Advance in Composite Gels

Advance in Composite Gels

Editor

Hiroyuki Takeno

MDPI • Basel • Beijing • Wuhan • Barcelona • Belgrade • Manchester • Tokyo • Cluj • Tianjin

Editor
Hiroyuki Takeno
Graduate School of Science
and Technology
Gunma University
Kiryu
Japan

Editorial Office
MDPI
St. Alban-Anlage 66
4052 Basel, Switzerland

This is a reprint of articles from the Special Issue published online in the open access journal *Gels* (ISSN 2310-2861) (available at: www.mdpi.com/journal/gels/special_issues/Composite_Gels).

For citation purposes, cite each article independently as indicated on the article page online and as indicated below:

LastName, A.A.; LastName, B.B.; LastName, C.C. Article Title. *Journal Name* **Year**, *Volume Number*, Page Range.

ISBN 978-3-0365-6427-2 (Hbk)
ISBN 978-3-0365-6426-5 (PDF)

Contents

About the Editor

Hiroyuki Takeno

Hiroyuki Takeno is currently an associate professor in the Division of Molecular Science at the Graduate School of Science and Technology, Gunma University, and at the Gunma University Center for Food Science and Wellness, Japan. He earned his bachelor's degree in 1992 and his doctoral degree from Kyoto University in 2000 under the supervision of Prof. Takeji Hashimoto.

His current research interests are 1) the mechanical and structural properties of tough composite gels, 2) the electro-responsive behavior of ion gels, and 3) the control of the mechanical properties of food gels.

gels

Editorial

Editorial on the Special Issue "Advances in Composite Gels"

Hiroyuki Takeno [1,2]

[1] Division of Molecular Science, Graduate School of Science and Technology, Gunma University, Kiryu 376-8515, Gunma, Japan; takeno@gunma-u.ac.jp
[2] Gunma University Center for Food Science and Wellness, 4-2 Aramaki, Maebashi 371-8510, Gunma, Japan

Polymer gels are soft materials composed of a large amount of solvent (water, organic solvent, and ionic liquid) and a polymer, and they are constructed using a three-dimensional network. The incorporation of other components into polymer gels gives birth to functional materials that cannot be attained in gels constituted of a single polymer. For example, polymer gels produced through conventional methods are mechanically brittle; by contrast, composite gels successfully incorporate inorganic nanoparticles and nanofibers into the polymer network, thus acquiring mechanically tough characteristics. Moreover, polymer composite hydrogels have potential applications in various fields such as tissue engineering, biomedical engineering, electrochemistry, and environmental chemistry. To fabricate smart composite gels, it is necessary to control the structural morphology of composite gels, and the interactions between the additive and the polymer network [1,2].

In addition to polymer composite gels, supramolecular gels formed via the self-assembly of small-molecular gelators and surfactants have raised much interest in terms of the production of smart nanomaterials [3]. A self-assembling structure is formed through non-covalent bonds such as hydrogen bonds, electrostatic interactions, π–π interactions, and van der Waals interactions. Factors such as molecular interactions, molecular architecture, and chirality significantly affect the self-assembling morphology. Recent studies highlight the importance of controlling the dimension of the supramolecular systems [4] and complex-ordered aggregates that cause aggregation-induced emission [5]. Moreover, two-component gelator systems including the combination of polymers and surfactants are also promising in the fabrication of smart nanomaterials [4].

In recent years, many advances have been made in the development of novel composite gels. This book aims to present the latest findings of composite gels by experts around the world in various fields. Each chapter in this book has been previously published in a Special Issue of the international journal, *Gels*, entitled "*Advances in Composite Gels*". These articles reveal the promising potential of composite gels as materials for cell scaffolds [6,7], opt-electrical devices [8], energy storage devices [9], catalysts [8], biomedicine [10], drug delivery systems [11], protein quantification [12], dental mold gypsum [13], excellent photoluminescence properties [5], and wastewater remediation [14]. Recent advances in composite hydrogels deal with biomaterials [15], as well as bioinspired and biomimetic materials [7]. Learning nature's strategy to efficiently fabricate smart materials proves valuable. The extracellular matrix is a three-dimensional network composed of macromolecular systems [7], which provide an adequate environment for cell adhesion, cell proliferation, and cell differentiation [6]. Furthermore, the eye lens having adaptive and focus-tunable characteristics may be regarded as a kind of biological gel [16]. Gel polymer electrolytes garner much attention due to having electrochemical applications such as portable electro devices and soft gel actuators [9,17]. The use of two-component additives produces composite gels with multi-functionality and excellent material properties, possibly resulting in a synergistic effect between the additives [2,18]. The incorporation of inorganic nanoparticles and nanofibers into a gel matrix provides robust and highly stretchable composite gels constructed by multi-crosslinking [2]. Additionally, the review articles included in this

Citation: Takeno, H. Editorial on the Special Issue "Advances in Composite Gels". *Gels* **2023**, *9*, 46. https://doi.org/10.3390/gels9010046

Received: 30 December 2022
Accepted: 1 January 2023
Published: 6 January 2023

collection introduce recent advances in drug delivery systems [19,20], microgels for oil recovery [21], and multi-layer hydrogels [22].

Finally, I would like to express my sincere gratitude to all the contributing authors who have made great contributions to the publication of this book.

Conflicts of Interest: The author declares no conflict of interest.

References

1. Nagai, D.; Isobe, N.; Inoue, T.; Okamoto, S.; Maki, Y.; Yamanobe, T. Preparation of Various Nanomaterials via Controlled Gelation of a Hydrophilic Polymer Bearing Metal-Coordination Units with Metal Ions. *Gels* **2022**, *8*, 435. [CrossRef] [PubMed]
2. Takeno, H.; Suto, N. Robust and Highly Stretchable Chitosan Nanofiber/Alumina-Coated Silica/Carboxylated Poly (Vinyl Alcohol)/Borax Composite Hydrogels Constructed by Multiple Crosslinking. *Gels* **2021**, *8*, 6. [CrossRef] [PubMed]
3. Azum, N.; Rub, M.A.; Khan, A.; Alotaibi, M.M.; Asiri, A.M. Synergistic Interaction and Binding Efficiency of Tetracaine Hydrochloride (Anesthetic Drug) with Anionic Surfactants in the Presence of NaCl Solution Using Surface Tension and UV–Visible Spectroscopic Methods. *Gels* **2022**, *8*, 234. [CrossRef] [PubMed]
4. Zhang, P.; Liu, Y.; Fang, X.; Ma, L.; Wang, Y.; Ji, L. Stoichiometric Ratio Controlled Dimension Transition and Supramolecular Chirality Enhancement in a Two-Component Assembly System. *Gels* **2022**, *8*, 269. [CrossRef]
5. Liu, X.; Li, C.; Wang, Z.; Zhang, N.; Feng, N.; Wang, W.; Xin, X. Luminescent Hydrogel Based on Silver Nanocluster/Malic Acid and Its Composite Film for Highly Sensitive Detection of Fe3+. *Gels* **2021**, *7*, 192. [CrossRef]
6. Vitale, M.; Ligorio, C.; Smith, I.P.; Richardson, S.M.; Hoyland, J.A.; Bella, J. Incorporation of Natural and Recombinant Collagen Proteins within Fmoc-Based Self-Assembling Peptide Hydrogels. *Gels* **2022**, *8*, 254. [CrossRef]
7. Riacci, L.; Sorriento, A.; Ricotti, L. Genipin-Based Crosslinking of Jellyfish Collagen 3D Hydrogels. *Gels* **2021**, *7*, 238. [CrossRef]
8. Özuğur Uysal, B.; Nayır, Ş.; Açba, M.; Çıtır, B.; Durmaz, S.; Koçoğlu, Ş.; Yıldız, E.; Pekcan, Ö. 2D Materials (WS2, MoS2, MoSe2) Enhanced Polyacrylamide Gels for Multifunctional Applications. *Gels* **2022**, *8*, 465. [CrossRef]
9. Velez, A.A.I.; Reyes, E.; Diaz-Barrios, A.; Santos, F.; Fernández Romero, A.J.; Tafur, J.P. Properties of the PVA-VAVTD KOH Blend as a Gel Polymer Electrolyte for Zinc Batteries. *Gels* **2021**, *7*, 256. [CrossRef]
10. Ekama, S.O.; Ilomuanya, M.O.; Azubuike, C.P.; Ayorinde, J.B.; Ezechi, O.C.; Igwilo, C.I.; Salako, B.L. Enzyme Responsive Vaginal Microbicide Gels Containing Maraviroc and Tenofovir Microspheres Designed for Acid Phosphatase-Triggered Release for Pre-Exposure Prophylaxis of HIV-1: A Comparative Analysis of a Bigel and Thermosensitive Gel. *Gels* **2021**, *8*, 15. [CrossRef]
11. Rub, M.A.; Azum, N.; Kumar, D.; Asiri, A.M. Interaction of TX-100 and Antidepressant Imipramine Hydrochloride Drug Mixture: Surface Tension, 1H NMR, and FT-IR Investigation. *Gels* **2022**, *8*, 159. [CrossRef] [PubMed]
12. Helbing, D.L.; Böhm, L.; Oraha, N.; Stabenow, L.K.; Cui, Y. A Ponceau S Staining-Based Dot Blot Assay for Rapid Protein Quantification of Biological Samples. *Gels* **2022**, *8*, 43. [CrossRef] [PubMed]
13. Ma, L.; Xie, Q.; Evelina, A.; Long, W.; Ma, C.; Zhou, F.; Cha, R. The Effect of Different Additives on the Hydration and Gelation Properties of Composite Dental Gypsum. *Gels* **2021**, *7*, 117. [CrossRef]
14. Sharma, G.; Kumar, A.; Ghfar, A.A.; García-Peñas, A.; Naushad, M.; Stadler, F.J. Fabrication and Characterization of Xanthan Gum-cl-poly(acrylamide-co-alginic acid) Hydrogel for Adsorption of Cadmium Ions from Aqueous Medium. *Gels* **2021**, *8*, 23. [CrossRef] [PubMed]
15. Tunit, P.; Thammarat, P.; Okonogi, S.; Chittasupho, C. Hydrogel Containing Borassus flabellifer L. Male Flower Extract for Antioxidant, Antimicrobial, and Anti-Inflammatory Activity. *Gels* **2022**, *8*, 126. [CrossRef]
16. Malik, A.; Khan, J.M.; Alhomida, A.S.; Ola, M.S. Modulation of the Structure and Stability of Novel Camel Lens Alpha-Crystallin by pH and Thermal Stress. *Gels* **2022**, *8*, 273. [CrossRef]
17. Sadiq, N.M.; Aziz, S.B.; Kadir, M.F.Z. Development of Flexible Plasticized Ion Conducting Polymer Blend Electrolytes Based on Polyvinyl Alcohol (PVA): Chitosan (CS) with High Ion Transport Parameters Close to Gel Based Electrolytes. *Gels* **2022**, *8*, 153. [CrossRef]
18. Safri, A.; Fletcher, A.J. Effective Carbon/TiO2 Gel for Enhanced Adsorption and Demonstrable Visible Light Driven Photocatalytic Performance. *Gels* **2022**, *8*, 215. [CrossRef]
19. Xie, Y.; Guan, Q.; Guo, J.; Chen, Y.; Yin, Y.; Han, X. Hydrogels for Exosome Delivery in Biomedical Applications. *Gels* **2022**, *8*, 328. [CrossRef]
20. Sastri, T.K.; Gupta, V.N.; Chakraborty, S.; Madhusudhan, S.; Kumar, H.; Chand, P.; Jain, V.; Veeranna, B.; Gowda, D.V. Novel Gels: An Emerging Approach for Delivering of Therapeutic Molecules and Recent Trends. *Gels* **2022**, *8*, 316. [CrossRef]
21. Rozhkova, Y.A.; Burin, D.A.; Galkin, S.V.; Yang, H. Review of Microgels for Enhanced Oil Recovery: Properties and Cases of Application. *Gels* **2022**, *8*, 112. [CrossRef] [PubMed]
22. Jin, L.; Xu, J.; Xue, Y.; Zhang, X.; Feng, M.; Wang, C.; Yao, W.; Wang, J.; He, M. Research Progress in the Multilayer Hydrogels. *Gels* **2021**, *7*, 172. [CrossRef] [PubMed]

gels

Review

Hydrogels for Exosome Delivery in Biomedical Applications

Yaxin Xie [1,†], Qiuyue Guan [2,†], Jiusi Guo [1], Yilin Chen [1], Yijia Yin [1] and Xianglong Han [1,*]

1 State Key Laboratory of Oral Diseases, Department of Orthodontics, National Clinical Research Center for Oral Diseases, West China Hospital of Stomatology, Sichuan University, Chengdu 610041, China; ashley_xieyaxin@163.com (Y.X.); jiusiguo@163.com (J.G.); dentistchenyl@163.com (Y.C.); yvonnist@163.com (Y.Y.)

2 Department of Geriatrics, People's Hospital of Sichuan Province, Chengdu 610041, China; qiuyueguan@126.com

* Correspondence: xhan@scu.edu.cn

† These authors contributed equally to this work.

Abstract: Hydrogels, which are hydrophilic polymer networks, have attracted great attention, and significant advances in their biological and biomedical applications, such as for drug delivery, tissue engineering, and models for medical studies, have been made. Due to their similarity in physiological structure, hydrogels are highly compatible with extracellular matrices and biological tissues and can be used as both carriers and matrices to encapsulate cellular secretions. As small extracellular vesicles secreted by nearly all mammalian cells to mediate cell–cell interactions, exosomes play very important roles in therapeutic approaches and disease diagnosis. To maintain their biological activity and achieve controlled release, a strategy that embeds exosomes in hydrogels as a composite system has been focused on in recent studies. Therefore, this review aims to provide a thorough overview of the use of composite hydrogels for embedding exosomes in medical applications, including the resources for making hydrogels and the properties of hydrogels, and strategies for their combination with exosomes.

Keywords: composite hydrogel; exosome; biomedical engineering

Citation: Xie, Y.; Guan, Q.; Guo, J.; Chen, Y.; Yin, Y.; Han, X. Hydrogels for Exosome Delivery in Biomedical Applications. *Gels* **2022**, *8*, 328. https://doi.org/10.3390/gels8060328

Academic Editor: Hiroyuki Takeno

Received: 19 April 2022
Accepted: 21 May 2022
Published: 24 May 2022

Publisher's Note: MDPI stays neutral with regard to jurisdictional claims in published maps and institutional affiliations.

1. Introduction

Hydrogels are three-dimensional macromolecular polymeric networks composed of hydrophilic polymer chains. They can generally be divided into three categories according to their origin: natural, synthetic, and hybrid. Hydrogels are degradable, with a high affinity for water, and can be fabricated under physiological conditions, resulting in excellent biocompatibility [1]. They can be formed chemically and/or physically upon initiation with crosslinking agents and produced with a certain viscosity and elasticity. The innovation of Wichterle and Lim pioneered a new approach to applying crosslinked hydroxyethyl methacrylate (HEMA) hydrogels as biomaterials in 1960 [2]. In the two decades following this discovery, Lim and Sun demonstrated calcium alginate hydrogels with applications in cell encapsulation [3]. It is not surprising that hydrogels, having mechanical and structural properties similar to those of many tissues and the extracellular matrix (ECM), have attracted great attention, and significant progress has been made in designing, synthesizing, and using these materials for many biological and biomedical applications [4].

Exosomes are small, single-membrane, secreted extracellular vesicles (EVs), enriched in certain proteins, nucleic acids, and lipids. Budding at both the plasma and endosomal membranes of all the mammalian cell types studied to date, they are produced to remodel the ECM and deliver signals and functional macromolecules to adjacent cells. Numerous surface molecules on exosomes enable them to be internalized via endocytosis by recipient cells, playing an important role in regulating cell–cell communication [5]. Therefore, the study of exosomes in the pathology of various diseases is an active area of research, and the exploration of therapeutic exosomes as delivery vesicles has offered new insights for clinical applications in recent years. However, the stability and retention of exosomes released in vivo are major hurdles, as they are rapidly cleared by the innate immune system or accumulate in the liver, spleen, and lungs via the blood circulation [6].

To overcome the rapid clearance and maintain the bioactivity of exosomes, hydrogels have been utilized to realize protection and controlled release by encapsulating these small vesicles. The excellent biodegradability of hydrogels allows them to be controlled by cell growth. Additionally, when they are used as scaffolds, exosomes can be loaded and released through their porous structure [7]. This review outlines the major applications of the hydrogel encapsulation of exosomes for physiological and pathological contexts, with a focus on the synthesis of, modification of, and exosome-loading strategies for hydrogels (Figure 1).

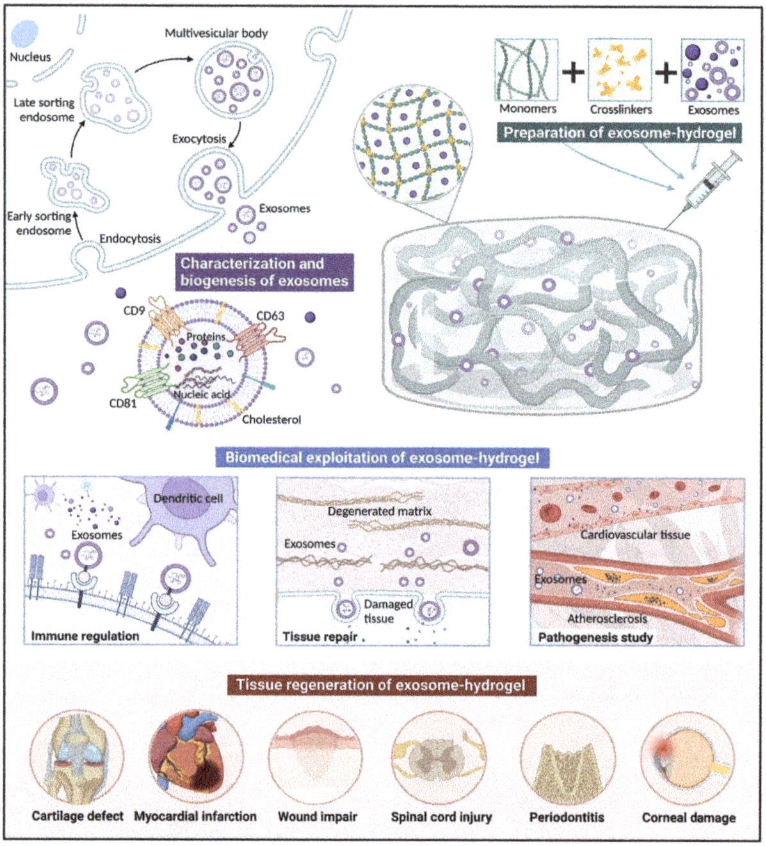

Figure 1. Hydrogels for exosome delivery in biomedical applications. Inward budding of the cellular plasma membrane results in the formation of endosomes, and the continuous inward invagination of the limiting membrane produces multivesicular bodies (MVBs). MVBs then fuse with the lysosome or plasma membrane, while the vesicles are released into the extracellular matrix to form exosomes. The secreted exosomes mainly contain proteins, nucleic acids, and lipids. The proteins contained in exosomes can be divided into two categories: one includes those commonly expressed in exosomes which can be used as markers (CD9, CD63, and CD81); the other includes the specific proteins from the parent cells. Hydrogels, as hydrophilic polymer networks, can encapsulate exosomes, overcoming the issue of low tissue retention and ensuring a controlled-release platform to localize their activity. Composite exosome–hydrogel systems have been applied in fields including tissue engineering and the study of pathogenesis. (Created with https://biorender.com/ accessed on 17 April 2022).

2. Hydrogels

Hydrogels, as hydrophilic polymer networks, absorb from 10–20% (an arbitrary lower limit) up to thousands of times their dry weights in water [2]. The high water content provides physical similarity to tissue, and gives the hydrogels excellent biocompatibility and the capability to easily encapsulate molecules [8]. The structural integrity of hydrogels depends on the crosslinks formed between the polymer chains, chemically and/or physically [9]. Naturally derived hydrogels are mostly formed by self-assembling physical crosslinks, including hydrogen bonds, van der Waals forces, and hydrophobic interactions, which cause macromolecules to fold and adopt well-defined structures and functionality [10]. Therefore, they can be synthesized in situ and used in injectable drug-delivery systems. Chemical crosslinking provides better stability because it allows substantially improved flexibility and spatiotemporal precision during gelation. The conventional mechanisms include radical polymerization, the chemical reaction of complementation groups, high-energy irradiation, and enzyme-enabled biochemistry, among others [11].

According to the different resources from which the polymers are sourced, hydrogels can be classified into natural, synthetic, and hybrid types. Recently, their network architectures, which can also be defined as conventional and unconventional polymer networks, such as interpenetrating and semi-interpenetrating polymer networks, have been extensively investigated. Concerning gel formation and drug release, a novel type of hydrogel capable of responding to a change in environmental conditions, such as temperature, pH, or the concentration of biomolecules, is the environment-sensitive hydrogel. As the design and preparation of hydrogels have been discussed in depth elsewhere, only a brief overview of common polymers is provided below [1,4,12].

2.1. Natural Hydrogels

Hydrogels derived from natural polymers tend to be highly compatible with biological tissues due to their similarity to the natural ECM or its components [13]. Therefore, the biodegradability and cell interactions in the tissue microenvironment of natural polymer networks mean that naturally derived hydrogels are widely used in tissue-engineering applications, and nearly all the hydrogels used for exosome encapsulation are based on naturally derived polymers [1]. The natural polymers for hydrogels can generally be divided into polysaccharides (alginate, hyaluronic acid, chitosan, agarose, and cellulose) and proteins (collagen, gelatin, and fibrin).

Alginate has been widely used as a scaffold in tissue engineering for cells, their encapsulation, and drug delivery; alginate is a linear polysaccharide copolymer of (1–4)-linked b-d-mannuronic acid (M) and a-l-guluronic acid (G) monomers and can be obtained from brown seaweeds and bacteria [14–16]. Alginate hydrogels are hypotoxic and easily available, while the dissociation of individual chains during gelation results in a loss of mechanical stiffness. Hyaluronic acid (HA) has been investigated for cell and molecule delivery, stem cell therapy, and tissue regeneration [17–19]. It is the simplest glycosaminoglycan (GAG), composed of a repeating disaccharide of (1–3)- and (1–4)-linked beta-D-glucuronic acid and N-acetyl-beta-D-glucosamine units, and is present in all mammals, especially in soft connective tissues [20,21]. HA hydrogels are nonimmunogenic, are biocompatible, and can be degraded by hyaluronidase for cell regulation. Chitosan is a linear polysaccharide, composed of β-(1–4)-linked D-glucosamine and N-acetyl-D-glucosamine, and obtained from arthropod exoskeletons [22]. Similar to naturally occurring GAGs, it has been applied in tissue engineering, showing excellent biocompatibility and biodegradability [23].

As the most abundant protein in animal bodies and the main component of the natural ECM, low-immunogenic collagen comprises three polypeptide chains wrapped around one another to form a three-stranded rope structure [24]. By introducing various chemical crosslinks, physical treatments, and other polymer modifications, collagen can be mechanically and stably enhanced, and is widely used in drug delivery and tissue reconstruction [25–27]. Gelatin is a single-stranded molecule naturally derived from breaking the triple-helix structure of collagen. Similar to collagen, gelatin requires covalent crosslinking, modifications, and interactions to further improve its physical properties [28,29]. GelMA hydrogels are hydrogels that are covalently crosslinked by introducing methacryloyl substituent groups to gelatin through photoinitiated radical polymerization [30]. A cargo-delivery platform can be created by mixing GelMA with nanoparticles such as laponite nanoclay to form a GelMA/nanoclay hydrogel with desirable combined mechanical and biological properties for specific biological applications [31]. Fibrin is a naturally derived polymer that is attractive for use in tissue sealants and adhesives for controlling bleeding in wound healing [32], as well as for scaffolds for tissue engineering [33,34]. It can be produced autologously from blood, thus possessing low antigenicity and being less likely to induce inflammatory responses [35].

However, the stability, mechanical properties, and cell adhesion of natural hydrogels need to be improved by extra crosslinking and modifications to realize specific biological and mechanical properties [13]. Covalent crosslinkers (e.g., glutaraldehyde and genipin) and physical treatments (e.g., UV irradiation and heating) have been applied to improve the mechanical properties of natural hydrogels [36–39]. A classic example of peptide modification is the introduction of the arginine–glycine–aspartic acid (RGD) sequence, which is used to enhance the cell-adhesion property [40].

2.2. Synthetic Hydrogels

Synthetic hydrogels can be fabricated with specific molecular weights, block structures, degradable linkages, and crosslinking modes to have tunable architectures at customized size scales and with controlled degradation rates. In addition, synthetic polymers are good in terms of cost, supply, and reproducible production. Examples of such synthetic materials discussed here are vinyl polymers (PHEMA and PVA), PEG, and polyesters (PLA).

Poly(hydroxyethyl methacrylate) (PHEMA) hydrogels can be prepared by the free-radical polymerization of HEMA. Copolymerization with acrylic or acrylamide monomers can achieve tunable swelling and mechanical properties for PHEMA hydrogels [41]. However, pure PHEMA requires extra biofunctional and bioactive motifs to realize cell adhesion and degradability in the tissue microenvironment [42]. PVA is mainly obtained from the partial or full hydrolysis of poly(vinyl acetate). Physically crosslinked PVA hydrogels exhibit high elasticity and fatigue resistance with low friction. PVA hydrogels, therefore, have been widely studied for cartilage tissue engineering [43]. Similar to PHEMA, pure PVA hydrogels need to be conjugated with several oligopeptide sequences to enhance their cellular interactions [44].

Hydrogels made from poly(ethylene glycol) (PEG) and the chemically similar poly (ethylene oxide) (PEO) are usually obtained from the polymerization of ethylene oxide [45]. Chemically crosslinked PEG hydrogels can be formed by photo-/UV-induced or radiation-induced free-radical polymerization with the modification of end chains with various chemical groups [46]. The physically crosslinked networks can also be generated by various motifs, which render the hydrogels reversible and stimulus-responsive [47]. Meanwhile, a triblock copolymer hydrogel has also been successfully manufactured and showed good performance for slow-release small EVs [48].

Poly(lactic acid) (PLA) is obtained from the ring-opening polymerization of lactide. The stability of PLA hydrogels can be improved via chemical crosslinking, such as photo-crosslinking to prevent autocatalytic decomposition [49]. Depending on the choice of lactide monomer, poly($_{\text{L}}$-lactic acid) (PLLA) and poly(D,$_{\text{L}}$-lactic acid) (PDLLA) can be generated as stereoisomers, and result in differing stiffnesses for hydrogels encapsulating hMSCs [50].

The limitation of synthetic hydrogels is the lack of native tissue topography and structure. Ergo, hybrid hydrogels comprising both natural and synthetic materials have recently attracted increasing attention, with the biological moieties of natural materials being combined with the benefits of tunable synthetic materials [7]. They are defined as polymers composed of hundreds of chemically or physically crosslinked nanogels, or systems combined with different polymers and/or with nanoparticles. The structural similarity to the natural ECM, tunable viscoelasticity and mechanical properties, high water contents, and permeability for oxygen and essential nutrients make hybrid hydrogels good candidates for tissue-engineering scaffolds [51].

3. Exosomes

3.1. Characterization and Biogenesis of Exosomes

Nearly all types of mammalian cells secrete extracellular vesicles (EVs), including mesenchymal stem cells [52], immune cells [53], neuronal cells [54], endothelial cells [55], and cancer cells [56]. As determined by their biogenesis, EVs can be broadly divided into three categories: exosomes, microvesicles, and apoptotic bodies [57]. Exosomes originate from endosomes with a size range of 40 to 160 nm (average ~100 nm) in diameter [58]. The inward budding of the cellular plasma membrane results in the formation of endosomes, and the continuous inward invagination of the limiting membrane produces multivesicular bodies (MVBs) [59]. Therefore, they can selectively incorporate cytosolic contents, and transmembrane and peripheral proteins, which contributes to the heterogeneity of exosomes. MVBs may then fuse with lysosomes or the plasma membrane, while the vesicles released into the extracellular matrix form exosomes [60,61].

Exosomes mainly contain proteins, nucleic acids, and lipids; the proteins contained in exosomes can be divided into two categories. One comprises proteins commonly expressed in exosomes that can be used as markers to identify exosomes, such as the CD9, CD63, and CD81 tetraspanin proteins, as well as TSG101, Alix, flotillin, and Rab [62]. The other comprises specific proteins from exosomes from different sources. For example, exosomes from T cells can carry CD3 molecules [63]. A major feature of exosomes that can distinguish them from other biological vesicles is that they contain a large number of nucleic acids, including DNA, RNA, miRNA, and noncoding RNA [55,64–66]. Moreover, exosomes can be engineered to deliver diverse therapeutic payloads. Small RNAs (sRNAs), particularly microRNAs, are transferred to mediate cell-to-cell communication and deliver genetic information [67,68].

Since the above biogenesis of exosomes is physiologic behavior, large-scale production for clinical studies and commercialization requires a higher yield of exosomes. There are some strategies used to stimulate EV shedding and enhance yield that can also be explored for exosomes. Wang et al. found that exosome secretion by MSCs could be enhanced by N-methyldopamine and norepinephrine without altering their modulatory capacity [69]. Other strategies such as pH variations or low-oxygen conditions may also stimulate an increase in exosome production [70].

3.2. Isolation and Analyses of Exosomes

The heterogeneity of exosomes originates from their size, molecular content, functional impact, and cellular origin. Therefore, the isolation and detection of exosomes are necessary for their embedding in hydrogels [71]. A variety of conventional isolation and enrichment methods have been developed, including ultracentrifugation, gradient ultracentrifugation, coprecipitation, size-exclusion chromatography, and field-flow fractionation. Ultracentrifugation is the current gold standard and most commonly used conventional approach for exosome isolation [72]. Sucrose-gradient centrifugation can further fractionate according to different vesicular densities and is more typically used to isolate exosomes. Coprecipitation is performed using commercial kits that rely on polymer coprecipitation, which are expensive for large-scale usage and lack specificity for exosomes [71]. Size-exclusion chromatography and field-flow fractionation separate exosomes and other molecules based on their sizes and molecular weights [73,74]. Compared to conventional methods, various new enrichment methods such as microfluidic filtering, contact-free sorting, and immunoaffinity enrichment have been developed to improve the isolation efficiency and specificity [75–78].

Since the enrichment methods are mainly based on the size, structure, and capture of some of the membrane proteins of exosomes, it is necessary to study exosomes by physical, chemical, and biological characterization to distinguish them from other vesicles and macromolecular protein complexes. Scanning electron microscopy (SEM) and transmission electron microscopy (TEM) are widely used to determine the morphology and structure of exosomes [79]. Dynamic light scattering (DLS) and nanoparticle-tracking analysis (NTA) are still attractive techniques for measuring the concentrations and size distributions of exosomes [80,81]. Conventional methods for the detection of exosomal proteins include Western blotting, enzyme-linked immunosorbent assay (ELISA), mass spectrometry, and flow cytometry [82–84], while novel methods include micro-nuclear magnetic resonance (μNMR) and exosome sensors [85,86]. It has been found that exosomes are enriched with tetraspanins (CD9, CD63, and CD81), membrane trafficking proteins (RAB proteins and annexins), and MVB-related proteins (ALIX, TSG101, and clathrin) [87]. The nucleic acids of exosomes, as potential circulating biomarkers, and intercellular regulators can be amplified through polymerase chain reactions (PCRs) and sequenced [71].

Despite these developments, some questions remain for subpopulations of EVs lacking precise definitions. It is still difficult to distinguish exosomes from other small vesicles with confidence. According to the updated guidelines for studies of EVs, researchers are encouraged to consider the use of operational terms for EV subtypes that refer to physical characteristics, biochemical composition, or descriptions of conditions or cells of origin [88]. Therefore, many studies have regarded different types of EVs as an entire cargo to deliver a packaged set of bioactive components [89]. For the further understanding of EVs' contents, single-EV analysis provides a benchmark by resolving EVs at a single-particle level [90]. Rogers et al. successfully detected EVs by using a single-EV microarray, which can assess EV proteins comprehensively and quantitatively [91].

3.3. Physiological Functions of Exosomes

Exosomes can be released under normal physiological conditions to regulate a range of biological processes. However, the precise roles of exosomes remain unclear due to the lack of physiological models in vitro and in vivo [90]. Ongoing experimental advances are likely to yield a thorough understanding of their heterogeneity and biological functions. The section below briefly discusses their main physiological functions.

1. Exosomes as mediators of intercellular communication. There are a variety of mechanisms that mediate cell–cell communication via exosomes. The phagocytosis-like uptake of exosomes by recipient cells enables them to transmit signals and molecules. Specific miRNA and protein cargoes in exosomes can contribute to tissue development and maintenance [92]. By directly fusing with the receptor cells, exosomes can exchange transmembrane proteins and lipids [93]. These properties mean that exosomes are involved in many physiological and pathological processes.
2. Exosomes as remodelers of the ECM. Cells can release exosomes into the ECM to manipulate its composition and function. Conversely, changes in the ECM affect cellular proliferation, migration, and organ morphogenesis. For example, exosomes can promote ECM synthesis by regulating matrix metalloproteinases (MMPs) [94], whereas some exosomes can inhibit the deposition of the ECM by suppressing collagen biosynthesis [95].
3. Exosomes as regulators of the immune response. Exosomes secreted by cells can modulate the immune response in various ways. Antigen-presenting cells can shed exosomes with the same cell-surface proteins such as MHC II and costimulatory signals [96]. An example of this is the release of exosomes containing bacterial mRNA by macrophages to activate the immune system [97]. MSC-derived exosomes can carry cytokines, miRNA, and other active molecules involved in proinflammatory and anti-inflammatory regulation [98].

4. Exosome-Loading Strategies

The stability and retention of exosomes are a major hurdle for clinical applications, as they are eliminated immediately by the immune system once injected in vivo [99]. Conventional delivery in cell-free exosome therapy includes intravenous, subcutaneous, and intraperitoneal injections. However, fluorescence imaging revealed that the majority of directly injected exosomes accumulated in various organs and tissues such as the liver and spleen [100]. Consequently, the method of administration should be optimized to achieve a high therapeutic efficacy and specificity, which requires delivering desirable exosomes to target tissues.

As the field rapidly evolves, biomaterials such as hydrogels allow exosomes to overcome the low tissue retention and ensure a controlled-release platform to localize their activity [7]. By embedding exosomes in a composite system, hydrogels play a dual role as carriers for cargo delivery and matrices for cellular interaction. Some of the first polymers used to synthesize hydrogels such as PHEMA and PEG are commonly used as cell culture materials. Much of the pioneering work with these hydrogels sought to elucidate the effects of the matrix stiffness on biological behavior [12]. However, these synthetic hydrogels are typically amorphous, homogeneous materials, considerably different from those of the native ECM. As progress has been made in 3D cell cultures, several strategies that permit cells and cellular molecules to spread and signal under physiological conditions have emerged. Hydrogels exhibiting passive hydrolytic degradation or cell-mediated enzymatic degradation have been considered, which enable the degradation rate of the matrices to be customized for the optimal release of the entrapped exosomes [4].

There are three common approaches for loading exosomes into a hydrogel matrix:

1. Polymers and exosomes are mixed and injected with crosslinkers in situ simultaneously. Exosomes are mixed with both polymers and crosslinkers simultaneously, and injected in situ with a dual-chamber syringe. After irradiation, ion exchanges, or environmental changes, polymerization can be achieved, inducing gelation [101]. In situ gelation can realize precise conformation to irregular cavities, and result in excellent integration and retention rates in the injection sites [102,103]. For example, entrapping effervescently generated CO_2 bubbles can help to form highly interconnected porous networks in injectable hydrogels in vivo, which is conducive to cellular attachment, infiltration, proliferation, and ECM deposition [104].
2. Polymers and exosomes are incorporated before the addition of crosslinkers for gelation. Exosomes are combined with polymers followed by crosslinkers for gelation. For example, Qin utilized a composite matrix (thiolated hyaluronic acid, heparin, and gelatin) to encapsulate bone marrow stem cell (BMSC)-derived exosomes, followed by the addition of poly(ethylene glycol) diacrylate (PEGDA) as a crosslinker [105]. The combination based on covalent crosslinking improves the retention and release rates for the exosomes embedded in the polymers. A problem that cannot be ignored is that residual unreacted crosslinkers can be cytotoxic, drawing attention to optimizing the reaction conditions, such as the gelling temperature, and choosing alternative nontoxic crosslinkers such as genipin [37,106].
3. Polymers and crosslinkers are gelated before their physical combination with exosomes. This method involves dehydrating the already-swollen hydrogel and soaking it in a solution containing exosomes. Due to the super-water-absorbent and swelling properties of the hydrogel, the exosomes are absorbed into the porous structure [107]. On account of the weak physical incorporation of exosomes, the pore size is pivotal; exosomes may easily leak from large pores or have difficulty in entering through small pores.

5. Biomedical Exploitation of Exosomes Delivered in Hydrogels

Exosomes functioning in the delivery of functional cargos are currently an active research hotspot. The biological features of exosomes make them suitable as potential therapeutics for the diagnosis and treatment of several diseases. There are generally three approaches to obtaining exosomes with therapeutic and diagnostic potential. (1) Naturally derived exosomes (e.g., MSC-Exos) have been verified to be therapeutic by themselves [108]. (2) Engineering exosomes by transferring molecules such as microRNAs has achieved targeted applications [109]. (3) Exosome mimetics have been exploited as promising biomaterials [67,110]. Below, the emerging roles of exosomes in tissue repair, immune modulation, and the study of pathogenesis are discussed.

5.1. Tissue Repair

Of the many classes of biomaterials that have been used in tissue repair, hydrogels have been regarded as one of the most prominent and versatile for supporting most cellular behaviors and nutrient transport. Protected by them, cellular secretions can maintain their biological activity and undergo controlled release in pathological environments (Table 1).

Table 1. Advances in tissue regeneration via the hydrogel encapsulation of EVs.

Composite Hydrogel Type	Exosome Source	Release Kinetics	Therapeutic Application	Reference
GelMA/nanoclay hydrogel	hUCMSCs	90% in a month	Cartilage regeneration	[31]
HA hydrogel	ECs	80% in a week	Fracture repair	[109]
GMOCS hydrogel	BMSCs	80% in 2 weeks	Repair of growth plate injuries	[111]
PEO–PPO–PEO hydrogel	PRP	80% in 20 days	Subtalar osteoarthritis	[112]
Pluronic F-127 hydrogel	Melanoma cells	Release peaked at 24 h	Chronic wound repair	[68]
HA@MnO hydrogel	M2	Over 80% in 21 days	Repair of chronic diabetic wounds	[113]
Methylcellulose–chitosan hydrogel	PMSCs	Not mentioned	Severe wound healing	[114]
HA hydrogel	iPS-CPCs and iPS-MSCs	Lasting over 2 weeks	Cardiac remodeling after MI	[115]
AT-EHBPE/HA-SH/CP05 hydrogel	hUCMSCs	Not mentioned	MI and reperfusion injury	[116]
Gelatin–laponite nanocomposite hydrogel	hADSCs	Not mentioned	Repair of peri-infarct myocardium	[117]
PDNP–PELA hydrogel	ADSCs	92.5 ± 5.7% in 2 weeks	Erectile dysfunction treatment	[118]
Peptide-modified HA hydrogel	hPAMMSCs	80% in a week	Recovery from spinal cord injury	[119]
Chitosan hydrogel	DPSCs	80% in a week	Periodontitis	[108]
Fibrin hydrogel	Rat BMSCs	Left over 2 weeks	Tendon regeneration	[120]

hUCMSC (human umbilical cord mesenchymal stem cell); EC (endothelial cell); BMSC (bone marrow mesenchymal stem cell); OCS (chondroitin sulfate); GM (gelatin macryloyl); PRP (platelet-rich plasma); M2 (M2 macrophage); PMSC (placental mesenchymal stem cell); iPS (induced pluripotent stem cell); CPC (cardiac progenitor cell); MI (myocardial infarction); AT (aniline tetramer); EHBPE (epoxy macromer); HA-SH (thiolated hyaluronic acid); hADSC (human adipose-derived stem cell); PDNP (polydopamine nanoparticle); PELA (poly(ethylene glycol)poly(ε-caprolactone-co-lactide)); hPAMMSC (human placenta amniotic membrane mesenchymal stem cell); DPSC (dental pulp stem cell).

5.1.1. Bone and Cartilage Defects

Overwhelming evidence shows that the exogenous transport of miRNAs by exosomes can regulate osteogenic and angiogenic differentiation. An example of this is a study carried out by Mi et al., who created a cocktail therapy by transferring miR-26a-5p into endothelial cell-derived exosomes (EC-Exos) in an HA hydrogel. The EC-Exos$^{miR-26a-5p}$ promoted osteogenic and osteoclast differentiation in mice with femoral fractures [109]. In another study, Hu et al. found that human umbilical cord MSC-derived small EVs (hUCMSC-sEVs) activated the PTEN/AKT signaling pathway by transferring miR-23a-3p when investigating the role and mechanism of cartilage regeneration [31]. Compared to increasing the specific miRNA in the target cells, the inhibition of miR-29a was verified to stimulate endogenous BMP/Smad signaling, which triggers subsequent osteogenic differentiation [67]. Therefore, the overexpression of miRNA can be an attractive method for improving the therapeutic effects. For example, miR-375 could be enriched in human adipose MSC (hASC)-derived exosomes by overexpressing the miRNA cargo in the parent cells [121].

Extensive research has shown that the essential properties of a bone and cartilage engineering scaffold are mechanical strength and a porous structure, to support the attachment and infiltration of osteogenic cells [122]. Hu et al. recently utilized an injectable and UV-crosslinked gelatin methacrylate (GelMA) to fabricate with nanoclay and achieved the sustained release of small EVs with the degradation of the hydrogel (Figure 2). The addition of laponite nanoclay significantly enhanced its ultimate strength for local administration in cartilage defects [31]. In addition to additives, 3D technology can also be applied to customize the shapes and sizes of porous scaffolds in accordance with bone defects. Fan et al. encapsulated umbilical MSC-derived exosomes (UMSC-Exos) in an HA hydrogel and combined them with 3D-printed nanohydroxyapatite/poly-ε-caprolactone (nHP) scaffolds [123]. Taken together, hydrogels can regulate extracellular matrix (ECM) formation, which provides a three-dimensional (3D) culture system for exosome secretion [89,124].

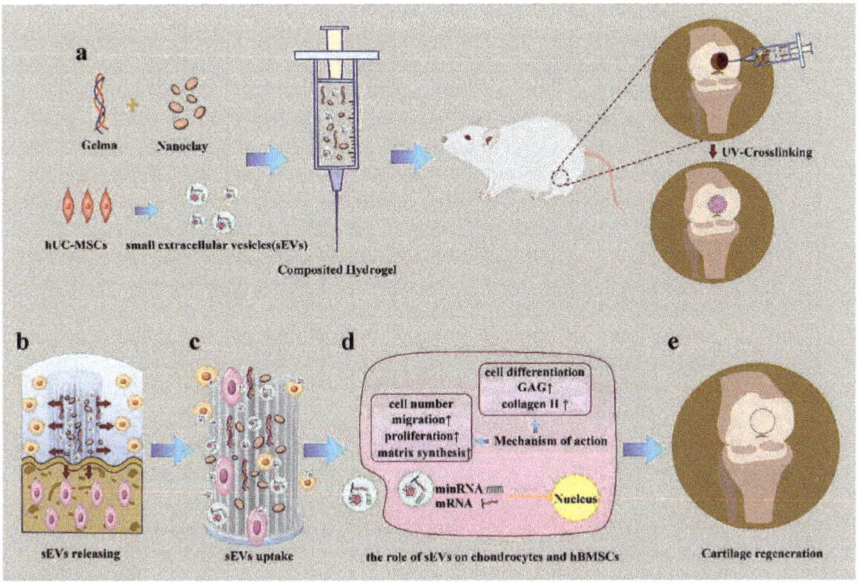

Figure 2. Schematic illustration of therapeutic sEVs released from a GelMA/nanoclay hydrogel for cartilage regeneration. (**a**) Preparation of a GelMA/nanoclay/sEV hydrogel and cartilage defect implantation. (**b**) Sustained release of sEVs with the degradation of the hydrogel. (**c**) Internalization of therapeutic sEVs by chondrocytes and hBMSCs. (**d**) The effect of the EVs on chondrocytes and hBMSCs. (**e**) Regeneration of a cartilage defect by the composite hydrogel. Copyright 2020, with permission from John Wiley and Sons [31].

5.1.2. Wound Repair

As a complicated biological process, wound healing consists of inflammation, proliferation, and remodeling [125]. The conventional treatment of chronic wounds includes regular wound debridement for stimulating skin regeneration and the protection of the wound using a specific dressing [126]. Recent interventions inspired by cell therapy approaches involve exosomes derived from MSCs, plasma, and cancer cells, while stem cell-derived exosomes are being developed for tissue recovery [68,127,128]. In a diabetes-impaired wound model, a wound dressing biomaterial was applied by combining antioxidant polyurethane (PUAO) for attenuating oxidative stress and adipose-derived stem cell (ADSC) exosomes for tissue remodeling [128]. Similarly, immobilizing ADSC-derived exosomes in a composite hydrogel that includes poly-ε-L-lysine (EPL), a natural cationic polypeptide from Streptomyces albulus, can help to realize antibacterial activity and adhesive ability [129]. Another study explored the feasibility of a composite hydrogel formed from silk fibroin (SF) and silk sericin (SS) due to the excellent mechanical properties of SF, and the cell-adhesion and biocompatibility properties of SS. After encapsulating and delivering UMSC-Exos, SF–SS hydrogels promoted wound healing and angiogenesis [130]. Additionally, the delivery of platelet-rich plasma exosomes in a composite chitosan–silk hydrogel sponge was found to upregulate collagen synthesis and deposition, as well as angiogenesis, at the wound site in diabetic rat models [127]. In addition, exosomes were enriched in miR-21, miR-23a, miR-125b, and miR-14, which can be blocked to reduce scar formation when they are laden in hydrogels [131]. Chitosan hydrogels functionalized with exosomes from synovium MSCs transduced to overexpress miR-126 promoted healing and angiogenesis in skin wounds [132].

5.1.3. Cardiovascular Diseases

Ischemic myocardial infarction (MI) results from the severe blockage of blood arteries, which, in turn, interrupts nutrient supply. However, clinical treatments may lead to further myocardial ischemia/reperfusion injury [133]. New findings have triggered studies investigating the potential of utilizing MSC-derived EVs after MI to promote angiogenesis and restore cardiac function [117,134–136]. For example, Zou at al. elaborated an exo-anchoring conductive hydrogel enabling electrical conduction within the myocardial fibrotic area and promoting the synchronous contraction of the myocardium. In this study, an aniline tetramer (AT) was employed as a crosslinker, and the researchers endowed it with electroconductibility. The CP05 peptide was applied for its capability of binding to CD63 on the exosomal surface, to anchor and capture exosomes from human UC-MSCs [116]. Based on the intended application, hydrogels can be synthesized with different preparations. A notable application is to encapsulate EVs from induced pluripotent stem cells in a hydrogel patch and apply them directly onto the rat myocardium. The hydrogel patch enabled sustainable release, which protected the acutely injured heart against pathological hypertrophy [89].

5.1.4. Spinal Cord Injury

Spinal cord injury (SCI) is among the most fatal diseases of the central nervous system, resulting in a temporary or permanent loss of sensation, movement, strength, and body functions [137]. To overcome the low cell survival resulting from the inhibitory environment at the lesion site, the local injection of exosomes protected by hydrogels is a promising therapeutic strategy. Li et al. improved the affinity of HA hydrogels and MSC-derived exosomes by a laminin modification, and successfully promoted spinal cord regeneration and the recovery of hindlimb motor function in vivo [119]. Surprisingly, plant (e.g., ginseng)-derived exosomes that can stimulate the neural differentiation of BMSCs have been demonstrated, and can be loaded in GelMA to fit the irregular shapes of injury defects [138]. The promotion of angiogenesis is beneficial for the regeneration of neuronal networks after SCI. Inspired by this, Luo et al. utilized a hybrid hydrogel system comprising GelMA, HA-NB, and a photoinitiator (LAP) to immobile exosomes from M2 macrophages. The hydrogel-mediated release system protected the exosomes from severe oxidative stress and inflammation [129].

5.1.5. Other Diseases

In addition to the aforementioned applications, exosomes have also played important roles in periodontal, endometrial, and corneal repairs. In the context of periodontitis, the incorporation of dental pulp stem cell-derived exosomes and chitosan hydrogels repolarized macrophages and accelerated periodontal regeneration [108]. The dynamic coordination of adipose stem cell-derived exosomes and PEG hydrogels via Ag^+–S resulted in outstanding injectable, self-healing, and antibacterial properties for endometrial and fertility restoration [113]. To effectively promote the repair of corneal damage, exosomes derived from MSCs were loaded in thermosensitive chitosan-based hydrogels [95].

5.2. Immune Regulation

Commonly, the adaptive immune response is regulated by antigen-presenting cells (APCs), such as dendritic cells (DCs), B cells, and macrophages, directly interacting with T cells and natural killer (NK) cells through cell-surface proteins [90]. Exosomes produced by APCs play an important role in the regulation of immunity, mediating immune stimulation or suppression, and driving inflammatory, autoimmune, and infectious disease pathology [96]. Inspired by dendritic cell-derived exosomes (DEXs), which improve cardiac function by activating $CD4^+$ T cells in the spleen and lymph nodes [139], Zhang et al. encapsulated DEXs in a simple alginate hydrogel and injected the DEX-Gel into the MI model. The DEXs significantly upregulated the infiltration of Treg cells and M2 macrophages, which resulted in better wound remodeling, and preserved systolic function after MI. Furthermore, the combined application of the hydrogel provides physical support to the infarcted area [140].

MSCs confer regenerative effects in different tissue injuries, while in some cases, MSCs have been confirmed to secrete immunosuppressive cytokines and other factors, resulting in anti-inflammatory effects from stem cells [141]. Notably, the analysis of MSC-derived EVs revealed that they also have immunosuppressive therapeutic effects [142]. To harness EVs' immunosuppressive properties, Fuhrmann et al. innovatively incorporated enzyme-loaded vesicles from MSCs into PVA hydrogels and applied this bioactive material for enzyme prodrug therapy. Once vesicles are released into the desired site, the injected nontoxic prodrugs are converted to anti-inflammatory drugs by enzymes [143]. The polarization of M2 macrophages, which can inhibit inflammation and induce tissue regeneration, has recently drawn great attention [108,109,144]. A classic cue is osteoimmunology, in which exosomes overexpressing miR-181 from human bone marrow-derived MSCs (hBM-MSCs) combined with a hydrogel were verified to significantly enhance osseointegration [144].

Tumor-derived EVs have been revealed to suppress tumor-specific and non-specific immune responses [96]. Metastatic melanoma releases a high level of exosomes carrying PD-L1 on their surfaces, which help in the evasion of immune surveillance. Based on how tumor cells suppress the immune system, Su et al. isolated exosomes from melanoma cells overexpressing PD-L1 to decrease T cell proliferation in a wound-healing model. The application of the thermoresponsive Pluronic F-127 hydrogel ensured that exosomes were released in a sustained manner [68].

5.3. Pathogenesis Study

Along with mediating physiological intercellular communication, exosomes also spread pathogenetic cargoes in diseases. Identifying the proteins and RNAs of exosomes can provide therapeutic targets. However, exosomal behavior can be dictated by the environment [4]. Therefore, hydrogels providing certain mechanical, structural, and compositional cues in the extracellular microenvironment are adopted as a novel strategy to recapitulate numerous physiologically relevant cell behaviors [145].

Tumor-derived exosomes can assist tumor growth and promote metastasis. To demonstrate the role of exosomes in ECM stiffness-triggered breast cancer invasiveness, Patwardhan et al. fabricated stiffness-tunable polyacrylamide (PA) gels as ECM mimics (Figure 3). Interestingly, stiff ECM cultures fostered exosome secretion by a series of changes in cell morphology, adhesion, and protrusion dynamics, which resulted in the invasion of breast cancer cells [146]. Aberrant cell behaviors can be induced by in vitro 2D culture, and the heterogeneity of exosomal behaviors also depends on the culture conditions [147]. Therefore, Millan et al. created 3D-engineered microtissues using the polysaccharides alginate and chitosan for the study of prostate cancer-derived EVs. Proteomics and RNA sequencing comparing 2D- and 3D-cultured cells revealed significantly differential expression of EV biomarkers. Some proteins known to be drivers of prostate cancer progression that were not detectable in the 2D conditions were enriched in the 3D cultures [148].

Exosomes from different cells such as endothelial cells and smooth muscle cells can contribute to atherosclerosis and cardiovascular disease when circulating in the blood [149,150]. In atherosclerosis-prone areas, EVs from smooth muscle cells (SMCs) and valvular interstitial cells (VICs) can cause a phospholipidic imbalance and, consequently, vascular and valvular calcification. Three-dimensional collagen hydrogels were utilized to produce a cardiovascular calcification model with which to observe the aggregation and microcalcification at the EV level [91]. Moreover, lesion macrophages can deliver exosomes that regulate vascular SMCs during the progression of atherosclerosis. In a study investigating the potential role of exosomes from nicotine-treated macrophages, Zhu et al. incorporated the above exosomes with chitosan hydrogels to stimulate release at the abdominal aorta [151].

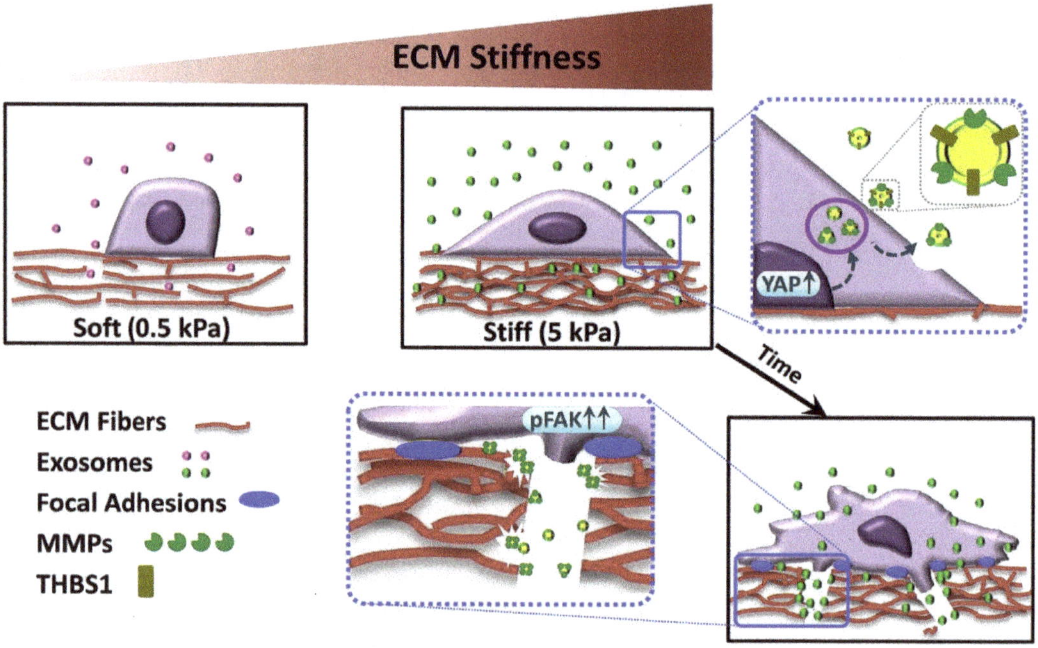

Figure 3. Proposed model of regulation of stiffness-dependent cancer invasiveness by stiffness-tuned exosomes. Breast cancer cells, in response to stiff substrates (5 kPa), secrete an excessive number of exosomes due to the activation of YAP/TAZ signaling. The stiffness-tuned exosomes confer an invasive, mesenchymal-like phenotype, accompanied by fostered focal adhesions and ECM remodeling. These changes are majorly elicited by exosomal THBS1 in concert with FAK and MMPs. Copyright 2021, with permission from Elsevier [146].

6. Conclusions and Outlook

The recent development of hydrogels as biomaterials has been aided by progress in material science, polymer physics, preparation techniques, and biomimetic characteristics. Despite these advances, there remain many challenges and clinical needs for biological and biomedical applications. Secreted from parent cells, exosomes can become components of the ECM. Therefore, hydrogels, as loading and release systems for maintaining the bioactivity of exosomes, need to mimic the matrix. Conventional forms of optimization such as the tuning of the pore size, degradability, and compatibility may greatly improve the retention and release profiles of exosomes in vivo. For instance, 3D printing has been applied to improve the functional porosities, pore shapes, and geometries of hydrogel scaffolds [152]. Tunable release and prolonged delivery can also be achieved by introducing materials such as integrins into synthetic hydrogels [153].

Compared to enhancing biomaterial characteristics, how to deliver exosomes to target cells is more challenging because the interaction between biomaterials and cellular behaviors on a relevant timescale needs to be considered. Recent advances in prolonging the half-lives and increasing the purity of exosomes could be exploited in order to overcome this hurdle. Design strategies for composite gels that combine different types of polymeric components to obtain unique properties are, therefore, common. Further study needs to be undertaken regarding smart hydrogels, such as CRISPR gel, which can be tailored and render programmable gels from traditional materials, thus capable of providing spatiotemporally defined interactions with exosomes for clinical translation [154,155].

Author Contributions: Writing—original draft, Y.X. and Q.G.; Visualization, J.G.; Writing—review & editing, Y.C. and Y.Y.; Project administration, X.H. All authors have read and agreed to the published version of the manuscript.

Funding: This research was funded by the National Natural Science Foundation of China (grant no. 81870803) and the Applied Fundamental Research Project of Sichuan Province (2022–2024).

Data Availability Statement: The figures used and analyzed during the current review are available from the publisher and corresponding author on reasonable request.

Conflicts of Interest: The authors declare no conflict of interest.

References

1. Zhao, X.H.; Chen, X.Y.; Yuk, H.; Lin, S.T.; Liu, X.Y.; Parada, G. Soft Materials by Design: Unconventional Polymer Networks Give Extreme Properties. *Chem. Rev.* **2021**, *121*, 4309–4372. [CrossRef] [PubMed]
2. Wichterle, O.; Lim, D. Hydrophilic gels for biological USE. *Nature* **1960**, *185*, 117–118. [CrossRef]
3. Lim, F.; Sun, A.M. Microencapsulated islets as bioartificial endocrine pancreas. *Science* **1980**, *210*, 908–910. [CrossRef] [PubMed]
4. Rosales, A.M.; Anseth, K.S. The design of reversible hydrogels to capture extracellular matrix dynamics. *Nat. Rev. Mater.* **2016**, *1*, 15012. [CrossRef] [PubMed]
5. Pegtel, D.M.; Gould, S.J. Exosomes. *Annu. Rev. Biochem.* **2019**, *88*, 487–514. [CrossRef]
6. Riau, A.K.; Ong, H.S.; Yam, G.H.F.; Mehta, J.S. Sustained Delivery System for Stem Cell-Derived Exosomes. *Front. Pharmacol.* **2019**, *10*, 1368. [CrossRef]
7. Brennan, M.A.; Layrolle, P.; Mooney, D.J. Biomaterials Functionalized with MSC Secreted Extracellular Vesicles and Soluble Factors for Tissue Regeneration. *Adv. Funct. Mater.* **2020**, *30*, 1909125. [CrossRef]
8. Li, J.; Mooney, D.J. Designing hydrogels for controlled drug delivery. *Nat. Rev. Mater.* **2016**, *1*, 16071. [CrossRef]
9. Drury, J.L.; Mooney, D.J. Hydrogels for tissue engineering: Scaffold design variables and applications. *Biomaterials* **2003**, *24*, 4337–4351. [CrossRef]
10. Zhang, Y.S.; Khademhosseini, A. Advances in engineering hydrogels. *Science* **2017**, *356*, eaaf3627. [CrossRef]
11. Hennink, W.E.; van Nostrum, C.F. Novel crosslinking methods to design hydrogels. *Adv. Drug Deliv. Rev.* **2012**, *64*, 223–236. [CrossRef]
12. Buwalda, S.J.; Boere, K.W.; Dijkstra, P.J.; Feijen, J.; Vermonden, T.; Hennink, W.E. Hydrogels in a historical perspective: From simple networks to smart materials. *J. Control Release* **2014**, *190*, 254–273. [CrossRef] [PubMed]
13. Pishavar, E.; Luo, H.; Naserifar, M.; Hashemi, M.; Toosi, S.; Atala, A.; Ramakrishna, S.; Behravan, J. Advanced Hydrogels as Exosome Delivery Systems for Osteogenic Differentiation of MSCs: Application in Bone Regeneration. *Int. J. Mol. Sci.* **2021**, *22*, 6203. [CrossRef] [PubMed]
14. Jana, P.; Shyam, M.; Singh, S.; Jayaprakash, V.; Dev, A. Biodegradable polymers in drug delivery and oral vaccination. *Eur. Polym. J.* **2021**, *142*, 110155. [CrossRef]
15. Kang, S.-W.; Cha, B.-H.; Park, H.; Park, K.-S.; Lee, K.Y.; Lee, S.-H. The Effect of Conjugating RGD into 3D Alginate Hydrogels on Adipogenic Differentiation of Human Adipose-Derived Stromal Cells. *Macromol. Biosci.* **2011**, *11*, 673–679. [CrossRef]
16. Toh, W.S.; Lee, E.H.; Guo, X.-M.; Chan, J.K.Y.; Yeow, C.H.; Choo, A.B.; Cao, T. Cartilage repair using hyaluronan hydrogel-encapsulated human embryonic stem cell-derived chondrogenic cells. *Biomaterials* **2010**, *31*, 6968–6980. [CrossRef]
17. Burdick, J.A.; Ward, M.; Liang, E.; Young, M.J.; Langer, R. Stimulation of neurite outgrowth by neurotrophins delivered from degradable hydrogels. *Biomaterials* **2006**, *27*, 452–459. [CrossRef]
18. Kim, I.L.; Mauck, R.L.; Burdick, J.A. Hydrogel design for cartilage tissue engineering: A case study with hyaluronic acid. *Biomaterials* **2011**, *32*, 8771–8782. [CrossRef]
19. Sabra, W.; Zeng, A.P.; Deckwer, W.D. Bacterial alginate: Physiology, product quality and process aspects. *Appl. Microbiol. Biotechnol.* **2001**, *56*, 315–325. [CrossRef]
20. Xu, X.; Jha, A.K.; Harrington, D.A.; Farach-Carson, M.C.; Jia, X. Hyaluronic acid-based hydrogels: From a natural polysaccharide to complex networks. *Soft. Matter.* **2012**, *8*, 3280–3294. [CrossRef]
21. Burdick, J.A.; Prestwich, G.D. Hyaluronic Acid Hydrogels for Biomedical Applications. *Adv. Mater.* **2011**, *23*, H41–H56. [CrossRef] [PubMed]
22. Suh, J.K.F.; Matthew, H.W.T. Application of chitosan-based polysaccharide biomaterials in cartilage tissue engineering: A review. *Biomaterials* **2000**, *21*, 2589–2598. [PubMed]
23. Li, Z.; Zhang, K.; Zhao, X.; Kong, D.; Zhao, Q.; Liu, N.; Ma, F. Enhanced Therapeutic Effects of MSC-derived Exosomes with an Injectable Hydrogel for Hindlimb Ischemia Treatment. *Circ. Res.* **2018**, *123*, A490. [CrossRef]
24. Lee, C.H.; Singla, A.; Lee, Y. Biomedical applications of collagen. *Int. J. Pharm.* **2001**, *221*, 1–22. [CrossRef]
25. Marelli, B.; Achilli, M.; Alessandrino, A.; Freddi, G.; Tanzi, M.C.; Fare, S.; Mantovani, D. Collagen-Reinforced Electrospun Silk Fibroin Tubular Construct as Small Calibre Vascular Graft. *Macromol. Biosci.* **2012**, *12*, 1566–1574. [CrossRef] [PubMed]
26. McGuigan, A.P.; Sefton, M.V. The thrombogenicity of human umbilical vein endothelial cell seeded collagen modules. *Biomaterials* **2008**, *29*, 2453–2463. [CrossRef]

27. Yuan, T.; Zhang, L.; Li, K.; Fan, H.; Fan, Y.; Liang, J.; Zhang, X. Collagen hydrogel as an immunomodulatory scaffold in cartilage tissue engineering. *J. Biomed. Mater. Res. Part B Appl. Biomater.* **2014**, *102*, 337–344. [CrossRef]
28. Choi, Y.S.; Hong, S.R.; Lee, Y.M.; Song, K.W.; Park, M.H.; Nam, Y.S. Study on gelatin-containing artificial skin: I. Preparation and characteristics of novel gelatin-alginate sponge. *Biomaterials* **1999**, *20*, 409–417. [CrossRef]
29. Kuijpers, A.J.; Engbers, G.H.M.; Feijen, J.; De Smedt, S.C.; Meyvis, T.K.L.; Demeester, J.; Krijgsveld, J.; Zaat, S.A.J.; Dankert, J. Characterization of the network structure of carbodiimide cross-linked gelatin gels. *Macromolecules* **1999**, *32*, 3325–3333. [CrossRef]
30. Yue, K.; Trujillo-de Santiago, G.; Moises Alvarez, M.; Tamayol, A.; Annabi, N.; Khademhosseini, A. Synthesis, properties, and biomedical applications of gelatin methacryloyl (GelMA) hydrogels. *Biomaterials* **2015**, *73*, 254–271. [CrossRef]
31. Hu, H.; Dong, L.; Bu, Z.; Shen, Y.; Luo, J.; Zhang, H.; Zhao, S.; Lv, F.; Liu, Z. miR-23a-3p-abundant small extracellular vesicles released from Gelma/nanoclay hydrogel for cartilage regeneration. *J. Extracell. Vesicles* **2020**, *9*, 1778883. [CrossRef] [PubMed]
32. Spotnitz, W.D.; Prabhu, R. Fibrin sealant tissue adhesive—Review and update. *J. Long Term Eff. Med. Implant.* **2005**, *15*, 245–270. [CrossRef] [PubMed]
33. Johnson, T.S.; Xu, J.W.; Zaporojan, V.V.; Mesa, J.M.; Weinand, C.; Randolph, M.A.; Bonassar, L.J.; Winograd, J.M.; Yaremchuk, M.J. Integrative repair of cartilage with articular and nonarticular chondrocytes. *Tissue Eng.* **2004**, *10*, 1308–1315. [CrossRef] [PubMed]
34. Chrobak, M.O.; Hansen, K.J.; Gershlak, J.R.; Vratsanos, M.; Kanellias, M.; Gaudette, G.R.; Pins, G.D. Design of a Fibrin Microthread-Based Composite Layer for Use in a Cardiac Patch. *Acs Biomater. Sci. Eng.* **2017**, *3*, 1394–1403. [CrossRef] [PubMed]
35. Ahmed, T.A.E.; Dare, E.V.; Hincke, M. Fibrin: A versatile scaffold for tissue engineering applications. *Tissue Eng. Part B Rev.* **2008**, *14*, 199–215. [CrossRef]
36. Chan, K.L.S.; Khankhel, A.H.; Thompson, R.L.; Coisman, B.J.; Wong, K.H.K.; Truslow, J.G.; Tien, J. Crosslinking of collagen scaffolds promotes blood and lymphatic vascular stability. *J. Biomed. Mater. Res. Part A* **2014**, *102*, 3186–3195. [CrossRef]
37. Li, Q.; Gong, S.; Yao, W.; Yang, Z.; Wang, R.; Yu, Z.; Wei, M. Exosome loaded genipin crosslinked hydrogel facilitates full thickness cutaneous wound healing in rat animal model. *Drug Deliv.* **2021**, *28*, 884–893. [CrossRef]
38. Lee, C.R.; Grodzinsky, A.J.; Spector, M. The effects of cross-linking of collagen-glycosaminoglycan scaffolds on compressive stiffness, chondrocyte-mediated contraction, proliferation and biosynthesis. *Biomaterials* **2001**, *22*, 3145–3154. [CrossRef]
39. Schoof, H.; Apel, J.; Heschel, I.; Rau, G. Control of pore structure and size in freeze-dried collagen sponges. *J. Biomed. Mater. Res.* **2001**, *58*, 352–357. [CrossRef]
40. Huang, C.C.; Kang, M.; Shirazi, S.; Lu, Y.; Cooper, L.F.; Gajendrareddy, P.; Ravindran, S. 3D Encapsulation and tethering of functionally engineered extracellular vesicles to hydrogels. *Acta Biomater.* **2021**, *126*, 199–210. [CrossRef]
41. Sefton, M.V.; May, M.H.; Lahooti, S.; Babensee, J.E. Making microencapsulation work: Conformal coating, immobilization gels and in vivo performance. *J. Control. Release* **2000**, *65*, 173–186. [CrossRef]
42. Meyvis, T.K.L.; De Smedt, S.C.; Demeester, J.; Hennink, W.E. Influence of the degradation mechanism of hydrogels on their elastic and swelling properties during degradation. *Macromolecules* **2000**, *33*, 4717–4725. [CrossRef]
43. Kumar, A.; Han, S.S. PVA-based hydrogels for tissue engineering: A review. *Int. J. Polym. Mater. Polym. Biomater.* **2017**, *66*, 159–182. [CrossRef]
44. Miao, T.; Miller, E.J.; McKenzie, C.; Oldinski, R.A. Physically crosslinked polyvinyl alcohol and gelatin interpenetrating polymer network theta-gels for cartilage regeneration. *J. Mater. Chem. B* **2015**, *3*, 9242–9249. [CrossRef]
45. Ostuni, E.; Chapman, R.G.; Holmlin, R.E.; Takayama, S.; Whitesides, G.M. A survey of structure-property relationships of surfaces that resist the adsorption of protein. *Langmuir* **2001**, *17*, 5605–5620. [CrossRef]
46. Sakai, T.; Matsunaga, T.; Yamamoto, Y.; Ito, C.; Yoshida, R.; Suzuki, S.; Sasaki, N.; Shibayama, M.; Chung, U.-i. Design and fabrication of a high-strength hydrogel with ideally homogeneous network structure from tetrahedron-like macromonomers. *Macromolecules* **2008**, *41*, 5379–5384. [CrossRef]
47. Dong, R.; Pang, Y.; Su, Y.; Zhu, X. Supramolecular hydrogels: Synthesis, properties and their biomedical applications. *Biomater. Sci.* **2015**, *3*, 937–954. [CrossRef]
48. Tao, S.C.; Huang, J.Y.; Gao, Y.; Li, Z.X.; Wei, Z.Y.; Dawes, H.; Guo, S.C. Small extracellular vesicles in combination with sleep-related circRNA3503: A targeted therapeutic agent with injectable thermosensitive hydrogel to prevent osteoarthritis. *Bioact Mater* **2021**, *6*, 4455–4469. [CrossRef]
49. Stanford, M.J.; Dove, A.P. Stereocontrolled ring-opening polymerisation of lactide. *Chem. Soc. Rev.* **2010**, *39*, 486–494. [CrossRef]
50. Sun, A.X.; Lin, H.; Fritch, M.R.; Shen, H.; Alexander, P.G.; DeHart, M.; Tuan, R.S. Chondrogenesis of human bone marrow mesenchymal stem cells in 3-dimensional, photocrosslinked hydrogel constructs: Effect of cell seeding density and material stiffness. *Acta Biomater.* **2017**, *58*, 302–311. [CrossRef]
51. Vasile, C.; Pamfil, D.; Stoleru, E.; Baican, M. New Developments in Medical Applications of Hybrid Hydrogels Containing Natural Polymers. *Molecules* **2020**, *25*, 1539. [CrossRef]
52. Keshtkar, S.; Azarpira, N.; Ghahremani, M.H. Mesenchymal stem cell-derived extracellular vesicles: Novel frontiers in regenerative medicine. *Stem Cell Res. Ther.* **2018**, *9*, 63. [CrossRef]
53. Veerman, R.E.; Akpinar, G.G.; Eldh, M.; Gabrielsson, S. Immune Cell-Derived Extracellular Vesicles—Functions and Therapeutic Applications. *Trends Mol. Med.* **2019**, *25*, 382–394. [CrossRef]
54. Deng, Z.; Wang, J.; Xiao, Y.; Li, F.; Niu, L.; Liu, X.; Meng, L.; Zheng, H. Ultrasound-mediated augmented exosome release from astrocytes alleviates amyloid-beta-induced neurotoxicity. *Theranostics* **2021**, *11*, 4351–4362. [CrossRef]

55. Balaj, L.; Lessard, R.; Dai, L.; Cho, Y.-J.; Pomeroy, S.L.; Breakefield, X.O.; Skog, J. Tumour microvesicles contain retrotransposon elements and amplified oncogene sequences. *Nat. Commun.* **2011**, *2*, 180. [CrossRef]
56. Chen, G.; Huang, A.C.; Zhang, W.; Zhang, G.; Wu, M.; Xu, W.; Yu, Z.; Yang, J.; Wang, B.; Sun, H.; et al. Exosomal PD-L1 contributes to immunosuppression and is associated with anti-PD-1 response. *Nature* **2018**, *560*, 382–386. [CrossRef]
57. El Andaloussi, S.; Maeger, I.; Breakefield, X.O.; Wood, M.J.A. Extracellular vesicles: Biology and emerging therapeutic opportunities. *Nat. Rev. Drug Discov.* **2013**, *12*, 348–358. [CrossRef]
58. Kalluri, R.; LeBleu, V.S. The biology, function, and biomedical applications of exosomes. *Science* **2020**, *367*, eaau6977. [CrossRef]
59. Thery, C.; Zitvogel, L.; Amigorena, S. Exosomes: Composition, biogenesis and function. *Nat. Rev. Immunol.* **2002**, *2*, 569–579. [CrossRef]
60. Hurley, J.H.; Hanson, P.I. Membrane budding and scission by the ESCRT machinery: It's all in the neck. *Nat. Rev. Mol. Cell Biol.* **2010**, *11*, 556–566. [CrossRef]
61. Luzio, J.P.; Gray, S.R.; Bright, N.A. Endosome-lysosome fusion. *Biochem. Soc. Trans.* **2010**, *38*, 1413–1416. [CrossRef]
62. Mathivanan, S.; Simpson, R.J. ExoCarta: A compendium of exosomal proteins and RNA. *Proteomics* **2009**, *9*, 4997–5000. [CrossRef]
63. Blanchard, N.; Lankar, D.; Faure, F.; Regnault, A.; Dumont, C.; Raposo, G.; Hivroz, C. TCR activation of human T cells induces the production of exosomes bearing the TCR/CD3/zeta complex. *J. Immunol.* **2002**, *168*, 3235–3241. [CrossRef]
64. Valadi, H.; Ekstrom, K.; Bossios, A.; Sjostrand, M.; Lee, J.J.; Lotvall, J.O. Exosome-mediated transfer of mRNAs and microRNAs is a novel mechanism of genetic exchange between cells. *Nat. Cell Biol.* **2007**, *9*, 654–659. [CrossRef]
65. Williams, C.; Rodriguez-Barrueco, R.; Silva, J.M.; Zhang, W.; Hearn, S.; Elemento, O.; Paknejad, N.; Manova-Todorova, K.; Welte, K.; Bromberg, J.; et al. Double-stranded DNA in exosomes: A novel biomarker in cancer detection. *Cell Res.* **2014**, *24*, 766–769. [CrossRef]
66. Wei, Z.; Batagov, A.O.; Schinelli, S.; Wang, J.; Wang, Y.; El Fatimy, R.; Rabinovsky, R.; Balaj, L.; Chen, C.C.; Hochberg, F.; et al. Coding and noncoding landscape of extracellular RNA released by human glioma stem cells. *Nat. Commun.* **2017**, *8*, 1145. [CrossRef]
67. Fan, J.; Lee, C.S.; Kim, S.; Chen, C.; Aghaloo, T.; Lee, M. Generation of Small RNA-Modulated Exosome Mimetics for Bone Regeneration. *ACS Nano* **2020**, *14*, 11973–11984. [CrossRef]
68. Su, D.; Tsai, H.I.; Xu, Z.; Yan, F.; Wu, Y.; Xiao, Y.; Liu, X.; Wu, Y.; Parvanian, S.; Zhu, W.; et al. Exosomal PD-L1 functions as an immunosuppressant to promote wound healing. *J Extracell Vesicles* **2019**, *9*, 1709262. [CrossRef]
69. Wang, J.; Bonacquisti, E.E.; Brown, A.D.; Nguyen, J. Boosting the Biogenesis and Secretion of Mesenchymal Stem Cell-Derived Exosomes. *Cells* **2020**, *9*, 660. [CrossRef]
70. Herrmann, I.K.; Wood, M.J.A.; Fuhrmann, G. Extracellular vesicles as a next-generation drug delivery platform. *Nat. Nanotechnol.* **2021**, *16*, 748–759. [CrossRef]
71. Shao, H.; Im, H.; Castro, C.M.; Breakefield, X.; Weissleder, R.; Lee, H. New Technologies for Analysis of Extracellular Vesicles. *Chem. Rev.* **2018**, *118*, 1917–1950. [CrossRef]
72. Gardiner, C.; Di Vizio, D.; Sahoo, S.; Thery, C.; Witwer, K.W.; Wauben, M.; Hill, A.F. Techniques used for the isolation and characterization of extracellular vesicles: Results of a worldwide survey. *J. Extracell. Vesicles* **2016**, *5*, 32945. [CrossRef] [PubMed]
73. Kang, D.; Oh, S.; Ahn, S.-M.; Lee, B.-H.; Moon, M.H. Proteomic analysis of exosomes from human neural stem cells by flow field-flow fractionation and nanoflow liquid chromatography-tandem mass spectrometry. *J. Proteome Res.* **2008**, *7*, 3475–3480. [CrossRef] [PubMed]
74. Boing, A.N.; van der Pol, E.; Grootemaat, A.E.; Coumans, F.A.W.; Sturk, A.; Nieuwland, R. Single-step isolation of extracellular vesicles by size-exclusion chromatography. *J. Extracell. Vesicles* **2014**, *3*, 23430. [CrossRef]
75. Lee, K.; Shao, H.; Weissleder, R.; Lee, H. Acoustic Purification of Extracellular Microvesicles. *Acs Nano* **2015**, *9*, 2321–2327. [CrossRef] [PubMed]
76. Reategui, E.; van der Vos, K.E.; Lai, C.P.; Zeinali, M.; Atai, N.A.; Aldikacti, B.; Floyd, F.P., Jr.; Khankhel, A.H.; Thapar, V.; Hochberg, F.H.; et al. Engineered nanointerfaces for microfluidic isolation and molecular profiling of tumor-specific extracellular vesicles. *Nat. Commun.* **2018**, *9*, 175. [CrossRef]
77. Rho, J.; Chung, J.; Im, H.; Liong, M.; Shao, H.; Castro, C.M.; Weissleder, R.; Lee, H. Magnetic Nanosensor for Detection and Profiling of Erythrocyte-Derived Microvesicles. *Acs Nano* **2013**, *7*, 11227–11233. [CrossRef]
78. Wunsch, B.H.; Smith, J.T.; Gifford, S.M.; Wang, C.; Brink, M.; Bruce, R.L.; Austin, R.H.; Stolovitzky, G.; Astier, Y. Nanoscale lateral displacement arrays for the separation of exosomes and colloids down to 20 nm. *Nat. Nanotechnol.* **2016**, *11*, 936–940. [CrossRef]
79. Sokolova, V.; Ludwig, A.-K.; Hornung, S.; Rotan, O.; Horn, P.A.; Epple, M.; Glebel, B. Characterisation of exosomes derived from human cells by nanoparticle tracking analysis and scanning electron microscopy. *Colloids Surf. B Biointerfaces* **2011**, *87*, 146–150. [CrossRef]
80. Gardiner, C.; Ferreira, Y.J.; Dragovic, R.A.; Redman, C.W.G.; Sargent, I.L. Extracellular vesicle sizing and enumeration by nanoparticle tracking analysis. *J. Extracell. Vesicles* **2013**, *2*, 19671. [CrossRef]
81. Zhang, Y.; Xie, Y.; Hao, Z.; Zhou, P.; Wang, P.; Fang, S.; Li, L.; Xu, S.; Xia, Y. Umbilical Mesenchymal Stem Cell-Derived Exosome-Encapsulated Hydrogels Accelerate Bone Repair by Enhancing Angiogenesis. *ACS Appl. Mater. Interfaces* **2021**, *13*, 18472–18487. [CrossRef] [PubMed]
82. Kreimer, S.; Belov, A.M.; Ghiran, I.; Murthy, S.K.; Frank, D.A.; Ivanov, A.R. Mass-Spectrometry-Based Molecular Characterization of Extracellular Vesicles: Lipidomics and Proteomics. *J. Proteome Res.* **2015**, *14*, 2367–2384. [CrossRef] [PubMed]

83. Lobb, R.J.; Becker, M.; Wen, S.W.; Wong, C.S.F.; Wiegmans, A.P.; Leimgruber, A.; Moller, A. Optimized exosome isolation protocol for cell culture supernatant and human plasma. *J. Extracell. Vesicles* **2015**, *4*, 27031. [CrossRef] [PubMed]
84. Pospichalova, V.; Svoboda, J.; Dave, Z.; Kotrbova, A.; Kaiser, K.; Klemova, D.; Ilkovics, L.; Hampl, A.; Crha, I.; Jandakova, E.; et al. Simplified protocol for flow cytometry analysis of fluorescently labeled exosomes and microvesicles using dedicated flow cytometer. *J. Extracell. Vesicles* **2015**, *4*, 25530. [CrossRef]
85. Jeong, S.; Park, J.; Pathania, D.; Castro, C.M.; Weissleder, R.; Lee, H. Integrated Magneto-Electrochemical Sensor for Exosome Analysis. *Acs Nano* **2016**, *10*, 1802–1809. [CrossRef]
86. Shao, H.; Yoon, T.-J.; Liong, M.; Weissleder, R.; Lee, H. Magnetic nanoparticles for biomedical NMR-based diagnostics. *Beilstein J. Nanotechnol.* **2010**, *1*, 142–154. [CrossRef]
87. Jeppesen, D.K.; Fenix, A.M.; Franklin, J.L.; Higginbotham, J.N.; Zhang, Q.; Zimmerman, L.J.; Liebler, D.C.; Ping, J.; Liu, Q.; Evans, R.; et al. Reassessment of Exosome Composition. *Cell* **2019**, *177*, 428–445. [CrossRef]
88. Thery, C.; Witwer, K.W.; Aikawa, E.; Alcaraz, M.J.; Anderson, J.D.; Andriantsitohaina, R.; Antoniou, A.; Arab, T.; Archer, F.; Atkin-Smith, G.K.; et al. Minimal information for studies of extracellular vesicles 2018 (MISEV2018): A position statement of the International Society for Extracellular Vesicles and update of the MISEV2014 guidelines. *J. Extracell. Vesicles* **2018**, *7*, 1535750. [CrossRef]
89. Liu, B.; Lee, B.W.; Nakanishi, K.; Villasante, A.; Williamson, R.; Metz, J.; Kim, J.; Kanai, M.; Bi, L.; Brown, K.; et al. Cardiac recovery via extended cell-free delivery of extracellular vesicles secreted by cardiomyocytes derived from induced pluripotent stem cells. *Nat. Biomed. Eng.* **2018**, *2*, 293–303. [CrossRef]
90. Cheng, L.; Hill, A.F. Therapeutically harnessing extracellular vesicles. *Nat. Rev. Drug Discov.* **2022**, *21*, 379–399. [CrossRef]
91. Rogers, M.A.; Buffolo, F.; Schlotter, F.; Atkins, S.K.; Lee, L.H.; Halu, A.; Blaser, M.C.; Tsolaki, E.; Higashi, H.; Luther, K.; et al. Annexin A1-dependent tethering promotes extracellular vesicle aggregation revealed with single-extracellular vesicle analysis. *Sci. Adv.* **2020**, *6*, eabb1244. [CrossRef] [PubMed]
92. Feng, D.; Zhao, W.-L.; Ye, Y.-Y.; Bai, X.-C.; Liu, R.-Q.; Chang, L.-F.; Zhou, Q.; Sui, S.-F. Cellular Internalization of Exosomes Occurs Through Phagocytosis. *Traffic* **2010**, *11*, 675–687. [CrossRef] [PubMed]
93. Prada, I.; Amin, L.; Furlan, R.; Legname, G.; Verderio, C.; Cojoc, D. A new approach to follow a single extracellular vesicle-cell interaction using optical tweezers. *Biotechniques* **2016**, *60*, 35–41. [CrossRef] [PubMed]
94. Xing, H.; Zhang, Z.; Mao, Q.; Wang, C.; Zhou, Y.; Zhou, X.; Ying, L.; Xu, H.; Hu, S.; Zhang, N. Injectable exosome-functionalized extracellular matrix hydrogel for metabolism balance and pyroptosis regulation in intervertebral disc degeneration. *J. Nanobiotechnol.* **2021**, *19*, 264. [CrossRef] [PubMed]
95. Tang, Q.; Lu, B.; He, J.; Chen, X.; Fu, Q.; Han, H.; Luo, C.; Yin, H.; Qin, Z.; Lyu, D.; et al. Exosomes-loaded thermosensitive hydrogels for corneal epithelium and stroma regeneration. *Biomaterials* **2022**, *280*, 121320. [CrossRef]
96. Robbins, P.D.; Morelli, A.E. Regulation of immune responses by extracellular vesicles. *Nat. Rev. Immunol.* **2014**, *14*, 195–208. [CrossRef]
97. Singh, P.P.; Li, L.; Schorey, J.S. Exosomal RNA from Mycobacterium tuberculosis-Infected Cells Is Functional in Recipient Macrophages. *Traffic* **2015**, *16*, 555–571. [CrossRef]
98. Wang, R.; Ji, Q.; Meng, C.; Liu, H.; Fan, C.; Lipkind, S.; Wang, Z.; Xu, Q. Role of gingival mesenchymal stem cell exosomes in macrophage polarization under inflammatory conditions. *Int. Immunopharmacol.* **2020**, *81*, 106030. [CrossRef]
99. Imai, T.; Takahashi, Y.; Nishikawa, M.; Kato, K.; Morishita, M.; Yamashita, T.; Matsumoto, A.; Charoenviriyakul, C.; Takakura, Y. Macrophage-dependent clearance of systemically administered B16BL6-derived exosomes from the blood circulation in mice. *J. Extracell. Vesicles* **2015**, *4*, 26238. [CrossRef]
100. Gupta, D.; Liang, X.; Pavlova, S.; Wiklander, O.P.B.; Corso, G.; Zhao, Y.; Saher, O.; Bost, J.; Zickler, A.M.; Piffko, A.; et al. Quantification of extracellular vesicles in vitro and in vivo using sensitive bioluminescence imaging. *J. Extracell Vesicles* **2020**, *9*, 1800222. [CrossRef]
101. Ruel-Gariepy, E.; Leroux, J.C. In situ-forming hydrogels—Review of temperature-sensitive systems. *Eur. J. Pharm. Biopharm.* **2004**, *58*, 409–426. [CrossRef] [PubMed]
102. Mathew, A.P.; Uthaman, S.; Cho, K.-H.; Cho, C.-S.; Park, I.-K. Injectable hydrogels for delivering biotherapeutic molecules. *Int. J. Biol. Macromol.* **2018**, *110*, 17–29. [CrossRef] [PubMed]
103. Piantanida, E.; Alonci, G.; Bertucci, A.; De Cola, L. Design of Nanocomposite Injectable Hydrogels for Minimally Invasive Surgery. *Acc. Chem. Res.* **2019**, *52*, 2101–2112. [CrossRef] [PubMed]
104. Griveau, L.; Lafont, M.; Le Goff, H.; Drouglazet, C.; Robbiani, B.; Berthier, A.; Sigaudo-Roussel, D.; Latif, N.; Le Visage, C.; Gache, V.; et al. Design and characterization of an in vivo injectable hydrogel with effervescently generated porosity for regenerative medicine. *Acta Biomater.* **2022**, *140*, 324–337. [CrossRef] [PubMed]
105. Qin, Y.; Wang, L.; Gao, Z.; Chen, G.; Zhang, C. Bone marrow stromal/stem cell-derived extracellular vesicles regulate osteoblast activity and differentiation in vitro and promote bone regeneration in vivo. *Sci. Rep.* **2016**, *6*, 21961. [CrossRef]
106. Nicodemus, G.D.; Bryant, S.J. Cell encapsulation in biodegradable hydrogels for tissue engineering applications. *Tissue Eng. Part B Rev.* **2008**, *14*, 149–165. [CrossRef]
107. Thomas, V.; Yallapu, M.M.; Sreedhar, B.; Bajpai, S.K. Breathing-In/Breathing-Out Approach to Preparing Nanosilver-Loaded Hydrogels: Highly Efficient Antibacterial Nanocomposites. *J. Appl. Polym. Sci.* **2009**, *111*, 934–944. [CrossRef]

108. Shen, Z.; Kuang, S.; Zhang, Y.; Yang, M.; Qin, W.; Shi, X.; Lin, Z. Chitosan hydrogel incorporated with dental pulp stem cell-derived exosomes alleviates periodontitis in mice via a macrophage-dependent mechanism. *Bioact. Mater.* **2020**, *5*, 1113–1126. [CrossRef]
109. Mi, B.; Chen, L.; Xiong, Y.; Yang, Y.; Panayi, A.C.; Xue, H.; Hu, Y.; Yan, C.; Hu, L.; Xie, X.; et al. Osteoblast/Osteoclast and Immune Cocktail Therapy of an Exosome/Drug Delivery Multifunctional Hydrogel Accelerates Fracture Repair. *Acs Nano* **2022**, *6*, 771–782. [CrossRef]
110. Xu, Z.; Tsai, H.-i.; Xiao, Y.; Wu, Y.; Su, D.; Yang, M.; Zha, H.; Yan, F.; Liu, X.; Cheng, F.; et al. Engineering Programmed Death Ligand-1/Cytotoxic T-Lymphocyte-Associated Antigen-4 Dual-Targeting Nanovesicles for Immunosuppressive Therapy in Transplantation. *Acs Nano* **2020**, *14*, 7959–7969. [CrossRef]
111. Guan, P.; Liu, C.; Xie, D.; Mao, S.; Ji, Y.; Lin, Y.; Chen, Z.; Wang, Q.; Fan, L.; Sun, Y. Exosome-loaded extracellular matrix-mimic hydrogel with anti-inflammatory property Facilitates/promotes growth plate injury repair. *Bioact. Mater.* **2022**, *10*, 145–158. [CrossRef] [PubMed]
112. Zhang, Y.; Wang, X.; Chen, J.; Qian, D.; Gao, P.; Qin, T.; Jiang, T.; Yi, J.; Xu, T.; Huang, Y.; et al. Exosomes derived from platelet-rich plasma administration in site mediate cartilage protection in subtalar osteoarthritis. *J Nanobiotechnol.* **2022**, *20*, 56. [CrossRef] [PubMed]
113. Lin, J.; Wang, Z.; Huang, J.; Tang, S.; Saiding, Q.; Zhu, Q.; Cui, W. Microenvironment-Protected Exosome-Hydrogel for Facilitating Endometrial Regeneration, Fertility Restoration, and Live Birth of Offspring. *Small* **2021**, *17*, e2007235. [CrossRef] [PubMed]
114. Wang, C.; Liang, C.; Wang, R.; Yao, X.; Guo, P.; Yuan, W.; Liu, Y.; Song, Y.; Li, Z.; Xie, X. The fabrication of a highly efficient self-healing hydrogel from natural biopolymers loaded with exosomes for the synergistic promotion of severe wound healing. *Biomater. Sci.* **2019**, *8*, 313–324. [CrossRef] [PubMed]
115. Zhu, D.; Li, Z.; Huang, K.; Caranasos, T.G.; Rossi, J.S.; Cheng, K. Minimally invasive delivery of therapeutic agents by hydrogel injection into the pericardial cavity for cardiac repair. *Nat. Commun.* **2021**, *12*, 1412. [CrossRef] [PubMed]
116. Zou, Y.; Li, L.; Li, Y.; Chen, S.; Xie, X.; Jin, X.; Wang, X.; Ma, C.; Fan, G.; Wang, W. Restoring Cardiac Functions after Myocardial Infarction-Ischemia/Reperfusion via an Exosome Anchoring Conductive Hydrogel. *Acs Appl. Mater. Interfaces* **2021**, *13*, 56892–56908. [CrossRef]
117. Waters, R.; Alam, P.; Pacelli, S.; Chakravarti, A.R.; Ahmed, R.P.H.; Paul, A. Stem cell-inspired secretome-rich injectable hydrogel to repair injured cardiac tissue. *Acta Biomater.* **2018**, *69*, 95–106. [CrossRef]
118. Liang, L.; Shen, Y.; Dong, Z.; Gu, X. Photoacoustic image-guided corpus cavernosum intratunical injection of adipose stem cell-derived exosomes loaded polydopamine thermosensitive hydrogel for erectile dysfunction treatment. *Bioact. Mater.* **2022**, *9*, 147–156. [CrossRef]
119. Li, L.; Zhang, Y.; Mu, J.; Chen, J.; Zhang, C.; Cao, H.; Gao, J. Transplantation of Human Mesenchymal Stem-Cell-Derived Exosomes Immobilized in an Adhesive Hydrogel for Effective Treatment of Spinal Cord Injury. *Nano Lett.* **2020**, *20*, 4298–4305. [CrossRef]
120. Yu, H.; Cheng, J.; Shi, W.; Ren, B.; Zhao, F.; Shi, Y.; Yang, P.; Duan, X.; Zhang, J.; Fu, X.; et al. Bone marrow mesenchymal stem cell-derived exosomes promote tendon regeneration by facilitating the proliferation and migration of endogenous tendon stem/progenitor cells. *Acta Biomater.* **2020**, *106*, 328–341. [CrossRef]
121. Chen, S.; Tang, Y.; Liu, Y.; Zhang, P.; Lv, L.; Zhang, X.; Jia, L.; Zhou, Y. Exosomes derived from miR-375-overexpressing human adipose mesenchymal stem cells promote bone regeneration. *Cell Prolif.* **2019**, *52*, e12669. [CrossRef] [PubMed]
122. De Witte, T.-M.; Fratila-Apachitei, L.E.; Zadpoor, A.A.; Peppas, N.A. Bone tissue engineering via growth factor delivery: From scaffolds to complex matrices. *Regen. Biomater.* **2018**, *5*, 197–211. [CrossRef] [PubMed]
123. Wu, D.; Qin, H.; Wang, Z.; Yu, M.; Liu, Z.; Peng, H.; Liang, L.; Zhang, C.; Wei, X. Bone Mesenchymal Stem Cell-Derived sEV-Encapsulated Thermosensitive Hydrogels Accelerate Osteogenesis and Angiogenesis by Release of Exosomal miR-21. *Front. Bioeng. Biotechnol.* **2021**, *9*, 829136. [CrossRef] [PubMed]
124. Yu, W.; Li, S.; Guan, X.; Zhang, N.; Xie, X.; Zhang, K.; Bai, Y. Higher yield and enhanced therapeutic effects of exosomes derived from MSCs in hydrogel-assisted 3D culture system for bone regeneration. *Mater. Sci. Eng. C Mater. Biol. Appl.* **2022**, 112646. [CrossRef] [PubMed]
125. Martin, P. Wound healing-Aiming for perfect skin regeneration. *Science* **1997**, *276*, 75–81. [CrossRef] [PubMed]
126. Frykberg, R.G.; Banks, J. Challenges in the Treatment of Chronic Wounds. *Adv. Wound Care* **2015**, *4*, 560–582. [CrossRef]
127. Xu, N.; Wang, L.; Guan, J.; Tang, C.; He, N.; Zhang, W.; Fu, S. Wound healing effects of a Curcuma zedoaria polysaccharide with platelet-rich plasma exosomes assembled on chitosan/silk hydrogel sponge in a diabetic rat model. *Int. J. Biol. Macromol.* **2018**, *117*, 102–107. [CrossRef]
128. Shiekh, P.A.; Singh, A.; Kumar, A. Exosome laden oxygen releasing antioxidant and antibacterial cryogel wound dressing OxOBand alleviate diabetic and infectious wound healing. *Biomaterials* **2020**, *249*, 120020. [CrossRef]
129. Shi, Q.; Qian, Z.; Liu, D.; Sun, J.; Wang, X.; Liu, H.; Xu, J.; Guo, X. GMSC-Derived Exosomes Combined with a Chitosan/Silk Hydrogel Sponge Accelerates Wound Healing in a Diabetic Rat Skin Defect Model. *Front. Physiol.* **2017**, *8*, 904. [CrossRef]
130. Han, C.; Liu, F.; Zhang, Y.; Chen, W.; Luo, W.; Ding, F.; Lu, L.; Wu, C.; Li, Y. Human Umbilical Cord Mesenchymal Stem Cell Derived Exosomes Delivered Using Silk Fibroin and Sericin Composite Hydrogel Promote Wound Healing. *Front. Cardiovasc. Med.* **2021**, *8*, 713021. [CrossRef]

131. Tao, S.C.; Guo, S.C.; Li, M.; Ke, Q.F.; Guo, Y.P.; Zhang, C.Q. Chitosan Wound Dressings Incorporating Exosomes Derived from MicroRNA-126-Overexpressing Synovium Mesenchymal Stem Cells Provide Sustained Release of Exosomes and Heal Full-Thickness Skin Defects in a Diabetic Rat Model. *Stem Cells Transl. Med.* **2017**, *6*, 736–747. [CrossRef] [PubMed]
132. Fang, S.; Xu, C.; Zhang, Y.T.; Xue, C.Y.; Yang, C.; Bi, H.D.; Qian, X.J.; Wu, M.J.; Ji, K.H.; Zhao, Y.P.; et al. Umbilical Cord-Derived Mesenchymal Stem Cell-Derived Exosomal MicroRNAs Suppress Myofibroblast Differentiation by Inhibiting the Transforming Growth Factor-beta/SMAD2 Pathway During Wound Healing. *Stem Cells Transl. Med.* **2016**, *5*, 1425–1439. [CrossRef] [PubMed]
133. Hashimoto, H.; Olson, E.N.; Bassel-Duby, R. Therapeutic approaches for cardiac regeneration and repair. *Nat. Rev. Cardiol.* **2018**, *15*, 585–600. [CrossRef] [PubMed]
134. Zhang, K.; Zhao, X.; Chen, X.; Wei, Y.; Du, W.; Wang, Y.; Liu, L.; Zhao, W.; Han, Z.; Kong, D.; et al. Enhanced Therapeutic Effects of Mesenchymal Stem Cell-Derived Exosomes with an Injectable Hydrogel for Hindlimb Ischemia Treatment. *ACS Appl Mater Interfaces* **2018**, *10*, 30081–30091. [CrossRef]
135. Monguio-Tortajada, M.; Prat-Vidal, C.; Moron-Font, M.; Clos-Sansalvador, M.; Calle, A.; Gastelurrutia, P.; Cserkoova, A.; Morancho, A.; Ramirez, M.A.; Rosell, A.; et al. Local administration of porcine immunomodulatory, chemotactic and angiogenic extracellular vesicles using engineered cardiac scaffolds for myocardial infarction. *Bioact. Mater.* **2021**, *6*, 3314–3327. [CrossRef]
136. Wang, Q.; Zhang, L.; Sun, Z.; Chi, B.; Zou, A.; Mao, L.; Xiong, X.; Jiang, J.; Sun, L.; Zhu, W.; et al. HIF-1alpha overexpression in mesenchymal stem cell-derived exosome-encapsulated arginine-glycine-aspartate (RGD) hydrogels boost therapeutic efficacy of cardiac repair after myocardial infarction. *Mater. Today Bio* **2021**, *12*, 100171. [CrossRef]
137. Ahuja, C.S.; Wilson, J.R.; Nori, S.; Kotter, M.R.N.; Druschel, C.; Curt, A.; Fehlings, M.G. Traumatic spinal cord injury. *Nat. Rev. Dis. Primers* **2017**, *3*, 17018. [CrossRef]
138. Xu, X.H.; Yuan, T.J.; Dad, H.A.; Shi, M.Y.; Huang, Y.Y.; Jiang, Z.H.; Peng, L.H. Plant Exosomes As Novel Nanoplatforms for MicroRNA Transfer Stimulate Neural Differentiation of Stem Cells In Vitro and In Vivo. *Nano Lett.* **2021**, *21*, 8151–8159. [CrossRef]
139. Liu, H.; Gao, W.; Yuan, J.; Wu, C.; Yao, K.; Zhang, L.; Ma, L.; Zhu, J.; Zou, Y.; Ge, J. Exosomes derived from dendritic cells improve cardiac function via activation of CD4(+) T lymphocytes after myocardial infarction. *J. Mol. Cell. Cardiol.* **2016**, *91*, 123–133. [CrossRef]
140. Zhang, Y.; Cai, Z.; Shen, Y.; Lu, Q.; Gao, W.; Zhong, X.; Yao, K.; Yuan, J.; Liu, H. Hydrogel-load exosomes derived from dendritic cells improve cardiac function via Treg cells and the polarization of macrophages following myocardial infarction. *J Nanobiotechnol.* **2021**, *19*, 271. [CrossRef]
141. van Koppen, A.; Joles, J.A.; van Balkom, B.W.M.; Lim, S.K.; de Kleijn, D.; Giles, R.H.; Verhaar, M.C. Human embryonic mesenchymal stem cell-derived conditioned medium rescues kidney function in rats with established chronic kidney disease. *PLoS ONE* **2012**, *7*, e38746. [CrossRef] [PubMed]
142. Cantaluppi, V.; Gatti, S.; Medica, D.; Figliolini, F.; Bruno, S.; Deregibus, M.C.; Sordi, A.; Biancone, L.; Tetta, C.; Camussi, G. Microvesicles derived from endothelial progenitor cells protect the kidney from ischemia-reperfusion injury by microRNA-dependent reprogramming of resident renal cells. *Kidney Int.* **2012**, *82*, 412–427. [CrossRef] [PubMed]
143. Fuhrmann, G.; Chandrawati, R.; Parmar, P.A.; Keane, T.J.; Maynard, S.A.; Bertazzo, S.; Stevens, M.M. Engineering Extracellular Vesicles with the Tools of Enzyme Prodrug Therapy. *Adv. Mater.* **2018**, *30*, 1706616. [CrossRef] [PubMed]
144. Liu, W.; Yu, M.; Chen, F.; Wang, L.; Ye, C.; Chen, Q.; Zhu, Q.; Xie, D.; Shao, M.; Yang, L. A novel delivery nanobiotechnology: Engineered miR-181b exosomes improved osteointegration by regulating macrophage polarization. *J Nanobiotechnol.* **2021**, *19*, 269. [CrossRef]
145. Hippler, M.; Lemma, E.D.; Bertels, S.; Blasco, E.; Barner-Kowollik, C.; Wegener, M.; Bastmeyer, M. 3D Scaffolds to Study Basic Cell Biology. *Adv. Mater.* **2019**, *31*, 1808110. [CrossRef]
146. Patwardhan, S.; Mahadik, P.; Shetty, O.; Sen, S. ECM stiffness-tuned exosomes drive breast cancer motility through thrombospondin-1. *Biomaterials* **2021**, *279*, 121185. [CrossRef]
147. Palviainen, M.; Saari, H.; Karkkainen, O.; Pekkinen, J.; Auriola, S.; Yliperttula, M.; Puhka, M.; Hanhineva, K.; Siljander, P.R.M. Metabolic signature of extracellular vesicles depends on the cell culture conditions. *J. Extracell. Vesicles* **2019**, *8*, 1596669. [CrossRef]
148. Millan, C.; Prause, L.; Vallmajo-Martin, Q.; Hensky, N.; Eberli, D. Extracellular Vesicles from 3D Engineered Microtissues Harbor Disease-Related Cargo Absent in EVs from 2D Cultures. *Adv. Healthc. Mater.* **2022**, *11*, e2002067. [CrossRef]
149. van Balkom, B.W.M.; de Jong, O.G.; Smits, M.; Brummelman, J.; den Ouden, K.; de Bree, P.M.; van Eijndhoven, M.A.J.; Pegtel, D.M.; Stoorvogel, W.; Wuerdinger, T.; et al. Endothelial cells require miR-214 to secrete exosomes that suppress senescence and induce angiogenesis in human and mouse endothelial cells. *Blood* **2013**, *121*, 3997–4006. [CrossRef]
150. Bobryshev, Y.V.; Killingsworth, M.C.; Orekhov, A.N. Increased Shedding of Microvesicles from Intimal Smooth Muscle Cells in Athero-Prone Areas of the Human Aorta: Implications for Understanding of the Predisease Stage. *Pathobiology* **2013**, *80*, 24–31. [CrossRef]
151. Zhu, J.M.; Liu, B.; Wang, Z.Y.; Di, W.; Ni, H.E.; Zhang, L.L.; Wang, Y. Exosomes from nicotine-stimulated macrophages accelerate atherosclerosis through miR-21-3p/PTEN-mediated VSMC migration and proliferation. *Theranostics* **2019**, *9*, 6901–6919. [CrossRef]
152. Collins, M.N.; Ren, G.; Young, K.; Pina, S.; Reis, R.L.; Oliveira, J.M. Scaffold Fabrication Technologies and Structure/Function Properties in Bone Tissue Engineering. *Adv. Funct. Mater.* **2021**, *31*, 2010609. [CrossRef]
153. Huang, C.C.; Narayanan, R.; Alapati, S.; Ravindran, S. Exosomes as biomimetic tools for stem cell differentiation: Applications in dental pulp tissue regeneration. *Biomaterials* **2016**, *111*, 103–115. [CrossRef] [PubMed]

154. Chen, M.; Luo, D. A CRISPR Path to Cutting-Edge Materials. *N. Engl. J. Med.* **2020**, *382*, 85–88. [CrossRef]
155. Khayambashi, P.; Iyer, J.; Pillai, S.; Upadhyay, A.; Zhang, Y.; Tran, S.D. Hydrogel Encapsulation of Mesenchymal Stem Cells and Their Derived Exosomes for Tissue Engineering. *Int. J. Mol. Sci.* **2021**, *22*, 684. [CrossRef] [PubMed]

gels

MDPI

Review

Novel Gels: An Emerging Approach for Delivering of Therapeutic Molecules and Recent Trends

Trideva K. Sastri, Vishal N. Gupta *, Souvik Chakraborty, Sharadha Madhusudhan, Hitesh Kumar, Pallavi Chand, Vikas Jain, Balamuralidhara Veeranna and Devegowda V. Gowda

Department of Pharmaceutics, JSS College of Pharmacy, JSS Academy of Higher Education & Research, Sri Shivarathreeshwara Nagar, Mysuru 570015, India; trideva.k@gmail.com (T.K.S.); souvik93pharmacist@gmail.com (S.C.); msharadha1996@gmail.com (S.M.); hitesh.sahu1921@gmail.com (H.K.); pallavichand1990@gmail.com (P.C.); vikasjain@jssuni.edu.in (V.J.); baligowda@jssuni.edu.in (B.V.); dvgowda@jssuni.edu.in (D.V.G.)

* Correspondence: vkguptajss@gmail.com

Abstract: Gels are semisolid, homogeneous systems with continuous or discrete therapeutic molecules in a suitable lipophilic or hydrophilic three-dimensional network base. Innovative gel systems possess multipurpose applications in cosmetics, food, pharmaceuticals, biotechnology, and so forth. Formulating a gel-based delivery system is simple and the delivery system enables the release of loaded therapeutic molecules. Furthermore, it facilitates the delivery of molecules via various routes as these gel-based systems offer proximal surface contact between a loaded therapeutic molecule and an absorption site. In the past decade, researchers have potentially explored and established a significant understanding of gel-based delivery systems for drug delivery. Subsequently, they have enabled the prospects of developing novel gel-based systems that illicit drug release by specific biological or external stimuli, such as temperature, pH, enzymes, ultrasound, antigens, etc. These systems are considered smart gels for their broad applications. This review reflects the significant role of advanced gel-based delivery systems for various therapeutic benefits. This detailed discussion is focused on strategies for the formulation of different novel gel-based systems, as well as it highlights the current research trends of these systems and patented technologies.

Keywords: hydrogels; in situ gels; emulsion gels; microgels; nanogels; vesicular gels

Citation: Sastri, T.K.; Gupta, V.N.; Chakraborty, S.; Madhusudhan, S.; Kumar, H.; Chand, P.; Jain, V.; Veeranna, B.; Gowda, D.V. Novel Gels: An Emerging Approach for Delivering of Therapeutic Molecules and Recent Trends. *Gels* **2022**, *8*, 316. https://doi.org/10.3390/gels8050316

Academic Editor: Hiroyuki Takeno

Received: 30 April 2022
Accepted: 17 May 2022
Published: 19 May 2022

1. Introduction

In recent years, novel drug delivery systems have proven very adept at delivering therapeutic molecules with site-specific and localized effects. Additionally, these systems facilitate drug release at desired rates and simultaneously lower the undesired effects [1]. Gels are three-dimensional, semi-solid systems consisting of polymeric matrices. These behave in the same way as solid systems; however, they consist of relatively higher liquid components than solid dispersions [2,3]. Gel systems comprise long, arbitrary chains, albeit with reversible links at precise points. These systems comprise minimum two components and are fundamentally coherent colloidal dispersion systems [4]. The system components, namely the dispersion medium and the dispersed constituent, are uniformly scattered throughout the system. Gels are usually transparent or translucent in appearance entailing higher amounts of solvent [5]. When a suitable solvent is employed, the gelling agents entangle to form a three-dimensional colloidal network that confines fluid movement by entrapment and achieves immobilization of solvent molecules [6]. The network governs the viscoelastic properties of the gel system by developing endurance against deformation. In other words, the thixotropic behavior is contributed by the matrix's structure [7]. Gels are prepared mainly by fusion technique or by employing gelling agents. Gel-based systems can be alienated into two categories, organogels and hydrogels, based on the physical state of the gelling agent dispersion [8]. Dispersible colloids and water-soluble components constitute hydrogels, while lipophilic oleaginous components are employed in organogels [9]. The systems are further classified into xerogels or aqueous gels based on the nature of the solvents. Xerogels are solid gels with a minimum solvent concentration obtained mainly by solvent evaporation, thereby attaining a gel network [10,11]. However, the gel state can be reinstated by incorporating an imbibing agent that swells the matrix. Novel gels are capable of controlled and sustained release of loaded therapeutic molecules. Figure 1 portrays common novel gel-based delivery systems [12]. Smart gels can be developed which respond to biological and external stimuli, such as temperature, pH, chemical, enzymes, electrical, light, antigens, etc. These systems are highly instrumental in lowering undesired effects and are biodegradable and biocompatible [13]. High drug loading can be achieved. Their size (nanogels) expedites high drug accumulation at the tissue level and enables stealth systems by evading phagocytic cells [14–16]. Their distinctive surface properties enable passive and active targeting. This review underlines the advances in gel-based delivery systems, their developments, and a current update in the delivery of therapeutic molecules.

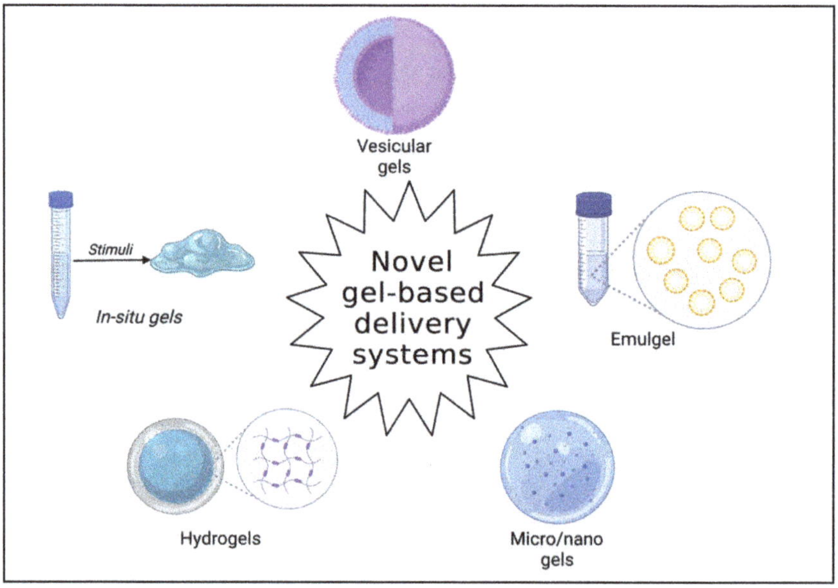

Figure 1. Common novel gel-based delivery systems (Created with BioRender.com, accessed on 28 April 2022).

2. Advances in Novel Gel-Based Delivery Systems

Novel gel-based drug delivery systems are classified by the nature of their structural network and by their response to stimuli [15]. The former is either a chemically aligned gel network or a physically aligned gel network system. At the same time, the latter category entails responsive, intelligent gel systems that imbibe solvents and swell on exposure to stimuli, such as temperature, pH, chemical, enzymes, electrical, light, antigens, etc. [17,18]. Novel gel systems are evaluated for rigorous characterizations to understand their efficacy as delivery systems. The most commonly employed evaluation parameters comprise swelling capacity, size and morphology, rheological properties, surface charge, etc. [19]. Further, they are scrutinized for physical appearance for compliance, physical state, homogeneity, and phase separation to understand their stability, extrudability, and spreading coefficient is significant for topical gels, as well as bioadhesive strength is a vital element for mucoadhesive gels [20]. Drug content, permeability, and release play a substantial role in any drug delivery system. The International Council on Harmonisation

(ICH) (Geneva, Switzerland) dictates the stability guidelines. The gels are subjected to various stress conditions and later scrutinized for drug content, release, and entrapment efficiency to assess their compliance [21]. Figure 2 illustrates various potential delivery routes for novel gel-based delivery systems.

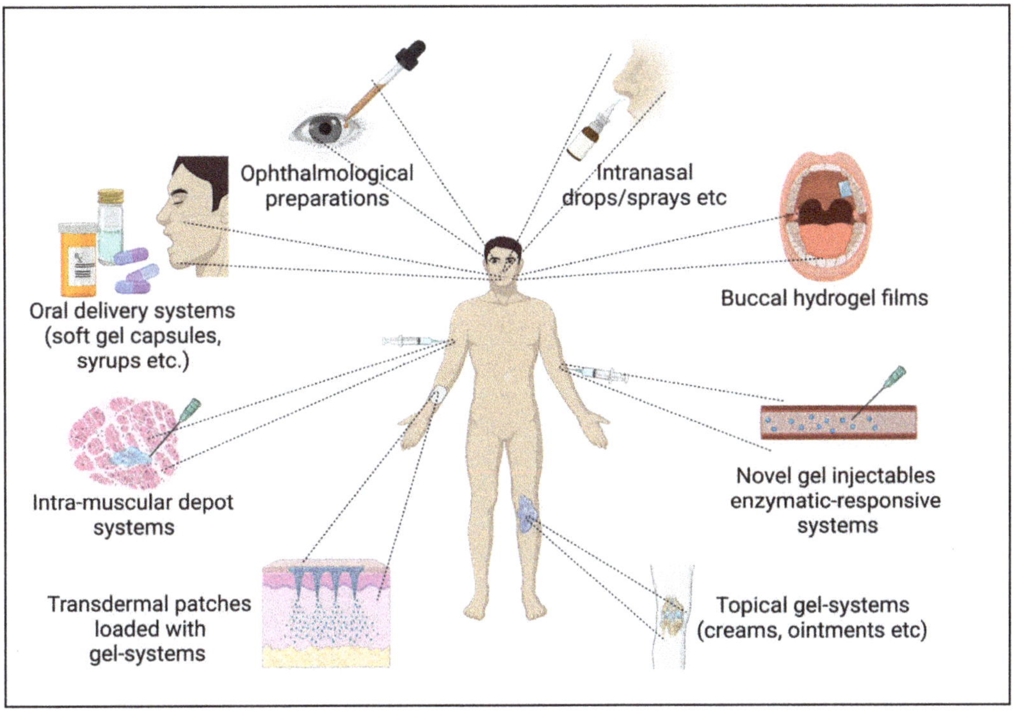

Figure 2. Potential delivery routes for novel gel-based delivery system. (Created with BioRender.com, accessed on 28 April 2022).

2.1. Intelligent Hydrogels

A three-dimensional network of hydrophilic polymers composes of hydrogels that inherently imbibe water and maintain the system's integrity [22]. These systems are one of the most versatile delivery systems among novel gels. Researchers have comprehensively developed several intelligent hydrogels that precisely retort to numerous physical stimuli such as temperature, light, electric fields, pressure, sound, and magnetic fields in recent years. Furthermore, stimuli pertain to pH, ions, enzymes, etc. These systems are beneficial for formulating controlled delivery systems [23–25].

Temperature-responsive hydrogels are triggered by a precisely established temperature range. These hydrogels are formulated with polymers that are capable of temperature-triggered phase transitions [26]. Regular polymers exhibit higher solubility with an increase in temperature; however, the polymers employed in temperature-responsive systems, such as poly (N, N-diethylacrylamide), poly (tertramethyleneether glycol), poly(N-isopropylacrylamide), and others, possess lower critical solution temperatures. Polymers with lower critical solution temperatures tend to shrink with the increase in temperature; these hydrogels are known as negative temperature responsive [22]. Poly(acrylic acid) and polyacrylamide polymers inherently imbibe at higher temperatures and shrink at lower temperatures; hence, hydrogels prepared from these polymers are positive temperature responsive. However, tetronics and pluronics are applied to formulate thermally reversible gels [27].

Electrical signal-responsive hydrogels endure swelling and contracting when subjected to an electric field. The demerit of some systems is that due to the charge orientation, one side swells while the other contracts, thus, comprising the stability [28]. Polyelectrolytes, such as poly(2-acrylamido-2-methylpropane sulphonic acid-co-n-butlymethacrylate), are usually employed to formulate these systems [29,30].

pH-responsive hydrogels tend to release or accept protons depending on the pH of the site. Researchers have studied these systems extensively and reported encouraging results. Poly(N,N′-diethylaminoethylmethacrylate) is ionized at a low pH, unlike poly(acrylic acid), which ionizes at a higher pH. However, polycations tend to swell less at neutral pH [31,32].

In enzyme-responsive systems, a suitable enzyme is considered to trigger a release or deliver at a precisely targeted site where the enzyme is operational at a specific temperature or pH [33]. Enzyme-responsive hydrogels are usually prepared from cellulose and other suitable polymers that facilitate the macromolecular networks and function in a controlled environment [34]. The most explored enzymatic stimuli-responsive system consists of a triggerable agent (usually a polymer or a lipid) into which a therapeutic molecule is incorporated. Indeed, this active agent is sensitive to swelling or degradation when it reaches the target site. Some reported enzymes include protease- and glycosidase-based catalyzed enzymatic reactions [35].

Other intelligent systems include light-responsive hydrogels that are functional in ophthalmic delivery systems. These systems are responsive to light and other stimuli, including pressure, thrombin, antigen, and so forth [36–38].

2.2. In Situ Gels

Over the years, in situ gels have exhibited tremendous benefits in controlled drug delivery systems and emerged as a significant intelligent drug delivery technology [39]. These incredible systems remain in a liquid state at room temperature and achieve a sol-gel state when exposed to any biological environments, such as altered pH and temperature. In other words, in situ gelling is a spontaneous gelation process at a specific site post-administration [40].

This system accommodates numerous routes of administration, namely ocular, oral, intranasal, vaginal, rectal, depot system, etc. In situ systems have proven their benefits, including prolonged residence time at the site of application. As a result, there is a marked reduction in dosage regimen. Good thixotropic properties expedite the flexibility for formulation development. Rapid absorption and onset can be easily achieved. As a result, the therapeutic benefits can be achieved at lower doses with minimal side effects. Besides, they expedite systemic circulation, avoiding localized hepatic circulation, and targeting can be accomplished [41,42].

In situ gel systems principally work on stimuli such as temperature alteration (chitosan, poloxamer), ion activation (sodium alginate), pH changes (carbopol), solvent exchange, and environmental factors. The gels are ideally dependent on physical or chemical mechanisms. The physical mechanism constitutes of imbibing liquids, mainly water and diffusion, and absorbing water by gel polymers in site-specific locations. While diffusion entails the solvent penetration from the polymer solution to the neighboring tissues, the polymer solidifies. Temperature–responsive, pH-responsive, enzymatic cross-linking, and ionic cross-linking are effective mechanisms that govern the precipitation of solids in gel systems [43].

The pH-responsive systems include fewer polymers such as cellulose acetate phthalate, carbopol, pseudolatexes, polyethylene glycol, and polymethacrilicacid. The temperature-responsive systems primarily form gels with temperature variations and they include polymers such as pluronics, chitosan, xyloglucans etc. [44,45]. Enzymatic cross-linking is governed by natural biological enzymes. The rate of gel formation is proportional to the enzyme concentration. Insulin delivery was studied with a smart stimuli-responsive delivery system and exhibited positive outcomes, e.g., in a study reported by Podual et.al., where the glucose oxidase enzyme was employed to facilitate the release [46–49]. In ionic cross-linking, different ions dictate the phase transition of polymers. Gellan gum, carrageenan, and alginic acid are a few ion-responsive polymers. Several natural polymers are available in nature, such as carrageenan, which transform to a gel state on exposure to ions. Gellan gum is predominantly available as Gelrite (commercially available). It is an anionic polysaccharide that instinctively forms a gel in the presence of Mg^{++}, Na^{+}, K^{+}, and Ca^{++}. Electromagnetic radiation is applied to facilitate in situ gels in photo-polymerization techniques [50,51]. This is achieved by injecting reactive macromers or a solution with initiators and monomers into the desired tissue site. To initiate photo-polymerization, specific polymerizable functional groups and acrylate or similar macromers that undergo dissociation in the presence of a photo-initiator are subjected to radiation. In ultraviolet photo-polymerization, a ketone, such as 2,2 dimethoxy-2-phenyl acetophenone, is used as the initiator, while camphorquinone and ethyl eosin initiators are used in visible light systems [52–55].

2.3. Emulsion Gels

An emulgel is an amalgamation of gel technology and emulsions. This system offers controlled release, especially in topical formulations, as the therapeutic molecules are loaded in a dual delivery system emulsion and a gel core. The fusion of these dual delivery systems overcomes the demerits of these conventional systems, such as stability and drug loading [56]. Emulgels are prepared by incorporating gelling agents in the continuous (usually water) emulsion phase. Compared to other gels, these systems facilitate higher entrapment efficiency, desired thixotropic behavior, and better patient compliance. The formulation of emulgels is achieved by loading the drug-loaded emulsion into a pre-gel and applying a shear to achieve a homogenous gel system [57]. Gelling agents, emulsifiers, water, and penetration enhancers (primarily for topical formulations) are the fundamental components of emulgels. Polyethylene glycol, tweens, spans, and so forth are utilized as emulsifiers, while carbopols, including Hydroxypropyl methylcellulose (HPMC), are used as gelling agents. Menthol, oleic acid, etc., are mostly effective penetration enhancers. Emulgels are amphiphilic and enable the load of both hydrophilic and lipophilic moieties. Furthermore, the bioavailability of specific molecules was enhanced by lowering the globule size to a micron (μm). Microemulsions are isotropic, clearer, and stable systems. As a result, the increased effective surface facilitates a higher bioavailability [58]. These systems are proven better than emulgels due to the significant increase in the penetration of topical formulations and better compliance [59].

2.4. Microgels

Microgels are described as gels that have a size range in microns (μm). In contrast to typical gels, these systems have cross-linking structures in microns (μm) and are colloidal dispersions [60]. The molecular arrangements are different in these systems, in comparison to regular gels. Electric charge, polymer–water bonding, and cross-link density are a few factors that underlie the swelling capacity of these systems [61]. The internal forces impart steadiness to the microgels. Similar to microgels, stimuli-responsive methods yield novelty. The colloidal nature of microgels augments characteristics such as controlled delivery of therapeutic molecules, high responsiveness to stimuli, extrudability, mass transport, and high therapeutic efficacy with minimal adversity [62]. The microgels can be created by reducing the size of macrogels by applying a shear or by employing monomers or polymers. The basic underlying principles for formulating are emulsification, where a pregel is prepared in the oil phase, followed by polymerization, yielding microgels. In nucleation, the adjacent particles (in the solvent) initiate the nucleation process and deliver homogeneous microgels [63]. No external forces are required. In contrast, complexation, where complexes are formed in water, can be achieved by adding two mild polymers to water. Another peculiar method can be adopted for developing microgels by the mere addition of polyelectrolytes, which are oppositely charged and at diluted concentrations yield colloidal dispersion [64].

2.5. Nanogels

Nanogels are mostly hydrogel polymeric network dispersions with particles lying in the nanometer range (nm). Nanogels allow chemical modifications to facilitate ligand targeting and triggered release. This system is a blend of colloidal dispersions and cross-linking networks of polymers [65]. Nanogels are capable of intracellular delivery with enhanced cellular concentrations without significant cellular toxicity [66]. Other advantages include rapid swelling, sharp responsiveness to stimuli, high bioavailability, accommodating more therapeutic moieties, avoiding renal clearance, and remaining undetected by opsonins. Besides, nanogels have excellent stability and accommodate highly lipophilic moieties [67]. Distinctly, nanogels can be prepared by chemical cross-linking (emulsion polymerization), pulse radiolysis, and photopolymerization. In the first method, the cross-linking agent, monomers, surfactant, and water are added to an organic phase. Later, this organic phase is irradiated and purified. In pulse radiolysis, ionizing radiation is irradiated onto polymers in suitable solvents to promote internal structural reorientations of polymer radicles to yield nanogels. In the photopolymerization technique, monomers are exposed to UV radiation. Initiators, cross-linking agents, or surfactants are not required for this method; hence, pure nanogels are produced [68,69].

Similarly, other procedures include heterogenous controlled radical polymerization. Recently, developed techniques, including reversible addition–fragmentation chain transfer and atom transfer radical polymerization, have been instrumental in developing polymer-conjugate systems. Chemical cross-linking is more commonly employed to formulate varied nanogels and is further classified into Michael addition reactions, carbodiimide coupling, and free radical polymerization [70]. Hydrophilic monomers are polymerized in the presence of varied cross linkers to yield synthetic nanogels. Self-assembly of polymers includes an aggregation of the hydrophilic polymers by electrostatic interactions, hydrophobic bonds, or hydrogen bonding in an aqueous medium. These systems accommodate large molecules; thus, they are effective in incorporating macromolecules, such as proteins and peptides [71].

2.6. Vesicular Gels

Vesicular gels are composed of carrier systems deliberate with amphiphilic molecules comprising lipids, surfactants, and co-polymers [72]. The carrier system consists of a hydrophilic core within an amphiphilic bilayer [73]. Vesicular drug delivery has been on the rise in recent studies for its versatility and broad application in drug delivery, cosmetics, etc. Vesicular gels have exhibited significant results in topical drug delivery. A hydrogel matrix can be incorporated with preformed vesicles in a simple technique [74,75].

2.6.1. Liposomal Gels

Liposomes are a very effective and successful novel delivery system. One significant advantage of these systems is that they possess biosimilar structures with desirable properties [76]. Their amphiphilic properties enable the incorporation of both lipophilic and hydrophilic therapeutic molecules. These carriers are biodegradable, exhibit no toxicity, and support localization and site-specific release. They are capable of infiltrating several bio-obstacles otherwise difficult to any conventional systems [77,78]. However, often the topical applications are limited due to rheological constraints. Traditionally, liposomes can be formulated by the standard technique of lipid film hydration. Other techniques include the French pressure cell method and solvent injection methods. The addition of a gelling agent in the aqueous phase incorporates liposomes into the gel system. Carbopols, xanthan gum, poloxamers, gellan gum, polyvinyl alcohol, and others are explored as gelling agents [79].

2.6.2. Niosomal Gels

Niosomes are similar to the liposomal system; however, these are prepared from non-ionic surfactants. This vesicular system resolves the instability of traditional liposomes due to phospholipids [80]. The bilayer, depending on the preparation technique, yields unilamellar or multilamellar niosomes. The system's integrity relies on the surfactant's chemical composition and on the hydrophilic–lipophilic balance (HLB) value. Primarily, proniosomes, emulsion niosomes, and aspasomes are different types of niosomes. Proniosomes are preliminary niosomes which are devoid of any aqueous phase [81]. These are reconstituted with a suitable buffer to yield niosomes. Proniosomes are advantageous over niosomes regarding dose dumping and stability. Niosomal gels comprise a therapeutic moiety, surfactant, and cholesterol that yield vesicles, which are later loaded into a gel base. However, broad applications of these systems are restricted to achieving controlled release, stealth systemic circulations, and enhanced stability [82,83].

2.6.3. Transferosome Gels

Transferosomes are improved versions of liposomes. These systems overcome various shortcomings of other vesicular systems, including aggregation, dose dumping, and poor permeability [84]. Similar to liposomes, these systems also possess an aqueous core enclosed by a lipid layer. However, these bilayers are modified with edge activators, which enables flexibility. Usually span 80, tween 80, and sodium cholate are proven effective edge activators. Phospholipids, edge activators, alcohol, and hydrating media (buffers) constitute transferosomes [85]. Hydrocolloids are incorporated into buffers to yield transferosome gels. The delivery of proteins and peptides, such as insulin, has shown promising results when incorporated into the transferosome gel system [85–87].

3. Novel Gel-Based Delivery Approaches for Delivering Therapeutic Molecules—Recent Trends

3.1. Hydrogel Systems

Various nanocarrier systems have been utilized in drug delivery applications for different diseases. The number of cancer patients is increasing day by day worldwide. However, the therapy for cancer treatment is still adapted as a conventional method, i.e., surgery and radiotherapy followed by chemotherapy. There are several limitations of conventional chemotherapy, which produces long- or short-term side effects and adverse effects. Such chemotherapy delivery to cancer patients has low bioavailability due to poor solubility issues, resulting in adverse effects on the biological systems. Hence, researchers are finding a better way to deliver chemotherapeutics with enhanced bioavailability to resolve these problems safely. The hydrogel drug delivery system is one of the carrier systems that can deliver hydrophilic chemotherapeutics agents and hydrophobic drugs in a sustained and controlled manner. Generally, hydrogels are hydrated in nature and possess self-shrinking and self-swelling characteristics in different biological conditions [88]. Their 3D structural properties enable the efficient encapsulation of chemotherapeutic agents into their internal structure, which protects the drug from degradation either in storage or enzymatically during circulation in biological systems. The advantage of the hydrogel drug delivery system in cancer therapy is that nanogels can be modified according to the response to the cell or the tumor microenvironments [89].

Hydrogels can be functionalized by targeting ligands for enhanced, prolonged, and specific drug delivery, making it a safer carrier system. In general, hydrogels could achieve high drug delivery efficiency in cancer therapy with conformational changes and degradation under specific conditions, such as temperature, pH, redox, and ultrasound [90].

The second advantage is that hydrogel synthesis can be done as per the demand of the drug delivery to particular tumor microenvironments. Internal and external stimuli have been utilized to design hydrogels for the delivery of chemotherapeutic drugs to manage cancer. The thermo-sensitive stimuli hydrogels are the most common used hydrogels. These hydrogels are usually in a gel state at room temperature and are attributed to a low critical solution temperature. Once it is administered into the body, it will change its form into a solution due to the cellular temperature. Usually, poly(N-isopropylacrylamide) or elastin-like polypeptides have been used in thermo-sensitive hydrogels recently [91,92].

Moreover, thermo-sensitive hydrogel in situ sites avoid the accumulation of chemotherapeutic drugs in the liver or spleen, which overcomes the biosafety limitations of drugs [93]. The chemotherapeutic agent, 7-ethyl-10-hydroxycamptothecin or SN-38, has a lower solubility in drug delivery applications. Bai et al. developed a thermo-sensitive liposomal system, which showed a better antitumor effect and reduced systematic toxicities in in vivo models at the same dose of the pure drug [94]. Similarly, alginate-based thermo-sensitive gels enhanced cisplatin's in vivo antitumoral effects through an in situ injection. They spurred the tumor growth up to 95% compared to the control group, along with an increased prolonged survival rate of the animals [95]. Furthermore, in another study, the interaction of ligand–receptor with the hydrogel enhances transcorneal permeability and precorneal retention of the drug activity. Upon topical instillation, dexamethasone and Arg-gly-asp supramolecular hydrogel increased the transcorneal permeability in rabbits' eyes to treat ocular inflammation [96]. The development of a copolymer from a mono-functional polymer exhibits a good sol-gel transition phase, thereby enhancing water solubility to prolong the mucoadhesive system. The effect of silsesquioxane thermo-responsive hydrogel of FK506 improved drug solubility, biocompatibility, and prolonged retention time by enhancing drug efficacy in a murine dry eye model [97].

These hydrogels can be used for single-drug therapy and combinatorial therapy by co-loading with other chemotherapeutics drugs [98,99]. For instance, Doxorubicin (DOX), IL-2, and IFN-g were delivered by a poly (g-ethyl-L glutamate)-poly (ethylene glycol)-poly(g-ethyl-L-glutamate) (PELG-PEG-PELG) hydrogel. The prepared hydrogel system showed long-term sustained drug release behavior for more than three weeks. The combination therapy enhanced the antitumor effect against B16F10 melanoma cells by inducing cell apoptosis and cell cycle arrest in the G2/S phase. The nanocarriers have not shown any systematic side effects in the xenograft mice model, suggesting an effective and promising approach to drug delivery in melanoma therapies [98].

3.2. *Thermosensitive Hydrogels*

Likewise, the thermo-sensitive poly(3-caprolactone)-10R5-PCL hydrogels, co-loaded with tannic acid and oxaplatin to manage colorectal cancer, restricted the CT26 colon cancer growth in a mice model. They improved the survival time of the animals (11). Moreover, the co-delivery of gemcitabine and cisplatin through the PDLLA-PEG-PDLLA hydrogels synergistically improved the anti-cancer efficacy against pancreatic cancer with sustained drug release. The dual drug-loaded hydrogels exhibited superior antitumor effects in the xenograft model than in single-drug therapies [100]. Similarly, when the thermo-sensitive stimuli hydrogel was utilized to co-deliver paclitaxel (PTX) and temozolomide (TMZ), it produced a synergistic effect against glioblastoma cells. The in vivo studies suggested that the combination therapy potentially reduced tumor growth and sustained drug release in mice brains for one month with no apparent side effects [101].

Chitosan-based thermo-sensitive hydrogels, including several polyols, have gained a critical identity that transforms to a hydrogel form upon contact with body temperature from a solution state. A study indicated that a hydrogel formulation that employs a green synthesis approach with chitosan, genepin, and poloxamer 407 has proved to sustain drug release. Brinzolamide-loaded nanostructured lipid carrier was entrapped into a hydrogel matrix using a hot-melt emulsification and sonication method that showed a sustained drug release for a longer duration (24 h) than marketed eye drops (8 h) in the management of glaucoma [102]. Moxifloxacin hydrochloride thermosensitive gel was formulated using chitosan-β-glycerophosphate to advance the ocular delivery. The drug release profile of the formulation was shown to be delivered in a sustained pattern and at a slower rate with a release of 53% in 1 h and 83.3% in 8 h due to the hydrogel's polymeric network, whereas a 75.6% release was identified from the drug solution. Hence, the formulated hydrogel was stated to be biodegradable, safe, and with a more significant drug loading to the administration site to manage bacterial infections [103]. The quaternized chitosan achieved a better swelling property as a therapeutic carrier. The thermo-sensitive transparent quaternized chitosan hydrogel was utilized to release timolol maleate in the management of glaucoma. Hemolysis and cytotoxicity profiles showed good biocompatibility. The in vitro release pattern of timolol maleate from the hydrogel showed a burst release initially and a linear release for one week, showing a sustained pattern. Hence, quaternized chitosan has a promising ability to sustain drug release in the anti-glaucoma model [104]. Similarly, chitosan has been an extensively used polysaccharide to prolong precorneal retention and corneal permeability, though it is poorly soluble in physiological solvents. Therefore, derivates of chitosan (glycol chitosan) have shown superior aqueous solubility in a broad pH range and are employed in ocular drug delivery systems, such as hydrogels, nanoparticles, and films. Hydrogel films for the topical ocular delivery of dexamethasone and levofloxacin were fabricated by utilizing many oxidation degrees of oxidized hyaluronic acid and glycol chitosan. The formulation showed potent activity in decreasing bacterial growth in different strains. Additionally, the formulation downregulated in vitro anti-inflammatory activities. Overall, the formulated hydrogel film would serve as a treatment for endophthalmitis with minimal corneal irritation and biocompatibility [105]. In another study, a combination of carbon dots and thermo-sensitive hydrogels was evaluated for their in vitro cellular toxicity after effectively delivering diclofenac sodium to an eye. The in vitro release showed characteristic biphasic release of the drug. Cellular toxicity studies revealed that the formulation has a better cytocompatibility with CD44 targeting and serves as a novel way for ocular delivery of drugs [106].

3.3. Light Stimuli Hydrogels

Furthermore, light stimuli hydrogels enhanced drug release from hydrogels. The mechanism of photosensitive hydrogels is that they can undergo structural and conformational changes under radiation, ultraviolet, and visible light sources and achieve sol-gel transition [107,108]. Photo stimuli hydrogels have been utilized in the past few years. For example, Fourniols and his co-worker developed a polyethylene glycol dimethacrylate-based photo-polymerized hydrogel for the local and sustained delivery of TMZ in the management of glioblastoma [109]. Similarly, azobenzene, α-cyclodextrin-functionalized hyaluronic acid, and gold nano-bipyramids-mesoporous silica nanoparticles-conjugated polymer-based in situ injectable hydrogels loaded with DOX and stimulated under near-infrared (NIR) radiation potentially improved the drug localization of drug into the nuclei of the tumor cells [110]. Another example of a NIR stimuli hydrogel was designed by Qui et al. in 2018. They synthesized DOX and black phosphorus-loaded agarose-based hydrogels. The drug release was enhanced after exposure to the light intensity of NIR, which therapeutically increased the anti-cancer efficacy for in vivo experiments [111].

3.4. pH Stimuli Hydrogels

Another most commonly used hydrogel for cancer therapy is the pH stimuli hydrogel. Due to the diverse microenvironments of tumor cells, the pH of the extracellular matrix of the tumor is in the range of 5.8–7.2 and the lysosomal or intracellular matrix pH is usually 5.5, which is acidic compared to normal cells (pH 7.4) [112,113]. Hence, both intracellular and extracellular acidic conditions stimulated the hydrogels for degradation and drug release at the tumor site. The pH-sensitivity could have been achieved by protonating the polymers' ionizable moiety or acid-cleavable bond break [114].

Usually, the pH stimuli hydrogels act as prodrugs, inactive in normal biological pH (7.4), but once they reach the tumor site, upon the release of the drug, they change to their active chemotherapeutic form, producing its effect. The FER-8 peptide hydrogel loaded with PTX exhibited high drug encapsulation and self-assembly of the peptide at pH 7.4. The hydrogel was stimulated by the acidic conditions of the tumor microenvironment. The PTX-loaded hydrogel significantly accumulated the drug in the tumor site and showed sustained and prolonged retention of PTX through an intratumoral injection. This hydrogel showed enhanced tumor inhibition [115]. The polyacrylic acid-based pH stimuli hydrogel for DOX delivery triggered the specific site of drug delivery with a less acidic microenvironment. The hydrogels showed improved pharmacokinetics and drug accumulation in a mice xenograft model with significant tumor growth regression and lowered adverse and side effects [116]. Various hydrogels have received attention in the last few years due to their biocompatibility, biodegradability, and low toxicity. The hydrogel-based drug delivery of chemotherapeutic agents has been attended in recent years [27]. Most chemotherapeutic drugs are associated with lower solubility. Hence, efficient drug delivery and significant therapeutic effects cannot be achieved by a lower dose and the involvement of a higher dose in cancer management includes side effects. Therefore, the hydrogel system is the emerging nanocarrier for the delivery of chemotherapeutic agents. Some recent significant research related to novel gel-based systems is shown in Table 1.

Table 1. Recent published research for novel gel-based delivery systems.

Sno.	Types of Hydrogels	Composition	Drug Used	Disease	References
1	Thermo-stimuli Hydrogel	7-ethyl-10-hydroxycamptothecin (SN-38) liposomal hydrogel	SN-38	Hepatocellular carcinoma	[94]
2		alginate nanogel co-loaded with cisplatin and gold nanoparticles	cisplatin	colorectal cancer	[95]
3		poly(γ-ethyl-L-glutamate)-poly(ethylene glycol)-poly(γ-ethyl-L-glutamate) (PELG-PEG-PELG) hydrogel	DOX/IL-2/IFN-γ	melanoma	[98]
4		PHB-b-PDMAEMA	paclitaxel and Temozolomide	glioblastoma	[99]
5		poly(3-caprolactone) (PCL)-10R5-PCL (PCLR) hydrogel	tannic acid/ oxaliplatin	colorectal cancer	[117]
6		Caprolactone-Polyethylene Glycol	Silibinin	melanoma	[118]
7		α-Cyclodextrin co-polymeric PEGylated iron oxide-based hydrogels	PTX/DOX	breast cancer	[119]
8		β- cyclodextrin complexed glycol chitosan hydrogel	PTX	Ovarian Cancer	[120]
9		mPEG-b-PELG	CA4P and cisplatin	colorectal cancer	[121]
10		Pluronic F127, Pluronic F68, and Hydroxy Propyl Methyl Cellulose.	Itraconazole	Fungal Keratitis	[122]
11		Sulfobutylether-β-cyclodextrin (SBE-β-CD)	Ketoconazole	Fungal Keratitis	[123]
12		Poloxamers (P407 and P188), Carbopol-93	Dipivefrin hydrochloride	Intraocular pressure	[124]
13		Triacetin, Transcutol-P, Poloxamer 407, Poloxamer188	Acyclovir	Ocular viral infections	[125]
14		Poloxamer 407, disodium EDTA	Chlorhexidine digluconate	*Acanthamoeba* keratitis	[126]
15		poloxamers, hyaluronic acid (HA), beta-lapachone (β Lap),	beta-lapachone (β Lap)	Restoring the synovial fluid	[127]
16		Poloxamers, D—(+)-GlcN hydrochloride, papain,	Glucosamine (GlcN)	controlling inflammation and promoting cartilage re-generation	[128]
17		Poloxamer, hyaluronic	Sulforaphane (SFN	SFN intra-articular release for OA treatment	[129]
18	Photosensitive Hydrogels	azobenzene and α-cyclodextrin-functionalized hyaluronic acid with gold nanobipyramids	DOX	human epider-mal keratinocyte	[110]
19		poly(N-isopropylacrylamide) hydrogel	Bortezomib and DOX	osteoblast	[130]
20		Agarose based hydrogel	black phosphorus and DOX	breast cancer	[111]
21		poly(*N*-phenylglycine)- poly (ethylene glycol) co-polymeric hydrogel	Cisplatin	breast cancer	[131]
23	pH-stimuli Hydrogels	poly (acrylic acid) complexed with stabilized amorphous calcium carbonate	DOX	hepatocarcinoma	[116]
24		amphiphilic hyaluronan (HA)-and cystamin-pyrenyl	Organoiridium (III)	lung cancer	[132]
25		Graphene oxide, L-arginine	5-fluorouracil	breast cancer	[133]

Table 1. *Cont.*

Sno.	Types of Hydrogels	Composition	Drug Used	Disease	References
26		FER-8 peptide	PTX	hepatocarcinoma	[115]
27		Dibenzaldehyde, poly (ethylene glycol)	DOX	hepatocarcinoma	[50]
28	Redox- stimuli Hydrogels	dextrin nanogel	DOX	Breast Cancer	[134]
29		Polydopamine, poly (ethylene glycol)	DOX	breast cancer	[135]
30		poly (ethylene glycol) monomethacrylate	Vorinostat and etoposide	cervical cancer	[136]
31		polyglycerol nanogel	DOX	cervical cancer	[137]
33		N-Isopropylacrylamide, Methacrylic acid, Benzalkonium chloride and poly (sulfobetaine methacrylate)	DOX and Indocyanine green	hepatocarcinoma	[138]
34	Magnetism-Responsive Hydrogels	ferromagnetic vortex-domain iron oxide, chitosan and poly (ethylene glycol)	DOX	breast cancer	[139]
35		methacrylic acid, ethylene glycol dimethacrylate, 2,2′-azobisisobutyronitrile and glycidyl methacrylate	sunitinib	cervical cancer, breast cancer and Human Thyroid Tumor	[140]
36		paramagnetic fullerene, DNA and Hyaluronic Acid	DOX	hepatocarcinoma	[141]
37	Proniosomal gel	Surfactant, lecithin and cholesterol	Curcumin	Ocular Inflammation	[142]
38	Liposomal gel	Lecithin: cholesterol, Carbopol 934	Travoprost	Glaucoma and ocular hypertension	[143]
39	Injectable hydrogel	horseradish peroxidase (HRP) and H2O, chitosan, hyaluronic acid (HA)	Dextrane Tyramine	cartilage tissue regeneration	[144]
40		bone marrow mesenchymal stem cell (MSC) spheroids, short fibre fillers, Kartogenin (KGN)	Celecoxib	cartilage regeneration, and inflammation removal	[145]
41		poly (ethylene glycol)-*b*-polythioketal-*b*-poly(ethylene glycol), micelles	dexamethasone acetate	preventing cartilage extracellular matrix degeneration	[146]
42		Gelatin, ulbecco's phosphate buffered saline (DPBS), methacrylic anhydride.	diclofenac sodium	preventing the development of degenerative changes in OA via the synergistical treatment of enhanced lubrication (COF reduction) and sustained drug release (inflammation down-regulation)	[147]
43		Hexachlorocyclotriphosphazene, Poly (dichlorophosphazene, Methoxy poly (ethylene glycol),	Triamcinolone acetonide	Effective prevention and long-term anti-OA treatment	[148]
44	Shear—sensitive hydrogels	Hyaluronic acid, aldehyde groups, amino Groups, HSPC lipid,	Celecoxib	Minimizing shear-induced cartilage damage and inflammation	[149]

4. Descriptive Patents Established for Novel Gel-Based Delivery Systems

Various publications illustrated the efficacy of novel gels over the last decades. The current section explains the established patents of novel gels. In patent 20210338211, aptamer and hydrogels cross-linked with DNAzyme are used for colorimetric identification of analytes in body fluids through an ocular device [150]. Patent 2021120395 tells that in situ hydrogels provide extended drug release by delivering to a tissue, usually agents with low water solubility [151]. Patent W/O/2021/113515 explains that hydrogels composed of Gelatin–hydroxyphenylpropionic acid (gelatin-HPA), hyaluronic acid–tyramine

(HA-Tyr), catalyzer, cross-linker, or other combinations can treat ocular disorders [152]. Patent 20210069496 describes polymeric formulations, including hydrogels formed by a UV cross-linking method. Here, the hydrogels act as a nasal stimulator that stimulates the lacrimal glands to mimic the production of tears electronically and to manage dry eye syndromes [153]. Patent WO/2021/038279 tells of the invention related to ion-exchange polymeric hydrogels for ocular treatment [154].

Patent 202121042889 talks about etoricoxib, which is formulated as a nanosponge hydrogel for the management of arthritis by the method of emulsion solvent diffusion using a polymeric organic solvent, ethylcellulose eudragit, and an aqueous phase. The formulated nanosponges were evaluated for differential scanning calorimetry (DSC), Fourier-transform infrared spectroscopy (FTIR), polydispersity index (PDI), scanning electron microscopy (SEM), zeta potential, drug content, entrapment efficiency, viscosity, spreadability, in vitro diffusion, irritation test, and in vivo antiarthritic effect. The synthesized formulation proved to be effective as a novel way for managing arthritic pain [155]. In patent 202031000910, a cross-linked protein matrix hydrogel was prepared for topical application in skin regeneration and wound healing [156].

Conductive hydrogels with an adhesiveness method of preparation were invented and discussed in patent 112442194. The process involves dopamine modification of carbon nanotubes and grafting to saccharides, followed by acrylamide mixing and formation of hydrogels in the presence of an initiator and a cross-linking agent. Hydrogel structure, electrical conductivity, adhesiveness, and biocompatibility can be improved by dispersing the modified carbon nanotubes in an aqueous solution to form hydrogen bonds and cross-link with the supramolecules of the hydrogel. Hence, conductive hydrogels can be used in biomedical fields, as well as for human body monitoring and electronic skin, etc. [157].

Patent 20210023121 discloses thrombin-responsive hydrogels for prolonged heparin delivery for auto-anticoagulant regulation in a controlled feedback mechanism. The formulated microneedle, containing a patch, can activate the thrombin and release heparin to avoid blood coagulation. The insertion of a microneedle patch containing hydrogel regulates blood coagulation sustainably in response to thrombin without leakage [158].

Patent 20210393780 discloses the effectiveness of thermo-sensitive polymer–protein-based hydrogels in the field of cancer therapeutics. The invention is enriched by photosensitizers, dyes, photothermal agents, and drugs. Hence, the invention proved to be less expensive, highly effective, and thermosensitive, resulting in a sustained drug release for targeted delivery [159]. Patent WO/2021/174021 describes a degradable hydrogel system for immunotherapy with an extended-release pattern of an anti-cancer drug linked with a hydrogel matrix synergistically for cancer treatment [160]. Patent 9758/CHENP/2012 explains self-assembling peptides and their use in hydrogels for the adhesion, proliferation, differentiation of neural stem cells, and their auto-healing properties. They are reported to be non-toxic in central nervous systems, as well as to avoid bleeding and have faster nervous regeneration [161]. Patent 20140286865 explores di-block co-polypeptide synthetic hydrogels in the central nervous system [162].

5. Conclusions

In this review, the authors have focused on recent trends in novel gel-based drug delivery systems and their applications. In recent times, these novel systems have exhibited proficient delivery of multiple therapeutic moieties and expressed desired properties and functions, such as selective targeting. The systems offer abundant benefits compared to conventional drug delivery approaches including controlled drug release, high drug loading, biocompatibility and biodegradability, and enriching patient compliance and comfort. The responsive gel technology is significant in formulating intelligent delivery systems; these systems respond to stimuli such as pH, temperature, enzymes, and so forth. Hence, these systems are site-specific and facilitate the controlled release of therapeutic molecules. Although, fundamentally, these systems have proven capabilities for effective drug delivery, there is a scope to explore new polymers to fabricate novel gels; therefore, the currently employed components can be modified. Furthermore, recent studies revealed that employing plant extracts to develop novel delivery systems has enabled the development of various drug delivery systems with non-toxic procedures. The formulation of substances of natural origin has advantages in various magnitudes on the environment. Over the years, the evolution of green chemistry has provided more eco-friendly procedures resulting in minor harm to nature. Current findings suggest promising results for green synthesised delivery systems over conventional systems. Green technology does not require common harmful chemicals. Instead, this technology uses biological and biocompatible reagents. Besides, reports suggest that green technology delivery systems have better stability than traditional methods. Formulations developed with green technology employing plant extracts and biomaterials, such as proteins or peptides, yielded non-toxic and highly biocompatible systems; thus, they have resolved the most concerning issue with traditional delivery systems, i.e., toxicity. Green technology will play a significant role in formulating novel delivery systems. However, we require further understanding of the development of systems with green technology.

Author Contributions: Conceptualization, T.K.S., V.N.G.; Formal analysis, V.N.G.; Resources, T.K.S., S.C., S.M., H.K., P.C.; Data curation, T.K.S., S.C., S.M., H.K., P.C.; Writing original draft preparation, T.K.S., S.C., S.M., H.K., P.C.; Writing review and editing, V.J., B.V. and D.V.G.; Supervision, V.N.G., V.J., B.V. and D.V.G. All authors have read and agreed to the published version of the manuscript.

Funding: This work received no external funding. The APC is supported by JSS Academy of Higher Education & Research, Mysuru.

Institutional Review Board Statement: Not applicable.

Informed Consent Statement: Not applicable.

Data Availability Statement: Not applicable.

Acknowledgments: The author(s) express deep sense of gratitude towards JSS College of Pharmacy, JSS Academy of Higher Education & Research (JSS AHER), Mysuru for their constant support and motivation.

Conflicts of Interest: The authors declare no conflict of interest.

References

1. Iriventi, P.; Gupta, N.V.; Osmani, R.A.M.; Balamuralidhara, V. Design & development of nanosponge loaded topical gel of curcumin and caffeine mixture for augmented treatment of psoriasis. *DARU J. Pharm. Sci.* **2020**, *28*, 489–506. [CrossRef]
2. Uzunalli, G.; Guler, M.O. Peptide gels for controlled release of proteins. *Ther. Deliv.* **2020**, *11*, 193–211. [CrossRef] [PubMed]
3. Nambiar, M.; Schneider, J.P. Peptide hydrogels for affinity-controlled release of therapeutic cargo: Current and potential strategies. *J. Pept. Sci.* **2021**, *28*, e3377. [CrossRef] [PubMed]
4. Sharma, S.; Tiwari, S. A review on biomacromolecular hydrogel classification and its applications. *Int. J. Biol. Macromol.* **2020**, *162*, 737–747. [CrossRef] [PubMed]
5. Micale, N.; Citarella, A.; Molonia, M.S.; Speciale, A.; Cimino, F.; Saija, A.; Cristani, M. Hydrogels for the Delivery of Plant-Derived (Poly)Phenols. *Molecules* **2020**, *25*, 3254. [CrossRef] [PubMed]
6. Op't Veld, R.C.; Walboomers, X.F.; Jansen, J.A.; Wagener, F.A.D.T.G. Design Considerations for Hydrogel Wound Dressings: Strategic and Molecular Advances. *Tissue Eng. Part B Rev.* **2020**, *26*, 230–248. [CrossRef]
7. Tavakoli, S.; Klar, A.S. Advanced Hydrogels as Wound Dressings. *Biomolecules* **2020**, *10*, 1169. [CrossRef]
8. Du, W.; Zong, Q.; Guo, R.; Ling, G.; Zhang, P. Injectable Nanocomposite Hydrogels for Cancer Therapy. *Macromol. Biosci.* **2021**, *21*, 2100186. [CrossRef]
9. Chen, T.; Hou, K.; Ren, Q.; Chen, G.; Wei, P.; Zhu, M. Nanoparticle-Polymer Synergies in Nanocomposite Hydrogels: From Design to Application. *Macromol. Rapid Commun.* **2018**, *39*, e1800337. [CrossRef]
10. Jiménez, G.; Venkateswaran, S.; López-Ruiz, E.; Peran, M.; Pernagallo, S.; Díaz-Monchón, J.J.; Canadas, R.; Antich, C.; Oliveira, J.M.; Callanan, A.; et al. A soft 3D polyacrylate hydrogel recapitulates the cartilage niche and allows growth-factor free tissue engineering of human articular cartilage. *Acta Biomater.* **2019**, *90*, 146–156. [CrossRef]
11. Zhang, Y.; Yu, J.-K.; Ren, K.; Zuo, J.; Ding, J.; Chen, X. Thermosensitive Hydrogels as Scaffolds for Cartilage Tissue Engineering. *Biomacromolecules* **2019**, *20*, 1478–1492. [CrossRef] [PubMed]
12. Barouti, G.; Liow, S.S.; Dou, Q.; Ye, H.; Orione, C.; Guillaume, S.M.; Loh, X.J. New Linear and Star-Shaped Thermogelling Poly([*R*]-3-hydroxybutyrate) Copolymers. *Chem. A Eur. J.* **2016**, *22*, 10501–10512. [CrossRef] [PubMed]

13. Yu, Y.; Cheng, Y.; Tong, J.; Zhang, L.; Wei, Y.; Tian, M. Recent advances in thermo-sensitive hydrogels for drug delivery. *J. Mater. Chem. B* **2021**, *9*, 2979–2992. [CrossRef] [PubMed]
14. Taghizadeh, B.; Taranejoo, S.; Monemian, S.A.; Moghaddam, Z.S.; Daliri, K.; Derakhshankhah, H.; Derakhshani, Z. Classification of stimuli–responsive polymers as anticancer drug delivery systems. *Drug Deliv.* **2015**, *22*, 145–155. [CrossRef]
15. Marques, A.C.; Costa, P.J.; Velho, S.; Amaral, M.H. Stimuli-responsive hydrogels for intratumoral drug delivery. *Drug Discov. Today* **2021**, *26*, 2397–2405. [CrossRef]
16. Khan, S.; Akhtar, N.; Minhas, M.U.; Badshah, S.F. pH/Thermo-Dual Responsive Tunable In Situ Cross-Linkable Depot Injectable Hydrogels Based on Poly(N-Isopropylacrylamide)/Carboxymethyl Chitosan with Potential of Controlled Localized and Systemic Drug Delivery. *AAPS PharmSciTech* **2019**, *20*, 119. [CrossRef]
17. Oliva, N.; Conde, J.; Wang, K.; Artzi, N. Designing Hydrogels for On-Demand Therapy. *Acc. Chem. Res.* **2017**, *50*, 669–679. [CrossRef]
18. Gao, W.; Zhang, Y.; Zhang, Q.; Zhang, L. Nanoparticle-Hydrogel: A Hybrid Biomaterial System for Localized Drug Delivery. *Ann. Biomed. Eng.* **2016**, *44*, 2049–2061. [CrossRef]
19. Wang, T.; Chen, L.; Shen, T.; Wu, D. Preparation and properties of a novel thermo-sensitive hydrogel based on chitosan/hydroxypropyl methylcellulose/glycerol. *Int. J. Biol. Macromol.* **2016**, *93*, 775–782. [CrossRef]
20. Raghuwanshi, V.S.; Garnier, G. Characterisation of hydrogels: Linking the nano to the microscale. *Adv. Colloid Interface Sci.* **2019**, *274*, 102044. [CrossRef]
21. Azeera, M.; Vaidevi, S.; Ruckmani, K. *Characterization Techniques of Hydrogel and Its Applications*; Springer: Cham, Switzerland, 2018; pp. 1–24. [CrossRef]
22. Alexander, A.; Ajazuddin; Khan, J.; Saraf, S.; Saraf, S. Polyethylene glycol (PEG)–Poly(N-isopropylacrylamide) (PNIPAAm) based thermosensitive injectable hydrogels for biomedical applications. *Eur. J. Pharm. Biopharm.* **2014**, *88*, 575–585. [CrossRef] [PubMed]
23. Qu, J.; Zhao, X.; Ma, P.X.; Guo, B. Injectable antibacterial conductive hydrogels with dual response to an electric field and pH for localized "smart" drug release. *Acta Biomater.* **2018**, *72*, 55–69. [CrossRef] [PubMed]
24. Malekmohammadi, S.; Aminabad, N.S.; Sabzi, A.; Zarebkohan, A.; Razavi, M.; Vosough, M.; Bodaghi, M.; Maleki, H. Smart and Biomimetic 3D and 4D Printed Composite Hydrogels: Opportunities for Different Biomedical Applications. *Biomedicines* **2021**, *9*, 1537. [CrossRef] [PubMed]
25. Kasiński, A.; Zielińska-Pisklak, M.; Oledzka, E.; Sobczak, M. Smart Hydrogels—Synthetic Stimuli-Responsive Antitumor Drug Release Systems. *Int. J. Nanomed.* **2020**, *15*, 4541–4572. [CrossRef]
26. Shi, J.; Yu, L.; Ding, J. PEG-based thermosensitive and biodegradable hydrogels. *Acta Biomater.* **2021**, *128*, 42–59. [CrossRef]
27. Sun, Z.; Song, C.; Wang, C.; Hu, Y.; Wu, J. Hydrogel-Based Controlled Drug Delivery for Cancer Treatment: A Review. *Mol. Pharm.* **2020**, *17*, 373–391. [CrossRef]
28. Le, T.M.D.; Duong, H.T.T.; Thambi, T.; Giang Phan, V.H.; Jeong, J.H.; Lee, D.S. Bioinspired pH- and Temperature-Responsive Injectable Adhesive Hydrogels with Polyplexes Promotes Skin Wound Healing. *Biomacromolecules* **2018**, *19*, 3536–3548. [CrossRef]
29. Basu, S.; Pacelli, S.; Paul, A. Self-healing DNA-based injectable hydrogels with reversible covalent linkages for controlled drug delivery. *Acta Biomater.* **2020**, *105*, 159–169. [CrossRef]
30. Rizzo, F.; Kehr, N.S. Recent Advances in Injectable Hydrogels for Controlled and Local Drug Delivery. *Adv. Healthc. Mater.* **2020**, *10*, 2001341. [CrossRef]
31. Stayton, P.; El-Sayed, M.; Murthy, N.; Bulmuş, V.; Lackey, C.; Cheung, C.; Hoffman, A. 'Smart' delivery systems for biomolecular therapeutics. *Orthod. Craniofacial Res.* **2005**, *8*, 219–225. [CrossRef]
32. Zhu, Y.J.; Chen, F. pH-Responsive Drug-Delivery Systems. *Chem. Asian J.* **2015**, *10*, 284–305. [CrossRef] [PubMed]
33. Chandrawati, R. Enzyme-responsive polymer hydrogels for therapeutic delivery. *Exp. Biol. Med.* **2016**, *241*, 972–979. [CrossRef] [PubMed]
34. Culver, H.R.; Clegg, J.R.; Peppas, N.A. Analyte-Responsive Hydrogels: Intelligent Materials for Biosensing and Drug Delivery. *Acc. Chem. Res.* **2017**, *50*, 170–178. [CrossRef] [PubMed]
35. Billah, S.M.R.; Mondal, I.H.; Somoal, S.H.; Pervez, M.N.; Haque, O. *Enzyme-Responsive Hydrogels*; Springer: Cham, Switzerland, 2019; pp. 309–330. [CrossRef]
36. Rafael, D.; Melendres, M.M.R.; Andrade, F.; Montero, S.; Martinez-Trucharte, F.; Vilar-Hernandez, M.; Durán-Lara, E.F.; Schwartz, S.; Abasolo, I. Thermo-responsive hydrogels for cancer local therapy: Challenges and state-of-art. *Int. J. Pharm.* **2021**, *606*, 120954. [CrossRef]
37. Andrade, F.; Roca-Melendres, M.M.; Durán-Lara, E.F.; Rafael, D.; Schwartz, S., Jr. Stimuli-Responsive Hydrogels for Cancer Treatment: The Role of pH, Light, Ionic Strength and Magnetic Field. *Cancers* **2021**, *13*, 1164. [CrossRef]
38. Griffin, D.R.; Kasko, A.M. Photoselective Delivery of Model Therapeutics from Hydrogels. *ACS Macro Lett.* **2012**, *1*, 1330–1334. [CrossRef]
39. Fakhari, A.; Subramony, J.A. Engineered in-situ depot-forming hydrogels for intratumoral drug delivery. *J. Control. Release* **2015**, *220*, 465–475. [CrossRef]
40. Fang, G.; Yang, X.; Wang, Q.; Zhang, A.; Tang, B. Hydrogels-based ophthalmic drug delivery systems for treatment of ocular diseases. *Mater. Sci. Eng. C* **2021**, *127*, 112212. [CrossRef]
41. Choi, B.; Loh, X.J.; Tan, A.; Loh, C.K.; Ye, E.; Joo, M.K.; Jeong, B. Introduction to in situ forming hydrogels for biomedical applications. In *In-Situ Gelling Polymers*; Springer: Singapore, 2015; pp. 5–35. [CrossRef]

42. Loh, X.J. *In-Situ Gelling Polymers: For Biomedical Applications*; Springer: Singapore, 2015.
43. Dimatteo, R.; Darling, N.J.; Segura, T. In situ forming injectable hydrogels for drug delivery and wound repair. *Adv. Drug Deliv. Rev.* **2018**, *127*, 167–184. [CrossRef]
44. Shim, W.S.; Kim, J.-H.; Kim, K.; Kim, Y.-S.; Park, R.-W.; Kim, I.-S.; Kwon, I.C.; Lee, D.S. pH- and temperature-sensitive, injectable, biodegradable block copolymer hydrogels as carriers for paclitaxel. *Int. J. Pharm.* **2007**, *331*, 11–18. [CrossRef]
45. Kang, J.H.; Turabee, H.; Lee, D.S.; Kwon, Y.J.; Ko, Y.T. Temperature and pH-responsive in situ hydrogels of gelatin derivatives to prevent the reoccurrence of brain tumor. *Biomed. Pharmacother.* **2021**, *143*, 112144. [CrossRef] [PubMed]
46. Podual, K.; Doyle, F.J.; Peppas, N.A. Dynamic behavior of glucose oxidase-containing microparticles of poly(ethylene glycol)-grafted cationic hydrogels in an environment of changing pH. *Biomaterials* **2000**, *21*, 1439–1450. [CrossRef]
47. Yu, J.; Zhang, Y.; Bomba, H.; Gu, Z. Stimuli-responsive delivery of therapeutics for diabetes treatment. *Bioeng. Transl. Med.* **2016**, *1*, 323–337. [CrossRef]
48. Xu, X.; Shang, H.; Zhang, T.; Shu, P.; Liu, Y.; Xie, J.; Zhang, D.; Tan, H.; Li, J. A stimuli-responsive insulin delivery system based on reversible phenylboronate modified cyclodextrin with glucose triggered host-guest interaction. *Int. J. Pharm.* **2018**, *548*, 649–658. [CrossRef] [PubMed]
49. Wells, C.M.; Harris, M.; Choi, L.; Murali, V.P.; Guerra, F.D.; Jennings, J.A. Stimuli-Responsive Drug Release from Smart Polymers. *J. Funct. Biomater.* **2019**, *10*, 34. [CrossRef]
50. Qu, J.; Zhao, X.; Ma, P.X.; Guo, B. pH-responsive self-healing injectable hydrogel based on N-carboxyethyl chitosan for hepatocellular carcinoma therapy. *Acta Biomater.* **2017**, *58*, 168–180. [CrossRef]
51. Jalalvandi, E.; Shavandi, A. In situ-forming and pH-responsive hydrogel based on chitosan for vaginal delivery of therapeutic agents. *J. Mater. Sci. Mater. Med.* **2018**, *29*, 158. [CrossRef]
52. Prajapati, V.D.; Jani, G.K.; Moradiya, N.G.; Randeria, N.P. Pharmaceutical applications of various natural gums, mucilages and their modified forms. *Carbohydr. Polym.* **2013**, *92*, 1685–1699. [CrossRef]
53. Bhardwaj, T.R.; Kanwar, M.; Lal, R.; Gupta, A. Natural Gums and Modified Natural Gums as Sustained-Release Carriers. *Drug Dev. Ind. Pharm.* **2000**, *26*, 1025–1038. [CrossRef]
54. Guo, J.-H.; Skinner, G.W.; Harcum, W.W.; Barnum, P.E. Pharmaceutical applications of naturally occurring water-soluble polymers. *Pharm. Sci. Technol. Today* **1998**, *1*, 254–261. [CrossRef]
55. Dodane, V.; Vilivalam, V.D. Pharmaceutical applications of chitosan. *Pharm. Sci. Technol. Today* **1998**, *1*, 246–253. [CrossRef]
56. Farjami, T.; Madadlou, A. An overview on preparation of emulsion-filled gels and emulsion particulate gels. *Trends Food Sci. Technol.* **2019**, *86*, 85–94. [CrossRef]
57. Singh, L.P.; Bhattacharyya, S.K.; Kumar, R.; Mishra, G.; Sharma, U.; Singh, G.; Ahalawat, S. Sol-Gel processing of silica nanoparticles and their applications. *Adv. Colloid Interface Sci.* **2014**, *214*, 17–37. [CrossRef]
58. Dickinson, E. Emulsion gels: The structuring of soft solids with protein-stabilized oil droplets. *Food Hydrocoll.* **2012**, *28*, 224–241. [CrossRef]
59. Zhao, X.; Chen, B.; Sun, Z.; Liu, T.; Cai, Y.; Huang, L.; Deng, X.; Zhao, M.; Zhao, Q. A novel preparation strategy of emulsion gel solely stabilized by alkaline assisted steam-cooking treated insoluble soybean fiber. *Food Hydrocoll.* **2022**, *129*, 107646. [CrossRef]
60. Smeets, N.M.B.; Hoare, T. Designing responsive microgels for drug delivery applications. *J. Polym. Sci. Part A Polym. Chem.* **2013**, *51*, 3027–3043. [CrossRef]
61. Imaz, A.; Forcada, J. New Biocompatible Microgels. *Macromol. Symp.* **2009**, *281*, 85–88. [CrossRef]
62. Martín-Molina, A.; Quesada-Pérez, M. A review of coarse-grained simulations of nanogel and microgel particles. *J. Mol. Liq.* **2019**, *280*, 374–381. [CrossRef]
63. Sung, B.; Kim, C.; Kim, M.-H. Biodegradable colloidal microgels with tunable thermosensitive volume phase transitions for controllable drug delivery. *J. Colloid Interface Sci.* **2015**, *450*, 26–33. [CrossRef]
64. Vinogradov, S.V. Colloidal Microgels in Drug Delivery Applications. *Curr. Pharm. Des.* **2006**, *12*, 4703–4712. [CrossRef]
65. Kousalová, J.; Etrych, T. Polymeric Nanogels as Drug Delivery Systems. *Physiol. Res.* **2018**, *67*, S305–S317. [CrossRef] [PubMed]
66. Varshosaz, J.; Taymouri, S.; Ghassami, E. Supramolecular Self-Assembled Nanogels a New Platform for Anticancer Drug Delivery. *Curr. Pharm. Des.* **2018**, *23*, 5242–5260. [CrossRef] [PubMed]
67. Ryu, J.-H.; Jiwpanich, S.; Chacko, R.; Bickerton, S.; Thayumanavan, S. Surface-Functionalizable Polymer Nanogels with Facile Hydrophobic Guest Encapsulation Capabilities. *J. Am. Chem. Soc.* **2010**, *132*, 8246–8247. [CrossRef] [PubMed]
68. Suhail, M.; Rosenholm, J.M.; Minhas, M.U.; Badshah, S.F.; Naeem, A.; Khan, K.U.; Fahad, M. Nanogels as drug-delivery systems: A comprehensive overview. *Ther. Deliv.* **2019**, *10*, 697–717. [CrossRef] [PubMed]
69. Müller, R. Biodegradability of Polymers: Regulations and Methods for Testing. In *Biopolymers Online*; American Cancer Society: Atlanta, GA, USA, 2005. [CrossRef]
70. Oh, J.K.; Drumright, R.; Siegwart, D.J.; Matyjaszewski, K. The development of microgels/nanogels for drug delivery applications. *Prog. Polym. Sci.* **2008**, *33*, 448–477. [CrossRef]
71. Ahmed, S.; Alhareth, K.; Mignet, N. Advancement in nanogel formulations provides controlled drug release. *Int. J. Pharm.* **2020**, *584*, 119435. [CrossRef]
72. Rajkumar, J.; Gv, R.; Sastri K, T.; Burada, S. Recent update on proniosomal gel as topical drug delivery system. *Asian J. Pharm. Clin. Res.* **2019**, *12*, 54–61. [CrossRef]
73. Brandl, M. Vesicular Phospholipid Gels: A Technology Platform. *J. Liposome Res.* **2007**, *17*, 15–26. [CrossRef]

74. Tian, W.; Schulze, S.; Brandl, M.; Winter, G. Vesicular phospholipid gel-based depot formulations for pharmaceutical proteins: Development and in vitro evaluation. *J. Control. Release* **2010**, *142*, 319–325. [CrossRef]
75. Breitsamer, M.; Winter, G. Vesicular phospholipid gels as drug delivery systems for small molecular weight drugs, peptides and proteins: State of the art review. *Int. J. Pharm.* **2018**, *557*, 1–8. [CrossRef]
76. Fetih, G.; Fathalla, D.; El-Badry, M. Liposomal Gels for Site-Specific, Sustained Delivery of Celecoxib: In Vitro and In Vivo Evaluation. *Drug Dev. Res.* **2014**, *75*, 257–266. [CrossRef] [PubMed]
77. Pavelić, Ž.; Skalko-Basnet, N.; Schubert, R. Liposomal gels for vaginal drug delivery. *Int. J. Pharm.* **2001**, *219*, 139–149. [CrossRef]
78. Pavelić, Z.; Skalko-Basnet, N.; Jalsenjak, I. Liposomal gel with chloramphenicol: Characterisation and in vitro release. *Acta Pharm.* **2004**, *54*, 319–330. [PubMed]
79. Singh, S.; Vardhan, H.; Kotla, N.G.; Maddiboyina, B.; Sharma, D.; Webster, T.J. The role of surfactants in the formulation of elastic liposomal gels containing a synthetic opioid analgesic. *Int. J. Nanomed.* **2016**, *11*, 1475–1482. [CrossRef]
80. Patel, K.K.; Kumar, P.; Thakkar, H.P. Formulation of Niosomal Gel for Enhanced Transdermal Lopinavir Delivery and Its Comparative Evaluation with Ethosomal Gel. *AAPS PharmSciTech* **2012**, *13*, 1502–1510. [CrossRef]
81. Muzzalupo, R.; Tavano, L. Niosomal drug delivery for transdermal targeting: Recent advances. *Res. Rep. Transdermal Drug Deliv.* **2015**, *4*, 23–33. [CrossRef]
82. Yoshida, H.; Lehr, C.-M.; Kok, W.; Junginger, H.E.; Verhoef, J.C.; Bouwstra, J.A. Niosomes for oral delivery of peptide drugs. *J. Control. Release* **1992**, *21*, 145–153. [CrossRef]
83. Bhardwaj, P.; Tripathi, P.; Gupta, R.; Pandey, S. Niosomes: A review on niosomal research in the last decade. *J. Drug Deliv. Sci. Technol.* **2020**, *56*, 101581. [CrossRef]
84. Das, B.; Sen, S.O.; Maji, R.; Nayak, A.K.; Sen, K.K. Transferosomal gel for transdermal delivery of risperidone: Formulation optimization and ex vivo permeation. *J. Drug Deliv. Sci. Technol.* **2017**, *38*, 59–71. [CrossRef]
85. Cevc, G.; Blume, G.; Schätzlein, A. Transfersomes-mediated transepidermal delivery improves the regio-specificity and biological activity of corticosteroids in vivo1Dedicated to the late Dr. Henri Ernest Bodde.1. *J. Control. Release* **1997**, *45*, 211–226. [CrossRef]
86. Malakar, J.; Sen, S.O.; Nayak, A.K.; Sen, K.K. Formulation, optimization and evaluation of transferosomal gel for transdermal insulin delivery. *Saudi Pharm. J.* **2012**, *20*, 355–363. [CrossRef] [PubMed]
87. Benson, H.A.E. Transfersomes for transdermal drug delivery. *Expert Opin. Drug Deliv.* **2006**, *3*, 727–737. [CrossRef] [PubMed]
88. Li, Y.; Maciel, D.; Rodrigues, J.; Shi, X.; Tomás, H. Biodegradable Polymer Nanogels for Drug/Nucleic Acid Delivery. *Chem. Rev.* **2015**, *115*, 8564–8608. [CrossRef] [PubMed]
89. Sahiner, N.; Godbey, W.T.; McPherson, G.L.; John, V.T. Microgel, nanogel and hydrogel–hydrogel semi-IPN composites for biomedical applications: Synthesis and characterization. *Colloid Polym. Sci.* **2006**, *284*, 1121–1129. [CrossRef]
90. Thambi, T.; Phan, V.H.G.; Lee, D.S. Stimuli-Sensitive Injectable Hydrogels Based on Polysaccharides and Their Biomedical Applications. *Macromol. Rapid Commun.* **2016**, *37*, 1881–1896. [CrossRef]
91. Yoshida, R.; Uchida, K.; Kaneko, Y.; Sakai, K.; Kikuchi, A.; Sakurai, Y.; Okano, T. Comb-type grafted hydrogels with rapid deswelling response to temperature changes. *Nature* **1995**, *374*, 240–242. [CrossRef]
92. Meyer, D.E.; Chilkoti, A. Purification of recombinant proteins by fusion with thermally-responsive polypeptides. *Nat. Biotechnol.* **1999**, *17*, 1112–1115. [CrossRef]
93. Li, Y.; Lin, T.-Y.; Luo, Y.; Liu, Q.; Xiao, W.; Guo, W.; Lac, D.; Zhang, H.; Feng, C.; Wachsmann-Hogiu, S.; et al. A smart and versatile theranostic nanomedicine platform based on nanoporphyrin. *Nat. Commun.* **2014**, *5*, 4712. [CrossRef]
94. Bai, R.; Deng, X.; Wu, Q.; Cao, X.; Ye, T.; Wang, S. Liposome-loaded thermo-sensitive hydrogel for stabilization of SN-38 via intratumoral injection: Optimization, characterization, and antitumor activity. *Pharm. Dev. Technol.* **2018**, *23*, 106–115. [CrossRef]
95. Mirrahimi, M.; Abed, Z.; Beik, J.; Shiri, I.; Dezfuli, A.S.; Mahabadi, V.P.; Kamrava, S.K.; Ghaznavi, H.; Shakeri-Zadeh, A. A thermo-responsive alginate nanogel platform co-loaded with gold nanoparticles and cisplatin for combined cancer chemo-photothermal therapy. *Pharmacol. Res.* **2019**, *143*, 178–185. [CrossRef]
96. Chen, L.; Deng, J.; Yu, A.; Hu, Y.; Jin, B.; Du, P.; Zhou, J.; Lei, L.; Wang, Y.; Vakal, S.; et al. Drug-peptide supramolecular hydrogel boosting transcorneal permeability and pharmacological activity via ligand-receptor interaction. *Bioact. Mater.* **2022**, *10*, 420–429. [CrossRef] [PubMed]
97. Han, Y.; Jiang, L.; Shi, H.; Xu, C.; Liu, M.; Li, Q.; Zheng, L.; Chi, H.; Wang, M.; Liu, Z.; et al. Effectiveness of an ocular adhesive polyhedral oligomeric silsesquioxane hybrid thermo-responsive FK506 hydrogel in a murine model of dry eye. *Bioact. Mater.* **2022**, *9*, 77–91. [CrossRef] [PubMed]
98. Lv, Q.; He, C.; Quan, F.; Yu, S.; Chen, X. DOX/IL-2/IFN-γ co-loaded thermo-sensitive polypeptide hydrogel for efficient melanoma treatment. *Bioact. Mater.* **2018**, *3*, 118–128. [CrossRef] [PubMed]
99. Zhao, D.; Song, H.; Zhou, X.; Chen, Y.; Liu, Q.; Gao, X.; Zhu, X.; Chen, D. Novel facile thermosensitive hydrogel as sustained and controllable gene release vehicle for breast cancer treatment. *Eur. J. Pharm. Sci.* **2019**, *134*, 145–152. [CrossRef] [PubMed]
100. Shi, K.; Xue, B.; Jia, Y.; Yuan, L.; Han, R.; Yang, F.; Peng, J.; Qian, Z. Sustained co-delivery of gemcitabine and cis-platinum via biodegradable thermo-sensitive hydrogel for synergistic combination therapy of pancreatic cancer. *Nano Res.* **2019**, *12*, 1389–1399. [CrossRef]
101. Zhao, M.; Bozzato, E.; Joudiou, N.; Ghiassinejad, S.; Danhier, F.; Gallez, B.; Préat, V. Codelivery of paclitaxel and temozolomide through a photopolymerizable hydrogel prevents glioblastoma recurrence after surgical resection. *J. Control. Release* **2019**, *309*, 72–81. [CrossRef]

102. Chakole, C.M.; Sahoo, P.K.; Pandey, J.; Chauhan, M.K. A green chemistry approach towards synthesizing hydrogel for sustained ocular delivery of brinzolamide: In vitro and ex vivo evaluation. *J. Indian Chem. Soc.* **2021**, *99*, 100323. [CrossRef]
103. Asfour, M.H.; El-Alim, S.H.A.; Awad, G.E.A.; Kassem, A.A. Chitosan/β-glycerophosphate in situ forming thermo-sensitive hydrogel for improved ocular delivery of moxifloxacin hydrochloride. *Eur. J. Pharm. Sci.* **2021**, *167*, 106041. [CrossRef]
104. Pakzad, Y.; Fathi, M.; Omidi, Y.; Mozafari, M.; Zamanian, A. Synthesis and characterization of timolol maleate-loaded quaternized chitosan-based thermosensitive hydrogel: A transparent topical ocular delivery system for the treatment of glaucoma. *Int. J. Biol. Macromol.* **2020**, *159*, 117–128. [CrossRef]
105. Bao, Z.; Yu, A.; Shi, H.; Hu, Y.; Jin, B.; Lin, D.; Dai, M.; Lei, L.; Li, X.; Wang, Y. Glycol chitosan/oxidized hyaluronic acid hydrogel film for topical ocular delivery of dexamethasone and levofloxacin. *Int. J. Biol. Macromol.* **2021**, *167*, 659–666. [CrossRef]
106. Wang, L.; Pan, H.; Gu, D.; Li, P.; Su, Y.; Pan, W. A composite System Combining Self-Targeted Carbon Dots and Thermosensitive Hydrogels for Challenging Ocular Drug Delivery. *J. Pharm. Sci.* **2022**, *111*, 1391–1400. [CrossRef] [PubMed]
107. Pereira, R.F.; Bártolo, P.J. Photopolymerizable hydrogels in regenerative medicine and drug delivery. In *Hot Topics in Biomaterials*; Future Science Ltd.: London, UK, 2014; pp. 6–28. [CrossRef]
108. Lim, H.L.; Hwang, Y.; Kar, M.; Varghese, S. Smart hydrogels as functional biomimetic systems. *Biomater. Sci.* **2014**, *2*, 603–618. [CrossRef] [PubMed]
109. Fourniols, T.; Randolph, L.D.; Staub, A.; Vanvarenberg, K.; Leprince, J.G.; Préat, V.; des Rieux, A.; Danhier, F. Temozolomide-loaded photopolymerizable PEG-DMA-based hydrogel for the treatment of glioblastoma. *J. Control. Release* **2015**, *210*, 95–104. [CrossRef] [PubMed]
110. Chen, X.; Liu, Z.; Parker, S.G.; Zhang, X.; Gooding, J.J.; Ru, Y.; Liu, Y.; Zhou, Y. Light-Induced Hydrogel Based on Tumor-Targeting Mesoporous Silica Nanoparticles as a Theranostic Platform for Sustained Cancer Treatment. *ACS Appl. Mater. Interfaces* **2016**, *8*, 15857–15863. [CrossRef] [PubMed]
111. Qiu, M.; Wang, D.; Liang, W.; Liu, L.; Zhang, Y.; Chen, X.; Sang, D.K.; Xing, C.; Li, Z.; Dong, B.; et al. Novel concept of the smart NIR-light–controlled drug release of black phosphorus nanostructure for cancer therapy. *Proc. Natl. Acad. Sci. USA* **2018**, *115*, 501–506. [CrossRef] [PubMed]
112. Ojugo, A.S.E.; McSheehy, P.M.J.; McIntyre, D.J.O.; McCoy, C.; Stubbs, M.; Leach, M.O.; Judson, I.R.; Griffiths, J.R. Measurement of the extracellular pH of solid tumours in mice by magnetic resonance spectroscopy: A comparison of exogenous19F and31P probes. *NMR Biomed.* **1999**, *12*, 495–504. [CrossRef]
113. van Dyke, R.W. *Acidification of Lysosomes and Endosomes*; Springer: Boston, MA, USA, 1996; pp. 331–360.
114. Hoffman, A.S. Stimuli-responsive polymers: Biomedical applications and challenges for clinical translation. *Adv. Drug Deliv. Rev.* **2013**, *65*, 10–16. [CrossRef]
115. Raza, F.; Zhu, Y.; Chen, L.; You, X.; Zhang, J.; Khan, A.; Khan, M.W.; Hasnat, M.; Zafar, H.; Wu, J.; et al. Paclitaxel-loaded pH responsive hydrogel based on self-assembled peptides for tumor targeting. *Biomater. Sci.* **2019**, *7*, 2023–2036. [CrossRef]
116. Xu, C.; Yan, Y.; Tan, J.; Yang, D.; Jia, X.; Wang, L.; Xu, Y.; Cao, S.; Sun, S. Biodegradable Nanoparticles of Polyacrylic Acid–Stabilized Amorphous $CaCO_3$ for Tunable pH-Responsive Drug Delivery and Enhanced Tumor Inhibition. *Adv. Funct. Mater.* **2019**, *29*, 1808146. [CrossRef]
117. Ren, Y.; Li, X.; Han, B.; Zhao, N.; Mu, M.; Wang, C.; Du, Y.; Wang, Y.; Tong, A.; Liu, Y.; et al. Improved anti-colorectal carcinomatosis effect of tannic acid co-loaded with oxaliplatin in nanoparticles encapsulated in thermosensitive hydrogel. *Eur. J. Pharm. Sci.* **2019**, *128*, 279–289. [CrossRef]
118. Makhmalzadeh, B.S.; Molavi, O.; Vakili, M.R.; Zhang, H.-F.; Solimani, A.; Abyaneh, H.S.; Loebenberg, R.; Lai, R.; Lavasanifar, A. Functionalized Caprolactone-Polyethylene Glycol Based Thermo-Responsive Hydrogels of Silibinin for the Treatment of Malignant Melanoma. *J. Pharm. Pharm. Sci.* **2018**, *21*, 143–159. [CrossRef] [PubMed]
119. Wu, H.; Song, L.; Chen, L.; Zhang, W.; Chen, Y.; Zang, F.; Chen, H.; Ma, M.; Gu, N.; Zhang, Y. Injectable magnetic supramolecular hydrogel with magnetocaloric liquid-conformal property prevents post-operative recurrence in a breast cancer model. *Acta Biomater.* **2018**, *74*, 302–311. [CrossRef] [PubMed]
120. Hyun, H.; Park, M.H.; Jo, G.; Kim, S.Y.; Chun, H.J.; Yang, D.H. Photo-Cured Glycol Chitosan Hydrogel for Ovarian Cancer Drug Delivery. *Mar. Drugs* **2019**, *17*, 41. [CrossRef] [PubMed]
121. Yu, S.; Wei, S.; Liu, L.; Qi, D.; Wang, J.; Chen, G.; He, W.; He, C.; Chen, X.; Gu, Z. Enhanced local cancer therapy using a CA4P and CDDP co-loaded polypeptide gel depot. *Biomater. Sci.* **2019**, *7*, 860–866. [CrossRef]
122. Permana, A.D.; Utami, R.N.; Layadi, P.; Himawan, A.; Juniarti, N.; Anjani, Q.K.; Utomo, E.; Mardikasari, S.A.; Arjuna, A.; Donnelly, R.F. Thermosensitive and mucoadhesive in situ ocular gel for effective local delivery and antifungal activity of itraconazole nanocrystal in the treatment of fungal keratitis. *Int. J. Pharm.* **2021**, *602*, 120623. [CrossRef]
123. Chaudhari, P.; Naik, R.; Mallela, L.S.; Roy, S.; Birangal, S.; Ghate, V.; Kunhanna, S.B.; Lewis, S.A. A supramolecular thermosensitive gel of ketoconazole for ocular applications: In silico, in vitro, and ex vivo studies. *Int. J. Pharm.* **2022**, *613*, 121409. [CrossRef]
124. Alkholief, M.; Kalam, M.A.; Almomen, A.; Alshememry, A.; Alshamsan, A. Thermoresponsive sol-gel improves ocular bioavailability of Dipivefrin hydrochloride and potentially reduces the elevated intraocular pressure in vivo. *Saudi Pharm. J.* **2020**, *28*, 1019–1029. [CrossRef]
125. Mahboobian, M.M.; Mohammadi, M.; Mansouri, Z. Development of thermosensitive in situ gel nanoemulsions for ocular delivery of acyclovir. *J. Drug Deliv. Sci. Technol.* **2020**, *55*, 101400. [CrossRef]

126. Cucina, A.; Filali, S.; Risler, A.; Febvay, C.; Salmon, D.; Pivot, C.; Pelandakis, M.; Pirot, F. Dual 0.02% chlorhexidine digluconate—0.1% disodium EDTA loaded thermosensitive ocular gel for Acanthamoeba keratitis treatment. *Int. J. Pharm.* **2019**, *556*, 330–337. [CrossRef]
127. Diaz-Rodriguez, P.; Mariño, C.; Vázquez, J.A.; Caeiro-Rey, J.R.; Landin, M. Targeting joint inflammation for osteoarthritis management through stimulus-sensitive hyaluronic acid based intra-articular hydrogels. *Mater. Sci. Eng. C* **2021**, *128*, 112254. [CrossRef]
128. Zhang, T.; Chen, S.; Dou, H.; Liu, Q.; Shu, G.; Lin, J.; Zhang, W.; Peng, G.; Zhong, Z.; Fu, H. Novel glucosamine-loaded thermosensitive hydrogels based on poloxamers for osteoarthritis therapy by intra-articular injection. *Mater. Sci. Eng. C* **2021**, *118*, 111352. [CrossRef] [PubMed]
129. Nascimento, M.H.M.D.; Ambrosio, F.N.; Ferraraz, D.C.; Windisch-Neto, H.; Querobino, S.M.; Nascimento-Sales, M.; Alberto-Silva, C.; Christoffolete, M.A.; Franco, M.K.K.D.; Kent, B.; et al. Sulforaphane-loaded hyaluronic acid-poloxamer hybrid hydrogel enhances cartilage protection in osteoarthritis models. *Mater. Sci. Eng. C* **2021**, *128*, 112345. [CrossRef] [PubMed]
130. GhavamiNejad, A.; Samarikhalaj, M.; Aguilar, L.E.; Park, C.H.; Kim, C.S. pH/NIR Light-Controlled Multidrug Release via a Mussel-Inspired Nanocomposite Hydrogel for Chemo-Photothermal Cancer Therapy. *Sci. Rep.* **2016**, *6*, 33594. [CrossRef] [PubMed]
131. Ruan, C.; Liu, C.; Hu, H.; Guo, X.-L.; Jiang, B.-P.; Liang, H.; Shen, X.-C. NIR-II light-modulated thermosensitive hydrogel for light-triggered cisplatin release and repeatable chemo-photothermal therapy. *Chem. Sci.* **2019**, *10*, 4699–4706. [CrossRef] [PubMed]
132. Cai, Z.; Zhang, H.; Wei, Y.; Wei, Y.; Xie, Y.; Cong, F. Reduction- and pH-Sensitive Hyaluronan Nanoparticles for Delivery of Iridium(III) Anticancer Drugs. *Biomacromolecules* **2017**, *18*, 2102–2117. [CrossRef] [PubMed]
133. Malekimusavi, H.; Ghaemi, A.; Masoudi, G.; Chogan, F.; Rashedi, H.; Yazdian, F.; Omidi, M.; Javadi, S.; Haghiralsadat, B.F.; Teimouri, M.; et al. Graphene oxide- l -arginine nanogel: A pH-sensitive fluorouracil nanocarrier. *Biotechnol. Appl. Biochem.* **2019**, *66*, 772–780. [CrossRef]
134. Zhang, F.; Gong, S.; Wu, J.; Li, H.; Oupicky, D.; Sun, M. CXCR4-Targeted and Redox Responsive Dextrin Nanogel for Metastatic Breast Cancer Therapy. *Biomacromolecules* **2017**, *18*, 1793–1802. [CrossRef]
135. Zhao, J.; Yang, Y.; Han, X.; Liang, C.; Liu, J.; Song, X.; Ge, Z.; Liu, Z. Redox-Sensitive Nanoscale Coordination Polymers for Drug Delivery and Cancer Theranostics. *ACS Appl. Mater. Interfaces* **2017**, *9*, 23555–23563. [CrossRef]
136. Kumar, P.; Wasim, L.; Chopra, M.; Chhikara, A. Co-delivery of Vorinostat and Etoposide Via Disulfide Cross-Linked Biodegradable Polymeric Nanogels: Synthesis, Characterization, Biodegradation, and Anticancer Activity. *AAPS PharmSciTech* **2018**, *19*, 634–647. [CrossRef]
137. Park, H.; Choi, Y.; Jeena, M.T.; Ahn, E.; Choi, Y.; Kang, M.-G.; Lee, C.G.; Kwon, T.-H.; Rhee, H.-W.; Ryu, J.-H.; et al. Reduction-Triggered Self-Cross-Linked Hyperbranched Polyglycerol Nanogels for Intracellular Delivery of Drugs and Proteins. *Macromol. Biosci.* **2018**, *18*, 1700356. [CrossRef]
138. Li, F.; Yang, H.; Bie, N.; Xu, Q.; Yong, T.; Wang, Q.; Gan, L.; Yang, X. Zwitterionic Temperature/Redox-Sensitive Nanogels for Near-Infrared Light-Triggered Synergistic Thermo-Chemotherapy. *ACS Appl. Mater. Interfaces* **2017**, *9*, 23564–23573. [CrossRef] [PubMed]
139. Gao, F.; Xie, W.; Miao, Y.; Wang, D.; Guo, Z.; Ghosal, A.; Li, Y.; Wei, Y.; Feng, S.; Zhao, L.; et al. Magnetic Hydrogel with Optimally Adaptive Functions for Breast Cancer Recurrence Prevention. *Adv. Healthc. Mater.* **2019**, *8*, 1900203. [CrossRef] [PubMed]
140. Parisi, O.I.; Morelli, C.; Scrivano, L.; Sinicropi, M.S.; Cesario, M.G.; Candamano, S.; Puoci, F.; Sisci, D. Controlled release of sunitinib in targeted cancer therapy: Smart magnetically responsive hydrogels as restricted access materials. *RSC Adv.* **2015**, *5*, 65308–65315. [CrossRef]
141. Wang, L.; Wang, Y.; Hao, J.; Dong, S. Magnetic Fullerene-DNA/Hyaluronic Acid Nanovehicles with Magnetism/Reduction Dual-Responsive Triggered Release. *Biomacromolecules* **2017**, *18*, 1029–1038. [CrossRef] [PubMed]
142. Aboali, F.A.; Habib, D.A.; Elbedaiwy, H.M.; Farid, R.M. Curcumin-loaded proniosomal gel as a biofreindly alternative for treatment of ocular inflammation: In-vitro and in-vivo assessment. *Int. J. Pharm.* **2020**, *589*, 119835. [CrossRef]
143. Shukr, M.H.; Ismail, S.; El-Hossary, G.G.; El-Shazly, A.H. Design and evaluation of mucoadhesive in situ liposomal gel for sustained ocular delivery of travoprost using two steps factorial design. *J. Drug Deliv. Sci. Technol.* **2021**, *61*, 102333. [CrossRef]
144. Jin, R.; Teixeira, L.S.M.; Dijkstra, P.J.; van Blitterswijk, C.A.; Karperien, M.; Feijen, J. Enzymatically-crosslinked injectable hydrogels based on biomimetic dextran–hyaluronic acid conjugates for cartilage tissue engineering. *Biomaterials* **2010**, *31*, 3103–3113. [CrossRef]
145. Wei, J.; Ran, P.; Li, Q.; Lu, J.; Zhao, L.; Liu, Y.; Li, X. Hierarchically structured injectable hydrogels with loaded cell spheroids for cartilage repairing and osteoarthritis treatment. *Chem. Eng. J.* **2022**, *430*, 132211. [CrossRef]
146. Zhou, T.; Xiong, H.; Wang, S.Q.; Zhang, H.L.; Zheng, W.W.; Gou, Z.R.; Fan, C.Y.; Gao, C.Y. An injectable hydrogel dotted with dexamethasone acetate-encapsulated reactive oxygen species-scavenging micelles for combinatorial therapy of osteoarthritis. *Mater. Today Nano* **2022**, *17*, 100164. [CrossRef]
147. Han, Y.; Yang, J.; Zhao, W.; Wang, H.; Sun, Y.; Chen, Y.; Luo, J.; Deng, L.; Xu, X.; Cui, W.; et al. Biomimetic injectable hydrogel microspheres with enhanced lubrication and controllable drug release for the treatment of osteoarthritis. *Bioact. Mater.* **2021**, *6*, 3596–3607. [CrossRef]
148. Seo, B.-B.; Kwon, Y.; Kim, J.; Hong, K.H.; Kim, S.-E.; Song, H.-R.; Kim, Y.-M.; Song, S.-C. Injectable polymeric nanoparticle hydrogel system for long-term anti-inflammatory effect to treat osteoarthritis. *Bioact. Mater.* **2022**, *7*, 14–25. [CrossRef] [PubMed]

149. Lei, Y.; Wang, X.; Liao, J.; Shen, J.; Li, Y.; Cai, Z.; Hu, N.; Luo, X.; Cui, W.; Huang, W. Shear-responsive boundary-lubricated hydrogels attenuate osteoarthritis. *Bioact. Mater.* **2022**, *16*, 472–484. [CrossRef] [PubMed]
150. Ocular Inserts with Analyte Capture and Release Agents. U.S. Patent 20,210,338,211, 3 May 2021. Available online: https://patentscope.wipo.int/search/en/detail.jsf?docId=US340278458&_cid=P20-L2LQFZ-64472-1 (accessed on 30 April 2022).
151. Drug Delivery from Hydrogels. JP Patent 2,021,120,395, 11 May 2021. Available online: https://patentscope.wipo.int/search/en/detail.jsf?docId=JP334768825&_cid=P20-L2LQIS-64737-1 (accessed on 30 April 2022).
152. Injectable Hydrogels for Cell Delivery to the Vitreous. WO Patent 2,021,113,515, 10 June 2020. Available online: https://patentscope.wipo.int/search/en/detail.jsf?docId=WO2021113515&_cid=P20-L2LQN0-65347-1 (accessed on 30 April 2022).
153. Polymer Formulations for Nasolacrimal Stimulation. U.S. Patent 20,210,069,496, 11 September 2020. Available online: https://patentscope.wipo.int/search/en/detail.jsf?docId=US319902595&_cid=P20-L2LQOA-65557-1 (accessed on 30 April 2022).
154. A Method for Obtaining Ion-Exchange Polymeric Hydrogels for Eye Treatment and Hydrogel Lenses Thereof. WO Patent 2,021,038,279, 4 March 2021. Available online: https://patentscope.wipo.int/search/en/detail.jsf?docId=WO2021038279&_cid=P20-L2LQQ0-65789-1 (accessed on 30 April 2022).
155. An Etoricoxib Loaded Nanosponge Hydrogel for Arthritis & Process to Prepare Thereof. IN Patent 202,121,042,889, 22 September 2021. Available online: https://patentscope.wipo.int/search/en/detail.jsf?docId=IN340501612&_cid=P20-L2LQRJ-65914-1 (accessed on 30 April 2022).
156. Protein Hydrogel for Topical Application and Formulation for Skin Regeneration/Wound Healing. IN Patent 202,031,000,910, 8 January 2020. Available online: https://patentscope.wipo.int/search/en/detail.jsf?docId=IN334769630&_cid=P20-L2LQSP-66090-1 (accessed on 30 April 2022).
157. Preparation Method of Conductive Adhesive Hydrogel. CN Patent 112,442,194, 4 September 2019. Available online: https://patentscope.wipo.int/search/en/detail.jsf?docId=CN320288668&_cid=P20-L2LQU0-66236-1 (accessed on 30 April 2022).
158. Thrombin-Responsive Hydrogels and Devices for Auto-Anticoagulant Regulation. U.S. Patent 20,210,023,121, 14 September 2020. Available online: https://patentscope.wipo.int/search/en/detail.jsf?docId=US316414241&_cid=P20-L2LQV8-66462-1 (accessed on 30 April 2022).
159. Thermosensitive Hydrogel for Cancer Therapeutics and Methods of Preparation Thereof. U.S. Patent 20,210,393,780, 20 May 2021. Available online: https://patentscope.wipo.int/search/en/detail.jsf?docId=US345561289&_cid=P20-L2LQWJ-66715-1 (accessed on 30 April 2022).
160. Tunable Extended Release Hydrogels. WO Patent 2,021,174,021, 2 September 2021. Available online: https://patentscope.wipo.int/search/en/detail.jsf?docId=WO2021174021&_cid=P20-L2LQXW-66853-1 (accessed on 30 April 2022).
161. Novel Self-Assembling Peptides and Their Use in the Formaiion of Hydrogels. IN Patent 9758/CHENP/2012, 19 November 2012. Available online: https://patentscope.wipo.int/search/en/detail.jsf?docId=IN211695000&_cid=P20-L2LQZK-67034-1 (accessed on 30 April 2022).
162. Synthetic Diblock Copolypeptide Hydrogels for Use in the Central Nervous System. U.S. Patent 20,140,286,865, 18 February 2014. Available online: https://patentscope.wipo.int/search/en/detail.jsf?docId=US123273430&_cid=P20-L2LR0O-67168-1 (accessed on 30 April 2022).

gels

Article

Interaction of TX-100 and Antidepressant Imipramine Hydrochloride Drug Mixture: Surface Tension, ^{1}H NMR, and FT-IR Investigation

Malik Abdul Rub [1,2,*], Naved Azum [1,2], Dileep Kumar [3,4,*] and Abdullah M. Asiri [1,2]

1 Center of Excellence for Advanced Materials Research, King Abdulaziz University, Jeddah 21589, Saudi Arabia; nhassan2@kau.edu.sa (N.A.); aasiri2@kau.edu.sa (A.M.A.)
2 Chemistry Department, Faculty of Science, King Abdulaziz University, Jeddah 21589, Saudi Arabia
3 Division of Computational Physics, Institute for Computational Science, Ton Duc Thang University, Ho Chi Minh City 700000, Vietnam
4 Faculty of Applied Sciences, Ton Duc Thang University, Ho Chi Minh City 700000, Vietnam
* Correspondence: aabdalrab@kau.edu.sa (M.A.R.); dileepkumar@tdtu.edu.vn (D.K.)

Abstract: Interfacial interaction amongst the antidepressant drug-imipramine hydrochloride (IMP) and pharmaceutical excipient (triton X-100 (TX-100-nonionic surfactant)) mixed system of five various ratios in dissimilar media (H_2O/50 mmol·kg^{-1} NaCl/250 mmol·kg^{-1} urea) was investigated through the surface tension method. In addition, in the aqueous solution, the ^{1}H-NMR, as well as FT-IR studies of the studied pure and mixed system were also explored and deliberated thoroughly. In NaCl media, properties of pure/mixed interfacial surfaces enhanced as compared with the aqueous system, and consequently the synergism/attractive interaction among constituents (IMP and TX-100) grew, whereas in urea (U) media a reverse effect was detected. Surface excess concentration (Γ_{max}), composition of surfactant at mixed monolayer (X_1^σ), activity coefficient ($f_1{}^\sigma$ (TX-100) and $f_2{}^\sigma$ (IMP)), etc. were determined and discussed thoroughly. At mixed interfacial surfaces interaction, parameter (β^σ) reveals the attractive/synergism among the components. The Gibbs energy of adsorption (ΔG^{o}_{ads}) value attained was negative throughout all employed media viewing the spontaneity of the adsorption process. The ^{1}H NMR spectroscopy was also employed to examine the molecular interaction of IMP and TX-100 in an aqueous system. FT-IR method as well illustrated the interaction amongst the component. The findings of the current study proposed that TX-100 surfactant could act as an efficient drug delivery vehicle for an antidepressant drug. Gels can be used as drug dosage forms due to recent improvements in the design of surfactant systems. Release mechanism of drugs from surfactant/polymer gels is dependent upon the microstructures of the gels and the state of the drugs within the system.

Keywords: amphiphilic drug; nonionic surfactant; surface property; thermodynamic; chemical shift; FT-IR

Citation: Rub, M.A.; Azum, N.; Kumar, D.; Asiri, A.M. Interaction of TX-100 and Antidepressant Imipramine Hydrochloride Drug Mixture: Surface Tension, ^{1}H NMR, and FT-IR Investigation. *Gels* **2022**, *8*, 159. https://doi.org/10.3390/gels8030159

Academic Editor: Hiroyuki Takeno

Received: 28 January 2022
Accepted: 2 March 2022
Published: 4 March 2022

Publisher's Note: MDPI stays neutral with regard to jurisdictional claims in published maps and institutional affiliations.

1. Introduction

Gels are used for various applications based on their drug-loading properties, rheological properties, and release mechanisms. Drugs can either be soluble in water with no interaction through any of the constituents, electrostatically/hydrophobically tied with polymer, or soluble within micelles and polymer/surfactant associates. The use of surfactant/polymer systems for gene therapy has a great deal of promise, and certain polymers can interact with the natural (nonionic) surfactant, which can be utilized to lock in bile salts for controlling cholesterol levels in the body. The interfacial/micellar characteristics of amphiphiles mixtures have been broadly studied due to their extensive applications, for instance, hydrate inhibitors, biologicals, foaming, in fabric moderating, pharmaceutics, improved oil recovery procedure, and so forth [1–3]. In aqueous/non-aqueous solvent, the surfactant monomers (comprising hydrophobic and hydrophilic parts into single molecules) were orientated into an associated form after surpassing a certain concentration into the solution (solvent) and formed the associate structure, called the micelle. The corresponding concentration is symbolized as the critical micelle concentration (*cmc*) [3–6]. Surfactant micelles revealed a considerable role in the solubilization of several hydrophobic materials including drugs [3,7]. Surfactant also acts as a drug carrier in combination with a specific additive, and therefore, extensive inspections of the influences of several additives (organic and inorganic) on the association performance of the drug are needed [3,7]. As compared with singular surfactant micelle formation, the mixed surfactants have substantial considerable properties in a variety of features [3]. Usually, a mixed surfactants system (ionic amphiphile with other ionic or nonionic amphiphiles) has smaller surface energy, higher solubilization capability, and smaller *cmc* along with higher surface activities as compared to the singular surfactants because of the attractive interaction/synergetic influence [3,8]. To diagnose osteoarthritis, Yin et al. [9] have made significant progress in eliminating major hurdles to using extracellular vesicles for delivery and as markers. Osteoarthritis therapeutics can be delivered effectively via extracellular vesicles because of their size, surface expression patterns, low immunogenicity, and low cytotoxicity.

Within various kinds of surfactants (cationic/anionic/nonionic), the non-ionic surfactant is valued as the best one for safe drug delivery, as they are physiologically more supportable than ionic surfactant [7]. TX-100 is one of the most applied surfactants in bio-chemical and chemical practices. The head groups of non-ionic surfactants consist of no electrical charge; therefore, they are generally soluble in water through H-bonding formation between the hydrophilic parts of the surfactant with water. Triton X-100 (TX-100) non-ionic surfactant has a huge industrial significance applied in the formulation of foams and found several applications in the pharmaceutical sciences for purpose of cleaning and as an ingredient in a few curative products [10,11]. TX-100 comprises a hydrophilic chain of 9 to 10 ethylene oxide units coupled with an aromatic ring, having a branched hydrocarbon chain. Different properties (interfacial, micellization, drugs solubilization ability, clouding property, etc.) of TX-100 in the occurrence of charge amphiphiles have been analyzed by means of experimental methods [3,12,13]. TX-100 varies from other conventional nonionic surfactants because their hydrophilic portion was found to be longer compared with the hydrophobic section of the monomer [14]. Herein, the interaction of TX-100 with antidepressant IMP was evaluated by means of different techniques. The mixed system of IMP+TX-100 reveals a compact packing at the surface as well as higher interfacial activity.

At a higher concentration, numerous amphiphilic drugs also formed a micellar structure in a similar manner to a conventional surfactant [15,16]. Pure amphiphilic drugs self-association studies, for any particular purpose are usually out-of-focus due to their high *cmc*, because of the use of a high amount of a drug, which might create numerous side effects [17]. Therefore, amphiphilic drugs are generally used in combination with additives such as surfactant, hydrotropes, bile salts, etc., as a drug carrier that generally forms mixed micelles [8,15]. As a mixture, the *cmc* value reduced more than 10 times. Hence, a very low quantity of drug is used along with a mixed micellar system to raise the absorption of numerous drugs [8].

Imipramine hydrochloride (IMP) is an amphiphilic tricyclic antidepressant drug that has two main parts, one is a large rigid tricyclic hydrophobic ring (tail) and the other one is a small alkyl amine part (head) and endures aggregation but higher concentration [15]. This drug color is white to off-white, odorless compound, and is employed to treat depression. The nature of IMP drug is protonated (cationic) at a lower pH range (below 7) and deprotonated at a high range (above 7) of pH ($pK_a = 9.5$) [15]. Apart from their uses to heal depression, this drug also indicated some unwanted impact. Consequently, to lessen the unwanted impact of IMP, mixed micellization investigation of IMP with TX-100 (as a drug carrier) (Scheme 1) was conducted in different media by means of the several methods.

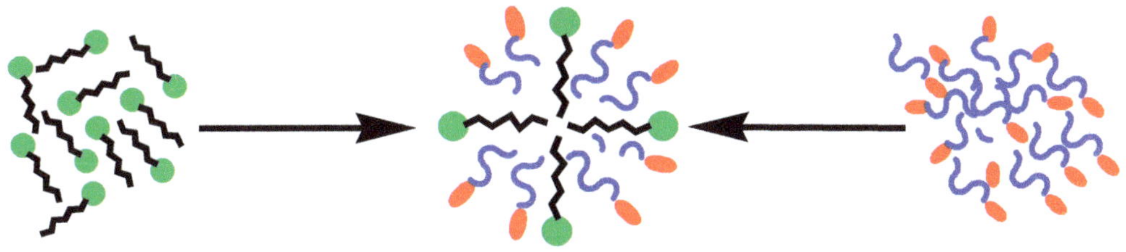

Scheme 1. Mixed micelle formation of the IMP+TX-100 mixed system.

Previously, our group have examined solution (bulk) properties (mixed micellization behavior) of pure and mixed system of IMP and TX-100 in water, NaCl, and urea media [18] and the current study is an extension of our previous work [18]. Herein, the interfacial properties of IMP and TX-100 mixture were evaluated by tenstiometic method in different media, along with ^{1}H NMR and FT-IR spectroscopy, which were also employed to evaluate the interaction amongst IMP and TX-100 in an aqueous system. Combining IMP with TX-100, might enhance drug characteristics, such as their solubility along with stability in living atmospheres [8,19]. Previously, in an aqueous solution, Alam and Siddiq [20] examined the association and surface behavior of an IMP drug and TX-100 mixed system by differing the mole fraction of a drug by tensiometric method. Irrespective of surface tension and ^{1}H NMR methods, the FTIR study of the akin system in aqueous media was also investigated to crisscross the reliability of the interaction between IMP drug and TX-100. ^{1}H NMR of IMP+TX-100 mixture in five different ratios has been investigated to explain the mechanism of IMP and TX-100 interactions. Several theoretical models regarding the interfacial behavior are employed to illustrate the mixed monolayer formation of the drug-surfactant mixed system in three different media. Various parameters, such as surface excess concentration (Γ_{max}), composition of constituent at mixed monolayer and the interaction parameter (β^{σ}) at interface, activity coefficient of employed ingredients ($f_1{}^{\sigma}$ (TX-100) and $f_2{}^{\sigma}$ (IMP)) at the boundary, packing parameter, etc., at the mixed monolayer, have been assessed and discussed [3,21]. Different thermodynamic functions (Gibbs's energy of adsorption (ΔG^{o}_{ad}), minimum free energy (G_{min}), excess free energy at mixed monolayer (ΔG^{σ}_{ex})), and chemical shifts by ^{1}H NMR study have also been thoroughly evaluated and debated. According to the current study, the results have relevance to model drug delivery, but no direct evidence can be drawn for drug delivery. As a result of this study, drugs and their possible carriers are examined physiochemically using various theoretical models, which is vital since the surfactant may also be utilized as a drug carrier. In addition, the choice of 50 mmol·kg^{-1} NaCl and 250 mmol·kg^{-1} urea concentration was not based on any specific reason other than to examine the effects of salt and urea that are normally found in human being. To provide knowledge (thermodynamic and additional) for the widely used drug-surfactant combinations in the absence and presence of NaCl and urea in drug delivery, our primary goal had been to exhibit how the two ingredients interacted in the aqueous system as well as in salt and urea media. Further enhancement of drug-surfactant conjugate delivery systems is possible if salt/urea are present as their presence increases/decreases the spontaneity of the mixture.

2. Results and Discussion

2.1. Characteristics at the Air-Interfacial Surfaces of Pure and Mixed System

Amphiphiles are likely to settle at the air-interfacial surface as compared with the bulk solution. Gibbs's adsorption equation [22] is employed to assess a variety of surface parameters of drug–surfactant mixed system. All interfacial parameters were evaluated by using the surface tension plot given in our previous work [18]. The adsorbed quantity of molecules in each unit area of the surface is computed through the assistance of Gibbs adsorption equation [22]. The surface excess concentration (Γ_{max}) along with minimum area per monomer (A_{min}) values in aqueous/non-aqueous media were determined utilizing the subsequent equations [3,22]:

$$\Gamma_{max} = -\frac{1}{2.303nRT}\left(\frac{\partial\gamma}{\partial log(C)}\right)\ (\text{mol}\cdot\text{m}^{-2}), \tag{1}$$

$$A_{min} = \frac{10^{20}}{N_A\,\Gamma_{max}}\ (\text{Å}^2). \tag{2}$$

Here, the γ, C, T, n, R, and N_A is the surface tension ($\text{mN}\cdot\text{m}^{-1}$), employed concentration of IMP, TX-100, or IMP+TX-100 mixtures, temperature, whole number of solute species obtained during adsorption, gas constant, and Avogadro number, respectively [3]. The n is considered 2 and 1 in the case of individual IMP and TX-100, respectively. However, in mixtures, n values were assessed using term: $n = n_1X_1^\sigma + n_2(1 - X_1^\sigma)$ [3], where n_1 = number of species in component 1 and n_2 = number of species in component 2 after ionization. X_1^σ = interfacial composition of component 1 at the mixed surface (Table 1). Throughout the study, the first component, or component 1, is used for TX-100 and the second component, or component 2, is used for IMP. The slope = $\partial\gamma/\partial log(C)$ value is attained from the γ vs. log(C) plot of any fixed concertation in all cases.

Table 1. Different interfacial parameters for IMP+TX-100 mixture in several media at 298.15 K [a].

α_1	X_1^σ	β^σ	f_1^σ	f_2^σ	Γ_{max} 10^7 ($\text{mol}\cdot\text{m}^{-2}$)	A_{min}/A^{id} (Å^2)	γ_{cmc} ($\text{mN}\cdot\text{m}^{-1}$)	π_{cmc} ($\text{mN}\cdot\text{m}^{-1}$)	pC_{20}	$\ln(C_1/C_2)$
					Aqueous solution					
0					12.78	129.95	42.58	28.42	1.95	
0.1	0.7334	−6.07	0.6496	0.0382	23.09	71.89/137.77	30.35	40.65	3.90	−6.04
0.3	0.7621	−7.77	0.6443	0.0110	23.22	71.51/139.94	29.94	41.06	4.36	−6.04
0.5	0.7741	−8.79	0.6385	0.0052	24.16	68.72/140.84	29.67	41.33	4.58	−6.04
0.7	0.7969	−9.34	0.6803	0.0027	25.73	64.54/142.56	29.36	41.64	4.68	−6.04
0.9	0.8421	−9.62	0.7868	0.0011	27.90	59.52/145.97	29.40	41.60	4.71	−6.04
1					36.02	46.10	29.31	41.69	4.57	
					50 $\text{mmol}\cdot\text{kg}^{-1}$ NaCl					
0					8.86	187.41	44.69	26.31	2.07	
0.1	0.8828	−2.76	0.9629	0.1168	27.25	60.93/149.04	30.66	40.34	3.88	−6.32
0.3	0.8712	−4.80	0.9235	0.0262	26.04	63.77/148.16	29.91	41.09	4.38	−6.32
0.5	0.9149	−4.76	0.9661	0.0186	28.65	57.95/151.46	29.83	41.17	4.57	−6.32
0.7	0.8819	−6.75	0.9101	0.0052	28.01	59.27/148.97	29.53	41.47	4.75	−6.32
0.9	0.9210	−7.23	0.9559	0.0022	31.30	53.05/151.91	29.45	41.55	4.82	−6.32
1					27.60	60.15	29.65	41.35	4.96	
					250 $\text{mmol}\cdot\text{kg}^{-1}$ U					
0					12.46	133.24	44.03	26.97	1.86	
0.1	0.7776	−4.69	0.7930	0.0587	25.19	65.91/141.11	30.21	40.79	3.70	−6.05
0.3	0.8085	−6.12	0.7989	0.0183	25.92	64.06/143.43	29.59	41.41	4.15	−6.05
0.5	0.8149	−7.26	0.7797	0.0080	25.42	65.31/143.92	29.31	41.69	4.38	−6.05
0.7	0.8189	−8.50	0.7567	0.0033	25.57	64.94/144.22	29.80	41.20	4.54	−6.05
0.9	0.8096	−10.99	0.6715	0.0007	24.60	67.49/143.52	29.71	41.29	4.71	−6.05
1					33.18	50.04	30.17	40.83	4.49	

[a] $A_1 = A_{min}$ of TX-100 and $A_2 = A_{min}$ of IMP. A_1 = 46.10 (in aqueous), 60.15 (in NaCl), 50.04 Å^2 (in urea). A_2 = 129.95 (in aqueous), 187.41 (in NaCl), 133.24 Å^2 (in urea).

In the ideal state, the minimum surface area per molecule (A^{id}) was evaluated by means of Equation (3):

$$A^{id} = X_1^\sigma A_1 + (1 - X_1^\sigma)A_2. \tag{3}$$

Here, A_1 and A_2 = per monomer minimum head group area of surfactant and IMP correspondingly. The assessed Γ_{max}, A_{min} and A^{id} value of individual and mixed components (IMP, TX-100, and IMP+TX-100) in the existence of different media were revealed in Table 1.

Table 1 showed the value of Γ_{max} and A_{min} of individual TX-100 in the aqueous system, which was found to be 36.02 mol m^{-2} and 46.10 Å^2 respectively, revealing that their value is in the same range with the previously reported value [23]. The parameter A_{min} value showed the opposite trend with the Γ_{max} value means, as each parameter was in reverse with each other. The Γ_{max} value of singular IMP obtained lesser than the Γ_{max} value of pure TX-100 means, and the A_{min} value showed the opposite behavior. This obtained behavior viewed that TX-100 molecules favored a compacted or strongly packed arrangement at the air-solvent interface as compared with IMP regardless of the media used, and therefore TX-100 showed more surface activity. The value of Γ_{max} of the IMP+TX-100 mixed system was found above the Γ_{max} value of singular IMP but was obtained below the Γ_{max} value of TX-100, so we can observe that mixed system surface activity was found higher than pure IMP but less than pure TX-100. In an aqueous system, the Γ_{max} value of the IMP+TX-100 mixed system was found to increase with an increase in α_1 of TX-100, observing that the mixed system surface activity increases with the increase of the composition of TX-100 in the solution mixture. However, in the presence of NaCl or U, the Γ_{max} value of the IMP+TX-100 mixture has not viewed a specific trend, nor did A_{min}, since A_{min} is inversely proportional to Γ_{max}.

The Γ_{max} value in NaCl media of IMP+TX-100 mixtures was achieved higher than other employed media (H_2O or U). The electrostatic repulsions between the ingredient's monomers decreased in NaCl media, observing that the efficiency of the molecules' existence at the interfacial surface increased and high compactness of IMP+TX-100 mixtures existed. However, in U solvent, pure IMP, and TX-100, Γ_{max} value found less but does not show any proper trend for mixed system.

The A^{id} value of IMP+TX-100 mixtures were observed to be higher than experimental A_{min}, implying that the space taken by apiece monomers was found below as expected for their ideal behavior. For mixtures (IMP+TX-100), the A_{min} value was obtained below the value of A_{min} of pure IMP. This result indicates that the introduction of TX-100 in the solution of IMP causes decreases in the repulsive force between IMP monomer molecules, and hence the value of mixture A_{min} decreased. Figure 1 showed the $\Gamma_{max}/A_{min}/A^{id}$ vs. α_1 plot for IMP+TX-100 mixture in diverse media (filled, open, and half-filled symbols represent Γ_{max}, A_{min}, and A^{id}, respectively), which shows the comparison of different surface parameters graphically.

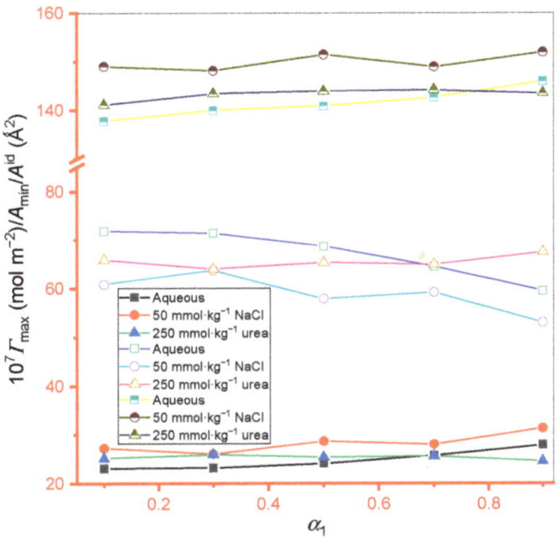

Figure 1. Plot of $\Gamma_{max}/A_{min}/A^{id}$ against α_1 plot for IMP+TX-100 mixture in different media (filled, open, and half-filled symbols represent Γ_{max}, A_{min}, and A^{id}, respectively).

The parameter surface tension value at the *cmc*, is symbolized via γ_{cmc} and the obtained value is depicted in Table 1. The γ_{cmc} value of singular and mixed system (IMP+TX-100) in aqueous and NaCl media were taken from the graph of our group's earlier published work [18]. For individual TX-100 and IMP+TX-100 mixtures, the value of γ_{cmc} was found close to each other's means in the same range, irrespective of the solvent employed. However, their value for individual IMP was found quite higher. The surface parameters-surface pressure at the *cmc* (π_{cmc}) and the adsorption efficiency, i.e., pC_{20}, was also exploited for individual ingredients and the IMP+TX-100 mixtures in all media. At the *cmc*, the surface pressure (π_{cmc}) parameter was explored by means of Equation (4) [3].

$$\pi_{cmc} = (\gamma_0 - \gamma_{cmc}). \tag{4}$$

In Equation (4), γ_0 signified the pure solvent surface tension, and γ_{cmc} indicated the γ at the *cmc* of the single and mixed components. All assessed values of γ_{cmc} and π_{cmc} are presented in Table 1. The obtained π_{cmc} is lowest for IMP irrespective of the media utilized but was found to be close to each other for individual TX-100 and the IMP+TX-100 mixture [24].

Another parameter, called pC_{20}, allowed the adsorption efficiency of the constituents at the interfacial surface. This parameter is demarcated as the negative logarithm of the concentration of monomer(s), as the individual solvent surface tension is lessened by 20 mN m^{-1} (C_{20}) [3]:

$$pC_{20} = -logC_{20}. \tag{5}$$

The higher the pC_{20} value, the larger the amphiphile efficiency for adsorption (higher surface activities) because a smaller amount (volume) of prepared solutions is needed to condense the solvent surface tension by 20 mN·m^{-1}. The obtained value of pC_{20} of IMP was considerably lower to a large extent, as compared to pC_{20} achieved for individual TX-100 regardless of the solvent used, which again confirmed that the IMP drug was less surface-active as compared with TX-100 (Table 1). This obtained phenomena showed that TX-100 has better adsorption ability along with being more effective in surface tension reduction of the solvent [24]. The pC_{20} value for IMP+TX-100 mixed systems was higher than individual IMP, observing that the mixed systems were more surface-active as compared with IMP, and their value increased with an enhancement in α_1 of TX-100, but blended systems pC_{20} value was found near the pC_{20} value TX-100 (Table 1). Pure species, as well as IMP+TX-100 mixtures pC_{20} value enhanced in the existence of NaCl because of the better surface activity in NaCl media as compared with aqueous system and the reverse trend, which was detected in the existence of U (Table 1).

2.2. Composition of Component and Interaction Parameters at the Air-Interfacial Surfaces

Before the start of micellization, at the interfacial surface, a mixed monolayer formation took place through adsorption phenomena. Rosen's theory [25] was applied to assess the composition of the constituent at mixed monolayer as well as the interaction parameter (β^σ) at the interface through subsequent equations.

$$\frac{(X_1^\sigma)^2 \ln(\alpha_1 C / X_1^\sigma C_1)}{(1 - X_1^\sigma)^2 \ln[(1 - \alpha_1)C/(1 - X_1^\sigma)C_2]} = 1, \tag{6}$$

$$\beta^\sigma = \frac{\ln(\alpha_1 C / X_1^\sigma C_1)}{(1 - X_1^\sigma)^2}. \tag{7}$$

In Equation (6), the X_1^σ = composition of the surfactant in the mixed monolayer (IMP+TX-100), and in Equations (6) and (7) C_1 = TX-100 concentration (first component), C_2 = IMP concentration (second component), and C = mixed monolayer concentration (IMP+TX-100) at different α_1, which is used for lessening the surface tension of a solvent of any selected value for all cases (pure and mixture).

The assessed values of X_1^σ and β^σ of all systems are itemized in Table 1. Herein, the X_1^σ (TX-100 composition at the mixed component surface) values were obtained amid 73% to 92% in all studied media, displaying that mainly TX-100 comprises the mixed monolayer. By increasing the α_1 value, the X_1^σ value was exhibiting any regular behavior (i.e., increase or decrease), but overall, their value was found to be higher at higher α_1. The X_1^σ value attained higher values in the NaCl system as compared with the aqueous solution at all α_1, displaying that salt diminished the repulsive forces existing amid components. Accordingly, there was an affection for early micellization prompted via the progressively hydrophobic atmosphere.

The β^σ values possess three possibilities: (1) $\beta^\sigma = 0$ for an ideal monolayer, which means no interaction among the mixture ingredients, (2) β^σ > zero for antagonistic interactions, whereas (3) β^σ < zero signifies the supremacy of attractive or synergistic interactions amongst mixture ingredients.

The β^σ values obtained was negative in all cases, revealing the existence of attractive interactions or synergism at the interfacial surface (Table 1) [26,27]. This occurred because of the closely packed formation of the mixed monolayer, owing to the clearly interactive forces amongst the ingredients at the interface in all utilized media. The decrease in electrostatic repulsion amongst molecules of both components applied its impact more at the planar interfacial surface, as compared in convex micelles [3]. The negative value of β^σ was revealed interactions amongst the constituent allocated to ion-ion dipole as well as hydrophobic interactions irrespective of the employed media. Consequently, the merger of these forces overwhelmed all electrostatic repulsion amongst the ingredients. In NaCl or U media, the β^σ value was not displaying a somewhat unique trend, but was found to be negative in the whole system (Table 1).

IMP+TX-100 mixtures display higher surface activity together with a much lesser *cmc* value as compared with individual IMP. The higher interaction amid the ingredients in the solution mixtures does not only serve as evidence for synergism in binary mixed system. Synergism in any mixed system at an interfacial surface occurs only if the subsequent circumstances are met [3]: (a) β^σ value should be below 0, and (b) $|\beta^\sigma|$ value should be more than $\ln(C_1/C_2)$ value, otherwise attractive mixed monolayers will be found. By viewing these results, it is shown that for all systems only first the circumstance was satisfied (Table 1) However, the second circumstance was not fulfilled in almost all cases. Therefore, attractive interactions were observed irrespective of the type of media employed for the surface tension reduction efficiency.

Akin to mixed micelles, the value of the activity coefficient of the employed ingredients ($f_1{}^\sigma$ (TX-100) and $f_2{}^\sigma$ (IMP)) at the boundary was also evaluated via subsequent equations [28]:

$$f_1^\sigma = \exp[\beta^\sigma(1 - X_1^\sigma)^2], \tag{8}$$

$$f_2^\sigma = \exp[\beta^\sigma(X_1^\sigma)^2]. \tag{9}$$

Table 1 shows that both $f_1{}^\sigma$ (TX-100) and $f_2{}^\sigma$ (IMP) values are obtained below one irrespective of the media employed [28]. Therefore, the system showed nonideal behavior as well as experienced attractive interactions amid the applied species at the boundary of the air-solvent. The results also showed that the $f_2{}^\sigma$ was found to be lower as compared to $f_1{}^\sigma$ (Table 1). This phenomenon showed that the involvement of IMP was much lower at the mixed monolayer than that of the TX-100. In NaCl or U media, no distinct behavior was detected.

2.3. Thermodynamic Parameters

Thermodynamic parameter, e.g., the Gibbs energy of adsorption (ΔG^o_{ad}) of the existing systems (pure and mixed), was obtained from Equation (10) [29,30]:

$$\Delta G^o_{ad} = \Delta G^o_m - \frac{\pi_{cmc}}{\Gamma_{max}}. \tag{10}$$

Table 2 showed the achieved ΔG^o_{ad} value of pure and mixed systems in different media. For the calculation of ΔG^o_{ad} of the current system, ΔG^o_m (Gibbs free energy) values were used from our previous article [18]. All ΔG^o_{ads} values were negative, which was symbolic of the spontaneity of the adsorption process at the air-solvent interface and their magnitude were higher than those of the previously calculated ΔG^{o}_{m} value [18] of the corresponding system. The occurrence of $\Delta G^{o}_{ad} > \Delta G^{o}_{m}$ hypothesized that adsorption phenomena were favored over the association process, meaning that after finishing the adsorption process, the micellization process starts, i.e., a slight effort is required to complete this phenomenon (energy supplied in micellization to bring the monomers from the surface to micellar state). The ΔG^{o}_{ads} value of the IMP+TX-100 mixture at all α_1 of the surfactant was more negative than the value associated with an individual component (IMP and TX-100) (Table 2). These obtained results showed that the adsorption phenomenon was additionally feasible in case of a mixed monolayer, as compared with the monolayer formed by a singular component. The ΔG^{o}_{ads} value did not view any specific trends in U or NaCl media in IMP+TX-100 mixtures. In the case of pure components, in NaCl/U media their negative value was found to increase/decrease, respectively.

Table 2. Various thermodynamic parameters along with packing parameter (P) for pure and IMP+TX-100 mixture in various media.

α_1	ΔG^{o}_{ad} (kJ·mol^{-1})	G_{min} (kJ·mol^{-1})	ΔG^{σ}_{ex} (kJ·mol^{-1})	P
		Aqueous system		
0	−40.06	33.33		0.34
0.1	−42.66	13.14	−2.94	0.60
0.3	−45.15	12.89	−3.49	0.61
0.5	−45.99	12.28	−3.81	0.63
0.7	−45.92	11.41	−3.74	0.67
0.9	−45.13	10.54	−3.17	0.73
1	−41.58	8.14		0.95
		50 mmol·kg^{-1} NaCl		
0	−47.80	50.44		0.24
0.1	−40.08	11.25	−0.71	0.71
0.3	−43.40	11.49	−1.33	0.68
0.5	−43.70	10.41	−0.92	0.75
0.7	−44.94	10.54	−1.74	0.73
0.9	−44.45	9.41	−1.30	0.82
1	−45.48	10.74		0.73
		250 mmol·kg^{-1} U		
0	−39.30	35.33		0.33
0.1	−40.40	11.99	−2.01	0.66
0.3	−42.41	11.42	−2.35	0.68
0.5	−44.11	11.53	−2.71	0.66
0.7	−45.0	11.66	−3.12	0.67
0.9	−46.25	12.08	−4.20	0.64
1	−41.30	9.09		0.87

One more thermodynamic parameter, named minimum free energy (G_{min}), which is attained at the outmost adsorption at equilibrium, is also used to determine the attractive interaction/synergism at the interfacial boundary via Equation (11) [31,32].

$$G_{min} = A_{min}\gamma_{cmc}N_A. \quad (11)$$

The value of the evaluated G_{min} value is given in Table 2. The value of G_{min} is usually correlated by the shipping of a component from the bulk system toward the interfacial boundary. The smaller magnitude of the G_{min} value detected in any studied case was characteristic of intensified stability of the air-solvent boundary [3]. The level through which the G_{min} value of the system is decreased is directly proportional to the extent of synergism allied through the system. The obtained G_{min} in our case was found to be lower in magnitude, showing the thermodynamic stable air-solvent boundary. The G_{min} seemed to be guileless in respect of any increase or decrease in value in any proper way by the occurrence of U/NaCl (Table 2).

An additional parameter of mixed monolayer called excess free energy (ΔG^{σ}_{ex}) of IMP+TX-100 was computed using Equation (12) [33–36].

$$\Delta G^{\sigma}_{ex} = RT[X^{\sigma}_1 \ln f^{\sigma}_1 + (1 - X^{\sigma}_1) \ln f^{\sigma}_2]. \tag{12}$$

The obtained value of ΔG^{σ}_{ex} was found to be negative in each solvent, observing that mixed monolayer formation is more stable than compared with a monolayer of either singular constituent (Table 2). Usually, at higher α_1, the ΔG^{σ}_{ex} value was found to be more negative, indicating that stability of the mixed monolayer was attained more at higher α_1, however, the ΔG^{σ}_{ex} value is not exhibiting a specific trend with the change of solvent (Table 2). Figure 2 showed the variation of $\Delta G^{o}_{ad}/G_{min}/\Delta G^{\sigma}_{ex}$ value with change in mole fraction (α_1) of TX-100 in different media (filled, open, and half-filled symbols represent ΔG^{o}_{ad}, G_{min}, and ΔG^{σ}_{ex} respectively) which depicted the comparison of different evaluated thermodynamic parameters graphically.

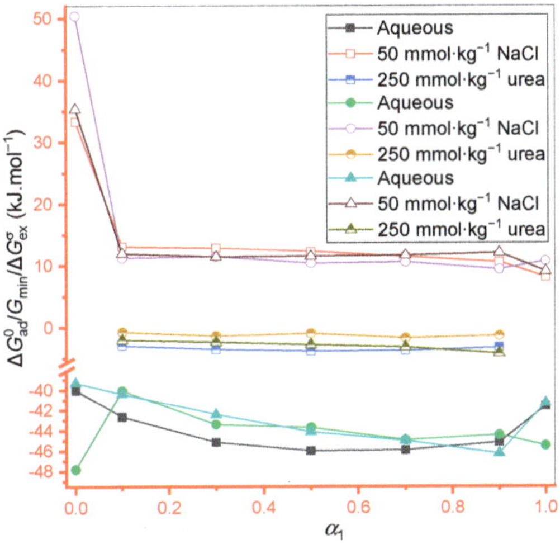

Figure 2. Variation of $\Delta G^{o}_{ad}/G_{min}/\Delta G^{\sigma}_{ex}$ value with change in mole fraction (α_1) of TX-100 in different media (filled, open, and half-filled symbols represent ΔG^{o}_{ad}, G_{min}, and ΔG^{σ}_{ex} respectively).

2.4. Packing Parameters

The structural geometry can be supposed via the packing parameter (P), i.e., the shape of micelles/mixed micelles in aqueous and non-aqueous solution was assessed through the following equation [37]:

$$P = \frac{V_0}{A_{\min} l_c}. \tag{13}$$

In Equation (13), l_c and V_0 are the effective chain length and volume of micellar interior, respectively, of the hydrophobic part of the employed monomers. Here, $A_{\min}$ value was used as achieved from the surface tension measurement. The V_0 and l_c value were computed by employing Tanford's theory [38].

$$V_0 = [27.4 + 26.9\ (n_c - 1)] \times 2\ (\text{Å}^3), \tag{14}$$

$$l_c = [1.54 + 1.26\ (n_c - 1)]\ (\text{Å}). \tag{15}$$

Here, n_c represents the whole sum of C-atoms in the C-chain length. The entire sum of C-atoms is measured one beneath the real count of C-atoms for the calculation of V_0 and l_c value, since the C-atom next to the head group is extremely solvated. Hence, the first corban is also considered as the head group portion [3]. Table 2 depicted the evaluated P

(packing parameter) value of the entire system. Micelles can be found in several shapes, depending on the obtained P value. As stated in literature [3,39] spherical micelles were detected as $P \leq 0.333$, cylinders or rods shapes micelles were noted for $0.333 < P < 0.5$, vesicles and bilayers shapes micelles were found for $0.5 < P < 1$, whereas inverted micelles were reported for $P > 1$. In our case, a P value for IMP was obtained for $0.333 < P < 0.5$ in the aqueous and U solvent, signifying that the micelles formed by IMP were cylinders or rods. In the NaCl solvent, the P value of IMP was found for 0.24, showing that IMP formed spherical micelles in the presence of NaCl (Table 2). For singular TX-100, the P was attained $0.5 < P < 1$ irrespective of the employed solvent, representing that the micellar shape of TX-100 were vesicles (Table 2). For the IMP+TX-100 mixture of the different ratio in the presence of a different solvent, the P value was achieved $0.5 < P < 1$, showing that a vesicle-shaped mixed micellar solution formed like pure TX-100, because mixed micelles consist of a maximum share of TX-100.

2.5. ^{1}H NMR Study

^{1}H NMR technique is one of the finest methods for confirming the structure and purity of compounds [40,41]. Currently, ^{1}H NMR is a very powerful method for examining an intermolecular interaction between both different compounds in their mixed micelles [42,43] and it gives us a great deal of information of interaction that is usually not available with other techniques. The present study also deals with the ^{1}H NMR study of the interaction among the drug IMP and TX-100 surfactant in different ratios in their mixed micellar solution of an aqueous system. The ^{1}H NMR signals of pure IMP, as well as TX-100, are clearly visible in D_2O. The ^{1}H NMR spectra of singular drug IMP and TX-100 is shown in Figure 3 with labeled hydrogen atoms attached to various carbons and obtained chemical shift value exposed in Table 3. Related data of pure TX-100 ^{1}H NMR have also been given in previously published work [24,44]. The spectra of pure IMP clearly show distinct six proton signals, and their corresponding proton numbers are allotted in the structure given in Scheme 2. The pure TX-100 spectra clearly show eight proton signals and the corresponding proton numbers are allocated in Scheme 3 [24,44]. Protons attached to -$N^+(CH_3)_2$ signals (I1 protons) are highly deshielded, that is, they resonate at high δ values because of the occurrence of N-atom in the drug IMP head group. All the NMR signals in both compounds drug IMP and non-ionic surfactant TX-100 (I1-I6 and T1-T8), in their pure form, show an increase in chemical shift δ values, which shows that each proton signal was highly deshielded. The proton signal I4 resonates at low δ values. This can be clearly observed, from the change in chemical shift values of I1, I3, I2, and I5, that the proton signals that present nearby to the head group are highly deshielded because of the occurrence of an adjacent N atom, whereas the proton signal I4 is highly shielded. No doubt, due to the combined electrostatic and hydrophobic effects, the interaction is stronger. In both drug and surfactant, the aromatic protons I6, T7, and T8 resonate at high δ values, i.e., they shift downfield.

Table 3. ^{1}H NMR chemical shifts (δ, ppm) of pure IMP and TX-100 in aqueous system.

Compound	Chemical Shifts (δ, ppm)							
Pure IMP	I1 2.478	I2 2.916	I3 2.790	I4 1.695	I5 3.564	I6 6.983		
Pure TX-100 [a]	T1 0.549	T2 1.124	T3 1.495	T4 3.591	T5 3.806	T6 3.865	T7 7.015	T8 7.097

[a] References [24,44].

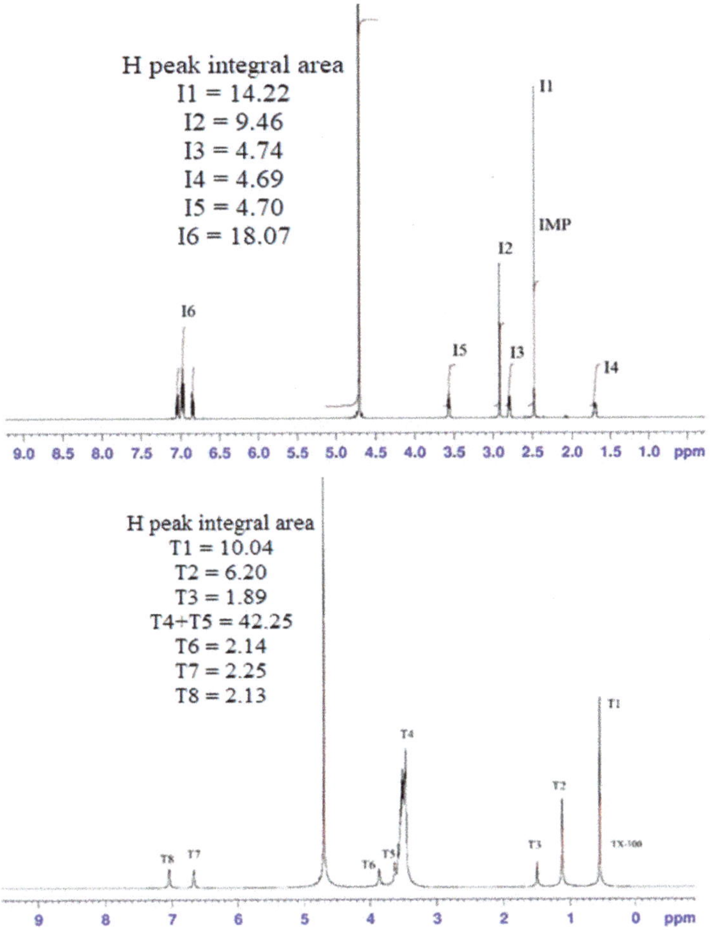

Figure 3. ^{1}H NMR (600 MHz) spectrum of singular compound IMP drug and TX-100 in D_2O.

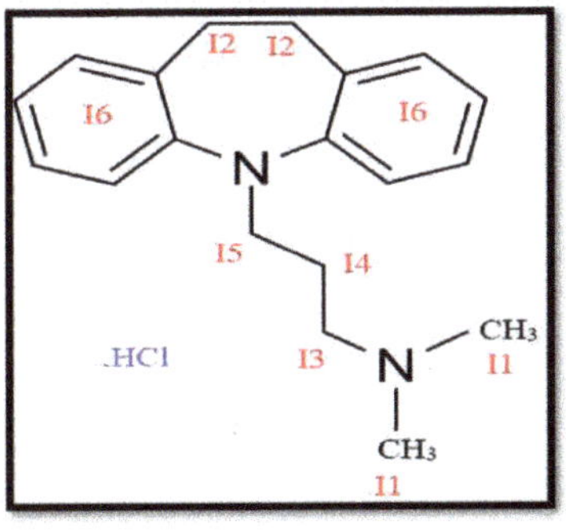

Scheme 2. Molecular model of drug IMP.

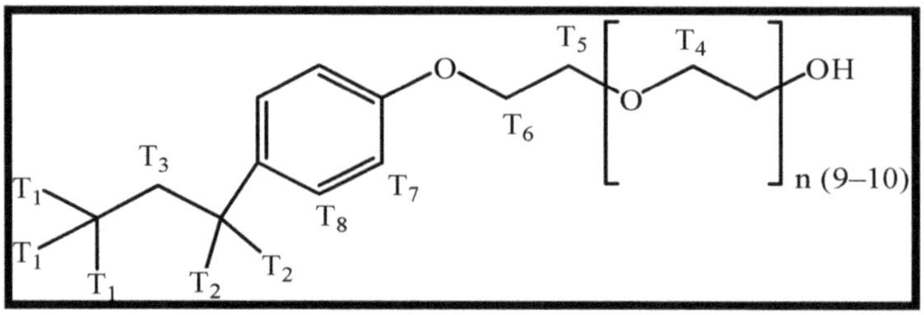

Scheme 3. Molecular model of TX-100.

Substantiation of complex formation for the drug–surfactant mixtures was obtained by NMR spectroscopy [45]. The proton signals of both the drug and surfactant show a significant change upon mixing (IMP+TX-100), which can be clearly understood from the chemical shift values given in Table 4 and also from the spectra presented in Figure 4. Table 4 depicted the addition of TX-100 in pure IMP solution cause noteworthy displacement in chemical shift values, which clearly point towards molecular interaction between IMP and TX-100. Chemical shifts are used to describe signals in NMR spectroscopy and the location and number of chemical shifts is symbolic of the structure of a compound.

Upon addition of TX-100 to pure IMP, a slight increase in chemical shift values is seen, i.e., they show a downfield shift, but not much interaction is seen at lower mixing ratios i.e., 0.1 TX-100 and 0.3 TX-100. However, as the mole fraction of TX-100 reaches 0.5, a prominent enhancement in δ values is seen through a rise in mole fraction (0.5–0.9), and from these values, it can be concluded that the extent of downfield shift is caused by the addition of TX-100; this depends upon the α_1 of surfactant in the solution of a drug and surfactant mixture. This increase in a downfield shift can be ascribed to an interaction of rigid tricyclic ring of IMP and polyoxyethylene chain of TX-100 structure.

Table 4. ^{1}H NMR chemical shifts (δ, ppm) of IMP+TX-100 mixtures in aqueous system.

	Chemical Shifts (δ, ppm)				
	α_1 = 0.1	α_1 = 0.3	α_1 = 0.5	α_1 = 0.7	α_1 = 0.9
T1	0.486	0.517	0.527	0.534	0.542
T2	1.01	1.06	1.08	1.096	1.111
T3	1.376	1.389	1.437	1.464	1.473
T4–T6	3.406	3.417	3.42	3.44	3.517
T7	6.82	6.831	6.944	6.999	7.004
T8	6.975	6.997	7.007	7.026	7.029
I1	2.496	2.515	2.524	2.526	2.685
I2	2.924	2.93	2.932	2.937	3.41
I3	2.834	2.845	2.86	2.874	2.903
I4	1.716	1.754	1.759	1.764	1.817
I5	3.57	3.582	3.596	3.628	3.65
I6	6.991	6.998	7.012	7.018	7.024

The changes in chemical shift values for the alkyl protons I1 to I5 upon addition of TX-100 are also given in Figure 4 and Table 4 and it is clear that upon mixing of both studied constituents, the resultant mixed micelles cause deshielding (a downfield shift) of all the hydrophobic tail protons of IMP. Upon mixing, the hydrophobic interactions, as well as electrostatic attractions, endorse spherically along with the compacted micelles, while steric repulsion sources the hindrance amongst the constituents, causing the exposure along with protons deshielding. Overall, the proton signals (I1–I6, T1–T8) for both drug IMP and surfactant TX-100 in mixed micelles resonating at high δ values show a downfield shift of protons. It is known that both electrostatic, as well as steric interactions, show the leading character during the mixed micelles formation. Therefore, through the rise in TX-100 mole fraction, all proton signals for drug-surfactant mixtures are highly deshielded, which point towards an increase in steric repulsion among the molecules, which leads to the formation of large micelles [45,46]. As compared to pure IMP, the length of peak I4 is increased in case of mixtures up to mole fractions 0.5, but as the mole fraction reaches 0.7 and 0.9, the length of the signal I4 is decreased, which shows that these mole fractions (0.7 and 0.9) of TX-100 are more effective as compared to IMP. Similar changes were recorded for other NMR signals, such as I1. It is clearly visible from the spectra as well, being stable, that the aromatic protons related to the tricyclic rigid ring in IMP as well as the protons related to the mono aromatic ring in TX-100 are highly deshielded and show high δ values. Therefore, a clear downfield shift is observed. In the case of mixtures, all peaks are showing a clear downfield shift for both compounds, I1–I6 as well as T1–T8, at different mixing ratios. The compactness of the micelles varies with the variation of mole fraction, which is clear from the chemical shift values [47,48]. This change in chemical shift values is attributed to the interplay of electrostatic and steric interactions.

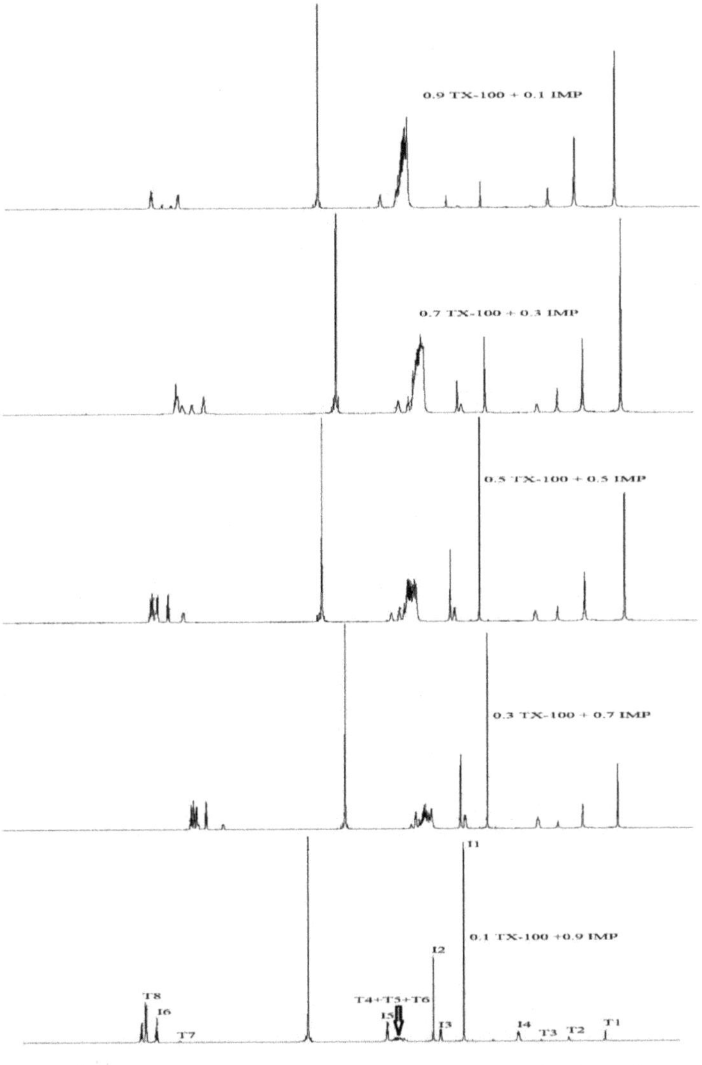

Figure 4. ^{1}H NMR (600 MHz) spectrum of IMP+TX-100 mixture having various α_1 of TX-100 in D_2O.

2.6. FT-IR Study

The interaction impact can also be qualitatively followed via FT-IR spectra [49]. FT-IR spectroscopy was utilized to describe diverse functional groups and to examine the interaction amongst unlikely groups existing in the binary mixed system. Background-deducted FT-IR spectra of a pure drug and IMP+TX-100 mixed system of equal ratio in an aqueous solution are depicted in Figure 5a,b. Amphiphilic compound head-groups along with hydrophobic portion frequencies give statistics on the structural change in the monomers of formed micelles [50,51]. The feasible interaction amongst IMP+TX-100 mixed system will possibly alter the C–H bending and stretching and C–N stretching frequency of the drug head group.

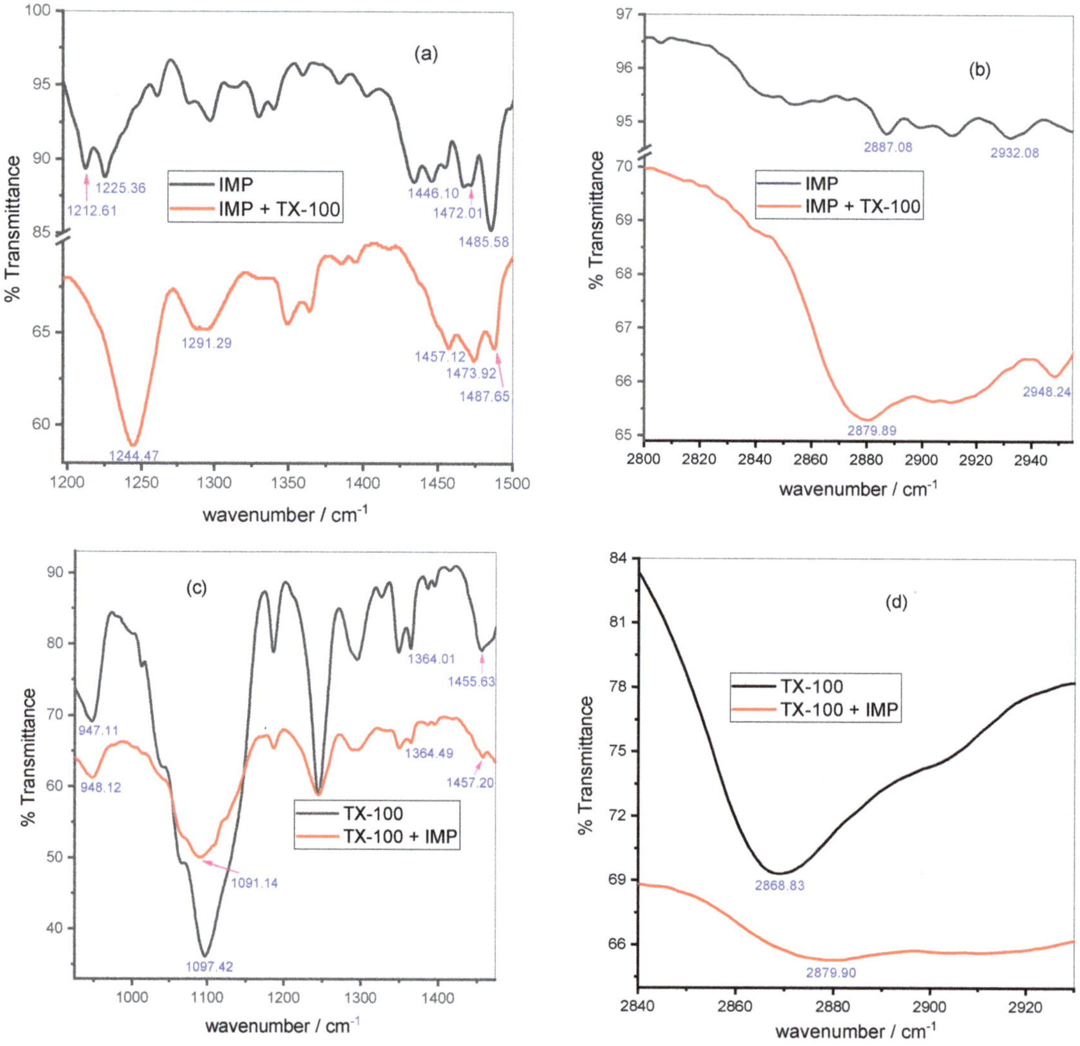

Figure 5. FTIR spectra of IMP (**a**,**b**) in the absence and existence of TX-100 and FTIR spectra of TX-100 (**c**,**d**) in the absence and existence of IMP in the selected wavenumber regions (cm^{-1}).

To view the effect of TX-100 on the aliphatic C–N bond stretching band as well as C–H bond bending band in the IMP molecule of IMP+TX-100 mixture, a frequency range of 1195 to 1500 cm^{-1} was selected (Figure 5a). As shown from Scheme 2, the nature of the employed drug IMP is cationic, as it keeps a positively charged N atom allied with three alkyl groups. As depicted in Figure 5a, the singular IMP spectra showed C–N bond stretching at two different frequencies: one at 1212.61 and the second one at 1225.36 cm^{-1}. However, in the occurrence of TX-100, the C–N stretching in IMP was shifted to a higher frequency. The first C–N bond stretching was shifted to 1244.47 from 1212.61 cm^{-1}, and the second one was shifted to 1291.29 from 1225.36 cm^{-1}. IMP showed C–H bending at three different frequencies (1446.10, 1472.01, and 1485.58 cm^{-1}) (Figure 5a) and in the occurrence of TX-100, the frequency of C–H bending in IMP was significantly shifted to a higher frequency from their initial position (1457.12 cm^{-1}, 1473.92 cm^{-1}, and 1487.65 cm^{-1}, correspondingly) because of the interaction of TX-100 with IMP. Shifting to a higher or lower frequency region is dependent on the environment of the interacting group of molecules. Through the addition of TX-100, the alteration in C–N stretching and C–H bending frequency in IMP showed the attractive interaction between constituents, owing to mixed micelles formation.

To investigate the C–H stretching in IMP, the frequency band region of 2800 to 2960 cm^{-1} was chosen to assess the effect of TX-100, and the achieved plotted graph is displayed in Figure 5b. As depicted in the graph, IMP showed a C–H bond stretching band at 2887.08 as well as 2932.08 cm^{-1} of the alkane methyl group. In presence of TX-100, the obtained C–H bond stretching band was shifted from 2887.08 to 2979.89 and 2932.08 to 2948.24 cm^{-1}. In the presence of TX-100, the occurrence of this shifting in the C–H stretching frequency band in the IMP functional group, reveals an interaction amongst both employed ingredients (IMP and TX-100) [52].

Figure 5c,d depicted the FT-IR spectra of a singular TX-100 and TX-100+IMP mixture with an identical ratio. Figure 5c showed the spectra of singular TX-100 between 940 and 1470 cm^{-1} frequency that showed the C–O stretching at 947.11 and 1097.42 cm^{-1}, O–H bending at 1364.01 cm^{-1}, and C–H bending band at 1455.63 cm^{-1}. Upon addition of IMP in the solution of TX-100, both C–O bond stretching was shifted from their original position. The first one shifted from 947.11 to 948.12 cm^{-1}, and the second one shifted from 1097.42 to 1091.14. In addition, in the presence of IMP, the shifting in the frequency band of O–H bending in TX-100 occurred from 1364.01 to 1364.49 cm^{-1} and the C–H bending band from 1455.63 to 1457.20 cm^{-1}, signaling the interaction amongst the constituents. Figure 5d showed the spectra of TX-100 as well as the TX-100+IMP mixture in the range 2840–2930 cm^{-1}. Pure TX-100 showed the medium and broad C-H stretching band (alkane) at 2868.83 cm^{-1}. The C–H stretching band (alkane) attained at 2868.83 cm^{-1} in TX-100 was moved to a higher frequency (2879.98 cm^{-1}) in the presence of IMP, signifying an interaction amongst TX-100+IMP mixture mixed micelles. The O–H stretching band was found in case of singular TX-100, but for the TX-100+IMP mixture, the O–H stretching band peak disappeared due to merging with the water peak (not shown graphically). Due to the interaction of the employed ingredients, the whole frequency band variation did not achieve much, but obtained to be reproducible. Overall, herein, the shifting in C–N stretching, O–H bending, along with C–H bending, and stretching frequency recommend the interaction between the employed ingredients [53–55].

3. Conclusions

Before a surfactant can be employed as an appropriate drug agent, a broad range analysis must be accomplished to examine the interaction of the surfactant through the proposed drug. Herein, ^{1}H NMR, FT-IR, and tensiometric studies were performed to explore the interaction of a TX-100 surfactant with the cationic drug IMP. Physiologically, the nonionic nature of surfactants is more suitable as compared with ionic ones (cationic/anionic) and, owing to their high surface activity, a nonionic surfactant, such as TX-10, is considered as an ideal nominee for drug delivery in comparison to other surfactants. The interfacial properties of IMP, TX-100 along with the IMP+TX-100 mixture of various ratios at the surface were evaluated using a tensiometric method in different solvents (H_2O/NaCl/urea). TX-100 decreases the surface thickness acquired by means of the water layer and enhances the hydrophobic film width of the studied systems. Interfacial composition (X_1^σ) and the β^σ values of the IMP+TX-100 system showed a much higher participation of TX-100 at the surface than IMP and attraction/synergism between the components at the surface, respectively. The obtained value of ΔG^o_{ad} specifies that the adsorption phenomena was a spontaneous process and the stability of the mixed monolayer. The P value of IMP+TX-100 was attained as $0.5 < P < 1$, showing that the micellar solution was vesicles-shaped. The value of Γ_{max} acquired more for the surfactant than IMP, confirming that the surfactant showed higher surface-activity as monomers of TX-100, favoring a compacted or strongly packed arrangement at the surface in all solvents. ^{1}H NMR study of solution mixed systems advocated that IMP and TX-100 interact with each other via hydrophobic interaction. FT-IR spectra showed that the frequency band of individual ingredients (IMP and TX-100) was shifted from the original position for the mixed system, proving the interaction amongst them. The conclusions of the current investigation contribute to the assessment of the implementation of the surfactant (as a capable drug delivery) with a drug mixed system and the supporting mechanisms, for a basic understanding required for the projected expansion of economical and efficient drug formulations.

4. Materials and Methods

4.1. Materials

Every material in the current study was used as received from their respective company. Drug IMP was obtained from Sigma (St. Louis, MO, USA) having purity $\geq$ 98.0%. Surfactant TX-100 was from Sigma (Taufkichen, Germany). Different additives such as NaCl was acquired from BDH (Poole, England), having a purity of 98.0%, and urea was obtained from Sigma (Taufkichen, Germany), with a purity of 98.0%. Deuterium oxide (D_2O) was purchased from Sigma (St. Louis, MO, USA) with a purity of 99%, which was used as the solution preparation for the ^{1}H NMR study only. For the rest of the study, distilled water was used for the solution preparation. Using calculated quantities of NaCl and urea dissolved in distilled water, the prepared solutions of these additives were used as solvents. In the aqueous system and in the occurrence of fixed NaCl/urea concentrations, the stock solutions of both employed constituents (IMP and TX-100) of fixed concertation were made separately, clearly above their corresponding *cmc*. Combinations of both components (IMP (drug) and TX-100 (surfactant as drug carrier) were readied by mixing the prepared stock solutions of both constituents (IMP and TX-100) in diverse mass ratios, varying the mole fraction of component 1 (TX-100 surfactant) from 0.1 to 0.9. These prepared solutions of diverse mass ratios were employed in the experiments, assuming that the density of the component's dilute solution at the experimental temperature is roughly constant.

4.2. Methods

4.2.1. Measurement of Surface Tension

For the surface tension (γ) measurement, an Attension tensiometer (Sigma 701, Darmstadt, Germany) working with the ring detachment process was applied for pure (IMP and TX-100) and mixed system (IMP+TX-100) in five ratios in aqueous/NaCl/U solvent. The γ of resultant system (IMP, TX-100, or IMP+TX-100) vs. *log* (*C* (conc.)) of pure IMP, TX-100, or IM+TX-100 were plotted, and each plot showed a break point that was termed *cmc* of the system [18]. Here, plots given for the IMP+TX-100 mixed system in different media in our previous work [18] were used for evaluation of different interfacial parameter evaluation. The error in γ and temperature was attained as $\pm$ 0.2 mNm^{-1} and $\pm$ 0.2 K, respectively.

4.2.2. ^{1}H NMR Study

For the ^{1}H NMR study, D_2O (as a solvent) was used rather than distilled water to prepare the solutions of the individual component (IMP, TX-100) and their mixtures (IMP+TX-100). The ^{1}H NMR spectra of IMP, the surfactant, and their mixture of various mole fractions in the aqueous system, were noted using a Bruker ultrashield plus 600 spectrometer, Billerica, MA, USA (600 MHz proton resonance frequency). Approximately 1 mL of every studied system is placed in a 5 mm tube for spectra measurements and chemical shifts were noted on the δ (ppm) scale. The reproducibility of δ was within 0.01 ppm. An organosilicon compound-tetramethylsilane was employed as an internal standard, which is recognized for calibrating a chemical shift.

4.2.3. FTIR Spectroscopy

In the aqueous system, the FTIR spectra (4000 to 400 cm^{-1} wavelength) of the singular components and IMP+TX-100 mixed system in an equal ratio were recorded by consuming a NICOLET iS50 FT-IR spectrometer possessing ATR accessory (Thermo Scientific, Madison, Waltham, MA, USA). Here, a particular part of the wavelength range is exposed in the graph for clarity purposes. From the entirely attained spectra of the chosen system, the water spectrum was consistently deducted. The concentration of IMP and TX-100 was maintained very well above their respective *cmc* value. Each spectrum was obtained at a resolution of 4.0 cm^{-1}.

Author Contributions: Conceptualization, M.A.R., N.A. and D.K.; investigation, M.A.R. and N.A.; validation, M.A.R., N.A., D.K. and A.M.A.; formal analysis, M.A.R., N.A. and D.K.; mthodology, M.A.R. and N.A.; project administration, M.A.R.; visualization, M.A.R. and N.A.; supervision, M.A.R.; writing—original draft, M.A.R., N.A., D.K. and A.M.A.; writing—review and editing, M.A.R., N.A., D.K. and A.M.A. All authors have read and agreed to the published version of the manuscript.

Funding: The authors extend their appreciation to the Deputyship for Research and Innovation, Ministry of Education in Saudi Arabia for funding this research work through the project number IFPIP: 63-130-1442 and King Abdulaziz University, DSR, Jeddah, Saudi Arabia.

Institutional Review Board Statement: Not applicable.

Informed Consent Statement: Not applicable.

Data Availability Statement: Not applicable.

Conflicts of Interest: The authors declare no conflict of interest.

References

1. Biswal, N.R.; Paria, S. Interfacial and wetting behavior of natural–synthetic mixed surfactant systems. *RSC Adv.* **2014**, *4*, 9182–9188. [CrossRef]
2. Sheng, Y.; Yan, C.; Li, Y.; Peng, Y.; Ma, L.; Wang, Q. Thermal stability of gel foams stabilized by xanthan gum, silica nanoparticles and surfactants. *Gels* **2021**, *7*, 179. [CrossRef] [PubMed]
3. Rosen, M.J. *Surfactants and Interfacial Phenomena*, 3rd ed.; John Wiley & Sons: New York, NY, USA, 2004.
4. Kumar, D.; Rub, M.A. Kinetic and mechanistic investigations of [Zn (II)-Trp]+ and ninhydrin in aqueous and cationic CTAB surfactant. *J. Phys. Org. Chem.* **2019**, *32*, e3997. [CrossRef]
5. Kumar, D.; Rub, M.A. Kinetic study of ninhydrin with chromium (III)-glycylleucine in aqueous–alkanediyl-α,ω-bis (dimethyl-cetylammonium bromide) gemini surfactants. *J. Phys. Org. Chem.* **2019**, *32*, e3946. [CrossRef]
6. de Molina, P.M.; Gradzielski, M. Gels obtained by colloidal self-assembly of amphiphilic molecules. *Gels* **2017**, *3*, 30. [CrossRef] [PubMed]
7. Attwood, D.; Florence, A.T. Surfactant systems. In *Their Chemistry, Pharmacy and Biology*; Chapman and Hall: New York, NY, USA, 1983.
8. Sachin, K.M.; Karpe, S.A.; Kumar, D.; Singh, M.; Dominguez, H.; Ríos-López, M.; Bhattarai, A. A simulation study of self-assembly behaviors and micellization properties of mixed ionic surfactants. *J. Mol. Liq.* **2021**, *336*, 116003. [CrossRef]
9. Yin, B.; Ni, J.; Witherel, C.E.; Yang, M.; Burdick, J.A.; Wen, C.; Wong, S.H.D. Harnessing Tissue-derived extracellular vesicles for osteoarthritis theranostics. *Theranostics* **2022**, *12*, 207–231. [CrossRef]
10. Hamed, F.A.; Zoveidavianpoor, M. The Foaming Behavior and Synergistic Effect in Aqueous CO_2 Foam by In Situ Physisorption of Alpha Olefin Sulfonate and Triton X-100 Surfactants and Their Mixture. *Petroleum Sci. Technol.* **2014**, *32*, 2376–2386.
11. Farcet, J.-B.; Kindermann, J.; Karbiener, M.; Kreil, T.R. Development of a Triton X-100 replacement for effective virus inactivation in biotechnology processes. *Eng. Rep.* **2019**, *1*, e12078. [CrossRef]
12. Thakkar, K.; Bharatiya, B.; Ray, D.; Aswal, V.; Bahadur, P. Molecular interactions involving aqueous Triton X-100 micelles and anionic surfactants: Investigations on, surface activity and morphological transitions. *J. Mol. Liq.* **2016**, *223*, 611–620. [CrossRef]

13. Dharaiya, N.; Bahadur, P.; Singh, K.; Marangoni, D.; Bahadur, P. Light scattering and NMR studies of Triton X-100 micelles in the presence of short chain alcohols and ethoxylates. *Colloids Surf. A Physicochem. Eng. Asp.* **2013**, *436*, 252–259. [CrossRef]
14. Ćirin, D.M.; Posa, M.M.; Krstonosic, V.S. Interactions between selected bile salts and Triton X-100 or sodium lauryl ether sulfate. *Chem. Central J.* **2011**, *5*, 89. [CrossRef] [PubMed]
15. Schreier, S.; Malheiros, S.V.; de Paula, E. Surface active drugs: Self-association and interaction with membranes and surfactants. Physicochemical and biological aspects. *Biochim. Biophys. Acta* **2000**, *1508*, 210–234. [CrossRef]
16. Kumar, D.; Hidayathulla, S.; Rub, M.A. Association behavior of a mixed system of the antidepressant drug imipramine hydrochloride and dioctyl sulfosuccinate sodium salt: Effect of temperature and salt. *J. Mol. Liq.* **2018**, *271*, 254–264. [CrossRef]
17. Alomar, M.J. Factors affecting the development of adverse drug reactions. *Saudi Pharm. J.* **2014**, *22*, 83–94. [CrossRef]
18. Rub, M.A.; Azum, N.; Alotaibi, M.M.; Asiri, A.M. Solution behaviour of antidepressant imipramine hydrochloride drug and non-ionic surfactant mixture: Experimental and theoretical study. *Polymers* **2021**, *13*, 4025. [CrossRef]
19. Bagheri, A. Comparison of the interaction between propranolol hydrochloride (PPL) with anionic surfactant and cationic surface active ionic liquid in micellar phase. *Colloids Surf. A* **2021**, *615*, 126183. [CrossRef]
20. Alam, M.S.; Siddiq, A.M. Self-association and mixed micellization of an amphiphilic antidepressant drug, 5-[3-(Dimethylamino)propyl]-10,11-dihydro-5H-dibenz[b,f]azepine hydrochloride and a nonionic surfactant, poly(ethylene glycol) t-octylphenyl ether: Evaluation of thermodynamics. *J. Mol. Liq.* **2018**, *252*, 321–328. [CrossRef]
21. Kumar, D.; Azum, N.; Rub, M.A.; Asiri, A.M. Aggregation behavior of sodium salt of ibuprofen with conventional and gemini surfactant. *J. Mol. Liq.* **2018**, *262*, 86–96. [CrossRef]
22. Pal, A.; Pillania, A. Modulations in surface and aggregation properties of non-ionic surfactant Triton X-45 on addition of ionic liquids in aqueous media. *J. Mol. Liq.* **2017**, *233*, 243–250. [CrossRef]
23. Kabir-ud-Din, K.A.; Naqvi, A.Z. Mixed micellization and interfacial properties of nonionic surfactants with the phenothiazine drug promazine hydrochloride at 30 °C. *J. Solut. Chem.* **2012**, *41*, 1587–1599. [CrossRef]
24. Rub, M.A.; Khan, F.; Sheikh, M.S.; Azum, N.; Asiri, A.M. Tensiometric, fluorescence and ^{1}H NMR study of mixed micellization of non-steroidal anti-inflammatory drug sodium salt of ibuprofen in the presence of non-ionic surfactant in aqueous/urea solutions. *J. Chem. Thermodyn.* **2016**, *96*, 196–207. [CrossRef]
25. Zhou, Q.; Rosen, M.J. Molecular interactions of surfactants in mixed monolayers at the air/aqueous solution interface and in mixed micelles in aqueous media: The regular solution approach. *Langmuir* **2003**, *19*, 4555–4562. [CrossRef]
26. Rub, M.A.; Azum, N.; Khan, F.; Asiri, A.M. Aggregation of sodium salt of ibuprofen and sodium taurocholate mixture in different media: A tensiometry and fluorometry study. *J. Chem. Thermodyn.* **2018**, *121*, 199–210. [CrossRef]
27. Dar, A.A.; Chatterjee, B.; Rather, G.M.; Das, A.R. Mixed micellization and interfacial properties of dodecyltrimethylammonium bromide and tetraethyleneglycol mono-n-dodecyl ether in absence and presence of sodium propionate. *J. Colloid Interface Sci.* **2006**, *298*, 395–405. [CrossRef]
28. Sharma, R.; Mahajan, R.K. An investigation of binding ability of ionic surfactants with trifluoperazine dihydrochloride: Insights from surface tension, electronic absorption and fluorescence measurements. *RSC Adv.* **2012**, *2*, 9571–9583. [CrossRef]
29. Rosen, M.J.; Aronson, S. Standard free energies of adsorption of surfactants at the aqueous solution/air interface from surface tension data in the vicinity of the critical micelle concentration. *Colloids Surf.* **1981**, *13*, 201–208. [CrossRef]
30. Fatma, N.; Panda, M.; Ansari, W.H. Environment-friendly ester bonded gemini surfactant: Mixed micellization of 14-E2-14 with ionic and nonionic conventional surfactants. *J. Mol. Liq.* **2015**, *211*, 247–255. [CrossRef]
31. Sugihara, G.; Miyazono, A.M.; Nagadome, S.; Oida, T.; Hayashi, Y.; Ko, J.S. Adsorption and micelle formation of mixed surfactant systems in water. II a combination of cationic gemini-type with MEGA-10. *J. Oleo Sci.* **2003**, *52*, 449–461. [CrossRef]
32. Kumar, D.; Azum, N.; Rub, M.A.; Asiri, A.M. Interfacial and spectroscopic behavior of phenothiazine drug/bile salt mixture in urea solution. *Chem. Pap.* **2021**, *75*, 3949–3956. [CrossRef]
33. Rodenas, E.; Valiente, M.; Villafruela, M.S. Different theoretical approaches for the study of the mixed tetraethylene glycol mono-n-dodecyl ether/hexadecyltrimethylammonium bromide micelles. *J. Phys. Chem. B* **1999**, *103*, 4549–4554. [CrossRef]
34. Khan, F.; Rub, M.A.; Azum, N.; Asiri, A.M. Mixtures of antidepressant amphiphilic drug imipramine hydrochloride and anionic surfactant: Micellar and thermodynamic investigation. *J. Phys. Org. Chem.* **2018**, *31*, e3812. [CrossRef]
35. Kumar, D.; Rub, M.A.; Azum, N.; Asiri, A.M. Mixed micellization study of ibuprofen (sodium salt) and cationic surfactant (conventional as well as gemini). *J. Phys. Org. Chem.* **2018**, *31*, e3730. [CrossRef]
36. Rub, M.A.; Azum, N.; Khan, F.; Asiri, A.M. Surface, micellar, and thermodynamic properties of antidepressant drug nortriptyline hydrochloride with TX-114 in aqueous/urea solutions. *J. Phys. Org. Chem.* **2017**, *30*, e3676. [CrossRef]
37. Israelashvili, J.N. *Intermolecular and Surface Forces*; Academic Press: London, UK, 1991.
38. Tanford, C. *The Hydrophobic Effect: Formation of Micelles and Biological Membranes*; Wiley: New York, NY, USA, 1980.
39. Azum, N.; Rub, M.A.; Asiri, A.M.; Bawazeer, W.A. Micellar and interfacial properties of amphiphilic drug–non-ionic surfactants mixed systems: Surface tension, fluorescence and UV–vis studies. *Colloids Surf. A* **2017**, *522*, 183–192. [CrossRef]
40. Khan, S.A.; Ullah, Q.; Almalki, A.S.A.; Kumar, S.; Obaid, R.J.; Alsharif, M.A.; Alfaifi, S.Y.; Hashmi, A.A. Synthesis and photophysical investigation of (BTHN) Schiff base as off-on Cd2+ fluorescent chemosensor and its live cell imaging. *J. Mol. Liq.* **2021**, *328*, 115407. [CrossRef]
41. Asad, M.; Khan, S.A.; Arshad, M.N.; Asiri, A.M.; Rehan, M. Design and synthesis of novel pyrazoline derivatives for their spectroscopic, single crystal X-ray and biological studies. *J. Mol. Struct.* **2021**, *1234*, 130131. [CrossRef]

42. Mahajan, R.K.; Mahajan, S.; Bhadani, A.; Singh, S. Physicochemical studies of pyridinium gemini surfactants with promethazine hydrochloride in aqueous solution. *Phys. Chem. Chem. Phys.* **2012**, *14*, 887–898. [CrossRef]
43. Jiang, Y.; Lu, X.-Y.; Chen, H.; Mao, S.-Z.; Liu, M.-L.; Luo, P.-Y.; Du, Y.-R. NMR study of the dynamics of cationic gemini surfactant 14-2-14 in mixed solution with conventional surfactants. *J. Phys. Chem. B* **2009**, *113*, 8357–8361. [CrossRef]
44. Rub, M.A.; Khan, F.; Azum, N.; Asiri, A.M.; Marwani, H.M. Micellization phenomena of amphiphilic drug and TX-100 mixtures: Fluorescence, UV-visible and ^{1} H NMR study. *J. Taiwan Inst. Chem. Eng.* **2016**, *60*, 32–43. [CrossRef]
45. Vautier-Giongo, C.; Bakshi, M.S.; Singh, J.; Ranganathan, R.; Joseph, H.; Bales, C.B.L. Effects of interactions on the formation of mixed micelles of 1,2-diheptanoyl-sn-glycero-3-phosphocholine with sodium dodecyl sulfate and dodecyltrimethylammonium bromide. *J. Colloid Interface Sci.* **2005**, *282*, 149–155. [CrossRef] [PubMed]
46. Lin, L.-T.; Tseng, M.-Y.; Chen, S.-H.; Roberts, M.F. Temperature dependence of the growth of diheptanoylphosphatidylcholine micelles studied by small-angle neutron scattering. *J. Phys. Chem.* **1990**, *94*, 7239–7243. [CrossRef]
47. Siddiqui, U.S.; Khan, F.; Khan, I.A.; Dar, A.A. Role of added counterions in the micellar growth of bisquaternary ammonium halide surfactant (14-s-14): 1H NMR and viscometric studies. *J. Colloid Interface Sci.* **2011**, *355*, 131–139. [CrossRef] [PubMed]
48. Rub, M.A. Effect of additives (TX-114) on micellization and microstructural phenomena of amphiphilic ibuprofen drug (sodium salt): Multi-technique approach. *J. Lumin.* **2018**, *197*, 252–265. [CrossRef]
49. Zhang, L.; Tu, Z.C.; Wang, H.; Kou, Y.; Wen, Q.H.; Fu, Z.F.; Chang, H.X. Response surface optimization and physicochemical properties of polysaccharides from Nelumbo nucifera leaves. *Int. J. Biol. Macromol.* **2015**, *74*, 103–110. [CrossRef]
50. Gaikar, V.G.; Padalkar, K.V.; Aswal, V.K. Characterization of mixed micelles of structural isomers of sodium butyl benzene sulfonate and sodium dodecyl sulfate by SANS, FTIR spectroscopy and NMR spectroscopy. *J. Mol. Liq.* **2008**, *138*, 155–167. [CrossRef]
51. Rub, M.A.; Azum, N. Association behavior of amphiphlic drug promethazine hydrochloride and sodium p-toluenesulfonate mixtures: Effect of additives. *J. Mol. Liq.* **2021**, *325*, 114654. [CrossRef]
52. Banjare, M.K.; Kurrey, R.; Yadav, T.; Sinha, S.; Satnami, M.L.; Ghosh, K.K. A comparative study on the effect of imidazolium-based ionic liquid on self-aggregation of cationic, anionic and nonionic surfactants studied by surface tension, conductivity, fluorescence and FTIR spectroscopy. *J. Mol. Liq.* **2017**, *241*, 622–632. [CrossRef]
53. Kumar, H.; Sharma, N.; Katal, A. Aggregation behaviour of cationic (cetyltrimethylammonium bromide) and anionic (sodium dodecylsulphate) surfactants in aqueous solution of synthesized ionic liquid [1-pentyl-3-methylimidazolium bromide] -Conductivity and FT-IR spectroscopic studies. *J. Mol. Liq.* **2018**, *258*, 285–294. [CrossRef]
54. Rub, M.A.; Azum, N.; Kumar, D.; Alotaibi, M.M.; Asiri, A.M. Impact of numerous media on association, interfacial, and thermodynamic properties of promethazine hydrochloride (PMT) + benzethonium chloride (BTC) mixture of various composition. *J. Mol. Liq.* **2022**, *346*, 118287. [CrossRef]
55. Rub, M.A.; Azum, N.; Kumar, D.; Khan, A.; Arshad, M.N.; Asiri, A.M.; Alotaibi, M.M. Aggregational behaviour of promethazine hydrochloride and TX-45 surfactant mixtures: A multi-techniques approach. *J. Mol. Liq.* **2021**, *342*, 117558.

gels

MDPI

Article

Synergistic Interaction and Binding Efficiency of Tetracaine Hydrochloride (Anesthetic Drug) with Anionic Surfactants in the Presence of NaCl Solution Using Surface Tension and UV–Visible Spectroscopic Methods

Naved Azum [1,2], Malik Abdul Rub [1,2,*], Anish Khan [1], Maha M. Alotaibi [2] and Abdullah M. Asiri [1,2]

1 Center of Excellence for Advanced Materials Research, King Abdulaziz University, Jeddah 21589, Saudi Arabia; nhassan2@kau.edu.sa (N.A.); akrkhan@kau.edu.sa (A.K.); aasiri2@kau.edu.sa (A.M.A.)
2 Chemistry Department, Faculty of Science, King Abdulaziz University, Jeddah 21589, Saudi Arabia; mmsalotaibi@kau.edu.sa
* Correspondence: aabdalrab@kau.edu.sa; Tel.: +966-563671946

Abstract: Surfactants are ubiquitous materials that are used in diverse formulations of various products. For instance, they improve the formulation of gel by improving its wetting and rheological properties. Here, we describe the effects of anionic surfactants on an anesthetic drug, tetracaine hydrochloride (TCH), in NaCl solution with tensiometry and UV–visible techniques. Various micellar, interfacial, and thermodynamic parameters were estimated. The outputs were examined by using different theoretical models to attain a profound knowledge of drug–surfactant mixtures. The presence of attractive interactions among drug and surfactant monomers (synergism) in mixed micelle was inferred. However, it was found that sodium dodecyl sulfate (SDS) showed greater interactions with the drug in comparison to sodium lauryl sarcosine (SLS). The binding of the drug with surfactants was monitored with a spectroscopic technique (UV–visible spectra). The results of this study could help optimize the compositions of these mixed aggregates and find the synergism between monomers of different used amphiphiles.

Keywords: tetracaine hydrochloride; sodium dodecyl sulfate; sodium lauroyl sarcosine; drug–surfactant mixed micelle; synergistic interaction

Citation: Azum, N.; Rub, M.A.; Khan, A.; Alotaibi, M.M.; Asiri, A.M. Synergistic Interaction and Binding Efficiency of Tetracaine Hydrochloride (Anesthetic Drug) with Anionic Surfactants in the Presence of NaCl Solution Using Surface Tension and UV–Visible Spectroscopic Methods. *Gels* **2022**, *8*, 234. https://doi.org/10.3390/gels8040234

Academic Editor: Hiroyuki Takeno

Received: 15 March 2022
Accepted: 7 April 2022
Published: 11 April 2022

Publisher's Note: MDPI stays neutral with regard to jurisdictional claims in published maps and institutional affiliations.

1. Introduction

It is often observed that the surfactant mixtures (e.g., surfactant–co-polymer, surfactant–drug, and surfactant–surfactant) exhibit better performance than single surfactants [1–6]. It is also common to use mixtures of surfactants and polymers to formulate gels that are used in drug-dosage forms to improve their properties or to improve their physical stability [7]. The anionic surfactant used in this study, sodium dodecyl sulfate, has been used to synthesize nanogels [8]. SDS has shown better activity in the formation of microgels based on poly(N-isopropylacrylamide) [8]. The synergistic or antagonistic effects of binary mixtures are produced by attraction or repulsion between surfactant monomers. Synergism is observed when the molecular interaction between the monomers of a mixture is greater than before mixing. The strength of synergism between different types of surfactants follows the order of anionic–cationic > nonionic–ionic > ionic–ionic > nonionic–nonionic. The interaction between oppositely charged head groups and the hydrophobic interaction between chains of amphiphiles are the two main factors that are responsible for strong synergistic effects inside cationic–anionic mixtures [9–11]. Ionic–anionic mixtures become turbid (precipitation) at some mole fractions, producing lamellar phases and rod-like morphologies.

A lesser water solubility and the dissolution characteristics of a drug usually limit its bioavailability and therapeutic efficacy. The poor water-solubility of drugs may also lead to disappointing and inconstant ingesting, which aggravates the complications of bioavailability and scarcity in the delivery of drugs. In addition, excessive dosages of drugs cause side effects such as vomiting, nausea, dizziness, and fatigue [12,13]. The development of increasing water solubility and improvements in encapsulation efficiency can enhance absorption, enhance bioavailability, and lower the required therapeutic dose [1,14,15]. Researchers have often studied different ways to increase solubilities, such as using small drug carriers, preparing nanoparticles, and using self-emulsifying formulations or amorphous formulations based on water-soluble polymers. A surfactant is a most-capable drug transporter in biomedical applications since it can be easily fabricated into different formulations such as micelles, hydrogels, and nanoparticles to enclose bioactive agents at several points of hydrophobicity [16–19]. Surfactants are polar molecules and contain both hydrophilic and hydrophobic components orientated at the surface to diminish the surface tension of water [20–22]. A micelle will only form when the concentration of the amphiphile is higher than a specific concentration (called the critical micelle concentration or cmc) that can be determined using diverse methods (surface tension, conductometry, fluorometry, UV–visible spectroscopy, cyclic voltammetry, and isothermal calorimetry) [23–26]. A valuable feature of these molecules is their cmc value. The cmc value depends on various aspects such as ionic strength, temperature, and the existence of additional compounds in the solution. Most chemical industries utilize surfactants, e.g., as pharmaceuticals, corrosion inhibitors, detergents, paints, and cosmetics [27–31].

Certain types of drugs, such as antidepressants, anticholinergics, antihistamines, and local anesthetics, are amphiphilic; they have surfactant-like properties and form micelles [32–35]. Invariably, their therapeutic activity is determined by how they interact with surfactants. Depending on their interactions in solution, any drug can be made more active. The mixed systems of many amphiphilic drugs have also been researched by our group using different techniques with different amphiphiles [36–45]. Tetracaine hydrochloride, TCH (Figure 1), is an amphiphilic compound that also possesses colloidal properties and is one of the most used local anesthetic drugs. It is used for stopping pain during surgery and eye infections. Since tetracaine is a poorly water-soluble compound, it is usually formulated as tetracaine hydrochloride. It has been hypothesized that the +ve charge on the drug, which is the functional component, interacts with the Na^+ channels on neuronal membranes and stops the transmission of the pain sensation along the nerve [46,47]. Furthermore, the cationic form provides an amphiphilic structure to such a drug, so it can be classified as a cationic tension-active molecule. Therefore, a TCH-like cationic surfactant undergoes an abrupt change above a critical concentration (cmc) and the Krafft temperature. The aqueous dissolution of tetracaine follows the same principle as all ionic surfactants (in that it is governed by both solubility and micellization). As a result, the nature of the surfactant, its counter ions, concentration, and temperature all affect the process. As the use of high concentrations of local anesthetic in spinal anesthesia is known to occasionally result in the sudden death of patients, it is important to understand how the micellization process occurs and what its phase diagram looks like.

In this work, surface tension and UV–visible measurements were carried out to examine the effects of anionic surfactants on a cationic drug. To the best of our knowledge, the mixed micellization of tetracaine hydrochloride (TCH) with sodium lauroyl sarcosine (SLS) and sodium dodecyl sulfate (SDS) in the presence of sodium chloride (NaCl) has not been previously described. Different theoretical approaches of mixed micellization (such as those by Clint, Rubingh, Rodenas, Rosen, and Motomura) were utilized to investigate the interactions of TCH + SDS/SLS mixtures. Various interfacial, micellization, and energetic parameters were analyzed. The output of this work can support the search for a surfactant-based carrier for drug delivery.

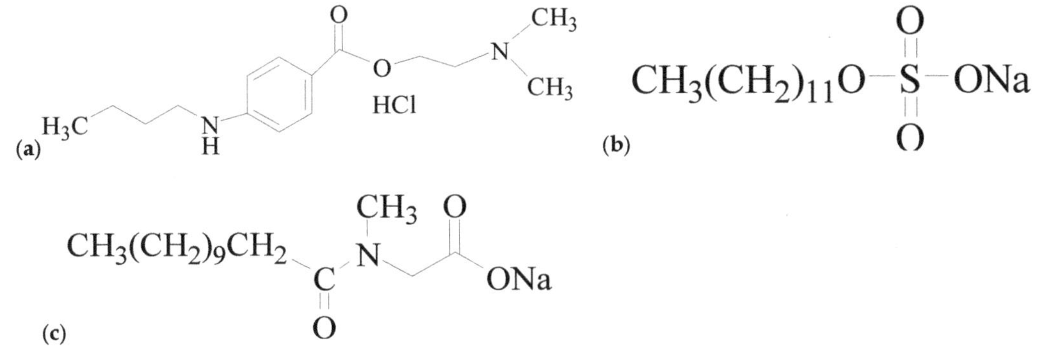

Figure 1. Chemical structures of (**a**) tetracaine hydrochloride (TCH), (**b**) sodium dodecyl sulfate (SDS), and (**c**) sodium lauryl sarcosine (SLS).

2. Result and Discussion

The stock solutions of numerous mole fractions (α_1) of component 1 (SDS/SLS) from 0 to 1 were prepared. As shown in Figure 2, the solution was turbid at some mole fractions (which barred the experiment), and we selected the mole fractions where no turbidity was observed. The surface tension (ST) measurements were used to estimate the cmc values of pure and binary mixtures of drugs and surfactants. Measurements of surface tension are widely used to provide authentic cmc values for all types of surfactants (cationic, anionic, and non-ionic). Illustrative ST graphs for the mixtures at different mole fractions of SLS in the presence of 100 mM NaCl at 298.15 K are displayed in Figure 3. The cmc values acquired via surface tension are listed in Table 1. As the surfactant molecules were mixed, a complex, which was more deeply adsorbed at the surface than single amphiphiles, was formed, thus suggesting an enhanced surface activity. The cmc values of single and mixed amphiphiles could be evaluated by the intersection of the linear fitting of the points (Figure 3). The cmc value of TCH was found to be 79.43 mM, which was lower than the values published by Miller et al. [48], who reported a value of nearly 100 mM without any salt. The cmc values of both employed surfactants in the existence of salt were also found to be less than those with a lack of salt. The values of cmc for currently employed surfactants in the presence of NaCl were in good agreement with the literature [49,50]. The obtained value of cmc for SDS in the presence of 100 mM NaCl was much lower than the cmc value computed by Thapa et al. [51] in an aqueous solution. When NaCl was added to the drug solution, the electrical atmosphere changed. The charge between the head group in the cationic drug became neutralized. Micelles could be formed at much lower concentrations in pure water because of the reduced electrostatic repulsion among the polar head groups. The cmc values for all mixtures unified in the center of two single amphiphiles, suggesting that the micellization of a drug was preferred in the company of surfactants. The observed decline in the cmc values of the mixture was due to the enrichment in the hydrophobic interaction among drugs and surfactants.

The whole study can be divided into two parts: (A) interactions of drugs with surfactants in the solution and (B) interactions of drugs with surfactants at the surface.

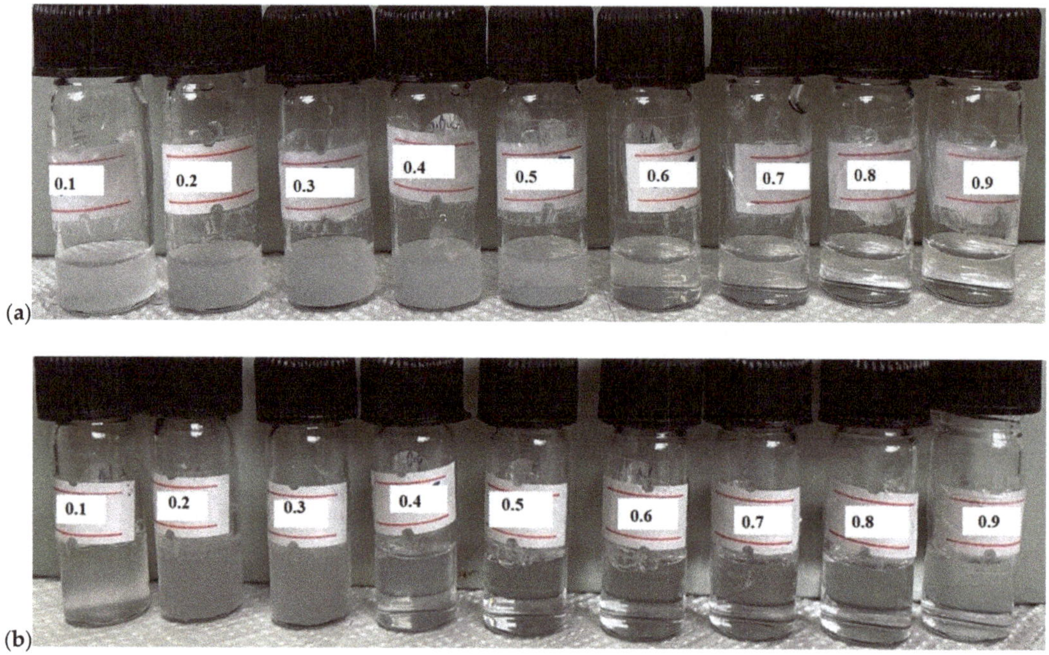

Figure 2. The physical appearance of TCH + SDS/SLS mixtures at different compositions: (**a**) SDS + TCH and (**b**) SLS + TCH.

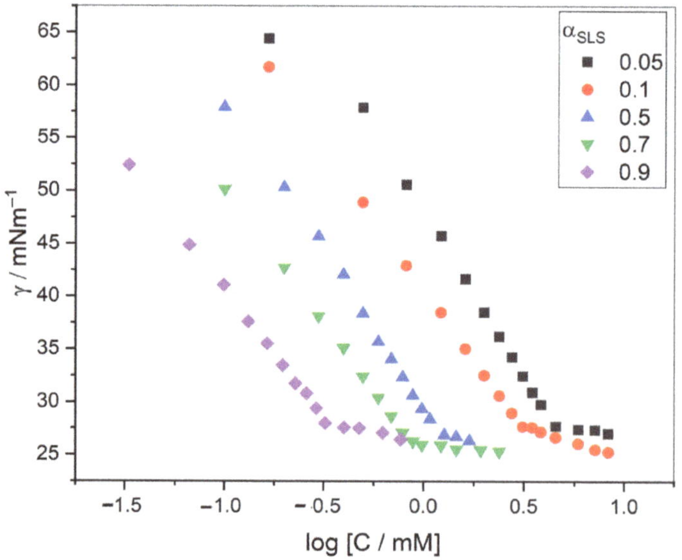

Figure 3. Graph of surface tension versus log molar concentration for SLS + TCH mixed systems.

Table 1. Physical parameters of TCH + SDS/SLS mixed systems in aqueous NaCl.

	cmc (mM)	*cmc** (mM)	X_1^{Rub}	X_1^{ideal}	$-\beta^{Rub}$	f_1^{Rub}	f_2^{Rub}
			SDS + TCH				
0.0	79.43	-	-	-	-	-	-
0.05	0.37	16.42	0.54	0.80	16.04	0.033	0.0095
0.1	0.31	9.16	0.56	0.90	15.20	0.055	0.0081
0.7	0.13	1.45	0.65	0.99	15.72	0.139	0.0014
0.8	0.15	1.27	0.66	0.99	15.44	0.174	0.0011
0.9	0.12	1.13	0.67	0.99	17.11	0.156	0.0004
1.0	1.02	-	-	-	-	-	-
			SLS + TCH				
0.0	79.43	-	-	-	-	-	-
0.05	4.49	41.24	0.50	0.51	8.86	0.110	0.1078
0.1	3.05	27.85	0.53	0.68	9.09	0.139	0.0742
0.5	1.24	7.74	0.62	0.95	9.96	0.243	0.0207
0.7	0.79	5.68	0.64	0.98	11.81	0.212	0.0082
0.9	0.33	4.49	0.64	0.99	16.53	0.115	0.0012
1.0	4.07	-	-	-	-	-	-

Relative standard uncertainties (u_r) are $u_r(cmc/cmc^*)$ = 0.03, u_r (X_1^{Rub}/X_1^{ideal}) = 0.02, u_r (β^{Ru}) = 0.03, and u_r (f_1^{Rub}/ f_2^{Rub}) = 0.04.

2.1. *Interactions of Drug with the Surfactants in the Mixed Micelle*

Using Rubingh's regular solution theory (RST) for mixtures of amphiphiles [52], the cmc of a mixed system (*cmc**) can be calculated via Equation (1):

$$\frac{1}{cmc^*} = \frac{\alpha_1}{f_1 cmc_1} + \frac{\alpha_2}{f_2 cmc_2} \tag{1}$$

where f_1 and f_2 are the activity coefficients of the surfactant (SDS/SLS) and drug in mixed micelles, respectively, and α_1 represents the mole fraction of surfactant (SDS/SLS) in the total mixed solution. The cmc values of surfactants and drugs are cmc_1 and cmc_2, respectively. $f_1 = f_2 = 1$ if we assume ideal behavior, so Equation (1) becomes:

$$\frac{1}{cmc^*} = \frac{\alpha_1}{cmc_1} + \frac{\alpha_2}{cmc_2} \tag{2}$$

Equation (2) was proposed by Clint [53]. Using the Clint equation, we could judge the ideality or non-ideality of a mixed system. Figure 4 displays a plot of *cmc* (experimentally determined)/*cmc** (calculated with Equation (2)) vs. α_1 (SDS/SLS). The *cmc* values of both mixtures were decreased with increases in the α_1. According to one possible explanation, the mixture was more favorable than expected under an ideal condition because of the interactions among hydrophobic chains of amphiphiles.

In contrast, for non-ideal mixtures, a new theory has been established and is referred to as the Rubingh model [52]. The Rubingh model uses RST to relate the activity coefficients of components with micellar mole fractions of component 1 as follows:

$$f_1^{Rub} = exp\left[\beta^{Rub}\left(1 - X_1^{Rub}\right)^2\right] \tag{3}$$

$$f_2^{Rub} = exp\left[\beta^{Rub}\left(X_1^{Rub}\right)^2\right] \tag{4}$$

where β^{Rub} and X_1^{Rub} are the interaction parameter and micellar mole fraction, respectively of component 1. If two variables have values of less than 1, the mixing components are not ideal. When computing the β^{Rub} values (parameter based on the *cmc* values of each amphiphile and their mixtures), the nature and strength of the interactions between the two surfactants are determined. Rubingh [52] derived the relationship shown in Equation (5) by considering the phase separation model for micellization.

$$\beta^{Rub} = \frac{\ln\left(\alpha_1 cmc / X_1^{Rub} cmc_1\right)}{\left(1 - X_1^{Rub}\right)^2} \tag{5}$$

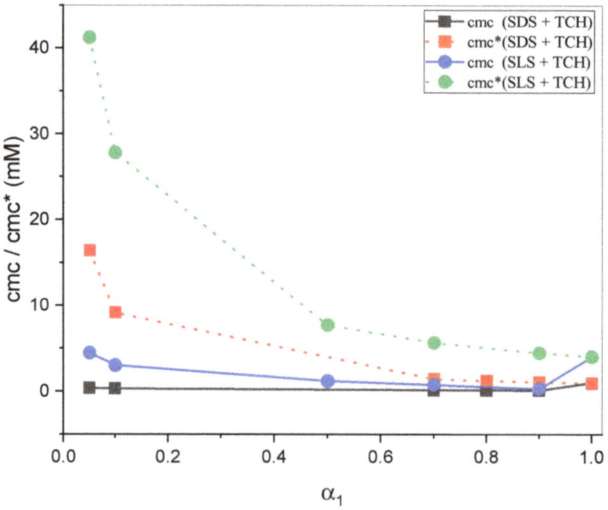

Figure 4. Experimentally determined critical micelle concentration (*cmc*) and ideal critical micelle concentration (*cmc**) against mole fraction of surfactants (SDS/SLS) in mixed systems at 298.15 K.

The micellar mole fraction of component 1 is represented by X_1^{Rub}, which is calculated by iteratively solving Equation (6):

$$\frac{\left(X_1^{Rub}\right)^2 \ln\left(\alpha_1 cmc / X_1^{Rub} cmc_1\right)}{\left(1 - X_1^{Rub}\right)^2 \ln\left[(1-\alpha_1) cmc / \left(1 - X_1^{Rub}\right) cmc_2\right]} = 1 \tag{6}$$

It is commonly believed that the deviation from zero of the interaction parameters (β^{Rub}) is due to interactions among the amphiphile head groups. Positive divergence from zero indicates antagonistic behavior, and negative deviation indicates synergistic interactions between two components. Free energy subsidies associated with amphiphile head groups have been found to be the main sources of mutual interaction. When positively and negatively charged amphiphiles are assorted in water, the most noteworthy feature of this mixture is its unusually huge drop in cmc values. A mixture of anionic and non-ionic surfactants usually yields a nonconformity from ideal behavior (less negative β^{Rub}) and synergistic effects in the mixed micelles of two non-ionic amphiphiles are even to a lesser extent. In most cases, experimentally computed values of β^{Rub} for mixtures of positively and negatively charged amphiphiles are higher. According to Table 1, there were considerable interactions (synergism) between the current mixed systems. The synergism was detected because of the electrostatic interaction among +ve and –ve charged head groups. The β^{Rub} average values were – 15.90 and – 11.25 for SDS + TCH and SLS + TCH, respectively. The positive and negatively charged amphiphiles were found to be firmly tied to one another through electrostatic and hydrophobic forces, consequently leading to ultimate attraction that promoted the growth of micellar aggregates. The synergism between two amphiphiles depends not only on the strength of the interaction but also on the individual amphiphile properties. The higher the hydrophobicity of an amphiphile, the easier it is to make micelles.

In a mixed system, the ideal micellar mole fraction of component 1 is represented by Equation (7) [54]

$$X_1^{ideal} = \frac{\alpha_1 cmc_2}{\alpha_1 cmc_2 + \alpha_2 cmc_1} \quad (7)$$

The values of X_1^{ideal} are given in Table 1. The values of X_1^{ideal} display nonconformity from the values of X_1^{Rub}, signifying non-ideality. The higher values of X_1^{ideal} for both binary mixtures at all mole fractions confirmed that added drug molecules replace some of the surfactant molecules from the mixed micelles, so the contribution of drug molecules is greater in mixed micelles than it should be in ideally mixed systems.

Thermodynamic Parameters for Drug–Surfactant Mixtures in the Mixed Micelle

Using RST, it is feasible to evaluate the free energy change for micellization in the following way [55–59]:

$$\Delta G_{mix} = RT[X_1^{Rub} ln(X_1^{Rub} f_1^{Rub}) + X_2^{Rub} ln\left(X_2^{Rub} f_2^{Rub}\right)] \quad (8)$$

If the values of activity coefficients (f_1^{Rub} and f_2^{Rub}) for an ideal mixed system are equal to unity, then Equation (8) becomes:

$$\Delta G_{mix}^{ideal} = RT[X_1^{Rub} ln X_1^{Rub} + X_2^{Rub} ln X_2^{Rub}] \quad (9)$$

where ΔG_{mix}^{ideal} is the free energy change for an ideal mixed system. Interestingly, the data (Table 2) show that the values were negative, implying that the micelles were spontaneously formed and were stable. If the values of ΔG_{mix}^{ideal} deviate from the values of ΔG_{mix}, rather than forming an ideal micelle, it then forms a real one. The literature confirms that previous investigators have observed the same behavior [60,61].

Table 2. Energetic constraints of TCH + SDS/SLS mixtures in aqueous NaCl [a].

α_1	$-G_{mix}^E$/$-\Delta H_m$ (kJmol^{-1})	$-\Delta G_{mix}$ (kJmol^{-1})	$-\Delta G_{mix}^{ideal}$ (kJmol^{-1})	$T\Delta S_m$ (kJmol^{-1})	$\lvert \frac{T\Delta S_m}{\Delta G_{mix}} \rvert$	$-\Delta G_m^o$ (kJmol^{-1})
		SDS + TCH				
0.0	-	-	-	-	-	16.23
0.05	9.87	11.78	1.71	6.40	0.54	29.48
0.1	9.27	11.15	1.69	6.32	0.57	29.93
0.7	8.90	10.69	1.61	6.01	0.56	32.08
0.8	8.54	10.29	1.58	5.88	0.57	31.79
0.9	9.36	11.12	1.57	5.89	0.53	32.34
1.0	-	-	-	-	-	27.01
		SLS + TCH				
0.0	-	-	-	-	-	16.23
0.05	5.49	7.33	1.72	6.17	0.84	23.34
0.1	5.60	7.44	1.71	6.16	0.82	24.30
0.5	5.79	7.56	1.64	5.93	0.78	26.54
0.7	6.76	8.52	1.62	5.92	0.69	27.66
0.9	9.45	11.26	1.62	6.07	0.54	29.77
1.0	-	-	-	-	-	23.59

[a] Relative standard uncertainties (u_r) are $u_r(G_{mix}^E/\Delta H_m)$ = 0.03, $u_r(\Delta G_{mix}/\Delta G_{mix}^{ideal})$ = 0.03, $u_r(\Delta S_m)$ = 0.03, and $u_r(\Delta G_m^o)$ = 0.03.

An excess thermodynamic function is a variation among the energetic function of the mixer for a non-ideal solution and the subsequent values for an ideal solution at a similar pressure and temperature [54]. The excess free energy of mixed micellization G_{mix}^E for a two-amphiphile mixtures can be computed with the help of equations 8 and 9 in form of Equation (10).

$$G^{E}_{mix} = \Delta H_m = RT[X_1^{Rub} ln f_1^{Rub} + X_2^{Rub} ln f_2^{Rub}] \quad (10)$$

From Table 2, we can observe that the values of G^{E}_{mix} were negative over the entire mole fraction range, confirming observations that the creation of the mixed micelles was thermodynamically more stable than the ideal state.

For the mixed system, Equations (9) and (10) were also used to calculate the entropy change as Equation (11):

$$\Delta S_m = \frac{\Delta H_m - \Delta G_m}{T} = -R\left[X_1^{Rub} ln X_1^{Rub} + X_2^{Rub} ln X_2^{Rub}\right] \quad (11)$$

Moreover, both binary and mixed micellization were found to be constrained by positive entropy values, which confirmed that entropy contribution drives mixed micellization. In the literature, the same results have previously been reported [55]. When we consider SDS + TCH mixed systems, the contributions to entropy were more significant at initial fractions. It was found to be an entropically favorable process when mixed micelles were formed, as the entropy/free energy change in this process was greater than 0.

Equation (12) was utilized to compute standard Gibbs free energy per mole of micellization using the mass-action model without considering counterion binding [58]:

$$\Delta G^{o}_{m} = RT ln X_{CMC} \quad (12)$$

In the above equation, X_{CMC} is the cmc value at mole fraction unit while R and T have their basic scientific meaning. The values of ΔG^{o}_{m} listed in Table 2 are negative for single and mixed amphiphiles. The negative values show that the micellization spontaneously occurred in the aqueous NaCl solution. The ΔG^{o}_{m} values of the drug were less than the single surfactants (SDS or SLS) and mixtures, confirming that mixed micelle formation of a drug with surfactants is more spontaneous compared to a drug alone. It is interesting to note here that the β^{Rub} values and ΔG^{o}_{m} values were directly proportional with respect to α_1, confirming that the higher interactions between amphiphile monomers cause more spontaneity in the process; the same results were reported by Bagheri et al. [54].

2.2. *Interfacial Properties of TCH + SDS/SLS Mixed System*

When amphiphiles are dissolved in water, the amphiphile monomers are adsorbed at the surface and the surface tension of water decreases, mainly due to the hydrophobic effects. The thermal motion and dynamic equilibrium determine the adsorption or desorption of monomers. Electrostatic interactions, hydrogen bonding, van der Waals interactions, and solvation/desolvation are factors that are less responsible for adsorption. Gibb's adsorption equation can be used to quantify the amount of amphiphiles adsorbed per unit area of the interface (surface excess, Γ_{max}) [62]:

$$\Gamma_{\max} = -\frac{1}{2.303nRT}\left(\frac{d\gamma}{dlogC}\right) \quad (13)$$

In Equation (13), $\frac{d\gamma}{dlogC}$ is the maximum slope, T is the absolute temperature in K, and $R = 8.314$ J mol^{-1} K^{-1}. Based on literature, the value of n was taken as 2 for pure amphiphiles and was calculated for mixtures with the following expression [62,63]

$$n = X_1^s n_1 X_2^s n_2 \quad (14)$$

The $\Gamma_{\max}$ values can be used to calculate the values of minimum area per molecule (A_{min}) with Equation (15) [64]

$$A_{min} = \frac{10^{20}}{N_A \Gamma_{\max}} \quad (15)$$

where N_A = 6.02214 × 10^{23} (Avogadro's number). The minimum area per molecule of an amphiphile suggests the packing (loose or close) and orientation of the amphiphile molecule at the surface. The low A_{min} (high Γ_{max}) values of the mixture at all mole fractions confirmed strong electrostatic interactions between cationic drugs and anionic surfactants (Table 3). This fact was also reflected in the negative interaction parameter values for the mixture. If there is no interaction between two amphiphiles in a mixed adsorbed film at the surface, the minimum area per molecule can be calculated with the following equation [62]:

$$A_{ideal} = \alpha_1 A_{min,\,1} + \alpha_2 A_{min,\,2} \tag{16}$$

Table 3. Interfacial and packing data of TCH + SDS/SLS mixed system in aqueous NaCl [a].

α_1	$10^6\ \Gamma_{max}$ (molm^{-2})	A_{min} (Å^2)	A_{ideal} (Å^2)	C_{20}	γ_{cmc} (mNm^{-1})	π_{cmc} (mNm^{-1})
			SDS + TCH			
0.0	1.64	1.01	-	19.36	39.57	31.43
0.05	1.77	0.94	1.01	0.03	27.79	43.21
0.1	2.44	0.68	1.01	0.05	28.55	42.45
0.7	3.10	0.53	0.99	0.03	29.88	41.12
0.8	3.39	0.49	0.98	0.04	30.17	40.83
0.9	3.28	0.51	0.97	0.03	30.68	40.32
1.0	1.71	0.97	-	0.09	30.60	40.40
			SLS + TCH			
0.0	1.64	1.01	-	19.36	39.57	31.43
0.05	2.73	0.61	1.01	0.80	27.88	43.11
0.1	2.28	0.73	1.01	0.41	27.72	43.28
0.5	2.57	0.65	1.03	0.19	26.89	44.11
0.7	2.14	0.77	1.04	0.09	27.11	43.89
0.9	2.01	0.83	1.05	0.04	28.04	42.96
1.0	1.57	1.05	-	0.18	23.80	47.20

[a] Relative standard uncertainties (u_r) are $u_r(\Gamma_{max})$ = 0.05, $u_r(A_{min}/A_{ideal})$ = 0.03, $u_r(C_{20})$ = 0.03, and $u_r(\gamma_{cmc}/\pi_{cmc})$ = 0.02.

The observed values (A_{min}) were lower than ideal values (A_{ideal}), indicating significant attractive interactions between the two components (Table 3). Water became 84–99% saturated following the adsorption of amphiphiles, which reduced its surface tension by approximately 20 dyn/cm. Adding an amphiphile to the water decreased the surface tension of H_2O by 20 mNm^{-1}, indicating the efficiency of its adsorption. Hence, it has the lowest concentration required to achieve saturation adsorption. By using Equation (17), we could calculate the adsorption efficiency (pC_{20}) as:

$$pC_{20} = -logC_{20} \tag{17}$$

where C_{20} is a measure of the adsorption efficiency of surfactants at the interface. The values of C_{20} are also listed in Table 3. It was concluded that the C_{20} values of SDS decreased with the addition of TCH. Decreasing C_{20} values of SDS with TCH were also shown by an earlier study [51]. In the case of SLS, the values of C_{20} only decreased at higher mole fractions. The C_{20} value of SDS in the presence NaCl has been found to be lower than in its absence [51], confirming that the surface activity of SDS is enhanced in the presence of NaCl.

Rosen and Hua modified Equations (5) and (6) for amphiphile adsorption to calculate the X_1^S and β^s with the following equations [64]

$$\frac{(X_1^s)^2 \ln(\alpha_1 C_{mix}/X_1^s C_1)}{(1 - X_1^S)^2 \ln[(1 - \alpha_1)C_{mix}/(1 - X_1^S)C_2]} = 1 \tag{18}$$

$$\beta^s = \frac{ln(\alpha_1 C_{mix}/X_1^S C_1)}{(1 - X_1^S)^2} \tag{19}$$

The interpretation of interaction parameter at the surface (β^s) is the same as in the case of bulk (β^{Rub}), with negative and positive β^s values that suggest synergism and antagonism, respectively. Here, the values of X_1^s were increased with the stoichiometric mole fraction (Table 4) and were always greater than X_1^{Rub}, showing amphiphiles contributed more to mixed monolayer formation than in the mixed micelle. Additionally, the contribution of SDS was greater than SLS in the mixed monolayer formation with the TCH. The β^s values were negative for both mixed systems, suggesting attractive interaction. The activity coefficients at the surface could be calculated by the following equations

$$ln f_1^S = \beta^s (X_2^S)^2 \tag{20}$$

$$ln f_2^S = \beta^s (X_1^S)^2 \tag{21}$$

Table 4. Thermodynamic and interfacial properties of TCH + SDS/SLS mixtures in aqueous NaCl [a].

α_1	X_1^s	$-\beta^s$	f_1^s	f_2^s	$-G_{ex}^s$ (kJmol^{-1})	$-\Delta G_{ads}$ (kJmol^{-1})	G_{min} (kJmol^{-1})
				SDS + TCH			
0.0	-	-	-	-	-	35.34	24.07
0.05	0.56	17.66	0.033	0.004	10.77	53.89	15.69
0.1	0.59	14.60	0.091	0.006	8.71	47.35	11.71
0.7	0.71	12.28	0.370	0.002	6.19	45.33	9.62
0.8	0.74	11.40	0.486	0.002	5.32	43.82	8.88
0.9	0.75	12.83	0.454	0.001	5.92	44.61	9.34
1.0	-	-	-	-	-	50.74	17.97
				SLS + TCH			
0.0	-	-	-	-	-	34.83	24.07
0.05	0.54	19.05	0.018	0.004	11.71	39.16	10.22
0.1	0.57	16.09	0.051	0.005	9.77	43.27	12.14
0.5	0.64	14.52	0.157	0.002	8.25	43.71	10.47
0.7	0.67	13.48	0.251	0.001	7.26	48.17	12.67
0.9	0.70	15.10	0.261	0.001	7.83	51.15	13.94
1.0		-	-	-	-	53.58	15.12

[a] Relative standard uncertainties (u_r) are $u_r(X_1^S) = 0.02$, $u_r(\beta^s) = 0.03$, $u_r(f_1^s/f_2^s) = 0.04$, $u_r(G_{ex}^s) = 0.03$, $u_r(\Delta G_{ads}) = 0.03$, and $u_r(G_{min}) = 0.03$.

The values of f_1^S and f_2^S are listed in Table 4 and were found to be less than unity, thus indicating non-ideality at the surface.

Thermodynamic Parameters for Drug–Surfactant Mixtures at the Surface

The standard free energy of interfacial adsorption (ΔG_{add}^o) can be computed by using the following relation [58]:

$$\Delta G_{add}^o = \Delta G_m^o - \left(\frac{\pi_{CMC}}{\Gamma_{max}}\right) \tag{22}$$

At the cmc, surface pressure is measured with the term π_{CMC}. Here in Equation (22), G_m^o is the standard Gibbs free energy previously computed with Equation (12). It was observed that the accomplished upsides of ΔG_{add}^o were –ve, similar to those of ΔG_m^o;

nonetheless, the extent was much more noteworthy, showing that adsorption was further unconstrained for this situation. f_1^S and f_2^S can be utilized to ascertain excess free energy (G_{exc}^s) at surface:

$$G_{exc}^s = RT\left[X_1^S lnf_1^S + \left(1 - X_1^S\right) lnf_2^S\right] \quad (23)$$

With negative values of G_{exc}^s, stability can be attained by the stable mixing at the surface, which is possible with the monolayer of surfactants or drugs alone. Negative G_{exc}^s values (Table 4) also indicate synergism at the surface. The degree of synergism for a mixed system can also be quantified by an energy parameter [65],

$$G_{min}^s = A_{min}\gamma_{CMC}N_A \quad (24)$$

The energy parameters that define the work required to create an interface per mole of the solution by transferring monomers from bulk to interface can be determined by the above-described energy parameters (G_{min}^s). According to Table 4, a lower value of G_{min}^s indicates a more stable surface, and this in turn results in increased surface activity.

3. UV–Visible Spectroscopic Study

The interaction of TCH with SDS and SLS was monitored with UV-visible absorption spectroscopy. The absorption spectrum of TCH (0.05 mM) in a 100 mM NaCl solution showed two absorption peaks at 226 and 310 nm due to the attendance of the aminobenzoate group. π–π^* and n–π^* transitions were involved in the first and second ones, respectively. When increasing concentrations of SDS and SLS were added to the TCH solution, the absorbance increased but the maximum absorbance at 310 nm was not changed (Figure 5). This spectral behavior indicates the electrostatic interactions between the positive charge of TCH molecules and the negative charge of surfactant monomers.

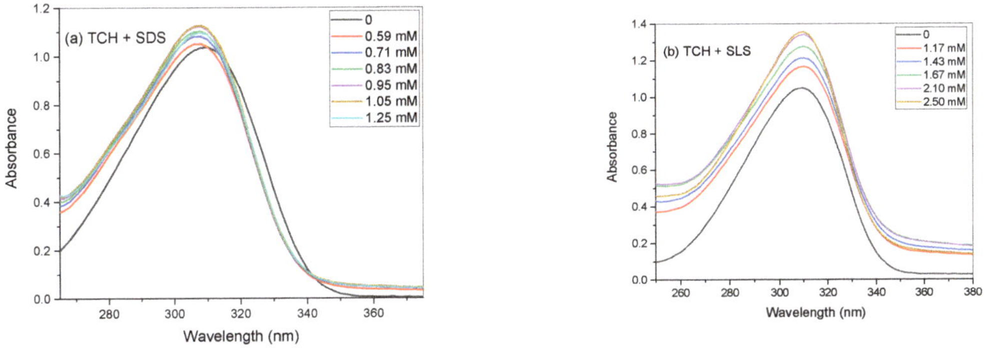

Figure 5. Absorption spectra of tetracaine hydrochloride in the presence of increased concentrations of (**a**) TCH + SDS and (**b**) TCH + SLS.

The binding constant and stoichiometric ratio were estimated with the differential absorbance method represented by the Benesi–Hildebrand equation [66]:

$$\frac{1}{A - A_0} = \frac{1}{K(A_{max} - A_0)[S]^n} + \frac{1}{A_{max} - A_0} \quad (25)$$

where the concentration of SDS/SLS is represented by [S], while A, A_0, and $A_{\max}$ represent values of absorbance due to the presence of surfactants, the absence of surfactants, and resulting absorbance due to the drug–surfactant complex, respectively. When plotting $1/(A - A_0)$ against $1/[SDS/SLS]^2$, a straight line is obtained (Figure 6), specifying the creation of the 1:2 complex. For an SDS + TCH mixed system without the addition of salt, Thapa et. al. reported a 1:1 complex [51]. However, for our system, a curvilinear fit was obtained, so the SDS + TCH complex was mainly 1:2. Using the Benesi–Hildebrand equation, the binding constant could be calculated (intercept/slope). We found values of K of 1.86×10^5 ($\pm$ 0.04) and 9.09×10^4 ($\pm$ 0.04) mol^{-1} dm^3 for the SDS + TCH and SLS + TCH mixed systems, respectively. The SLS + TCH mixed system had lesser binding constant values than the SDS + TCH system. In comparison, SDS has one functional group and SLS has two functional groups. The localized positive charge on the nitrogen atom on the TCH interacts with the negative charge on the sulphonic group, thus enhancing the electrostatic attraction between the guest and host. SLS, however, has methylated amide nitrogen, so the amide bond cannot be a hydrogen bond donor, which inhibits intermolecular attraction between SLS and TCH at the palisade layer. Furthermore, the steric hindrance of the N-methyl group of SLS may make it difficult to tightly align the amphiphiles. All these behaviors of SLS are responsible for its lesser binding constant compared to SDS.

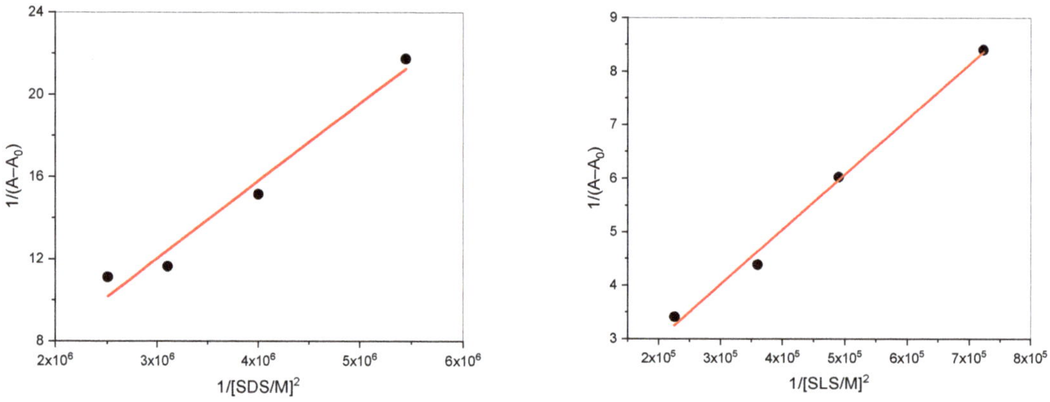

(a) (b)

Figure 6. Benesi–Hildebrand plots for the interaction of TCH (**a**) SDS and (**b**) SLS.

By using binding constant (K) values, free energy change of binding could be attained with Equation (26):

$$\Delta G_K = -RTlnK \tag{26}$$

The binding free energies were –30.08 ($\pm$ 0.2) $Jmol^{-1}$ for SDS + TCH and – 28.30($\pm$ 0.2) $kJmol^{-1}$ for SLS + TCH. In both mixed systems, the G values were negative, indicating that the binding process was spontaneous.

4. Conclusions

The synergistic interaction of TCH (+ve charged head group) with SDS and SLS (–ve charged head group) surfactants in the presence of salt (100 mM NaCl) was analyzed with both tensiometry and UV–visible spectroscopic techniques. The following conclusions can be derived:

1. The negative deviation of experimentally determined cmc values with hypothetical values confirms the nonideality of current mixtures.
2. The interaction parameter at the interface and in solution was determined to be –ve, thus validating synergism between monomers of two species at the surface and in bulk.
3. The higher values of the ideal mole fraction of component 1 (X_1^{ideal}) for both binary mixtures at all mole fractions indicate the strong ability of the drug to form of mixed micelles.

4. Energetics parameters confirm the spontaneity, stability, and entropic favorability of drug–surfactant mixtures.
5. The TCH with SLS had smaller binding constant values than SDS, possibly because SLS has a methylated amide nitrogen so the amide bond cannot be a hydrogen bond donor, which inhibits the intermolecular attraction between SLS and TCH at the palisade layer. Furthermore, the steric hindrance of the N-methyl group of SLS may make it difficult to tightly align the amphiphiles. All these behaviors of SLS are responsible for its smaller binding constant in comparison to SDS.

5. Experimental

5.1. Materials

Tetracaine hydrochloride (TCH, 99%), an anesthetic amphiphilic drug, and sodium lauroyl sarcosine (SLS, >95%) were supplied by Molecules On (Switzerland) and used as received. Sodium chloride (NaCl, 99%) and sodium dodecyl sulfate (SDS, 98.5%) were acquired from Sigma-Aldrich (St. Louis, MO, USA). At 298.15 K, all experiments were performed using ultra-pure, double-distilled de-ionized water with a conductivity between 1 and 2 μScm^{-1}. To prepare standard solutions for experiments, amphiphiles (both pure and mixed) were dissolved and accurately weighed in a 100 mM NaCl solution. The stock solutions for both techniques (surface tension and UV–vis spectrophotometer measurements) were prepared in aqueous 100 mM NaCl solutions.

5.2. Methods

5.2.1. Surface Tension Measurements

The surface tension experiments were conducted with a digital tensiometer (Sigma 700, Attention, Darmstadt, Germany) by using a platinum ring. The instrument was occasionally calibrated with ultra-pure distilled water. In tensiometric titration, an amphiphile stock solution was titrated into a static volume of H_2O. Throughout all experiments, water was circulated from a thermostatically controlled water bath through the outer jacket to keep the temperature at 298.15 K.

5.2.2. UV–Vis Spectrophotometer Measurements

We measured the spectra of the aqueous solutions of the drug and the drug–surfactant binary mixtures to determine the level of the binding of the drug with surfactants. As a first step, TCH in water was prepared as a stock solution in a volumetric flask. The desired concentration of surfactant solution was prepared from the aqueous TCH solution. Finally, a suitable volume of surfactant solution was added to the H_2O solution of TCH in a quartz cell. We measured the absorption spectra of TCH solutions with surfactants and plotted them against the wavelengths. For the measurement of the absorption spectrum of TCH solutions over the range of 200–400 nm, an Evolution 300 spectrophotometer from Thermo Scientific, Tokyo, Japan was used to record UV–visible spectra (Figure 2).

Author Contributions: N.A., M.A.R. & A.K. were involved in experimental planning, interpreting data and writing the manuscript; M.M.A., A.M.A. reviewed and edited the manuscript. All authors have read and agreed to the published version of the manuscript.

Funding: This research work was funded by Institutional Fund Projects under grant no (IFPRC-174-130-2020). Therefore, the authors gratefully acknowledge technical and financial support from the Ministry of Education and King Abdulaziz University, Jeddah, Saudi Arabia.

Institutional Review Board Statement: Not applicable.

Informed Consent Statement: Not applicable.

Data Availability Statement: Not applicable.

Conflicts of Interest: The authors declare no conflict of interest.

References

1. Van Eerdenbrugh, B.; Vermant, J.; Martens, J.A.; Froyen, L.; Van Humbeeck, J.; Van den Mooter, G.; Augustijns, P. Solubility Increases Associated with Crystalline Drug Nanoparticles: Methodologies and Significance. *Mol. Pharm.* **2010**, *7*, 1858–1870. [CrossRef]
2. Ruso, J.M.; Attwood, D.; Rey, C.; Taboada, P.; Mosquera, V.; Sarmiento, F. Light Scattering and NMR Studies of the Self-Association of the Amphiphilic Molecule Propranolol Hydrochloride in Aqueous Electrolyte Solutions. *J. Phys. Chem. B* **1999**, *103*, 7092–7096. [CrossRef]
3. Awang, N.; Ismail, A.F.; Jaafar, J.; Matsuura, T.; Junoh, H.; Othman, M.H.D.; Rahman, M.A. Functionalization of polymeric materials as a high performance membrane for direct methanol fuel cell: A review. *React. Funct. Polym.* **2015**, *86*, 248–258. [CrossRef]
4. Azum, N.; Rub, M.A.; Khan, A.; Alotaibi, M.M.; Asiri, A.M.; Rahman, M.M. Mixed Micellization, Thermodynamic and Adsorption Behavior of Tetracaine Hydrochloride in the Presence of Cationic Gemini/Conventional Surfactants. *Gels* **2022**, *8*, 128. [CrossRef] [PubMed]
5. Rub, M.A.; Azum, N.; Kumar, D.; Asiri, A.M. Interaction of TX-100 and Antidepressant Imipramine Hydrochloride Drug Mixture: Surface Tension, 1H NMR, and FT-IR Investigation. *Gels* **2022**, *8*, 159. [CrossRef]
6. Ahmed, M.F.; Abdul Rub, M.; Joy, M.T.R.; Molla, M.R.; Azum, N.; Anamul Hoque, M.; Rub, M.A.; Azum, N.; Kumar, D.; Asiri, A.M.; et al. Influences of NaCl and Na2SO4 on the Micellization Behavior of the Mixture of Cetylpyridinium Chloride + Polyvinyl Pyrrolidone at Several Temperatures. *Gels* **2022**, *8*, 62. [CrossRef] [PubMed]
7. Alvarez-Lorenzo, C.; Concheiro, A. Effects of Surfactants on Gel Behavior. *Am. J. Drug Deliv.* **2003**, *1*, 77–101. [CrossRef]
8. Wedel, B.; Brändel, T.; Bookhold, J.; Hellweg, T. Role of Anionic Surfactants in the Synthesis of Smart Microgels Based on Different Acrylamides. *ACS Omega* **2017**, *2*, 84–90. [CrossRef] [PubMed]
9. Maiti, K.; Mitra, D.; Mitra, R.N.; Panda, A.K.; Das, P.K.; Rakshit, A.K.; Moulik, S.P. Self-Aggregation of Synthesized Novel Bolaforms and Their Mixtures with Sodium Dodecyl Sulfate (SDS) and Cetyltrimethylammonium Bromide (CTAB) in Aqueous Medium. *J. Phys. Chem. B* **2010**, *114*, 7499–7508. [CrossRef] [PubMed]
10. Jafari-Chashmi, P.; Bagheri, A. The strong synergistic interaction between surface active ionic liquid and anionic surfactant in the mixed micelle using the spectrophotometric method. *J. Mol. Liq.* **2018**, *269*, 816–823. [CrossRef]
11. Mal, A.; Bag, S.; Ghosh, S.; Moulik, S.P. Physicochemistry of CTAB-SDS interacted catanionic micelle-vesicle forming system: An extended exploration. *Colloids Surf. A Physicochem. Eng. Asp.* **2018**, *553*, 633–644. [CrossRef]
12. Tozuka, Y.; Imono, M.; Uchiyama, H.; Takeuchi, H. A novel application of α-glucosyl hesperidin for nanoparticle formation of active pharmaceutical ingredients by dry grinding. *Eur. J. Pharm. Biopharm.* **2011**, *79*, 559–565. [CrossRef]
13. Shen, S.; Ng, W.K.; Chia, L.; Dong, Y.; Tan, R.B.H. Stabilized Amorphous State of Ibuprofen by Co-Spray Drying With Mesoporous SBA-15 to Enhance Dissolution Properties. *J. Pharm. Sci.* **2010**, *99*, 1997–2007. [CrossRef] [PubMed]
14. Sigfridsson, K.; Lundqvist, A.J.; Strimfors, M. Particle size reduction for improvement of oral absorption of the poorly soluble drug UG558 in rats during early development. *Drug Dev. Ind. Pharm.* **2009**, *35*, 1479–1486. [CrossRef]
15. Sugano, K.; Okazaki, A.; Sugimoto, S.; Tavornvipas, S.; Omura, A.; Mano, T. Solubility and Dissolution Profile Assessment in Drug Discovery. *Drug Metab. Pharmacokinet.* **2007**, *22*, 225–254. [CrossRef] [PubMed]
16. Schreier, S.; Malheiros, S.V.P.; de Paula, E. Surface active drugs: Self-association and interaction with membranes and surfactants. Physicochemical and biological aspects. *Biochim. Et Biophys. Acta (BBA)-Biomembr.* **2000**, *1508*, 210–234. [CrossRef]
17. YOKOYAMA, S.; FUJINO, Y.; KAWAMOTO, Y.; KANEKO, A.; FUJIE, T. Micellization of an Aqueous Solution of Piperidolate Hydrochloride in the Presence of Acetylcholine Chloride. *Chem. Pharm. Bull.* **1994**, *42*, 1351–1353. [CrossRef]
18. Attwood, D.; Tolley, J.A. Self-association of analgesics in aqueous solution: Association models for codeine, oxycodone, ethylmorphine and pethidine. *J. Pharm. Pharmacol.* **2011**, *32*, 761–765. [CrossRef]
19. Kumar, D.; Azum, N.; Rub, M.A.; Asiri, A.M. Interfacial and spectroscopic behavior of phenothiazine drug/bile salt mixture in urea solution. *Chem. Pap.* **2021**, *75*, 3949–3956. [CrossRef]
20. Ghosh, S.; Krishnan, A.; Das, P.K.; Ramakrishnan, S. Determination of Critical Micelle Concentration by Hyper-Rayleigh Scattering. *J. Am. Chem. Soc.* **2003**, *125*, 1602–1606. [CrossRef]
21. Zhu, Q.; Huang, L.; Su, J.; Liu, S. A sensitive and visible fluorescence-turn-on probe for the CMC determination of ionic surfactants. *Chem. Commun.* **2014**, *50*, 1107–1109. [CrossRef] [PubMed]
22. Tadros, T.F. *Applied Surfactants*; Wiley: Hoboken, NJ, USA, 2005; ISBN 9783527306299.
23. Chiu, Y.C.; Kuo, C.Y.; Wang, C.W. Using electrophoresis to determine zeta potential of micelles and critical micelle concentration. *J. Dispers. Sci. Technol.* **2000**, *21*, 327–343. [CrossRef]
24. Priev, A.; Zalipsky, S.; Cohen, R.; Barenholz, Y. Determination of Critical Micelle Concentration of Lipopolymers and Other Amphiphiles: Comparison of Sound Velocity and Fluorescent Measurements. *Langmuir* **2002**, *18*, 612–617. [CrossRef]
25. Romani, A.P.; da Hora Machado, A.E.; Hioka, N.; Severino, D.; Baptista, M.S.; Codognoto, L.; Rodrigues, M.R.; de Oliveira, H.P.M. Spectrofluorimetric Determination of Second Critical Micellar Concentration of SDS and SDS/Brij 30 Systems. *J. Fluoresc.* **2009**, *19*, 327–332. [CrossRef]
26. Pérez-Rodríguez, M.; Prieto, G.; Rega, C.; Varela, L.M.; Sarmiento, F.; Mosquera, V. A Comparative Study of the Determination of the Critical Micelle Concentration by Conductivity and Dielectric Constant Measurements. *Langmuir* **1998**, *14*, 4422–4426. [CrossRef]

27. Karsa, D.R. *Industrial Applications of Surfactants*; Elsevier: Amsterdam, The Netherlands, 1999.
28. Atta, A.M.; Abdullah, M.M.S.; Al-Lohedan, H.A.; Ezzat, A.O. Demulsification of heavy crude oil using new nonionic cardanol surfactants. *J. Mol. Liq.* **2018**, *252*, 311–320. [CrossRef]
29. Shaban, S.M.; Kang, J.; Kim, D.-H. Surfactants: Recent advances and their applications. *Compos. Commun.* **2020**, *22*, 100537. [CrossRef]
30. Hegazy, M.A.; Abdallah, M.; Ahmed, H. Novel cationic gemini surfactants as corrosion inhibitors for carbon steel pipelines. *Corros. Sci.* **2010**, *52*, 2897–2904. [CrossRef]
31. Torchilin, V.P. Structure and design of polymeric surfactant-based drug delivery systems. *J. Control. Release* **2001**, *73*, 137–172. [CrossRef]
32. King, S.-Y.P.; Basista, A.M.; Torosian, G. Self-Association and Solubility Behaviors of a Novel Anticancer Agent, Brequinar Sodium. *J. Pharm. Sci.* **1989**, *78*, 95–100. [CrossRef]
33. Matsuki, H.; Hashimoto, S.; Kaneshina, S.; Yamanaka, M. Surface Adsorption and Volume Behavior of Local Anesthetics. *Langmuir* **1994**, *10*, 1882–1887. [CrossRef]
34. Atherton, A.D.; Barry, B.W. Photon correlation spectroscopy of surface active cationic drugs. *J. Pharm. Pharmacol.* **2011**, *37*, 854–862. [CrossRef] [PubMed]
35. Sarmiento, F.; López-Fontán, J.L.; Prieto, G.; Mosquera, V.; Attwood, D. Mixed micelles of structurally related antidepressant drugs. *Colloid Polym. Sci.* **1997**, *275*, 1144–1147. [CrossRef]
36. Rub, M.A.; Azum, N.; Khan, F.; Asiri, A.M. Surface, micellar, and thermodynamic properties of antidepressant drug nortriptyline hydrochloride with TX-114 in aqueous/urea solutions. *J. Phys. Org. Chem.* **2017**, *30*, e3676. [CrossRef]
37. Abdul Rub, M.; Azum, N.; Asiri, A.M. Binary Mixtures of Sodium Salt of Ibuprofen and Selected Bile Salts: Interface, Micellar, Thermodynamic, and Spectroscopic Study. *J. Chem. Eng. Data* **2017**, *62*, 3216–3228. [CrossRef]
38. Azum, N.; Naqvi, A.Z.; Rub, M.A.; Asiri, A.M. Multi-technique approach towards amphiphilic drug-surfactant interaction: A physicochemical study. *J. Mol. Liq.* **2017**, *240*, 189–195. [CrossRef]
39. Azum, N.; Rub, M.A.; Asiri, A.M.; Bawazeer, W.A. Micellar and interfacial properties of amphiphilic drug–non-ionic surfactants mixed systems: Surface tension, fluorescence and UV–vis studies. *Colloids Surf. A Physicochem. Eng. Asp.* **2017**, *522*, 183–192. [CrossRef]
40. Kumar, D.; Rub, M.A.; Azum, N.; Asiri, A.M. Mixed micellization study of ibuprofen (sodium salt) and cationic surfactant (conventional as well as gemini). *J. Phys. Org. Chem.* **2018**, *31*, e3730. [CrossRef]
41. Khan, F.; Rub, M.A.; Azum, N.; Asiri, A.M. Mixtures of antidepressant amphiphilic drug imipramine hydrochloride and anionic surfactant: Micellar and thermodynamic investigation. *J. Phys. Org. Chem.* **2018**, *31*, e3812. [CrossRef]
42. Azum, N.; Rub, M.A.; Asiri, A.M. Interaction of antipsychotic drug with novel surfactants: Micellization and binding studies. *Chin. J. Chem. Eng.* **2018**, *26*, 566–573. [CrossRef]
43. Kumar, D.; Azum, N.; Rub, M.A.; Asiri, A.M. Aggregation behavior of sodium salt of ibuprofen with conventional and gemini surfactant. *J. Mol. Liq.* **2018**, *262*, 86–96. [CrossRef]
44. Rub, M.A.; Azum, N.; Khan, F.; Asiri, A.M. Aggregation of sodium salt of ibuprofen and sodium taurocholate mixture in different media: A tensiometry and fluorometry study. *J. Chem. Thermodyn.* **2018**, *121*, 199–210. [CrossRef]
45. Azum, N.; Ahmed, A.; Rub, M.A.; Asiri, A.M.; Alamery, S.F. Investigation of aggregation behavior of ibuprofen sodium drug under the influence of gelatin protein and salt. *J. Mol. Liq.* **2019**, *290*, 111187. [CrossRef]
46. Srivastava, A.; Thapa, U.; Saha, M.; Jalees, M. Aggregation behaviour of tetracaine hydrochloride with Gemini surfactants and the formation of silver nanoparticles using drug-Gemini surfactants mixture. *J. Mol. Liq.* **2019**, *276*, 399–408. [CrossRef]
47. Zhou, S.; Huang, G.; Chen, G. Synthesis and biological activities of local anesthetics. *RSC Adv.* **2019**, *9*, 41173–41191. [CrossRef]
48. Miller, K.J.; Goodwin, S.R.; Westermann-Clark, G.B.; Shah, D.O. Importance of molecular aggregation in the development of a topical local anesthetic. *Langmuir* **1993**, *9*, 105–109. [CrossRef]
49. Ray, G.B.; Ghosh, S.; Moulik, S.P. Physicochemical Studies on the Interfacial and Bulk Behaviors of Sodium N-Dodecanoyl Sarcosinate (SDDS). *J. Surfactants Deterg.* **2009**, *12*, 131–143. [CrossRef]
50. Umlong, I.M.; Ismail, K. Micellization behaviour of sodium dodecyl sulfate in different electrolyte media. *Colloids Surf. A Physicochem. Eng. Asp.* **2007**, *299*, 8–14. [CrossRef]
51. Thapa, U.; Kumar, M.; Chaudhary, R.; Singh, V.; Singh, S.; Srivastava, A. Binding behaviour of hydrophobic drug tetracaine hydrochloride used as organic counterion on ionic surfactants. *J. Mol. Liq.* **2021**, *335*, 116564. [CrossRef]
52. Holland, P.M.; Rubingh, D.N. Nonideal multicomponent mixed micelle model. *J. Phys. Chem.* **1983**, *87*, 1984–1990. [CrossRef]
53. Clint, J.H. Micellization of mixed nonionic surface active agents. *J. Chem. Soc. Faraday Trans. 1 Phys. Chem. Condens. Phases* **1975**, *71*, 1327. [CrossRef]
54. Motomura, K.; Yamanaka, M.; Aratono, M. Thermodynamic consideration of the mixed micelle of surfactants. *Colloid Polym. Sci.* **1984**, *262*, 948–955. [CrossRef]
55. Negm, N.A.; Tawfik, S.M. Studies of Monolayer and Mixed Micelle Formation of Anionic and Nonionic Surfactants in the Presence of Adenosine-5-monophosphate. *J. Solut. Chem.* **2012**, *41*, 335–350. [CrossRef]
56. Ren, Z.H.; Huang, J.; Zheng, Y.C.; Lai, L.; Yu, X.R.; Chang, Y.L.; Li, J.G.; Zhang, G.H. Mixed micellization of binary mixture of amino sulfonate amphoteric surfactant with octadecyltrimethyl ammonium bromide in water/isopropanol solution: Comparison with that in aqueous solution. *J. Dispers. Sci. Technol.* **2019**, *40*, 1353–1359. [CrossRef]

57. Das, S.; Ghosh, S.; Das, B. Formation of Mixed Micelle in an Aqueous Mixture of a Surface Active Ionic Liquid and a Conventional Surfactant: Experiment and Modeling. *J. Chem. Eng. Data* **2018**, *63*, 3784–3800. [CrossRef]
58. Rosen, M.J.; Cohen, A.W.; Dahanayake, M.; Hua, X.Y. Relationship of structure to properties in surfactants. 10. Surface and thermodynamic properties of 2-dodecyloxypoly(ethenoxyethanol)s, C12H25(OC2H4)xOH, in aqueous solution. *J. Phys. Chem.* **1982**, *86*, 541–545. [CrossRef]
59. Bagheri, A.; Abolhasani, A. Binary mixtures of cationic surfactants with triton X-100 and the studies of physicochemical parameters of the mixed micelles. *Korean J. Chem. Eng.* **2015**, *32*, 308–315. [CrossRef]
60. Ren, Z.H.; Luo, Y.; Zheng, Y.C.; Wang, C.J.; Shi, D.P.; Li, F.X. Micellization behavior of the mixtures of amino sulfonate amphoteric surfactant and octadecyltrimethyl ammonium bromide in aqueous solution at 40 °C: A tensiometric study. *J. Mater. Sci.* **2015**, *50*, 1965–1972. [CrossRef]
61. Ren, Z.H. Interacting behavior between amino sulfonate amphoteric surfactant and octylphenol polyoxyethylene ether (7) in aqueous solution and pH effect. *J. Ind. Eng. Chem.* **2014**, *20*, 3649–3657. [CrossRef]
62. Zhou, Q.; Rosen, M.J. Molecular Interactions of Surfactants in Mixed Monolayers at the Air/Aqueous Solution Interface and in Mixed Micelles in Aqueous Media: The Regular Solution Approach. *Langmuir* **2003**, *19*, 4555–4562. [CrossRef]
63. Rosen, M.J.; Hua, X.Y. Surface concentrations and molecular interactions in binary mixtures of surfactants. *J. Colloid Interface Sci.* **1982**, *86*, 164–172. [CrossRef]
64. Ananda, K.; Yadav, O.P.; Singh, P.P. Studies on the surface and thermodynamic properties of some surfactants in aqueous and water+1,4-dioxane solutions. *Colloids Surf.* **1991**, *55*, 345–358. [CrossRef]
65. Oida, T.; Nakashima, N.; Nagadome, S.; Ko, J.-S.; Oh, S.-W.; Sugihara, G. Adsorption and Micelle Formation of Mixed Surfactant Systems in Water. III. A Comparison between Cationic Gemini/Cationic and Cationic Gemini/Nonionic Combinations. *J. Oleo Sci.* **2003**, *52*, 509–522. [CrossRef]
66. Benesi, H.A.; Hildebrand, J.H. A Spectrophotometric Investigation of the Interaction of Iodine with Aromatic Hydrocarbons. *J. Am. Chem. Soc.* **1949**, *71*, 2703–2707. [CrossRef]

gels

Article

Hydrogel Containing *Borassus flabellifer* L. Male Flower Extract for Antioxidant, Antimicrobial, and Anti-Inflammatory Activity

Prakairat Tunit [1], Phanit Thammarat [2], Siriporn Okonogi [2,3] and Chuda Chittasupho [2,3,*]

1 Thai Traditional Medicine Program, Faculty of Nursing and Allied Health Sciences, Phetchaburi Rajabhat University, Phetchaburi 76000, Thailand; prakairat.tun@mail.pbru.ac.th
2 Department of Pharmaceutical Sciences, Faculty of Pharmacy, Chiang Mai University, Chiang Mai 50200, Thailand; phanit.thamma@cmu.ac.th (P.T.); siriporn.okonogi@cmu.ac.th (S.O.)
3 Research Center of Pharmaceutical Nanotechnology, Faculty of Pharmacy, Chiang Mai University, Chiang Mai 50200, Thailand
* Correspondence: chuda.c@cmu.ac.th

Abstract: *Borassus flabellifer* L. is a plant in Arecaceae family, widely distributed and cultivated in tropical Asian countries. The purpose of this study was to identify the bioactive compounds of *B. flabellifer* L. male flower ethanolic extract and investigate the antioxidant, anti-inflammatory, and antibacterial activities against *Cutibacterium acnes*. Total phenolic compounds and total flavonoids in *B. flabellifer* L. male flower ethanolic extract were determined by the Folin–Ciocalteu method and aluminum chloride colorimetric assay, respectively. Active substances in the extract and their quantities were analyzed by liquid chromatography and mass spectrometry (LC–MS/MS). The antioxidant evaluation was carried out using DPPH, ABTS free radical scavenging assays, and FRAP assay. *C. acnes* inhibitory activity was performed by the broth microdilution method. Anti-inflammatory activity was determined by the protein denaturation assay. In addition, gel containing different amounts of *B. flabellifer* L. male flower extract was formulated. The physical stability of the gel was observed by measuring viscosity and pH after six heating and cooling cycles, as well as 1-month storage at 4, 30, and 45 °C. The total phenolic content in the extract was 268.30 ± 12.84 mg gallic acid equivalent/g crude dry extract. The total flavonoid contents in the extract were 1886.38 ± 55.86 mg quercetin equivalent/g extract and 2884.88 ± 128.98 mg EGCG equivalent/g extract, respectively. The LC–MS/MS analysis revealed the presence of gallic acid, coumarin, and quercetin and the concentrations of quercetin, coumarin, and gallic acid in *B. flabellifer* male flower ethanolic extract were 0.912, 0.021, and 1.610 µg/mL, respectively. DPPH and ABTS antioxidant assays indicated that the *B. flabellifer* L. male flower extract had IC_{50} values of 31.54 ± 0.43 and 164.5 ± 14.3 µg/mL, respectively. FRAP assay revealed that the *B. flabellifer* male flower extract had high ferric ion reducing power. The extract was able to inhibit *C. acnes* bacteria with a minimum inhibitory concentration (MIC) of 250 mg/mL. At 250 and 500 µg/mL, the extract demonstrated the highest anti-inflammatory activity. The gel containing 31.25% *w/w* and 62.5% *w/w* showed good physical stability after six heating and cooling cycles, as well as 1-month storage.

Keywords: *Borassus flabellifer* L.; antioxidant activity; antibacterial activity; *Cutibacterium acnes*; gel

Citation: Tunit, P.; Thammarat, P.; Okonogi, S.; Chittasupho, C. Hydrogel Containing *Borassus flabellifer* L. Male Flower Extract for Antioxidant, Antimicrobial, and Anti-Inflammatory Activity. *Gels* **2022**, *8*, 126. https://doi.org/10.3390/gels8020126

Academic Editor: Hiroyuki Takeno

Received: 30 January 2022
Accepted: 12 February 2022
Published: 17 February 2022

1. Introduction

Acne vulgaris is a common skin disorder in which the skin's pores are blocked by sebum, bacteria, and dead cells. Acne is associated with inflammation of the pilosebaceous unit. Previous studies reported that oxidative stress components, such as reactive oxygen species and lipid peroxide are involved in the pathogenesis and progression of the disease [1]. In addition, acne vulgaris is triggered by *Cutibacterium acnes* (*C. acnes*) under the influence of normal circulating dehydroepiandrosterone. *C. acnes* is a Gram-positive bacteria that lives in the sebaceous follicle. It is one of the main causes of acne vulgaris by inducing inflammation and follicular epidermal proliferation [2]. Inhibition of *C. acnes* by topical antibacterial medications reduces its entry into the skin and prevents acne from spreading to other areas.

The natural antioxidant systems in the skin, including superoxide dismutase and catalase enzymes, maintain the balance of cellular redox reactions. Elevated reactive oxygen species (ROS) levels and decreased antioxidant levels can lead to oxidative stress and cause cellular membrane. Al-Shobaili showed a significant elevation of plasma lipid peroxide levels in acne patients compared with healthy volunteers [3]. In addition, the activities of superoxide dismutase and catalase enzymes and the level of total antioxidant capacity significantly diminished in acne patients [3].

Various acne vulgaris treatments target different steps in the pathogenesis of acne, including reducing *C. acnes* proliferation, suppressing androgens, and decreasing sebum production to prevent follicular occlusion [4]. The treatment of acne using topical antibacterial agents, such as clindamycin and erythromycin, is effective and faster than hormonal adjustments and laser treatments. However, the long-term use of topical antibiotics to treat acne vulgaris can cause undesirable side effects and induce *C. acnes* resistance. *C. acnes* resistance to antibiotics, such as erythromycin and clindamycin has been detected with high prevalence in Mediterranean countries mainly due to antibiotic abuse [5–8]. Side effects of topical antibiotic use are dryness of the treated area, skin irritation, and contact dermatitis, including red, dry, and itchy skin. Many plants have shown antibacterial and anti-inflammatory activities. Natural active compounds are attractive for use as a combination or replacement of antibacterial drugs to treat acne vulgaris since they possess fewer side effects and have multiple mechanisms of action [9].

Borassus flabellifer L. (*B. flabellifer*) (Arecaceae) is a plant widely grown in Southeast Asia [10]. Folk medicine uses various parts of *B. flabellifer* as a diuretic, antimicrobial, tonic, laxative, and wound healing agent [11]. A study on the pharmacological activity of the male flower found that the ethanol extracts at concentrations of 150 and 300 mg/kg had anti-inflammatory [12], analgesic, and antipyretic effects in rats [13]. The root and male flower parts extracted with methanol were found to have antioxidant activity [14]. However, the bioactive compounds, antioxidant, and anti-*C. acnes* activities of male flower ethanolic extract have never been reported. The objective of this study was to quantify phenolic compounds and flavonoids, as well as the active substances in the ethanol extract from the male flower of *B. flabellifer*. The antioxidant, anti-*C. acne*, and anti-inflammatory potential were evaluated. Gel containing different concentrations of *B. flabellifer* extract was developed and investigated for its physical stability.

2. Results and Discussion

2.1. Yield of *B. flabellifer* Male Flower Ethanolic Extract and Total Phenolic Content in *B. flabellifer* Male Flower Ethanolic Extract

The yield of ethanolic extract obtained from dried *B. flabellifer* male flower was measured. The yield was reported as 5.06 ± 1.35% *w/w*. The total phenolic content in *B. flabellifer* male flower ethanolic extract using the Folin–Ciocalteu reagent was expressed in terms of gallic acid equivalent. The standard curve equation was $y = 0.0147x + 0.1009$, $r^2 = 0.9990$ (Figure 1A). The results showed that total phenolic content increased with the concentration of *B. flabellifer* male flower extract with a correlation coefficient of 0.9979 (Figure 1B). The average total phenolic content in the extract was 268.30 ± 12.84 mg gallic acid equivalent/g crude dry extract, calculated from 31.25–250 μg/mL crude extract. The total phenolic content in plant extracts of *B. flabellifer* male flower ethanolic extract depends on the polarity of solvent used for extraction [15]. Tusskorn et al. reported that *B. flabellifer* flowers extracted with methanol and ethyl acetate had phenolic contents of 159.3 ± 0.3 and 90.0 ± 0.0 mg GAE/g extract, respectively [16]. The results suggested that the total phenolic content in *B. flabellifer* male flower's ethanolic extract was significantly greater than the extracts from methanol and ethyl acetate. The total phenolic content of *B. flabellifer* root extract from ethanol was significantly higher than the petroleum ether extract [17]. These results agreed with our results and previous reports, showing that the total phenolic compound in terms of the gallic acid equivalent amount was significantly higher when *B. flabellifer* was extracted with ethanol.

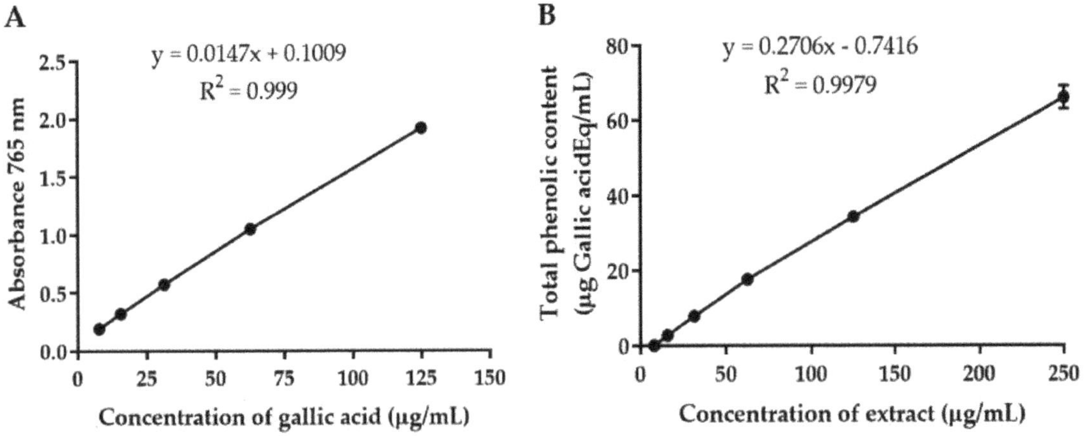

Figure 1. (**A**) Calibration curve of gallic acid. (**B**) Total phenolic content of *B. flabellifer* male flower ethanolic extract.

2.2. Total Flavonoid Content in B. flabellifer Male Flower Ethanolic Extract

The total flavonoid content in *B. flabellifer* male flower ethanolic extract determined by the aluminum chloride colorimetric method was expressed in quercetin and epigallocatechin (EGCG) equivalent amounts. The quercetin standard curve equation was $y = 0.0012x + 0.0276$, $r^2 = 0.9991$ (Figure 2A). The EGCG standard curve equation was $y = 0.0008x + 0.0217$, $r^2 = 0.999$ (Figure 3A). The concentration of total flavonoids in the extract increased with the concentration of the extract (Figures 2B and 3B). The average total flavonoid contents in the extract were 1886.38 ± 55.86 mg quercetin equivalent/g extract and 2884.88 ± 128.98 mg EGCG equivalent/g extract, calculated from 62.5–500 μg/mL of crude extract. These results indicated that the ethanolic extract of male flowers contained a high amount of flavonoids. The total flavonoid contents found in each part of *B. flabellifer* were different. The lower polar solvent could extract higher total flavonoid content from *B. flabellifer*. The root and fruit extracts from ethanol yielded higher total flavonoid concentrations at 3.57 ± 1.26 and 7.02 ± 0.61 mg quercetin equivalent/g of crude extract, respectively. In contrast, the root extract from petroleum ether contained 17.41 ± 1.89 mg quercetin equivalent/g of crude extract. In addition, part of *B. flabellifer* contained different amounts of total flavonoids. The total flavonoid content in quercetin equivalent in *B. flabellifer* flower was significantly higher than the root [18].

2.3. Analysis of B. flabellifer Male Flower Ethanolic Extract Phytochemical Component by LC–MS/MS

A total of four bioactive compounds in *B. flabellifer* male flower ethanolic extract were identified and characterized by a correlation of the molecular (precursor) ions and the fragmentation patterns (product ions) acquired by LC–MS/MS analysis (Figure 4). The LC–MS/MS data were compared with the molecular ions and fragmentation patterns of the reference standards. The results showed that the bioactive compounds identified in *B. flabellifer* male flower ethanolic extract were quercetin, coumarin, and gallic acid, as shown in Figure 5.

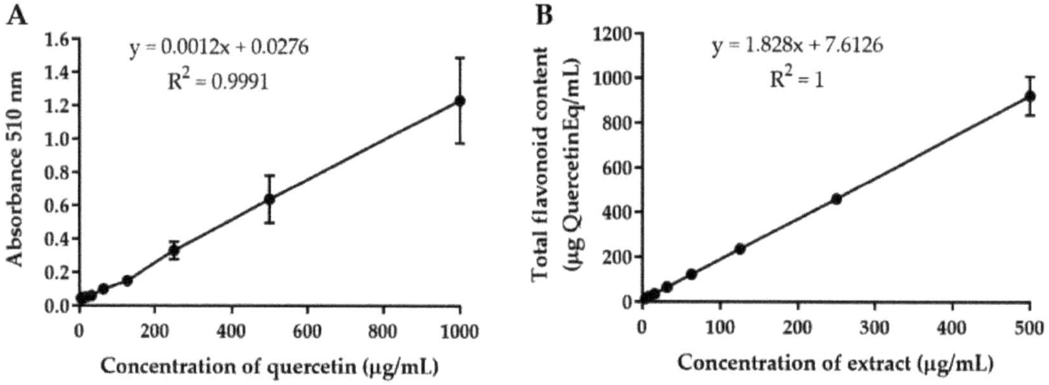

Figure 2. (**A**) Calibration curve of quercetin. (**B**) Total flavonoid content of *B. flabellifer* male flower ethanolic extract.

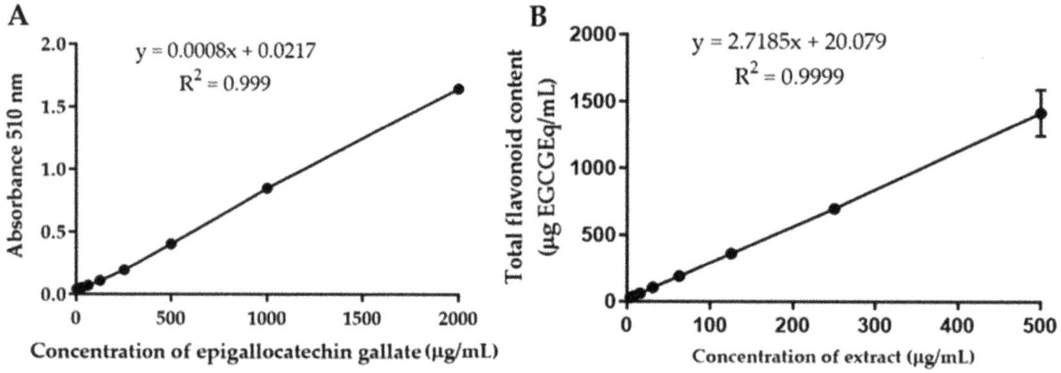

Figure 3. (**A**) Calibration curve of epigallocatechin gallate. (**B**) Total flavonoid content of *B. flabellifer* male flower ethanolic extract.

The main peaks of quercetin, coumarin, and EGCG reference standards were revealed at retention times of 4.173, 4.636, and 3.686 min corresponding to ion transitions of the precursor to the protonated ions [M-H]+ at *m/z* 303.15 → 153.15, 147.05 → 91.15, and 459.10 → 139.10, respectively. The main peak of gallic acid reference standard was shown at a retention time of 1.700 min. The ion transition of the precursor to the product ion was deprotonated ions [M-H] at *m/z* 169.05 → 125.05. The results of the LC–MS/MS analysis of *B. flabellifer* male flower ethanolic extract are presented in Figure 6 and Table 1. The LC–MS/MS analysis confirmed the presence of quercetin, coumarin, and gallic acid in *B. flabellifer* male flower ethanolic extract. In addition, the concentrations of quercetin, coumarin, and gallic acid in *B. flabellifer* male flower ethanolic extract were 0.912, 0.021, and 1.610 μg/mL, respectively.

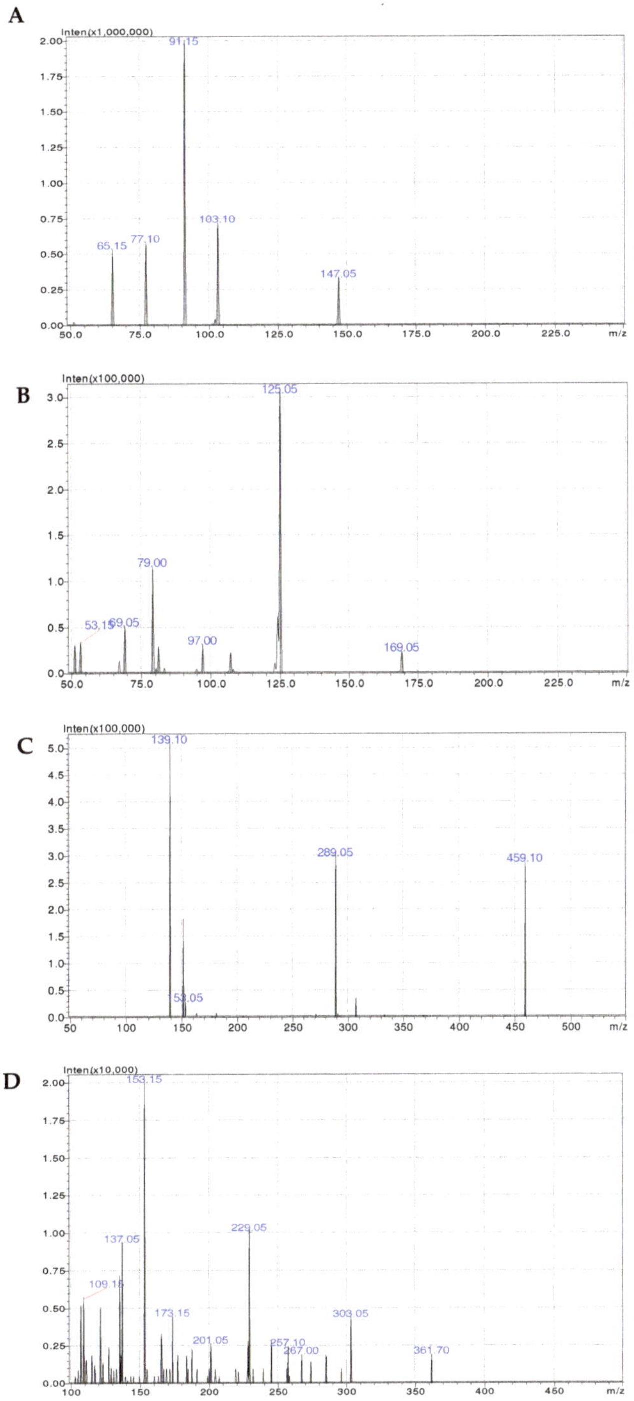

Figure 4. LC–MS/MS analysis presenting the precursor ion and product ion spectra of (**A**) coumarin, (**B**) gallic acid, (**C**) EGCG, and (**D**) quercetin.

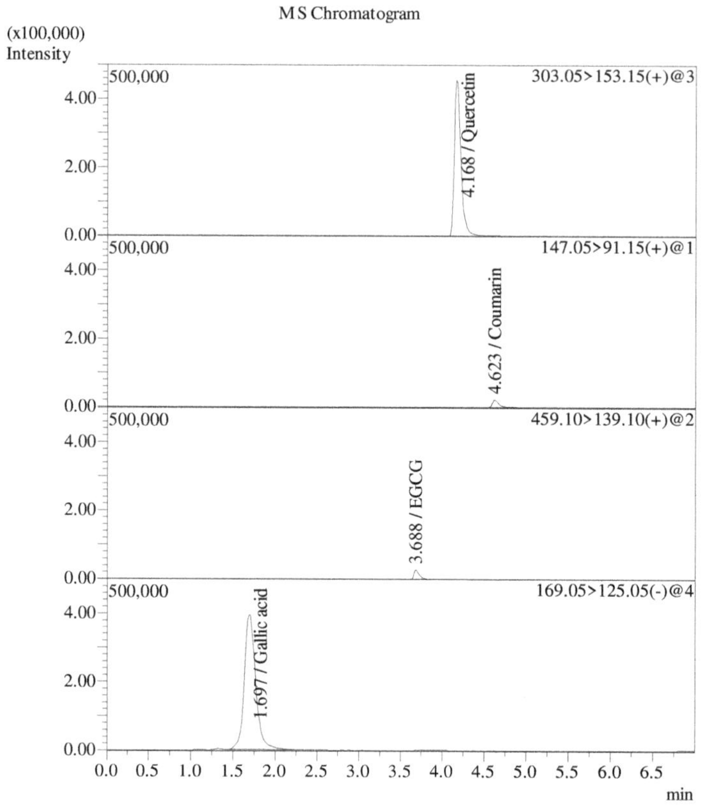

Figure 5. Chromatograms of standard compounds determined by LC–MS/MS.

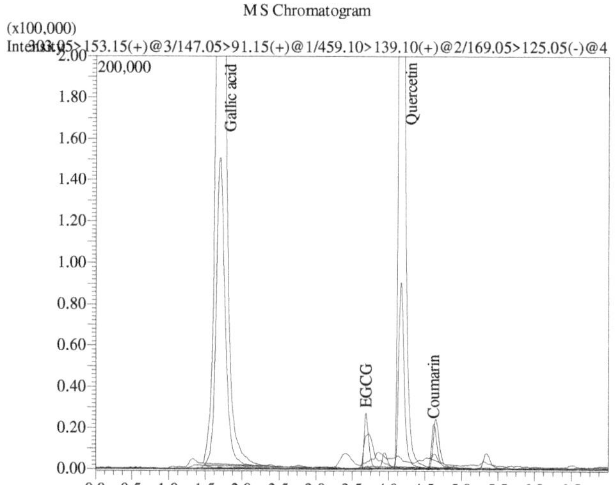

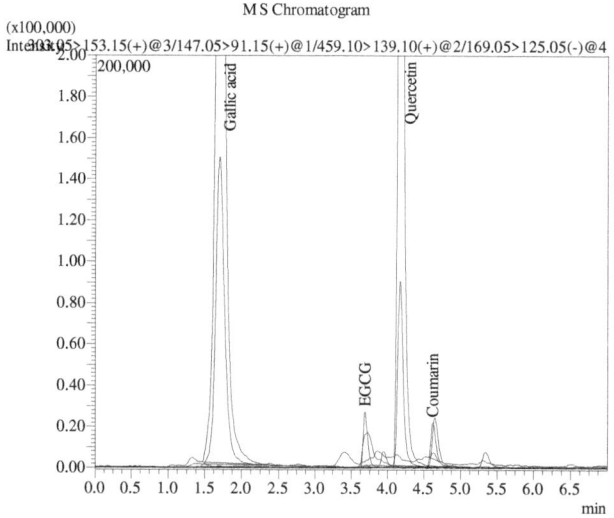

Figure 6. Overlay of LC–MS/MS extracted ion chromatograms (EIC) for *B. flabellifer* male flower ethanolic extract.

Table 1. Identification and quantification of compounds of ethanol extract of *B. flabellifer* male flower by LC–MS/MS.

Compound	Precursor Ions (m/z)	Product Ions		Retention Time (min)	Equation	R^2	Linearity (µg/mL)	Quantification of *B. flabillifer* Male Flower (µg/mL)
		Target Ion	Reference Ion					
Gallic acid	169.05	125.05	79.05	1.69	Y = (762,191)x + (50,587.9)	R^2 = 0.9989	0.250–5.000	1.610
EGCG	459.10	139.10	289.05	3.68	Y = (605,393)x + (−1614.28)	R^2 = 0.9980	0.005–0.200	Not detected
Quercetin	303.05	153.15	229.10	4.16	Y = (508,792)x + (24,227.3)	R^2 = 0.9992	0.250–5.000	0.912
Coumarin	147.05	91.15	103.10	4.62	Y = (593,159)x + (−37.6301)	R^2 = 0.9999	0.005–0.200	0.021

2.4. The 2,2-Diphenyl-1-picrylhydrazyl (DPPH) Free Radical Scavenging Activity of B. flabellifer Male Flower Ethanolic Extract

The *B. flabellifer* male flower extract showed DPPH free radical scavenging activity with the IC_{50} value of 31.54 ± 0.43 µg/mL. The IC_{50} values of gallic acid, quercetin, EGCG, and ascorbyl glucoside were <3.9, 13.77 ± 2.80, 6.25 ± 0.19, and 38.68 ± 2.00 µg/mL, respectively. These results indicated that the potency of antioxidants to scavenge the DPPH radical was in the following order: Gallic acid > EGCG > quercetin > *B. flabellifer* male flower extract > ascorbyl glucoside (Figure 7A). The DPPH scavenging activities of *B. flabellifer* male flower extracted by different types of solvents have been reported. Tusskorn et al. showed that the male flower of *B. flabellifer* extracted by ethyl acetate and chloroform had the IC_{50} values of 287.03 ± 6.90 and >1000 µg/mL, respectively [16]. Kavatagimath et al. reported that the *B. flabellifer* male flower extracted by 95% ethanol and butylated hydroxytoluene had the IC_{50} values of 460 and 419 µg/mL, respectively [18]. The IC_{50} values of *B. flabellifer* male flower ethanolic extract reported in our study showed a significantly lower value indicating that the male flower ethanolic extract had a superior antioxidant activity compared with the other solvents.

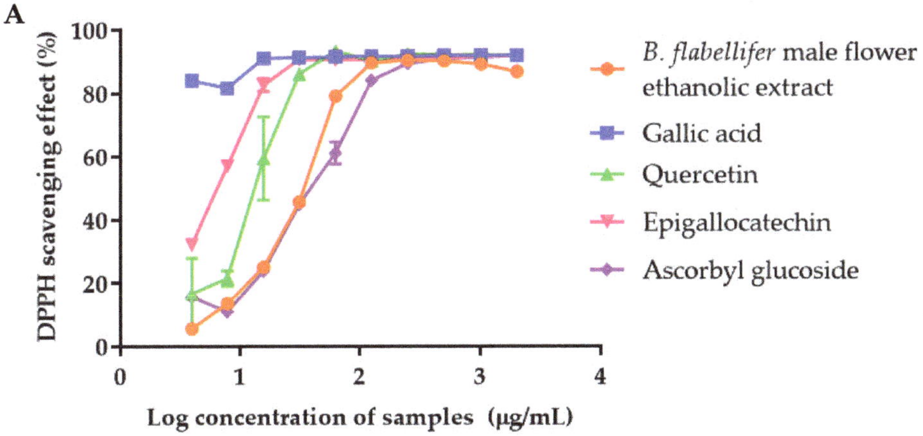

Figure 7. *Cont.*

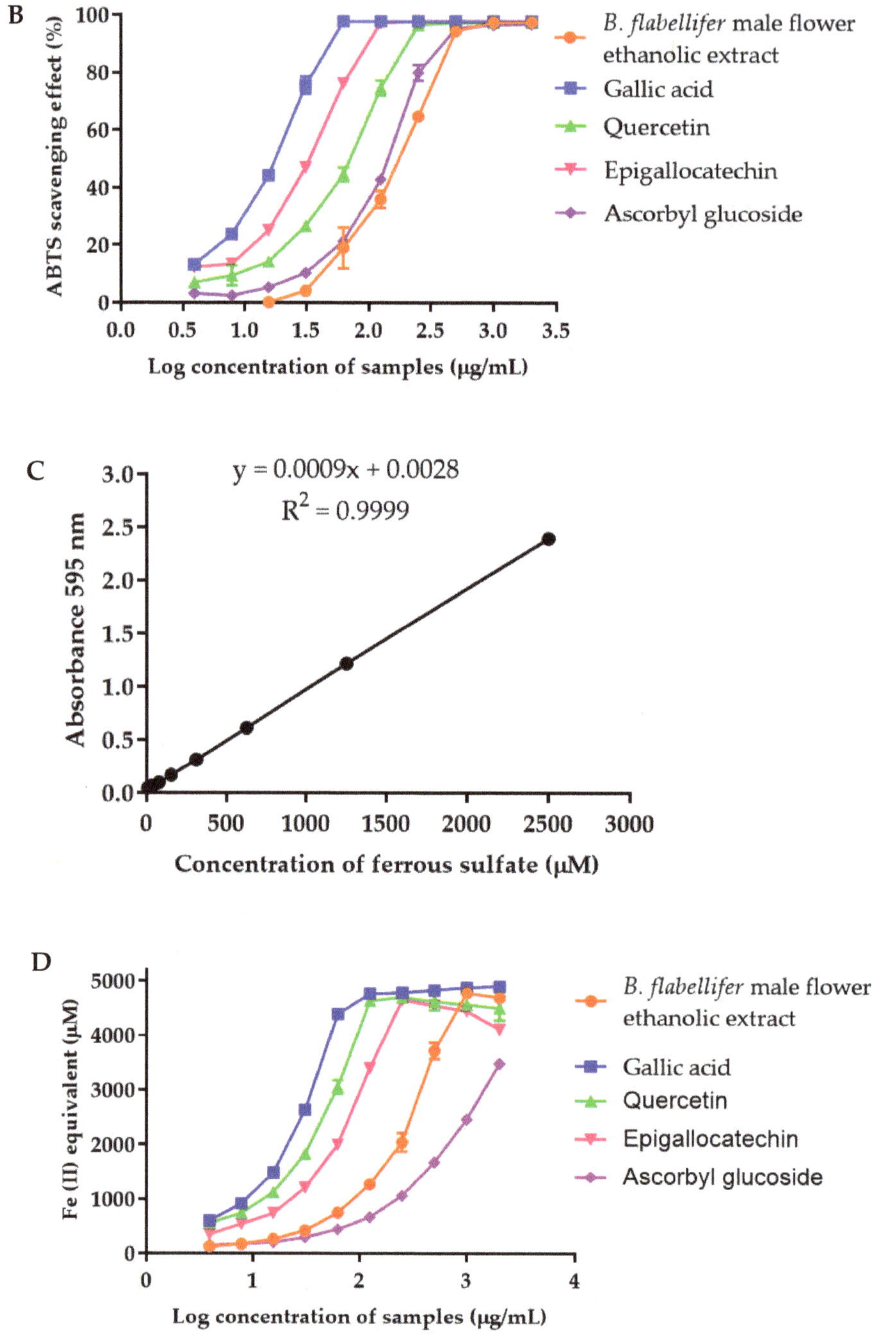

Figure 7. The antioxidant activity of gallic acid, quercetin, epigallocatechin, ascorbyl glucoside, and *B. flabellifer* flower ethanolic extract determined by (**A**) DPPH free radical scavenging assay, (**B**) ABTS free radical scavenging assay, (**C**) standard curve of ferric reducing antioxidant power assay using ferrous sulfate, (**D**) FRAP assay.

The superior antioxidant activity of *B. flabellifer* male flower extract obtained from our study might be due to the lower degradation extent of phytochemical compounds, i.e., polyphenolic compounds (phenolic compounds and flavonoids) in the extract during the extraction process. The polyphenolic compounds can degrade upon drying, extraction, and long-term storage [19]. According to the Kavatagimath et al. report, the extraction temperature might affect the stability of phenolic compounds, which are the active pharmaceutical ingredient for antioxidant activity [18]. The high temperature during continuous hot extraction using Soxhlet apparatus may result in degradation of the phenolic compounds and reduced antioxidant activity observed by the DPPH method [20]. Kotsiou et al. discussed that the main reason for phenolic compound degradation under high temperature was oxidation and hydrolysis reactions [21]. In addition, the age of *B. flabellifer* flower and the storage time of flower extract had been shown to influence the amount and stability of phenolic compounds in the extract, respectively [20,22]. Therefore, using mature (6 month-old) *B. flabellifer* flower, cold maceration, and freshly prepared extract as an active ingredient in the formulation was recommended to achieve the best antioxidant activity.

2.5. The 2,2′-Azino-bis(3-ethylbenzothiazoline-6-sulfonic acid) (ABTS) Free Radical Scavenging Activity of B. flabellifer Male Flower Ethanolic Extract

The *B. flabellifer* male flower ethanolic extract showed the scavenging activity of the $ABTS^+$ radical. The IC_{50} value of *B. flabellifer* male flower extract for ABTS free radical scavenging activity was 164.5 ± 14.3 μg/mL. The IC_{50} values of gallic acid, quercetin, EGCG, and ascorbyl glucoside were 16.11 ± 0.33, 61.92 ± 3.01, 27.21 ± 6.00, and 131.20 ± 3.91 μg/mL, respectively. The results indicated that the potency of antioxidants to scavenge ABTS radical was in the following order: Gallic acid > EGCG > quercetin > ascorbyl glucoside > *B. flabellifer* male flower extract (Figure 7B). The IC_{50} values of *B. flabellifer* male flower methanolic and ethyl acetate extract, determined by the ABTS free radical scavenging assay, were 10.82 ± 0.78 and 462.92 ± 25.70 μg/mL, respectively [16]. These results suggested that the *B. flabellifer* male flower extracted in this study yielded higher antioxidant activity compared with the previous report.

Do et al. reported the effect of plant extraction solvent on total phenolic content, total flavonoids content, and antioxidant activity [23]. They concluded that the best extracting solvent to yield the highest total phenolic compound and total flavonoids content in the extract was ethanol. The different solvents used in extraction resulted in the differences in compositions and antioxidant activities of the extracts. Ethanol (95%) is suitable for extracting bioactive compounds with a broader range of polarity, when compared with methanol (more polarity) and ethyl acetate (less polarity). Since the Pearson correlation showed that both phenolic and flavonoid contents were directly proportional to their antioxidant activity, the decrease in polyphenolic compounds resulted in the decrease in antioxidant activity. Therefore, these results imply that 95% of ethanol may be the most appropriate solvent to extract polyphenolic compounds from *B. flabellifer* male flower.

2.6. Ferric Reducing Antioxidant Potential (FRAP) of B. flabellifer Male Flower Ethanolic Extract

The FRAP assay was performed to determine the reducing capacity of *B. flabellifer* male flower extract in a redox reaction. The results revealed good linearity of ferrous sulfate obtained within the range of 9.8–2500 μM ($r^2 = 0.9999$) (Figure 7C). The results of FRAP assay were expressed as Fe^{2+} equivalent. Gallic acid, quercetin, EGCG, ascorbyl glucoside, and *B. flabellifer* male flower extract at 1 mg/mL exhibited 4877.13 ± 34.43, 4566.46 ± 59.81, 4442.79 ± 42.72, 2458.42 ± 41.93, and 4778.25 ± 56.20 μM Fe^{2+} equivalent, respectively (Figure 7D). The results indicated that the potency of antioxidants to reduce ferric ion was in the following order: Gallic acid > *B. flabellifer* male flower extract > quercetin > EGCG > ascorbyl glucoside. These results indicated that compared with the other mechanisms, the antioxidant activity of *B. flabellifer* male flower extract was mainly based on the reducing power of the compounds in the extract, which reduced ferric ion (Fe^{3+}) to the ferrous ion (Fe^{2+}).

2.7. Pearson Correlation of Total Phenolic and Flavonoid Contents with Antioxidant Activities of B. flabellifer Male Flower Ethanolic Extract

Pearson correlation coefficients for total phenolic and flavonoid contents with *B. flabellifer* male flower ethanolic extract antioxidant activities were shown in Table 2. Total phenolic and flavonoid contents positively and significantly correlated with the antioxidant activities measured by ABTS and FRAP assays. In addition, the results showed a correlation between the total phenolic content and antioxidant activity determined by the DPPH scavenging assay. There was no correlation between the total flavonoid content and DPPH free radical scavenging activity for *B. flabellifer* male flower ethanolic extract. These results indicated that the total phenolic and flavonoid contents contributed to the antioxidant activities of *B. flabellifer* male flower ethanolic extract.

Table 2. Pearson correlation coefficients of total phenolic and flavonoid contents and antioxidant activities of *B. flabellifer* male flower ethanolic extract measured by DPPH, ABTS, and FRAP assays.

	Total Phenolic Content (Gallic Acid Equivalent)	Total Flavonoid Content (Quercetin Equivalent)	Total Flavonoid Content (EGCG Equivalent)
DPPH assay	0.8256 *	0.7001	0.6962
ABTS assay	0.9964 ***	0.9932 ***	0.9932 ***
FRAP assay	0.9972 ****	0.9970 ****	0.9965 ****

* Indicated $p < 0.05$, *** indicated $p < 0.001$, and **** indicated $p < 0.0001$.

The polyphenols and flavonoids present in the extract from the male flower of *B. flabellifer* exert several antioxidant properties through several mechanisms, including free radical scavenging, reduction potential, and metal chelation. The antioxidant activity of polyphenols was associated with the capability to inactivate reactive radical species due to the neutralization of free radicals. Neutralization occurs when polyphenols transfer their electrons and/or hydrogen atoms to radicals. The reduction potential of polyphenols resulted from the reduction of Fe^{3+} to Fe^{2+}. The reduction potential of polyphenols depends on the presence of double bonds in the rings and the number of OH groups. In addition, chelation of Fe^{3+} may be another possible antioxidant mechanism of polyphenolic compounds containing two or more groups of OH groups [24].

The antioxidant capacity of a plant crude extract could be a combined effect of phenolic, flavonoids, and other reducing compounds, including gallic acid, coumarin, and quercetin. In the present study, the antioxidant activity was measured by DPPH, ABTS radical scavenging assay, and FRAP assay. The extract from the male flower of *B. flabellifer* showed a significantly higher radical scavenging activity and ferric chloride reducing power compared with the previous reports. In addition, the radical scavenging activity of the extract was in accordance with the phenolic and flavonoid contents in the extract. The higher value of total phenolic and flavonoid contents might be responsible for the potent antioxidant activity of this extract. This might be due to the drying process and extraction method of the plant that preserve the bioactive compounds.

The selective HPLC–DPPH post-column methodology might be another interesting option to investigate the antioxidant activity of the bioactive compounds containing *B. flabellifer* L. male flower extract [25].

2.8. Anti-Cutibacterium Acnes Activity of B. flabellifer Male Flower Ethanolic Extract

The minimum inhibitory concentration (MIC) of *B. flabellifer* male flower ethanolic extract against *C. acnes* was 250 mg/mL. In comparison, the standard antibacterial drug, i.e., clindamycin, had the minimum inhibitory concentration (MIC) of *C. acnes* as 0.781 µg/mL. The results suggested the use of *B. flabellifer* male flower ethanolic extract to treat acne vulgaris caused by *C. acnes* infection. Wongmanit et al. reported the antibacterial activity of *B. flabellifer* male flower ethanolic extract against *Enterococcus faecalis* and *Staphylococcus aureus*, but did not report the activity against *Escherichia coli* [22]. Alamelumangai et al. reported that the methanolic and ethanolic seed coat of *B. flabellifer* extract at 50 mg/mL could inhibit the growth of *Aspergillus brasiliensis* and *Bacillus subtilis* [26]. Reshma et al. isolated 2,3,4-trihydroxy-5-methylacetophenone from palm juice extract, which showed a broad-spectrum antibacterial activity against *Escherichia coli, Mycobacterium smegmatis, Staphylococcus aureus,* and *Staphylococcus simulans* [27].

Some phenolic compounds in *B. flabellifer* L. male flower extract, including gallic acid, quercetin, and coumarin have shown antibacterial activity against *C. acnes* [28]. Gallic acid is a hydroxybenzoic acid that inhibited bacterial growth by inducing irreversible changes in membrane properties of both Gram-positive and Gram-negative bacteria [29]. Gallic acid has shown a strong anti-*C. acnes* activity by the inhibition of lipase enzyme [30]. Coumarin and quercetin showed a significant antibacterial effect against *C. acnes* [31]. The mechanism of antibacterial activity of coumarin was inhibiting the bacterial fatty acid synthesis pathway [32]. The antibacterial mechanism of quercetin included destroying bacterial cell wall, changing cell permeability, affecting protein synthesis and expression, reducing enzyme activities, and inhibiting nucleic acid synthesis [33].

2.9. Effect of B. flabellifer Male Flower Ethanolic Extract on Inhibition of Protein Denaturation

The anti-inflammatory activity of *B. flabellifer* male flower ethanolic extract was evaluated against denaturation of egg albumin. The *B. flabellifer* male flower ethanolic extract at 250, 500, and 1000 µg/mL inhibited albumin denaturation of 47.55 $\pm$ 1.30, 47.79 $\pm$ 3.53, and 27.40 $\pm$ 5.08%, respectively. Diclofenac diethylammonium, a positive control, inhibited albumin denaturation to a greater extent than the extract with the IC_{50} of 0.26 $\pm$ 0.02 mg/mL. Denaturation of protein results in autoantigen generation, leading to inflammation. Inhibition of protein denaturation might inhibit an inflammatory activity. The results suggested that the *B. flabellifer* male flower ethanolic extract at 500 µg/mL was the optimal concentration for anti-inflammatory activity. The inhibition of *B. flabellifer* male flower ethanolic extract and diclofenac diethylammonium at higher than 500 and 2000 µg/mL, respectively, showed a decreasing inhibitory activity. These results agreed with the previous report showing a decreasing inhibition rate of ibuprofen [34].

Protein denaturation may cause auto-antigens, leading to inflammatory responses. The substances that can protect protein from denaturation may be used for developing an anti-inflammatory agent. The inhibition of albumin denaturation assay has been widely used for determining an in vitro anti-inflammatory assay [35–38]. *B. flabellifer* male flower ethanolic extract at 250 and 500 µg/mL was shown to inhibit albumin protein denaturation. The inhibition decreased with the increasing concentration of *B. flabellifer* male flower ethanolic extract. The mechanism of protein denaturation inhibition of the flower extract might be due to the interaction of polyphenolic compounds in the extract and albumin protein, resulting in improved thermal stability of proteins. Binding of gallic acid with protein increased protein intramolecular packing and induced higher thermal stability [39]. Coumarin and its derivatives exhibited an inhibition of heat-induced protein denaturation at specific concentrations [40,41]. The anionic form of quercetin was reported to bind a folded protein and increased the thermal stability of the planar conformation [42]. The interaction of polyphenols and protein may result in secondary structural changes in protein due to the reduction of surface hydrophobicity. Therefore, increasing the extract concentration containing an higher amount of polyphenols might result in a negative effect on the structural and thermal properties of protein [43].

2.10. Evaluation of Physical Characteristics and Stability of Gel Containing B. flabellifer Male Flower Ethanolic Extract

The freshly prepared gel containing *B. flabellifer* male flower ethanolic extract was characterized as a light brown gel with a smooth texture and the distinctive odor of

B. flabellifer male flower ethanolic extract. The gel had a darker color when the concentration of the flower extract increased (Table 3). The gel containing 25, 31.25, and 62.5% *w/w* of *B. flabellifer* male flower ethanolic extract had pH values of 6.05 ± 0.01, 6.31 ± 0.02, and 6.59 ± 0.02, respectively.

Table 3. Appearance of *B. flabellifer* male flower ethanolic extract and gel containing 25, 31.25, and 62.5 %*w/w* *B. flabellifer* male flower extract.

Concentration of *B. flabellifer* Male Flower Ethanolic Extract	**25% *w/w***	**31.25% *w/w***	**62.5% *w/w***
Extract (dissolved in 50% ethanol)			
Gel containing extract			

A plot of viscosity versus shear rate for the gel base and containing *B. flabellifer* male flower ethanolic extract is shown in Figure 8. The results clearly indicated that the apparent viscosity of the gel base and gel containing *B. flabellifer* male flower ethanolic extract decreased significantly with the increasing shear rate, i.e., from around 100,000 cP at a shear rate of 0.9 s^{-1} to below 1000 cP at a shear rate of 200 s^{-1}, indicating that the gel exhibited a pseudoplastic character. The viscosity values of gel containing *B. flabellifer* male flower ethanolic extract determined by a viscometer were 366.45 ± 20.01, 341.66 ± 28.91, and 326.2 ± 22.64 kcps, respectively. The viscosity of gel containing *B. flabellifer* male flower ethanolic extract, measured using a rheometer and a viscometer, decreased with the increasing extract concentration. This result might be due to the mild acidity of *B. flabellifer* male flower ethanolic extract (pH = 5.08 ± 0.3), which might reduce the neutralization capacity of triethanoloamine to carbopol.

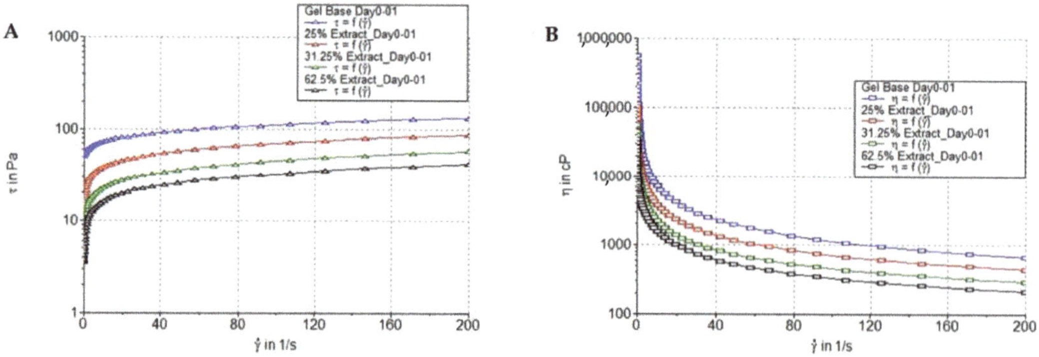

Figure 8. Rheology measurement. (**A**) Flow curves of gel base, gel containing 25, 31.25, and 62.5% *w/w* extract expressed as shear rate and applied shear stress. (**B**) Flow curves of gel base, gel containing 25, 31.25, and 62.5% *w/w* extract expressed as viscosity and shear rate.

The physical stability of all gel formulations containing *B. flabellifer* male flower ethanolic extract was confirmed by the maintained appearance, odor, color, and other physical characteristics. Gel separation and non-homogeneity were not observed in all formulations. After accelerated stability tests at all six cycles and 1-month storage at 4, 30, and 45 °C, the appearance, color, odor, and texture of the gel containing flower extract were not changed. The viscosity of the gel containing 31.25 and 62.5% *w/w* extract was not significantly changed, except the gel containing the lowest concentration of the extract (25% *w/w*) (Figure 9A). The viscosity of the gel formulation containing 25% *w/w* extract decreased significantly after the 6th heating–cooling cycle and after 1-month storage at 45 °C (Figure 9B). The pH values of all the formulations after they were freshly prepared and after the heating–cooling cycles and 1-month stability tests were not significantly different (Figure 10A,B).

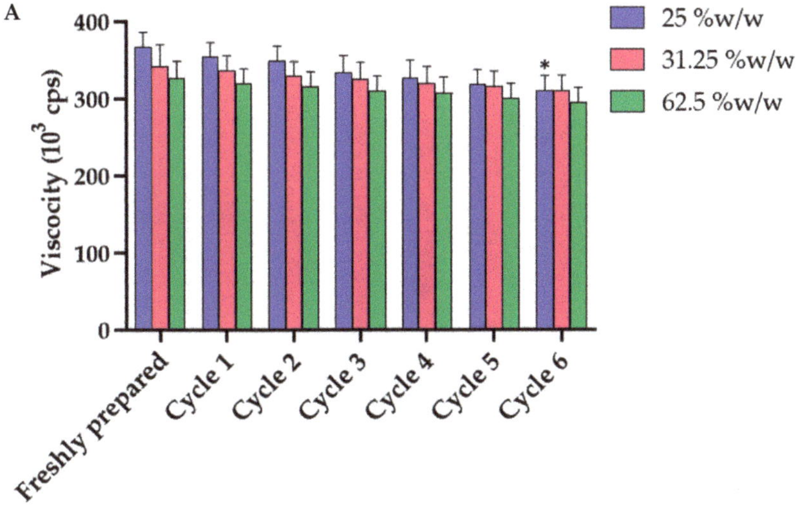

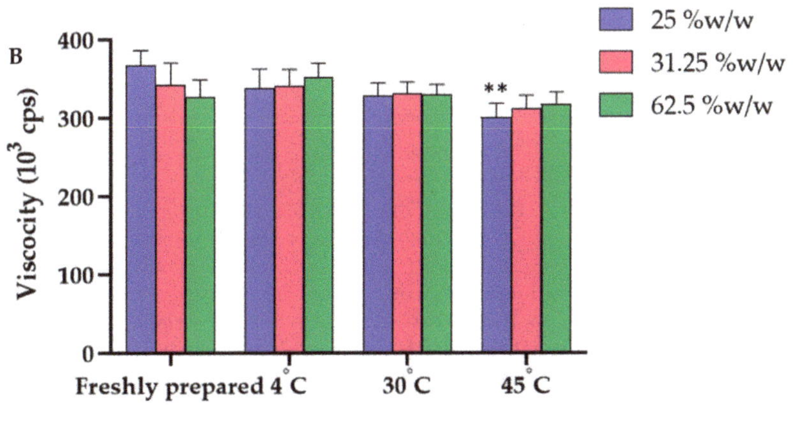

Figure 9. (**A**) Effects of six heating–cooling cycles on the viscosity of gel containing 25, 31.25 and 62.5% *w/w* of *B. flabellifer* male flower ethanolic extract. (**B**) Effect of temperature on the viscosity of gel containing 25, 31.25 and 62.5% *w/w* of *B. flabellifer* male flower ethanolic extract after storage at 4, 30, and 45 °C for 1 month. * Indicated $p < 0.05$ and ** indicated $p < 0.01$.

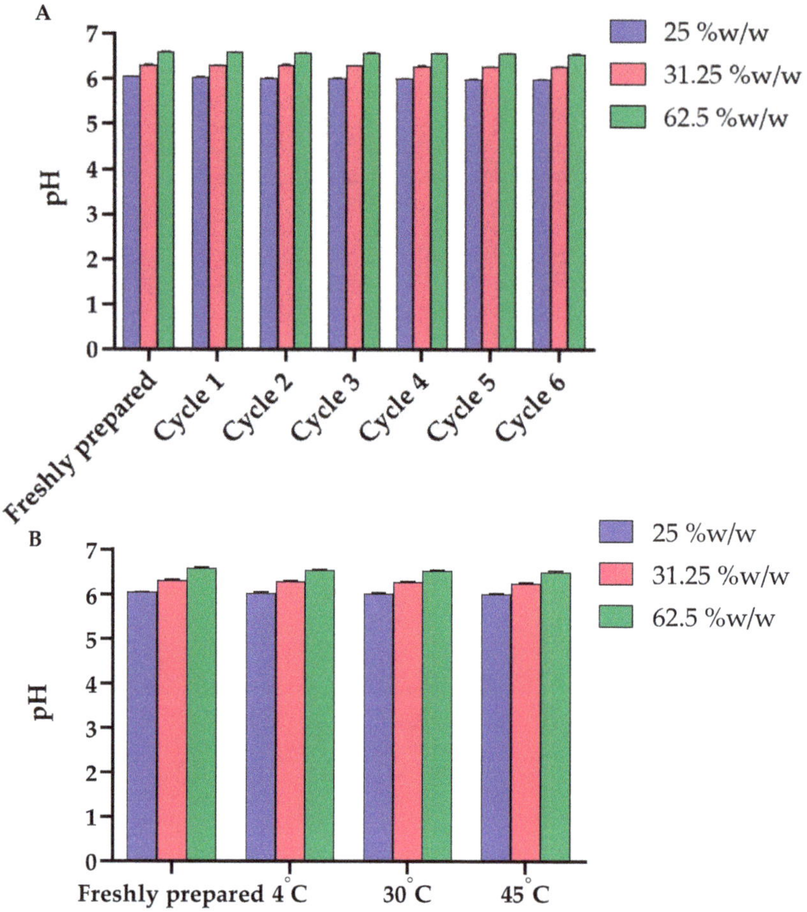

Figure 10. (**A**) Effects of six heating–cooling cycles on the pH of gel containing 25, 31.25 and 62.5% *w/w* of *B. flabellifer* male flower ethanolic extract. (**B**) Effect of temperature on the pH of gel containing 25, 31.25 and 62.5% *w/w* of *B. flabellifer* male flower ethanolic extract after storage at 4, 30 and 45 °C for 1 month.

3. Materials and Methods

3.1. Materials

Gallic acid, Griess reagent, DPPH (2,2-diphenyl-1-picrylhydrazyl), TPTZ (2,4,6-Tris(2-pyridyl)-s-triazine, ABTS (2,2′-azino-bis(3-ethylbenzothiazoline-6-sulfonic acid)), and coumarin were purchased from Sigma-Aldrich, St. Louis, USA. Absolute ethanol, dimethyl sulfoxide, sodium bicarbonate, sodium nitrate, and sodium hydroxide were purchased from RCI Lab-scan, Bangkok, Thailand. The Folin–Ciocalteu phenol reagent, aluminum chloride, sodium acetate trihydrate, ferrous sulfate heptahydrate (99% purity), and potassium persulfate were obtained from Loba Chemie, Mumbai, India. Quercetin (98% purity), epigallocatechin (EGCG) (98% purity), ascorbyl glucoside, phenoxyethanol, and triethanolamine were purchased from Chanjao Longevity Co., Ltd., Bangkok, Thailand. Clindamycin and diclofenac diethylammonium were purchased from RPC International Co., Ltd., Bangkok, Thailand. Carbopol Ultrez-21 polymer, propylene glycol, and glycerin were purchased from Namsiang, Bangkok, Thailand. Iron (III) chloride hexahydrate and 37% hydrochloric acid were purchased from Qrec, New Zealand.

3.2. Plant Material

The male flower of *B. flabellifer* was collected from Mueang District, Phetchaburi Province, Thailand. The 6-month-old *B. flabellifer* male flowers were harvested from a 5–20 year-old *B. flabellifer* tree at the end of April 2021 at 10:00–12:00 pm. The pictures of fresh and dried cut *B. flabellifer* male flowers were included in Figure 11. The sample has been identified by a botanist of the Forest Herbarium, Royal Forest Department, Bangkok, Thailand.

Figure 11. (**A**) Fresh male flower of *B. flabellifer.* (**B**) *B. flabellifer* male flowers dried extract.

3.3. Preparation of Extraction of B. flabellifer Male Flower Ethanolic Extract

The male flowers of *B. flabellifer* were washed, cut into pieces, and dried at room temperature for 20 days. *B. flabellifer* male flowers were dried in a room with good ventilation to protect from direct exposure to light and heat from sunlight. The flowers were flipped several times during the day. The dried flowers were powdered and extracted by maceration of 300 g of powdered plant with 1700 mL of 90% ethanol at room temperature for 72 h. Then, the extract was filtered using filter paper and concentrated by evaporating the solvent using a rotary vacuum evaporator. The dried crude extract was weighed, and the yield was calculated by Equation (1).

$$Yield\ (\%) = \frac{Weight\ of\ dried\ crude\ extract\ (g)}{Weight\ of\ dried\ plant} \times 100\% \quad (1)$$

3.4. Quantitative Analysis of Phenolic Compounds in the Ethanolic Extract of Male Flower of B. flabellifer by Folin–Ciocalteu Method

Total phenolic content was determined by the Folin–Ciocalteu reaction. Gallic acid solution at concentrations ranging from 7.8–125 μg/mL and *B. flabellifer* male flower extract solution at concentrations ranging from 7.8–250 μg/mL were added into 96-well plates (50 μL/well). Then, 10% *v/v* of Folin–Ciocalteu phenol reagent (100 μL) was added and mixed for 1 min. After 4 min of incubation, 7.5% *w/v* Na_2CO_3 solution (50 μL) was subsequently added to the mixture and incubated at room temperature in the dark for 2 h. Then, the absorbance was measured using a UV–visible spectrophotometer (Spectramax M3, Thermo Scientific, Waltham, MA, USA) at a wavelength of 765 nm. Total phenolic contents were calculated from the gallic acid standard curve. Data were expressed as μg gallic acid equivalents per mL of crude extract and μg/mg gallic acid equivalents of dry crude extract.

3.5. Quantitative Analysis of Total Flavonoid Content in B. flabellifer Male Flower Ethanolic Extract by the Aluminum Chloride Colorimetric Method

The total flavonoid contents in *B. flabellifer* male flower extract were determined by the aluminum chloride colorimetric method. Quercetin solution (3.9–1000 μg/mL), epigallocatechin solution (3.9–2000 μg/mL), and extract solution at concentrations of 3.9–500 μg/mL (100 μL/well) were added into 96-well plates. Then, 5% $NaNO_2$ (30 μL) was added to the well and incubated for 5 min. Aluminum chloride (2% *w/v*, 50 μL) was added and incubated for 6 min followed by 10 min of incubation with 1 N NaOH (50 μL). The absorbance was measured at a wavelength of 510 nm with a UV–Vis spectrophotometer (Spectramax M3, Thermo Scientific, Waltham, MA, USA). Total flavonoid contents were calculated from quercetin and EGCG standard curves, and data were expressed as μg quercetin and EGCG equivalents per mL of crude extract and μg/mg quercetin and EGCG equivalents of dry crude extract.

3.6. Phytochemical Screening and Quantitative Determination of Bioactive Compounds in B. flabellifer Male Flower Ethanolic Extract by LC–MS/MS

The phytochemical screening of B. flabellifer male flower ethanolic extract was conducted using an LC–MS/MS instrument (Shimadzu Corporation, Kyoto, Japan) with ESI interface. The LC–MS/MS comprises a Column Oven: CTO-20A, Autosampler: SIL-20ACXR, Pump: LC-20ADXR, Degasser unit: DGU-20A3R, and Valve unit: FCV-20AH2. MS detection was performed using the LCMS 8040 model triple quadrupole mass spectrometer equipped with an ESI source operating in both positive and negative ionization modes and the data were processed by Lab solution software, version 5.82 SP1. The operation parameters were as follows: Heat block temperature of 400 °C, DL temperature of 250 °C, and autosampler temperature of 15 °C.

HPLC separation was performed on ACE Excel 5 Super C18 column (150 × 2.1 mm). Mobile phase consisted of two solvents: Solvent *(A)* deionized water with 0.1% formic acid and solvent (B) acetonitrile. Gradient elution was performed at a flow rate of 0.4 mL/min at 40 °C. A 5 μL injection volume of the sample was injected on to the column, separated, and eluted using the following gradients: 0.01 min (10% B); 2.00 min (50% B), and allowed for a column stabilization of 2 min with initial conditions.

3.7. Determination of DPPH Free Radical Scavenging Activity of B. flabellifer Male Flower Ethanolic Extract

The DPPH free radical scavenging capacity of the *B. flabellifer* male flower extract was determined and compared with gallic acid, quercetin, EGCG, and ascorbyl glucoside [44,45]. Gallic acid, EGCG, ascorbyl glucoside, and *B. flabellifer* male flower extract were diluted in deionized water, and quercetin was diluted in 95% ethanol at concentrations of 3.9–2000 μg/mL. DPPH was dissolved in absolute ethanol at a concentration of 500 μM. Gallic acid, quercetin, EGCG, ascorbyl glucoside, and *B. flabellifer* male flower extract (100 μL) were mixed with DPPH solution (100 μL). The mixture was incubated at room temperature in the dark for 30 min. The absorbance was measured using a UV–Vis spectrophotometer microplate reader at a maximum wavelength of 517 nm. The percentage of radical scavenging activity was calculated by Equation (2). The 50% of scavenging (IC_{50}) was calculated from the non-linear regression analysis of the graph plotted between the percentages of DPPH free radical scavenging and the sample concentrations.

$$DPPH\ Free\ radical\ scavenging\ (\%) = \frac{A - B}{A} \times 100\% \quad (2)$$

where A is the absorbance of the reaction with solvent control and B is the absorbance of the reaction with the extract.

3.8. Determination of ABTS Free Radical Scavenging Activity of B. flabellifer Male Flower Ethanolic Extract

Gallic acid, EGCG, ascorbyl glucoside, and *B. flabellifer* male flower extract were diluted in deionized water, and quercetin was diluted in 95% ethanol at concentrations of 3.9–2000 μg/mL. The ABTS scavenging assay was modified from a previous report [46]. ABTS was dissolved in absolute ethanol to a concentration of 7 mM. ABTS radical cation ($ABTS^+$) was produced by the reacting ABTS stock. The solution with 2.45 mM potassium persulfate was dissolved in and remained in the dark at room temperature for 24 h prior to use. The $ABTS^+$ solution was diluted with absolute ethanol to an absorbance of 0.700 (±0.02) at 734 nm. Gallic acid, quercetin, EGCG, ascorbyl glucoside, and *B. flabellifer* male flower extract (20 µL) were mixed with the ABTS solution (180 µL). The mixture was incubated at room temperature in the dark for 15 min. The absorbance was measured using a UV–Vis spectrophotometer microplate reader at a wavelength of 734 nm. The percentage of radical scavenging activity was calculated by Equation (3). The 50% of scavenging (IC_{50}) was calculated from the non-linear regression analysis of the graph plotted between the percentages of $ABTS^+$ free radical scavenging and the sample concentrations.

$$ABTS\ Free\ radical\ scavenging\ (\%) = \frac{A - B}{A} \times 100\% \quad (3)$$

where A is the absorbance of the reaction with solvent control (deionized water) and B is the absorbance of the reaction with the extract.

3.9. Determination of Ferric Reducing Antioxidant Power of *B. flabellifer* Male Flower Ethanolic Extract

Gallic acid, EGCG, ascorbyl glucoside, and *B. flabellifer* male flower extract were diluted in deionized water, and quercetin was diluted in 95% ethanol at concentrations of 3.9–2000 µg/mL. The FRAP reagent was prepared from an acetate buffer (300 mM, pH 3.6), a solution of 10 mM TPTZ in 40 mM HCl, and 20 mM $FeCl_3$ at 10:1:1 (*v/v*). The FRAP reagent (180 µL) and sample solutions (20 µL) were added to each well and mixed thoroughly. The mixture was incubated at 37 °C for 30 min. Then, the absorbance was read at 595 nm. The standard curve was constructed using a ferrous sulfate solution (9.8–5000 µM), and the results were expressed as µmol Fe (II) equivalent.

3.10. Antibacterial Activity of *B. flabellifer* Male Flower Ethanolic Extract against *C. acnes*

The study of the minimum inhibitory concentration (MIC) of *C. acnes* bacteria was performed by the broth microdilution method [47]. Samples were prepared by dissolving the extract in 95% ethanol at a concentration range of 0.244–500 mg/mL. The culture medium was added into 96-well plates, with 100 µL in each well. The prepared sample (100 µL) was added to the well and mixed with the culture. Clindamycin was used as a positive control at a concentration range of 0.024–50 µg/mL. *C. acnes* with turbidity 0.5 McFarland standard (100 µL) was added to the well. The culture was incubated at 37 ± 1 °C for 48 h in anaerobic conditions. The concentration of clindamycin and the extract that inhibited *C. acnes* growth completely (the first clear well) were taken as the MIC value.

3.11. Anti-Inflammatory Activity of *B. flabellifer* Male Flower Ethanolic Extract by the Albumin Denaturation Method

B. flabellifer male flower ethanolic extract was diluted with deionized water to obtain the concentrations of 0.25, 0.5, 1, 2, and 4 mg/mL. The extract solution (2 mL) was mixed with 0.2 mL of egg albumin (from fresh hen's egg) and 2.8 mL of phosphate-buffered saline (PBS, pH 6.4). The reaction mixture was incubated at 37 ± 2 °C for 15 min and then heated for 5 min at 70 ± 2 °C. After cooling, the absorbance was measured at 660 nm using an UV–Vis spectrophotometer. Diclofenac sodium at the final concentration of (0.25, 0.5, 1, 2, and 4 mg/mL) was used as a positive control and treated similarly to determine absorbance. The percentage inhibition of protein denaturation was calculated using the following equation.

$$Inhibition\ (\%) = \frac{V_1}{V_0 - V_1} \times 100\% \quad (4)$$

where V_1 is the absorbance of a test sample and V_0 is the absorbance of control.

3.12. Formulation of Gel Containing *B. flabellifer* Male Flower Ethanolic Extract

The gel containing *B. flabellifer* male flower ethanolic extract was prepared by dispersing Carbopol Ultrez-21® (0.7% *w/w*) in an aqueous medium containing propylene glycol (1.5% *w/w*) and glycerin (2.5% *w/w*). Then, the flower extract (25, 31.25, and 62.5% *w/w*) and phenoxyethanol (1% *w/w*) were added into the dispersion. Triethanolamine (0.8% *w/w*) was added dropwise until the gel was formed. The concentration of *B. flabellifer* male flower ethanolic extract used in this formulation was 1-, 1.25-, and 2.5-fold of MIC value against *C. acnes*. The extract was diluted with deionized water and filtered through a 0.45 µm syringe filter prior to use.

3.13. Characterization of Gel Containing *B. flabellifer* Male Flower Ethanolic Extract

The gel formulations containing *B. flabellifer* male flower ethanolic extract were inspected visually for appearance, color, odor, homogeneity, and consistency. The gel's viscosity and pH were determined using a viscometer (Yanhe, China) and pH meter (Eutech Instruments, Singapore), respectively.

3.14. Rheological Measurement

The measurement of gel containing *B. flabellifer* male flower ethanolic extract rheology was performed using a Thermo Scientific HAAKE RheoStress 1 rheometer equipped with a plate and plate geometry (1.0 mm gap, 60 mm diameter). The temperature of the gel was set at 25 °C. The range of shear rate was from 0.01 to 200 s^{-1} with the frequency of 1 Hz.

3.15. Accelerated Stability of Gel Containing *B. flabellifer* Male Flower Ethanolic Extract

The accelerated stability study of gel formulations containing the flower extract was performed by heating–cooling cycles. The test was performed in 12 days with six cycles. In each cycle, the freshly prepared gel containing extract was kept at 4 °C for 24 h and 45 °C for another 24 h [48]. The pH and viscosity of the gel after each cycle were measured every cycle. In addition, the stability of all gel formulations was determined by maintaining the products at 4, 30, and 45 °C for 1 month. Then, the pH and viscosity of gels were measured.

3.16. Statistical Analysis

Data were presented as mean ± standard deviation (SD). The one-way analysis of variance (ANOVA) and Tukey post hoc test were performed for statistical analysis. To compare the significance of the difference between the means of two groups, a *t*-test was performed. A *p*-value < 0.05 was considered statistically significant.

4. Conclusions

The *B. flabellifer* L. male mature flower extract contained high concentrations of total phenolic and flavonoid contents. The bioactive compounds found in the extract were gallic acid, coumarin, and quercetin. The phytoconstituent of *B. flabellifer* L. male flower extract was shown to be correlated with the antioxidant activity of the extract. The *B. flabellifer* L. male flower extract had antibacterial activity against *C. acnes* with the MIC value of 250 mg/mL. At 250 and 500 µg/mL, the extract demonstrated the anti-inflammatory activity by preventing albumin denaturation. The anti-acne gel containing the *B. flabellifer* male flower ethanolic extract was initially and successfully developed with acceptable physicochemical properties and stability.

Author Contributions: Conceptualization, P.T. (Prakairat Tunit) and C.C.; methodology, P.T. (Prakairat Tunit), C.C. and P.T. (Phanit Thammarat); formal analysis, P.T. (Prakairat Tunit), C.C. and P.T. (Phanit Thammarat); investigation, P.T. (Prakairat Tunit), C.C. and P.T. (Phanit Thammarat); resources, S.O.; data curation, C.C.; writing—original draft preparation, P.T. (Prakairat Tunit), C.C. and P.T. (Phanit Thammarat); writing—review and editing, P.T. (Prakairat Tunit), C.C., P.T. (Phanit Thammarat) and S.O.; funding acquisition, P.T. (Prakairat Tunit) and S.O. All authors have read and agreed to the published version of the manuscript.

Funding: This work was supported by Phetchaburi Rajabhat University (Contract No. 15/2563). This work was partially supported by Chiang Mai University.

Institutional Review Board Statement: Not applicable.

Informed Consent Statement: Not applicable.

Conflicts of Interest: The authors declare no conflict of interest.

References

1. Mills, O.H.; Criscito, M.C.; Schlesinger, T.E.; Verdicchio, R.; Szoke, E. Addressing Free Radical Oxidation in Acne Vulgaris. *J. Clin. Aesthet. Derm.* **2016**, *9*, 25–30.
2. Alexeyev, O.A.; Dekio, I.; Layton, A.M.; Li, H.; Hughes, H.; Morris, T.; Zouboulis, C.C.; Patrick, S. Why we continue to use the name Propionibacterium acnes. *Br. J. Derm.* **2018**, *179*, 1227. [CrossRef]
3. Al-Shobaili, H.A. Oxidants and anti-oxidants status in acne vulgaris patients with varying severity. *Ann. Clin. Lab. Sci.* **2014**, *44*, 202–207. [PubMed]
4. Kraft, J.; Freiman, A. Management of acne. *CMAJ* **2011**, *183*, E430–E435. [CrossRef] [PubMed]
5. Walsh, T.R.; Efthimiou, J.; Dréno, B. Systematic review of antibiotic resistance in acne: An increasing topical and oral threat. *Lancet Infect. Dis.* **2016**, *16*, e23–e33. [CrossRef]
6. Abdel Fattah, N.S.; Darwish, Y.W. In vitro antibiotic susceptibility patterns of Propionibacterium acnes isolated from acne patients: An Egyptian university hospital-based study. *J. Eur. Acad. Derm. Venereol.* **2013**, *27*, 1546–1551. [CrossRef]
7. Jappe, U. Pathological mechanisms of acne with special emphasis on Propionibacterium acnes and related therapy. *Acta. Derm. Venereol.* **2003**, *83*, 241–248. [CrossRef]
8. Fanelli, M.; Kupperman, E.; Lautenbach, E.; Edelstein, P.H.; Margolis, D.J. Antibiotics, acne, and Staphylococcus aureus colonization. *Arch. Derm.* **2011**, *147*, 917–921. [CrossRef]
9. Nasri, H.; Bahmani, M.; Shahinfard, N.; Moradi Nafchi, A.; Saberianpour, S.; Rafieian Kopaei, M. Medicinal Plants for the Treatment of Acne Vulgaris: A Review of Recent Evidences. *Jundishapur J. Microbiol.* **2015**, *8*, e25580. [CrossRef]
10. Pipatchartlearnwong, K.; Swatdipong, A.; Vuttipongchaikij, S.; Apisitwanich, S. Genetic evidence of multiple invasions and a small number of founders of Asian Palmyra palm (*Borassus flabellifer*) in Thailand. *BMC Genet.* **2017**, *18*, 88. [CrossRef]
11. Nisha, R.; Parthasarathi, G.; Chandrakumar, M.; Arunachalam, R.; Krishnaveni, T.R.S. Potential Review on Palmyra (*Borassus flabellifer* L.). *Adv. Res.* **2020**, *21*, 29–40. [CrossRef]
12. Paschapur, M.S.; Patil, M.B.; Kumar, R.; Patil, S.R. Evaluation of anti-inflammatory activity of ethanolic extract of *Borassus flabellifer* L. male flowers (inflorescences) in experimental animals. *J. Med. Plants* **2009**, *3*, 49–54.
13. Paschapur, M.; Patil, S.; Patil, S.; Kumar, R.; Patil, M.B. Evaluation of the analgesic and antipyretic activities of ethanolic extract of male flowers (inflorescence) of *Borassus flabellifer* L. (Arecaceae). *Int. J. Pharm. Pharm. Sci.* **2009**, *1*, 98–106.
14. Talluri, M.R.; Gummadi, V.; Battu, G. Antioxidant activity of borassus flabellifer. *Int. Res. J. Pharm.* **2017**, *8*, 18–22. [CrossRef]
15. Mohsen, S.M.; Ammar, A.S.M. Total phenolic contents and antioxidant activity of corn tassel extracts. *Food Chem.* **2009**, *112*, 595–598. [CrossRef]
16. Tusskorn, O.; Pansuksan, K.; Machana, K. *Borassus flabellifer* L. crude male flower extracts alleviate cisplatin-induced oxidative stress in rat kidney cells. *Asian Pac. J. Trop. Biomed.* **2021**, *11*, 81–88. [CrossRef]
17. Lina, S.M.M.; Mahbub, K.M.M.; Ashab, I.; Al-Faruk, M.; Atanu, H.S.; Alam, M.J.; Sahriar, M. Antioxidant and cytotoxicity potential of alcohol and petroleum ether extract of *Borassus flabellifer* L. *Int. J. Pharm. Sci.Res.* **2013**, *4*, 1852.
18. Kavatagimath, S.A.; Jalalpure, S.S.; Hiremath, R.D. Screening of Ethanolic Extract of Borassus flabellifer Flowers for its Antidiabetic and Antioxidant Potential. *J. Nat. Remedies* **2016**, *16*, 22–32. [CrossRef]
19. Mediani, A.; Abas, F.; Tan, C.P.; Khatib, A. Effects of Different Drying Methods and Storage Time on Free Radical Scavenging Activity and Total Phenolic Content of Cosmos Caudatus. *Antioxidants* **2014**, *3*, 358–370. [CrossRef]
20. Ali, A.; Chong, C.H.; Mah, S.H.; Abdullah, L.C.; Choong, T.S.Y.; Chua, B.L. Impact of Storage Conditions on the Stability of Predominant Phenolic Constituents and Antioxidant Activity of Dried Piper betle Extracts. *Molecules* **2018**, *23*, 484. [CrossRef]
21. Kotsiou, K.; Tasioula-Margari, M. Monitoring the phenolic compounds of Greek extra-virgin olive oils during storage. *Food Chem.* **2016**, *200*, 255–262. [CrossRef] [PubMed]
22. Wongmanit, P.; Yasiri, A.; Pansuksan, K. Antibacterial Activities Derived from *Borrasus Flabellifer* L. Male Flower Extracts. *Int. J. Adv. Sci. Eng. Technol.* **2019**, *7*, 64–68.
23. Do, Q.D.; Angkawijaya, A.E.; Tran-Nguyen, P.L.; Huynh, L.H.; Soetaredjo, F.E.; Ismadji, S.; Ju, Y.H. Effect of extraction solvent on total phenol content, total flavonoid content, and antioxidant activity of Limnophila aromatica. *J. Food Drug Anal.* **2014**, *22*, 296–302. [CrossRef] [PubMed]
24. Olszowy, M. What is responsible for antioxidant properties of polyphenolic compounds from plants? *Plant Physiol. Biochem.* **2019**, *144*, 135–143. [CrossRef]
25. Baranauskaite, J.; Kubiliene, A.; Marksa, M.; Petrikaite, V.; Vitkevičius, K.; Baranauskas, A.; Bernatoniene, J. The Influence of Different Oregano Species on the Antioxidant Activity Determined Using HPLC Postcolumn DPPH Method and Anticancer Activity of Carvacrol and Rosmarinic Acid. *Biomed Res. Int.* **2017**, *2017*, 1681392. [CrossRef]

26. Alamelumangai, M.; Dhanalakshmi, J.; Mathumitha, M.; Renganayaki, R.S.; Muthukumaran, P.; Saraswathy, N. In vitro studies on phytochemical evaluation and antimicrobial activity of Borassus flabellifer Linn against some human pathogens. *Asian Pac. J. Trop. Med.* **2014**, *7*, S182–S185. [CrossRef]
27. Reshma, M.V.; Jacob, J.; Syamnath, V.L.; Habeeba, V.P.; Dileep Kumar, B.S.; Lankalapalli, R.S. First report on isolation of 2,3,4-trihydroxy-5-methylacetophenone from palmyra palm (*Borassus flabellifer* L.) syrup, its antioxidant and antimicrobial properties. *Food Chem.* **2017**, *228*, 491–496. [CrossRef]
28. Faridha Begum, I.; Mohankumar, R.; Jeevan, M.; Ramani, K. GC-MS Analysis of Bio-active Molecules Derived from Paracoccus pantotrophus FMR19 and the Antimicrobial Activity against Bacterial Pathogens and MDROs. *Indian J. Microbiol.* **2016**, *56*, 426–432. [CrossRef]
29. Li, K.; Guan, G.; Zhu, J.; Wu, H.; Sun, Q. Antibacterial activity and mechanism of a laccase-catalyzed chitosan–gallic acid derivative against Escherichia coli and Staphylococcus aureus. *Food Control.* **2019**, *96*, 234–243. [CrossRef]
30. Muddathir, A.M.; Yamauchi, K.; Mitsunaga, T. Anti-acne activity of tannin-related compounds isolated from Terminalia laxiflora. *J. Wood Sci.* **2013**, *59*, 426–431. [CrossRef]
31. Kim, S.; Oh, S.; Noh, H.B.; Ji, S.; Lee, S.H.; Koo, J.M.; Choi, C.W.; Jhun, H.P. In Vitro Antioxidant and Anti-Propionibacterium acnes Activities of Cold Water, Hot Water, and Methanol Extracts, and Their Respective Ethyl Acetate Fractions, from *Sanguisorba officinalis* L. Roots. *Molecules* **2018**, *23*, 3001. [CrossRef]
32. Katsori, A.M.; Hadjipavlou-Litina, D. Coumarin derivatives: An updated patent review (2012–2014). *Expert Opin. Pat.* **2014**, *24*, 1323–1347. [CrossRef]
33. Yang, D.; Wang, T.; Long, M.; Li, P. Quercetin: Its Main Pharmacological Activity and Potential Application in Clinical Medicine. *Oxid. Med. Cell Longev.* **2020**, *2020*, 8825387. [CrossRef] [PubMed]
34. Dharmadeva, S.; Galgamuwa, L.S.; Prasadinie, C.; Kumarasinghe, N. In vitro anti-inflammatory activity of Ficus racemosa L. bark using albumin denaturation method. *Ayu* **2018**, *39*, 239–242.
35. Banerjee, S.; Chanda, A.; Adhikari, A.; Das, A.; Biswas, S. Evaluation of Phytochemical Screening and Anti Inflammatory Activity of Leaves and Stem of *Mikania scandens* (L.) Wild. *Ann. Med. Health Sci. Res.* **2014**, *4*, 532–536. [CrossRef]
36. Gambhire, M.N.; Juvekar, A.R.; Wankhede, S.S. Evaluation of anti-inflammatory activity of methanol extract of Barleria Cristata leaves by in vivo and in vitro methods. *Internet. J. Pharmacol.* **2008**, *7*, 1–6.
37. Gunathilake, K.D.P.P.; Ranaweera, K.K.; Rupasinghe, H.P.V. In Vitro Anti-Inflammatory Properties of Selected Green Leafy Vegetables. *Biomedicines* **2018**, *6*, 107. [CrossRef]
38. Shunmugaperumal, T.; Kaur, V. In Vitro Anti-inflammatory and Antimicrobial Activities of Azithromycin After Loaded in Chitosan- and Tween 20-Based Oil-in-Water Macroemulsion for Acne Management. *AAPS PharmSciTech.* **2016**, *17*, 700–709. [CrossRef] [PubMed]
39. Precupas, A.; Leonties, A.R.; Neacsu, A.; Sandu, R.; Popa, V.T. Gallic acid influence on bovine serum albumin thermal stability. *New J. Chem.* **2019**, *43*, 3891–3898. [CrossRef]
40. Naik, M.D.; Bodke, Y.D.; Bc, R. An efficient one-pot synthesis of coumarin-amino acid derivatives as potential anti-inflammatory and antioxidant agents. *Synth. Commun.* **2020**, *50*, 1210–1216. [CrossRef]
41. Babu, S.; Ambikapathy, V.; Panneerselvam, A. Determination of Anti-Inflammatory Activity of Coumarin Compounds of *Coldenia Procumbens* L. *Int. J. Sci. Res. Biol. Sci.* **2018**, *5*, 5. [CrossRef]
42. Precupas, A.; Sandu, R.; Popa, V.T. Quercetin Influence on Thermal Denaturation of Bovine Serum Albumin. *J. Phys. Chem. B* **2016**, *120*, 9362–9375. [CrossRef] [PubMed]
43. Malik, M.A.; Sharma, H.K.; Saini, C.S. Effect of removal of phenolic compounds on structural and thermal properties of sunflower protein isolate. *J. Food Sci. Technol.* **2016**, *53*, 3455–3464. [CrossRef] [PubMed]
44. Tadtong, S.; Kanlayavattanakul, M.; Lourith, N. Neuritogenic and neuroprotective activities of fruit residues. *Nat. Prod. Commun.* **2013**, *8*, 1583–1586. [CrossRef] [PubMed]
45. Athikomkulchai, S.; Tunit, P.; Tadtong, S.; Jantrawut, P.; Sommano, S.R.; Chittasupho, C. Moringa oleifera Seed Oil Formulation Physical Stability and Chemical Constituents for Enhancing Skin Hydration and Antioxidant Activity. *Cosmetics* **2021**, *8*, 2. [CrossRef]
46. Chiangnoon, R.; Samee, W.; Uttayarat, P.; Jittachai, W.; Ruksiriwanich, W.; Sommano, S.R.; Athikomkulchai, S.; Chittasupho, C. Phytochemical Analysis, Antioxidant, and Wound Healing Activity of Pluchea indica L. (Less) Branch Extract Nanoparticles. *Molecules* **2022**, *27*, 635. [CrossRef]
47. Wongsukkasem, N.; Soynark, O.; Suthakitmanus, M.; Chongdiloet, E.; Chairattanapituk, C.; Vattanikitsiri, P.; Hongratanaworakit, T.; Tadtong, S. Antiacne-causing Bacteria, Antioxidant, Anti-Tyrosinase, Anti-Elastase and Anti-Collagenase Activities of Blend Essential Oil comprising Rose, Bergamot and Patchouli Oils. *Nat. Prod. Commun.* **2018**, *13*. [CrossRef]
48. Chittasupho, C.; Thongnopkoon, T.; Burapapisut, S.; Charoensukkho, C.; Shuwisitkul, D.; Samee, W. Stability, permeation, and cytotoxicity reduction of capsicum extract nanoparticles loaded hydrogel containing wax gourd extract. *Saudi Pharm. J.* **2020**, *28*, 1538–1547. [CrossRef]

gels

MDPI

Article

Enzyme Responsive Vaginal Microbicide Gels Containing Maraviroc and Tenofovir Microspheres Designed for Acid Phosphatase-Triggered Release for Pre-Exposure Prophylaxis of HIV-1: A Comparative Analysis of a Bigel and Thermosensitive Gel

Sabdat Ozichu Ekama [1,2,*], Margaret O. Ilomuanya [1], Chukwuemeka Paul Azubuike [1], James Babatunde Ayorinde [2], Oliver Chukwujekwu Ezechi [2], Cecilia Ihuoma Igwilo [1] and Babatunde Lawal Salako [2]

1 Department of Pharmaceutics and Pharmaceutical Technology, Faculty of Pharmacy, University of Lagos, Surulere, Lagos P.M.B 12003, Nigeria; milomuanya@unilag.edu.ng (M.O.I.); cazubuike@unilag.edu.ng (C.P.A.); cigwilo@unilag.edu.ng (C.I.I.)

2 Nigerian Institute of Medical Research, 6 Edmund Crescent, Yaba, Lagos P.M.B 12003, Nigeria; ayorindejames@gmail.com (J.B.A.); oezechi@nimr.gov.ng (O.C.E.); tundesalako@nimr.gov.ng (B.L.S.)

* Correspondence: sabdatekama@nimr.gov.ng; Tel.: +234-81-3476-1356

Abstract: The challenges encountered with conventional microbicide gels has necessitated the quest for alternative options. This study aimed to formulate and evaluate a bigel and thermosensitive gel, designed to combat the challenges of leakage and short-residence time in the vagina. Ionic-gelation technique was used to formulate maraviroc and tenofovir microspheres. The microspheres were incorporated into a thermosensitive gel and bigel, then evaluated. Enzyme degradation assay was used to assess the effect of the acid phosphatase enzyme on the release profile of maraviroc and tenofovir microspheres. HIV efficacy and cytotoxicity of the microspheres were assessed using HIV-1-BaL virus strain and HeLa cell lines, respectively. Maraviroc and tenofovir release kinetics followed zero-order and Higuchi model kinetics. However, under the influence of the enzyme, maraviroc release was governed by first-order model, while tenofovir followed a super case II transport-mechanism. The altered mode of release and drug transport mechanism suggests a triggered release. The assay of the microspheres suspension on the HeLa cells did not show signs of cytotoxicity. The thermosensitive gel and bigel elicited a progressive decline in HIV infectivity, until at concentrations of 1 µg/mL and 0.1 µg/mL, respectively. The candidate vaginal gels have the potential for a triggered release by the acid phosphatase enzyme present in the seminal fluid, thus, serving as a strategic point to prevent HIV transmission.

Keywords: HIV prevention; pre-exposure prophylaxis; HIV/AIDS; vaginal gels microbicides; acid phosphatase; microparticles

Citation: Ekama, S.O.; Ilomuanya, M.O.; Azubuike, C.P.; Ayorinde, J.B.; Ezechi, O.C.; Igwilo, C.I.; Salako, B.L. Enzyme Responsive Vaginal Microbicide Gels Containing Maraviroc and Tenofovir Microspheres Designed for Acid Phosphatase-Triggered Release for Pre-Exposure Prophylaxis of HIV-1: A Comparative Analysis of a Bigel and Thermosensitive Gel. *Gels* **2022**, *8*, 15. https://doi.org/10.3390/gels8010015

Academic Editor: Hiroyuki Takeno

Received: 30 November 2021
Accepted: 22 December 2021
Published: 24 December 2021

1. Introduction

Preventing new human immunodeficiency virus (HIV) infections using microbicides as a pre-exposure prophylactic measure has gained attention in recent times because of the disproportionate prevalence among women globally. HIV microbicides are medicaments, containing antiretroviral drugs that are administered via the vagina or rectum to prevent HIV infection.

Male to female heterosexual intercourse is a major mode of HIV transmission, and this has necessitated a focus on prevention strategies targeting viral infection at the point of sexual intercourse [1,2].

Pre-exposure prophylaxis is the use of antiretroviral drugs among HIV negative individuals to prevent HIV infection before possible contact with an infected individual.

This is commonly offered to individuals at high risk of HIV infection, such as couples in a serodiscordant relationship [3].

Microbicides are dosage forms used to administer pre-exposure prophylactic treatment. They have been designed to target HIV at the point of sexual intercourse, using different mechanisms; however, the early first-generation vaginal microbicide gels (SAVVY, PRO 2000, SLS, N-9) were non-specific in their mode of action on the HIV life cycle and were all ineffective in clinical trials [4].

The CAPRISA trial employed an antiretroviral drug, which is specific in its mode of action, formulated as a 1% tenofovir conventional microbicide gel. The trial reported a reduction in the risk of HIV acquisition by 39% among the study participants.

However, challenges associated with the conventional microbicide gel tested in the CAPRISA clinical trial, which include leakage, messiness, short vaginal residence time, and coital dependent use, which affected adherence to its use, as well as its efficacy, have necessitated the exploration of novel targeted and smart drug delivery techniques [5].

These novel drug delivery techniques are designed to target the active ingredients to the point of drug action, and one of the several ways to achieve targeted drug delivery is the use of microspheres (microparticles). They are free-flowing particles made from polymer, which could be modified to have mucoadhesive properties, such that it becomes adhesive to the vagina mucosa upon administration, thus increasing its residence time [6].

Stimuli sensitive systems, designed to trigger drug release in the presence of a stimulus, such as a change in pH, temperature, or enzymes, are also novel approaches to achieve smart drug delivery [7].

The stimulus of interest in this study is the enzyme acid phosphatase, which is a component of the male seminal fluid. Acid phosphatase is a glycoprotein with a molecular weight of 100,000 to 120,000, found in high concentrations in the male seminal fluid.

Seminal fluid is a rich source of enzymes, such as prostatic acid phosphatase and hyaluronidase. Prostatic acid phosphatase is an enzyme constituent of the seminal fluid, usually released from the prostate gland; it is also used as a tool in forensic examination to determine the presence of semen in cases of rape or sexual abuse [8,9].

Acid phosphatases are known to hydrolyze phosphate groups, and they are synthesized by epithelial cells that line the ductal elements within the prostrate. Therefore, its presence in the semen has been utilized in this study as a stimulus for a drug polymer microparticulate system that the enzyme can hydrolyze to give rise to polymer breakdown and a resultant drug release.

The candidate antiretroviral drugs, maraviroc and tenofovir, employed in this study are entry inhibitors and nucleotide reverse transcriptase inhibitors, respectively [10], which are designed and formulated as microspheres to be incorporated into a thermosensitive gel and bigel, which will serve as carriers for the microspheres. The polymer chitosan is a natural mucoadhesive polymer [11], which has been selected for the formulation of microspheres via a crosslinking process utilizing tripolyphosphate as a crosslinking agent.

These gels will be administered via the vaginal route to offer controlled and sustained release, with the goal of serving as a prophylactic agent against HIV-1 infection in women. Gels are of different types, the composition of the external liquid component, which could be either water or organic solvents, determines its classification, either as a hydrogel or organogel, respectively.

Thermosensitive gels are hydrogels that are designed to be liquid at room temperature but becomes viscous upon administration at vaginal temperature to enhance its retention, prevent leakage, and increase residence time [12].

Bigels are a hybrid of an organogel, and a hydrogel designed to allow for simultaneous incorporation of drugs that differ in hydrophilicity, which allows for better spreadability and retention [13,14].

This study seeks to explore the use of the polymer chitosan, crosslinked with tripolyphosphate, as a carrier for the antiretroviral drugs maraviroc and tenofovir, in the form of mucoadhesive microspheres. Furthermore, a stimulus-triggered release of the active ingredients is anticipated. This could occur because of the hydrolyzing effect of the acid phosphatase enzyme on the phosphate group of tripolyphosphate, leading to hydrolysis and breakdown of the microsphere, thus triggering the release of the drugs in the presence of semen during heterosexual intercourse. The hypothesis that a triggered release of the active drugs from the microspheres will occur in the presence of acid phosphatase enzyme was tested in this study.

This study, therefore, aims to formulate and evaluate vaginal microbicide gels containing maraviroc and tenofovir microspheres, designed for acid phosphatase-triggered release for pre-exposure prophylaxis of HIV-1.

2. Results and Discussion

2.1. Enzyme Degradation Assay

The cumulative percentage of drug released from the microspheres, under the influence of acid phosphatase enzyme and without it, is portrayed in Table 1, with the coefficient of correlation (R^2) and drug transport mechanism (n) of the kinetic models shown in Table 2. In the presence of the enzyme, the release of microspheres containing maraviroc started after 12 h, but there was an observed increase in the percentage of drug released which stopped at the end of 72 h. In contrast to the scenario without the enzymes, the percentage of drug released was less, and drug release did not end at 72 h (Table 1).

Table 1. Cumulative percentage of drug released from microspheres, with and without acid phosphatase enzyme.

Time (h)	Maraviroc (Without Enzymes)	Maraviroc (With Enzymes)	Tenofovir (Without Enzymes)	Tenofovir (With Enzymes)
0	0	0	0	0
1	0	0	30.84	34.88
3	0	0	33.80	40.13
6	0	0	35.60	54.36
9	0	0	39.80	58.86
12	31.02	46.3	42.76	59.66
24	42.31	66.2	49.20	99.65
48	60.65	75.4	79.92	-
72	94.70	100	81.40	-

In the absence of the enzyme, maraviroc release from the microspheres followed a zero-order (R^2 = 0.9409) kinetic release via a super case II transport mechanism (n = 1.1329), while under the influence of the enzyme it followed a first-order (R^2 = 0.8922) kinetic release model via a super case II transport mechanism (n = 1.239), as shown in Figure 1A,B and Table 2. The amount of drug released for the microspheres containing tenofovir was higher, and its rate of release increased from 1 h till 24 h, under the influence of acid phosphatase enzyme, with a completion of drug release at 24 h.

In comparison to the amount released without the enzymes, the percentage drugs released was less and drug release was not completed as at 72 h (Table 1). The release kinetic model for tenofovir followed a Higuchi model kinetics of release with and without the enzyme (Figure 2A,B); however, the release mechanism of transport was via a non-Fickian transport system without the enzyme (n = 0.5973) and a super case II transport mechanism (n = 0.8971) under the influence of the enzyme [15].

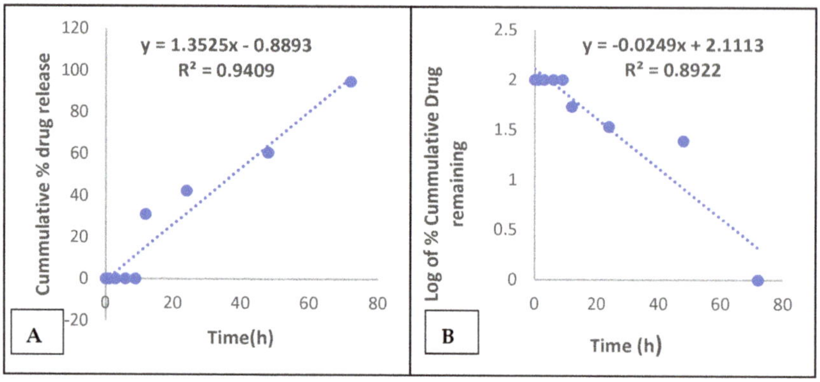

Figure 1. (**A**) Zero-order kinetic release of maraviroc from microspheres. (**B**) First-order kinetic release of maraviroc from microspheres facilitated with enzymes.

Table 2. Mathematical models for maraviroc and tenofovir release kinetic data.

Mathematical Models	R^2 Values without Enzyme *	R^2 Values with Enzyme *
Maraviroc		
Zero-order	0.9409	0.8460
First-order	0.8966	0.8922
Higuchi model	0.8953	0.8687
Korsmeyer–Peppas model	0.7543 n = 1.1329 **	0.7292 n = 1.239 **
Tenofovir		
Zero-order	0.8086	0.8475
First-order	0.8590	0.8470
Higuchi model	0.9034	0.9473
Korsmeyer–Peppas model	0.4907 n = 0.5973 **	0.5039 n = 0.8971 **

* R^2 = coefficient of correlation and the highest value best describes a drug kinetic model of release. ** Depicts the drug transport mechanism(n).

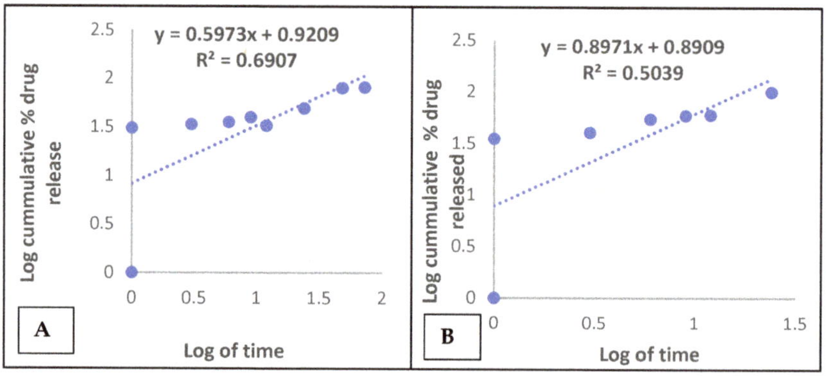

Figure 2. (**A**) Korsmeyer-peppas model kinetic release of tenofovir from microspheres. (**B**) Korsmeyer-peppas model kinetic release for tenofovir from microspheres facilitated with enzymes.

2.2. Characterization of Thermosensitive Gel Containing Maraviroc and Tenofovir

The varying pluronic combinations had gelation temperatures varying from 33.4 °C to 42.6 °C, while combinations with pluronic F127 below 2g showed no gelation. The gelation time ranged from 37 to 120 s. The pH of the various gel combinations was acidic, ranging from 5.69 to 5.97 (Table 3). The optimal thermosensitive gel had a gelation temperature and time of 36.4 °C and 36 s, respectively, with a pH of 5.83 (Figure 3). The viscosity of the gel ranged from 24 cps and 88 cps to 62,887 cps at 4 °C, 25 °C, and 37 °C, respectively, with an osmolality value of 991 mOsm/kg (Table 4).

Table 3. Composition of a 10 mL thermosensitive gel and characterization results.

Code	PF127 (g)	PF68 (g)	Gelation Temperature* (°C)	Gelation Time * (secs)	pH
T1	2.0	0.08	37.6 ± 0.6	37 ± 1.4	5.69
T2	2.0	0.1	36.4 ± 0.8	36 ± 1.1	5.83
T3	1.8	0.1	No gelation	-	5.97
T4	1.5	0.1	No gelation	-	5.94
T5	2.0	0.2	39.4 ± 1.3	39 ± 1.6	5.84
T6	1.8	0.2	42.6 ± 2.2	54 ± 1.8	5.89
T7	1.5	0.2	No gelation	-	5.97
T8	2.0	-	33.4 ± 0.9	96 ± 1.5	5.02
T9	1.8	-	39.8 ± 0.8	120 ± 2.2	5.89

* Each value is presented as mean ± SD, $n = 3$.

Figure 3. Thermosensitive gel, showing gelation at 36.4 °C.

Table 4. Characterization for thermosensitive gel containing maraviroc and tenofovir microspheres.

Parameter	Values
pH	5.83
Osmolality	991 mOsm/kg.
Viscosity at 4 °C	24 cps
Viscosity at 25 °C	88 cps
Viscosity at 37 °C	62,887 cps
Release kinetic model of maraviroc from the gel	Zero-order (R^2 = 0.9051)
Release kinetic model of tenofovir from the gel	Higuchi model (R^2 = 0.9163)

2.3. Characterization of the Organogel and Bigel

The minimum gelation concentration of Span 60 in the organogel was 10%. Organogels were not formed at 2% or 5%. Furthermore, F3 was the optimal formulation selected for the preparation of the bigel (Table 5). The organogel and hydrogel formulations were combined at varying ratios to form a bigel mix, with a corresponding pH assessment (Table 6). The ratio 1:1 of organogel to hydrogel was used for the bigel formulation (Figure 4). The pH of the formed bigel was acidic, and its values ranged from 3.5–5.6. The optimal bigel had a pH of 3.65, with viscosity and osmolality of 8840 cps and 628 mOsm/kg, respectively (Table 7).

Table 5. Formulation chart for varying organogel combinations and gelation characteristics.

Code	Span 60% *w/v*	Tween 80 *v/v*	Soya Bean Oil % *v/v*	Gel Characteristics
F1	2	1	97	No gelation
F2	5	1	94	No gelation
F3	10	1	89	Gelation
F4	15	1	84	Gelation
F5	18	1	81	Gelation
F6	20	1	79	Gelation
F7	25	1	74	Gelation
F8	2	2	96	No gelation
F9	5	2	93	No gelation
F10	10	2	88	Gelation
F11	15	2	83	Gelation
F12	18	2	80	Gelation
F13	20	2	78	Gelation
F14	25	2	73	Gelation

Table 6. Varying combinations of organogel–hydrogel mix for bigel formation.

Code	Ratio (Organogel: Hydrogel)	pH of Bigel Mix
T1	1:1	3.65
T2	2:1	4.43
T3	3:1	3.8
T4	4:1	4.8
	Ratio (Hydrogel : Organogel)	
T5	1:2	3.5
T6	1:3	3.7
T7	1:4	5.6

Table 7. Characterization parameters for the optimal bigel containing Maraviroc and Tenofovir microspheres.

Parameter	Value
pH	3.65
Osmolality	628 mOsm/kg
Viscosity	8840 cps
Release kinetic model of maraviroc from the bigel	Zero-order (R^2= 0.9431)
Release kinetic model of tenofovir from the bigel	Higuchi model(R^2 = 0.9206)

Figure 4. Organogel, hydrogel, and bigel mix.

2.4. In-Vitro Cytotoxicity of Microspheres

Assuming a significance level (α) of 0.05 and 95% confidence interval, a two- sample *t*-test, comparing the means of absorbance between the test versus the controls, was conducted to test for significance (Table 8). The percentage viability of the cells exposed to the varying maraviroc/tenofovir concentrations ranged from 71.2% to 98.4%, which is a measure of the live HeLa cells. The HeLa cell exposed to the positive control had a percentage viability of 13.5%.

There was no statistically significant difference between the absorbance of the negative control and the maraviroc/tenofovir concentrations of 10 μg/mL, 1 μg/mL, and 0.1 μg/mL (p = 0.054, 0.069, and 0.098, respectively).

There was a statistically significant difference between the absorbance of the positive control and all varying concentrations of the maraviroc/tenofovir concentrations ($p < 0.001$).

Furthermore, as shown in Figure 5, the percentage cytotoxicity of the various maraviroc and tenofovir concentrations were minimal (<8%) from concentrations of 10 μg/mL and below, compared to higher concentrations and the positive control.

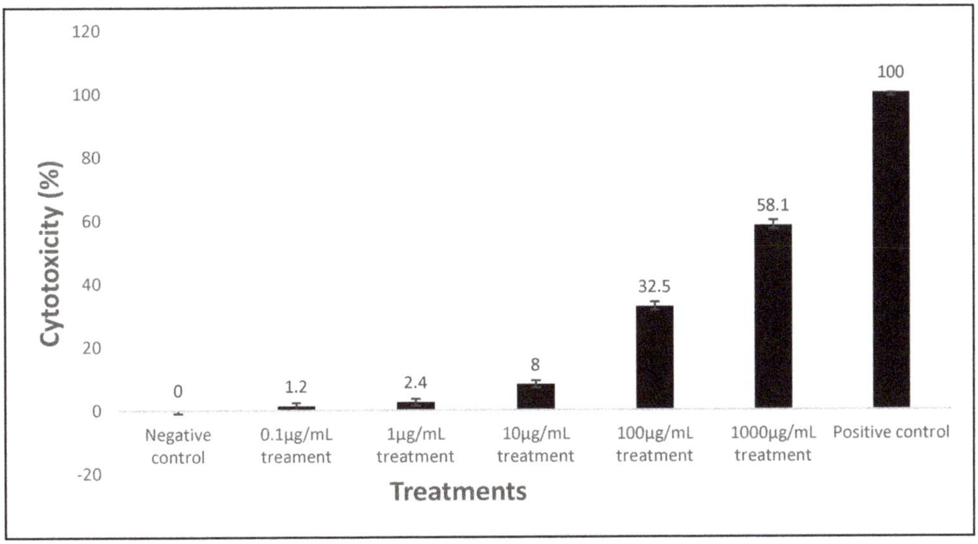

Figure 5. Cytotoxic effect of the varying concentrations of maraviroc and tenofovir microspheres, negative and positive control on the HeLa cell lines. Results are given as mean ± SD, $n = 3$.

Table 8. Independent sample *t*-test, comparing the means of absorbance of the test agent and controls to test for significance in cytotoxicity assay.

S/N	Test Agent	N	Viability ** (%)	Mean ± SD (Absorbance)	*p* Value *
1	MVC/TFV (1000 μg/mL)	3	71.2	0.371 ± 0.014	0.001
	Negative control	3	100	0.521 ± 0.01	
2	MVC/TFV (100 μg/mL)	3	83.5	0.435 ± 0.02	0.006
	Negative control	3	100	0.521 ± 0.01	
3	MVC/TFV (10 μg/mL)	3	95.3	0.497 ± 0.011	0.054
	Negative control	3	100	0.521 ± 0.01	
4	MVC/TFV (1 μg/mL)	3	98.1	0.511 ± 0.001	0.069
	Negative control	3	100	0.521 ± 0.01	
5	MVC/TFV (0.1 μg/mL)	3	98.4	0.513 ± 0.002	0.098
	Negative control	3	100	0.521 ± 0.01	
6	MVC/TFV (1000 μg/mL)	3	71.2	0.371 ± 0.014	<0.001
	Positive control	3	13.5	0.071 ± 0.01	
7	MVC /TFV (100 μg/mL)	3	83.5	0.435 ± 0.02	<0.001
	Positive control	3	13.5	0.071 ± 0.01	
8	MVC/TFV (10 μg/mL)	3	95.3	0.497 ± 0.011	<0.001
	Positive control	3	13.5	0.071 ± 0.01	
9	MVC/TFV (1 μg/mL)	3	98.1	0.511 ± 0.001	<0.001
	Positive control	3	13.5	0.071 ± 0.01	
10	MVC/TFV (0.1 μg/mL)	3	98.4	0.513 ± 0.002	<0.001
	Positive control	3	13.5	0.071 ± 0.01	
11	Negative control	3	100	0.521 ± 0.01	<0.001
	Positive control	3	13.5	0.071 ± 0.01	

* Significance level = 0.05. ** 95% confidence interval. Maraviroc = MVC, Tenofovir = TFV.

2.5. HIV Efficacy and TZM-bl Assay

Results from the in-vitro concentration response curve of the thermosensitive gel and bigel showed an initial threshold dose at a concentration of 0.0001g/mL. This indicates the onset of action of a decline in HIV infectivity, after which there was a progressive decline until at maximal concentration of 0.1 and 1.0 μg/mL, respectively.

There was a 10-fold difference between the maximal effective dose of the gels, with the bigel showing better efficacy, due to its lower maximal effective dose, compared to the thermosensitive gel.

Furthermore, there was an observed contrasting difference between the response of the negative control (nonoxynol-9 gel), compared to the thermosensitive gel and bigel, as shown in Figure 6.

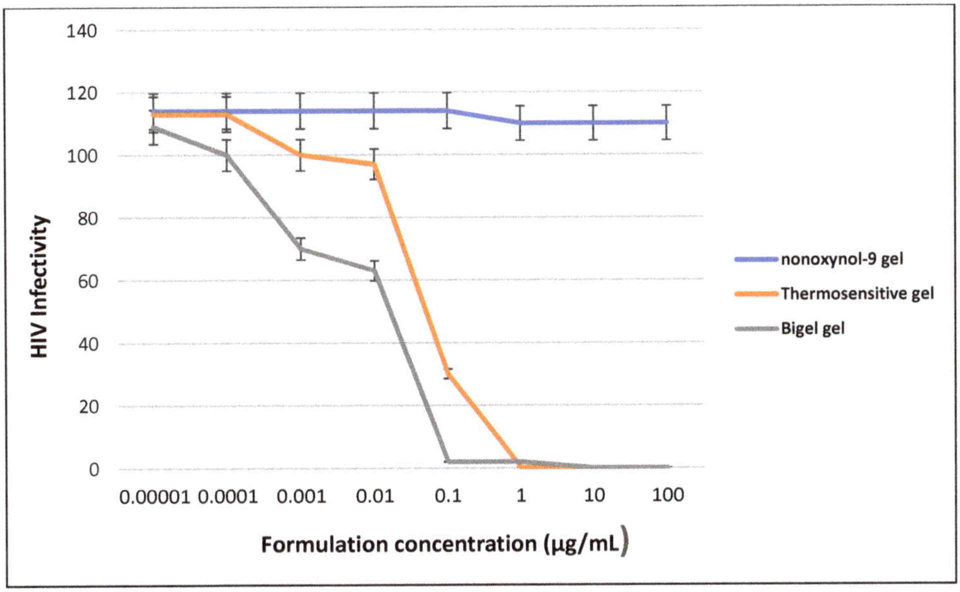

Figure 6. HIV infectivity dose-response curve for the optimal thermosensitive and bigel, containing maraviroc and tenofovir microspheres and nonoxynol-9 gel, incubated with HIV-1 indicator TZM-bl cells at different concentrations ($n = 5 \pm$ SD). No significant difference between the mean HIV infectivity values of the thermosensitive and bigel ($p > 0.05$).

2.6. Enzyme Degradation Assay

Exploiting enzymes as a trigger in the arena of smart drug delivery is valuable because of enzyme specific selectivity for its substrate [16]. Acid phosphatase enzymes are known to hydrolyze phosphate groups [17], and the optimal microspheres in this study were formulated using ionic gelation technique via a crosslinking of chitosan with tripolyphosphate.

The hypothesis that the crosslink formed by tripolyphosphate with chitosan might be hydrolyzed using acid phosphatase enzyme to influence and trigger the release of the drugs encapsulated in the microspheres was examined in this study. The rate and duration of drug release from the microspheres was faster under the influence of acid phosphatase enzyme. This is an indication that there was a breakdown of the polymer matrix to release the drugs faster. The release kinetics for maraviroc transformed from zero-order to first-order model kinetics. This implies that the release kinetics changed from being constant to a concentration-dependent release. The rate of tenofovir release was faster in the presence of the enzyme, resulting in a complete release at the end of 24 h, compared to its gradual release pattern beyond 72 h without enzyme influence. The drug transport mechanism for tenofovir changed from non-Fickian transport system to a super case II transport mechanism. Acid phosphatase enzyme is one of the enzyme compositions of the seminal fluid; therefore, these results suggest the possibility of a breakdown of the microsphere polymer matrix when the seminal fluid encounters the microspheres during sexual intercourse, thus triggering release of the drugs. Researchers have explored the possibility of other enzymes present in the seminal fluid serving as a trigger to facilitate the release of drugs by exploring polymers that can be hydrolyzed in the presence of these enzymes, which has informed the concept of this study. Some researchers [18] demonstrated the ability of hyaluronidase enzyme to hydrolyze the drug–polymer matrix, formulated using hyaluronic acid; a polymer that is hydrolyzed by hyaluronidase.

2.7. Evaluation of the Thermosensitive Gel and Bigel

A suitable vehicle to deliver the microspheres into the vagina is desirable and necessary. The goal of developing a microbicide formulation, with prolonged residence time in the vagina, has necessitated the use of gelling agents, such as poloxamers.

The combination of poloxamers 407 and 68, at different ratios, have been exploited for ocular, transdermal, and vaginal drug delivery [19,20].

Micelle formation by poloxamers in aqueous solution is reversible, and this is enhanced by the amphiphilic nature of the copolymers. The mechanism by which temperature sensitive hydrogels achieve thermogelation is either by cooling below the upper critical temperature (UCGT) or heating above the lower critical gelation temperature [21].

The poloxamer combination used in this study transitioned from solution to gel by heating above the lower critical gelation temperature, falling within the range of 33 °C to 37 °C.

The different combinations of pluronics F127 and F68, used for the formulation of the thermosensitive gel in this study, had varying effects on the gelation characteristics of the thermosensitive gel. It was observed that concentrations of PF127 below 2 g did not become viscous, even at high temperatures. The optimal thermosensitive gel showed varying viscosity values at different temperatures. Viscosity increased with increasing temperatures, with a quantum leap in viscosity values between 25 °C and 37 °C (88 to 62,887 cps). At 25 °C, the gel was still in liquid form, which is desired, as this is necessary to allow for easy handling and administration. However, at 37 °C, which is the vaginal temperature, the gel becomes viscous, which is equally desired because the goal is to prevent the gel from leaking from the vagina and combat the challenges of leakage experienced with conventional vaginal gels; this explains the high viscosity at this temperature.

Osmolality is an important parameter for an ideal microbicide gel [22]. Acceptable values below 1000 mOsm/kg are recommended to prevent mucosal damage and epithelial stripping of the vaginal tissues. The thermosensitive gel had an osmolality of 991mOsm/kg, which is within acceptable range. One of the setbacks of the 1% tenofovir gel, used for the CAPRISA 004 study, was the hyperosmolality (3111 mOsm/kg) of the gel [5].

A bigel is ideal for the simultaneous incorporation of hydrophilic and lipophilic drugs, such as tenofovir and maraviroc. The optimal formulated bigel had a pH of 3.65, which favors the protective acidic pH condition of a healthy vaginal environment.

Its viscosity value taken at room temperature was 8840 cps. Bigels are a blend of a highly a viscous organogel and less viscous hydrogel, which results in a gel with good spreadability and retention that should surmount the problems of messiness and leakage associated with previous conventional gels [23].

Osmolality is an important parameter for an ideal microbicide gel. Acceptable values below 1000 mOsm/kg are recommended to prevent mucosal damage and epithelial stripping of the vaginal tissues. The bigel had an osmolality of 628 mOsm/kg, which is within the acceptable range. One of the setbacks of the 1% tenofovir gel, used for the CAPRISA 004 study, was the hyperosmolality (3111 mOsm/kg) of the gel [5,22].

2.8. Comparative Biophysical Analysis of the Thermosensitive Gel and Bigel

The two smart gels formulated in this study belong to different category of gels; however, the goal is to design vaginal gels that will combat the challenge of short residence time in the vagina, leakage, and coital dependent administration. Both gels are suitable candidates for vaginal delivery of microbicide but differ, with respect to various rheological parameters and efficacy.

With respect to gel viscosity, which is a function of the residence time and leakage, both gel types demonstrated improved viscosity, compared to regular conventional hydrogels; however, thermosensitive gel has a higher viscosity, with a better tendency for retention in the vagina.

The osmolality values of both gels were below 1000 mOsm/kg, with a lower osmolality value recorded for the bigel, indicating a lower tendency for vaginal mucosa damage and epithelial stripping, compared to the thermosensitive gel.

The pH values of both gels were acidic, as is required for maintaining a conducive vaginal environment, although the pH value of the bigel was lower, compared to the thermosensitive gel. There was an adequate release of tenofovir and maraviroc from the thermosensitive and bigels, with no significant difference in their release profile. Tenofovir release followed Higuchi model kinetics from the thermosensitive and bigel and maraviroc, followed a zero-order kinetics order release from both gels, thus eliciting a similar pattern in the HIV infectivity decline rate.

The thermosensitive and bigel demonstrated suitable properties, which makes them potential appropriate microbicide candidates.

The pre-exposure prophylactic ability of the combined optimal tenofovir and maraviroc microspheres were evaluated via an in-vitro concentration-response procedure, using nonoxynol-9 as the control. Nonoxynol-9 is a spermicidal agent that was previously one of the early nonspecific antiviral agents that were tested as microbicides but failed in clinical trials [24,25].

The HIV infectivity of the thermosensitive gel and bigel, compared with nonoxynol-9, showed a marked difference in the HIV infectivity rate against the CCR5 tropic HIV-1 BaL strain.

The observed continuous decline in HIV infectivity of the microspheres, indicates adequate release of the drugs from the polymer and anti-HIV activity.

In comparing the efficacy of the bigel with the thermosensitive gel, the former appears to have a better potency, as demonstrated in it achieving a lower maximum effective dose than that of the thermosensitive gel. This might be attributed to the fact that bigel accommodates both lipophilic drugs and hydrophilic simultaneously and might allow better release, in comparison to the thermosensitive gel that will favor the release of hydrophilic drugs better than lipophilic drugs. Tenofovir is a hydrophilic drug, while maraviroc is lipophilic; therefore, a system that will favor the release of both drugs should demonstrate better potency as observed.

Ibrahim et al. [26], in a study of drugs incorporated into a hydrogel, organogel, and bigel, were able to demonstrate the potency of the formulation incorporated into the hydrogel, attributing it to the hydrophilicity of the drug, which was favored in the hydrogel.

2.9. In-Vitro Cytotoxicity Assay

The safety of any microbicide formulation on vagina cells is an important aspect of microbicide evaluation. A disruption or toxicity on vaginal cells will negatively impact adherence to its use and microbicide acceptance, regardless, of its efficacy. Lackman-Smith et al. [27], evaluated natural product-based microbicides containing lemon, lime, and vinegar, and the study reported the cytotoxicity of this microbicide to HeLa cells. Although this product demonstrated anti-HIV activity, its cytotoxicity did not fulfill the criteria for an ideal microbicide.

Maraviroc and tenofovir microspheres were exposed to HeLa cells, in order to evaluate cell viability and cytotoxicity in-vitro, to assess the effect on these cells over a 48-h period.

Maraviroc and tenofovir microspheres did not show significant signs of cytotoxicity on the HeLa cells at a concentration of 10 μg/mL. The cells showed a percentage viability of 95.3%, upon exposure at this concentration.

The absorbance values, which are a measure of the viable cells, did not have a statistically significant value at this concentration, when compared to the negative control ($p = 0.054$).

Similarly, some studies [28,29] demonstrated the safety of a 1mg/mL tenofovir microsphere suspension on vaginal and epithelial cell lines, which showed no significant reduction in cell viability.

3. Conclusions

Vaginal microbicide gels containing maraviroc and tenofovir microspheres were successfully developed, with adequate in-vitro release profiles that suggest good formulation properties.

Maraviroc and tenofovir microspheres have potential for a triggered release by acid phosphatase enzyme present in the seminal fluid, thus serving as a strategic point to prevent HIV transmission, when used by women during heterosexual intercourse. Its efficacy on the HIV-1 BaL virus strain and safety on vaginal lactic acid bacteria and HeLa cells suggests further in vivo studies might pave the way for promising results.

4. Materials and Methods

4.1. Materials

Poloxamer 407 (Pluronic F127), poloxamer 188 (Pluronic F68), and chitosan were supplied by Dideu Medichem Ltd. (China). Maraviroc was purchased from Shanghai Yudiao Chemistry Technology Co Ltd. (Shanghai, China). Tenofovir was supplied by Macklin Biochemical Co. Ltd. (Shanghai, China). Sodium tripolyphosphate, Span 60, and Hydroxy propyl methyl cellulose were procured from Shanghai Macklin Biochemical Co. Ltd. (Shanghai, China). Soya bean oil was purchased from Sunola Foods Ltd. (Lagos, Nigeria).

Milli-Q water (Milli-Q water purification system, Merck, Germany) was used for all the experiments. All other chemicals were of analytical grade and used as obtained from the suppliers, without further modification.

Cell lines: HeLa cells and fetal bovine serum were purchased from Cell Lines Service GmbH, Germany. HIV-1 BaL and TZM-bl cells were obtained from Fisher Bioservices /NIH-ARP German town, MD, USA.

4.2. Methods

4.2.1. Microsphere Preparation and Characterization

Maraviroc and tenofovir microspheres were prepared using an ionic gelation technique, whereby chitosan was crosslinked with sodium tripolyphosphate to form microparticle carriers for the drugs. Varying concentrations of chitosan (0.1 and 0.2 g) were dissolved in 10 mL of acetic acid (4% v/v), with a subsequent addition of 10 mg of the individual antiretroviral drugs. The resultant dispersion was added in droplets into a (2% w/v and 4% w/v) solution of sodium tripolyphosphate, using a 23 G syringe nozzle.

The resultant wet microsphere beads were allowed a curing time of 10 to 15 min, then dried at 45 °C for 12 h. The dried microspheres were further characterized for particle size, encapsulation efficiency, drug excipient compatibility, mucoadhesion, and in-vitro release, with details in [30].

4.2.2. Enzyme Degradation Assay

A phosphatase assay, using the enzyme recombinant human acid phosphatase, was used to analyze the effect of the enzyme on the release of the drugs from the drug–polymer matrix of the microspheres. A combination of seminal fluid simulant (SFS) and vaginal fluid simulant (VFS) were mixed at a ratio of 1:4, and the pH adjusted to 7.1 using 0.1 N HCl. The amount of acid phosphatase enzyme (200 U) present in an average volume of seminal ejaculate was added to 3 mL of the SFS and VFS mixture.

The formulated maraviroc and tenofovir microspheres were each resuspended in 3 mL of a mixture of seminal fluid simulant and vaginal fluid simulant (pH 7.1), and the dispersion was transferred into a 10 mL capacity dialysis membrane (Spectra/Por, Float-A-Lyzer G2, MWCO 3.5-5KD, Spectrum Laboratories Inc., Rancho Dominguez, CA, USA).

The membrane was maintained in a 20 mL release medium (SFS and VFS, pH 7.1), contained in the dialysis tube, and the whole system was placed in a conical flask, then put in a shaking water bath (Julabo SW-21C, Germany) and maintained at a temperature of 37 °C. At periodic time intervals of 0, 1, 3, 6, 24, 48, and 72 h, a 1 mL sample was withdrawn from the release medium in the dialysis tube for analysis. At each point of withdrawal an equivalent volume of the release medium was replaced to maintain sink condition [31]. The amount of maraviroc and tenofovir released was determined using HPLC (Agilent 1200 series, Germany), under the same conditions as described above for encapsulation efficiency. The above procedure was repeated, without addition of the recombinant human acid phosphatase enzyme, and this served as the control.

Mathematical models, which include the zero-order, first-order, Higuchi, and Korsmeyer–Peppas models, were used to interpret the release kinetics of the drugs, and the model with highest coefficient of correlation (R^2) best describes the pattern of drug release.

A zero-order model describes a process of constant drug release, which is determined by a plot of cumulative percentage drug released versus time (h). A first-order model depicts a process in which drug release is dependent on the drug concentration; it is represented graphically by a plot of the log of cumulative percentage drug remaining against time (h).

Diffusion controlled release is the prime mechanism of drug release for a Higuchi model, which is obtained from a plot of cumulative percentage drug released against square root of time. Furthermore, the Korsmeyer–Peppas model helps to ascertain the type of drug transport mechanism, which is characterized by the n value. The coefficient of X in the straight-line equation of the graphs represents (n), which depicts the drug transport mechanism, which is usually a range of values [32].

4.2.3. Preparation of Thermosensitive Gel

A citrate buffer of pH 4.5 was prepared by dissolving 42 mg of citric acid and 59 mg of trisodium citrate dihydrate in 1000 mL of Mili Q water, and pH was adjusted to 4.5 using 0.1 N HCl. A combination of the optimal maraviroc and tenofovir microspheres was added to 10 mL of this buffer, and then varying quantities of the polymers PF127 (Poloxamer 407) and PF68 (Poloxamer 188) were added to the dispersion, in a sample bottle, to identify the optimal combination that will give the desired gelation temperature. Methylparaben sodium (0.1%) and propyl paraben sodium (0.01%) were added as preservatives. This dispersion was kept in a refrigerator (Haier Thermocool, HRF350), overnight at 2–8 °C, to dissolve the polymers [5,33]. The resultant solution was assessed for gelation temperature, gelation time, and rheological properties.

4.2.4. Preparation of Bigel

The bigel was prepared from a hydrogel and organogel mix. The hydrogel was prepared by weighing 15 g of hydroxypropyl methylcellulose into 150 mL of Milli-Q water, then allowed to soak and dissolve overnight. The dispersion was stirred for 30 min with a magnetic stirrer, and the pH was adjusted to a pH of 2.01, using 0.1 N HCl. In preparing the organogel varying concentrations of Span 60 were prepared in soya bean oil ranging from 2%, 5%, 10%, 15%, 18%, 20%, and 25%. Tween 80 was added as surfactant. The combination F3, was used for organogel preparation. To prepare 100 g of the organogel, 10 g of Span 60 was weighed into a beaker containing 96 mL of soya bean oil and 1.8 mL of Tween 80. This was placed in a water bath at 60 °C and stirred until there is a homogenous mix. The dispersion was allowed to set at room temperature, after which it is stored at 2–8 °C overnight. The bigel was then prepared by mixing at a ratio of 1:1, 1:2, 2:1, 1:3, and 3:1 of the hydrogel and organogel to form a homogenous mix, with subsequent determination of the optimal gel mix. The tenofovir and maraviroc microspheres were then incorporated into the formulation for the optimal hydrogel and organogel (1:1), respectively, and then combined to formulate the optimal bigel mix [34].

4.3. Characterization of the Thermosensitive Gel and Bigel

4.3.1. Gelation Temperature and Time

A 25 mL beaker containing 10 mL of the pluronics dispersion was placed in a water bath, and the probe of a digital thermometer was inserted in the gel. The gel was heated with continuous stirring and temperature monitoring. The temperature at which the dispersion turned into a gel and did not flow upon inversion of the beaker was the gelation temperature. The time interval between when heat was applied and the dispersion became viscous is known as the gelation time [35].

4.3.2. Rheological Study and pH Determination

The viscosity and osmolality of the optimal gel (T2) was evaluated. A viscometer was used to determine the viscosity of T2 gel at varying temperatures. At a speed of 30 rpm, and with a spindle of size of 63, the viscosity of the gel at 4 °C and 25 °C was determined. While at a speed of 6 rpm, and with a spindle size of 64, the viscosity of the gel at 37 °C was a determined. The osmolality of the gel was also analyzed using an osmometer. The pH of the various sample combinations was determined using an Adwa pH/mV and temperature bench top pH meter (AD1040 series, Hungary). After calibration, the probe of the pH meter was inserted into the gel samples, and pH readings were recorded.

4.3.3. In-Vitro Release of Microspheres from the Thermosensitive Gel

The in-vitro release of the drugs from the gels were evaluated using a dialysis membrane. The individual gels containing optimal microspheres of tenofovir and maraviroc were introduced in the dialysis membrane, as well as the procedure described above (in the methodology section) for enzyme degradation assay.

4.3.4. Rheological Study and pH Determination

The viscosity and osmolality of the optimal bigel mix was evaluated. A viscometer (Brookfield viscometer, DV1) was used to determine the viscosity of the gel at a speed of 12 rpm and with a spindle of size of 64, the viscosity of the gel at room temperature was determined. The osmolality of the gel was also analyzed using an osmometer.

The pH of the various bigel sample combinations was determined using an Adwa pH/mV and temperature bench top pH meter (AD1040 series, Hungary). After calibration, the probe of the pH meter was inserted into the gel samples, and pH readings were recorded.

4.3.5. In-Vitro Release of Microspheres from the Bigel

The in-vitro release of the drugs from the bigel were evaluated using a dialysis membrane. The individual bigel containing optimal microspheres of tenofovir and maraviroc were introduced in the dialysis membrane and procedure described in above (in the methodology section) for enzyme degradation assay.

4.3.6. In-Vitro Cytotoxicity of the Microspheres on Vaginal HeLa Cells

The in-vitro cytotoxicity of the microspheres was evaluated using HeLa cells [18]. The cells, contained in a T25 flask, were allowed to grow until about 80% confluence was achieved. The media from the plates was drained using a 10 mL serological pipette and pipette gun (Dragon Med hero plus). A 100 μL solution of Trypsin EDTA (0.25%), at a pH of 7.21, was added to the flask to dislodge the cells. The flask containing the cells was placed in an incubator at 37 °C for 5 min. A 20 mL solution of 5% Dulbecco's modified Eagle media (DMEM) was added to the cells in the flask, then the whole suspension of cells was collected using a serological pipette into a falcon tube and centrifuged (MRC centrifuge) at 400 *g* for 10 min.

The sedimented cell suspension was collected after decanting the supernatant media. A total of 50 μL of trypan blue was added to 50 μL of the cell suspension, and this was transferred to a hemocytometer. The hemocytometer was viewed under the microscope, and the number of cells in each quadrant of the hemocytometer was counted using a tally counter, and an average of the cell count was determined. The viable cells were determined by the following formula:

$$\text{Viable cells/mL} = \text{Cell count} \times \text{scaling factor} \times \text{dilution factor} \quad (1)$$

(Where scaling factor = 104 and dilution factor = 2)

Using the formula $C_1V_1 = C_2V_2$, the volume of cell suspension (V_1) needed to achieve a cell suspension V_2 of 30 mL (capacity of 96-well plates for the test and controls) and 10^5 cells/mL concentration (C_2) was calculated from the viable cell/mL of

2.555×10^6 cells/mL (C_1). V_1 was calculated to be 1.17 mL. A volume of the cell suspension of 1.17 mL was measured into a plastic trough and made up to 30 mL using 5% DMEM.

A multichannel pipette was used to transfer 100 μL of the cell suspension (at a density of 10^5 cell/mL) into each well of the 96-well plates, and it was covered, sealed on all sides, and incubated at 37 °C for 24 h.

After incubation, the media was drained from each well, serial dilutions of concentrations of maraviroc and tenofovir microsphere at 1000 μg/mL to 0.1 μg/mL were added in triplicates to the wells containing the cells.

Camptothecin, at a concentration of 1000 μg/mL, was used as the positive control, while the media was used as the negative control. The 96-well plates were covered and sealed, then incubated at 37 °C for 24 h.

After 24 h, the media was drained from each well and 10 μL of water-soluble tetrazolium salt (WST) was added to each well. A Biochrom EZ microplate reader was used to determine the absorbance of the cells, at a wavelength of 450 nm.

The percentage viability was determined from the Equation:

$$\text{Viability (\%)} = \frac{\text{Absorbance(test)}}{\text{Absorbance (control)}} \times 100 \quad (2)$$

4.3.7. Efficacy Testing

TZM-bl cells assay was described by [20]. TZM-bl cells were plated, and 100 μL of the 10-fold serial dilution of the thermosensitive and bigel-containing maraviroc and tenofovir microspheres, in a 50:50 ratio, was applied. Nonoxynol 9 gel was used as the control. For efficacy testing, 100 μL of medium containing HIV-1BaL, without and with 12% simulated seminal fluid, was added to each well. After 48 h, 100 μL of medium was removed and replaced with 100 μL of Bright-Glo, and the luminescence measured. Inhibition was determined based on deviations from the HIV-1-only or HIV-1/semen controls.

Author Contributions: Conceptualization, S.O.E.; Methodology, S.O.E., M.O.I. and J.B.A., Investigation S.O.E., M.O.I., J.B.A. and C.P.A.; Resources, S.O.E. and M.O.I.; Original draft preparation, S.O.E. and M.O.I.; Writing review and editing, S.O.E., M.O.I., C.P.A., J.B.A., O.C.E., C.I.I. and B.L.S.; Data analysis, S.O.E. and M.O.I.; Project administration, C.I.I. and O.C.E.; Funding Acquisition, B.L.S. and O.C.E.; Supervision, C.I.I., M.O.I. and C.P.A. All authors have read and agreed to the published version of the manuscript.

Funding: This project was partly funded by the Nigerian Institute of Medical Research under Award number NIMR-PRDG-0003-18-01. The APC was funded by the Emory-Nigeria HIV Research Training Program under Award number D43TW010934.

Institutional Review Board Statement: This study was conducted in accordance with the Declaration of Helsinki and approved by the Institutional Review Board of the Nigerian Institute of Medical Research with Protocol number IRB/18/064 issued on the 4 February 2019.

Informed Consent Statement: Written Informed consent was obtained from study participants.

Acknowledgments: This project was supported by the Fogarty International Centre of the National Institutes of Health, under award number D43TW010934. The content is solely the responsibility of the authors and does not necessarily represent the official views of the National Institutes of Health.

Conflicts of Interest: The authors declare no conflict of interest in this work.

References

1. Baeten, J.M.; Hendrix, C.W.; Hillier, S.L. Topical Microbicides in HIV Prevention: State of the Promise. *Annu. Rev. Med.* **2020**, *71*, 361–377. [CrossRef]
2. Melo, M.G.; Sprinz, E.; Gorbach, P.M.; Santos, B.; Rocha, T.D.; Simon, M.; Almeida, M.; Lira, R.; Chaves, M.C.; Kerin, T.; et al. HIV-1 Heterosexual Transmission and Association with Sexually Transmitted Infections in the Era of Treatment as Prevention. *Int. J. Infect. Dis.* **2019**, *87*, 128–134. [CrossRef] [PubMed]

3. Chan, S.S.; Chappel, A.R.; Maddox, K.E.J.; Hoover, K.W.; Huang, Y.-L.A.; Zhu, W.; Cohen, S.M.; Klein, P.W.; De Lew, N. Pre-Exposure Prophylaxis for Preventing Acquisition of HIV: A Cross-Sectional Study of Patients, Prescribers, Uptake, and Spending in the United States, 2015–2016. *PLoS Med.* **2020**, *17*, e1003072. [CrossRef]
4. Romano, J.W.; Robbiani, M.; Doncel, G.F.; Moench, T. Non-Specific Microbicide Product Development: Then and Now. *Curr. HIV Res.* **2012**, *10*, 9–18. [CrossRef] [PubMed]
5. Date, A.; Shibata, A.; Goede, M.; Sanford, B.; La Bruzzo, K.; Belshan, M.; Destache, C.J. Development and Evaluation of a Thermosensitive Vaginal Gel Containing raltegravir + efavirenz Loaded Nanoparticles for HIV Prophylaxis. *Antivir. Res.* **2012**, *96*, 430–436. [CrossRef]
6. Kirtane, A.R.; Verma, M.; Karandikar, P.; Furin, J.; Langer, R.; Traverso, G. Nanotechnology Approaches for Global Infectious Diseases. *Nat. Nanotechnol.* **2021**, *16*, 369–384. [CrossRef]
7. Bhardwaj, A.; Kumar, L.; Mehta, S.; Mehta, A. Stimuli-Sensitive Systems—An Emerging Delivery System for Drugs. *Artif. Cells Nanomed. Biotechnol.* **2015**, *43*, 299–310. [CrossRef]
8. Xu, H.; Wang, F.; Li, H.; Ji, J.; Cao, Z.; Lyu, J.; Shi, X.; Zhu, Y.; Zhang, C.; Guo, F.; et al. Prostatic Acid Phosphatase (PAP) Predicts Prostate Cancer Progress in a Population-Based Study: The Renewal of PAP? *Dis. Markers* **2019**, *2019*, 7090545–10. [CrossRef] [PubMed]
9. Sakurada, K.; Watanabe, K.; Akutsu, T. Current Methods for Body Fluid Identification Related to Sexual Crime: Focusing on Saliva, Semen, and Vaginal Fluid. *Diagnostics* **2020**, *10*, 693. [CrossRef] [PubMed]
10. Woolard, M.S.; Kanmogne, G.D. Maraviroc: A Review of Its Use in HIV Infection and Beyond. *DDDT* **2015**, *9*, 5447–5468.
11. Karava, A.; Lazaridou, M.; Nanaki, S.; Michailidou, G.; Christodoulou, E.; Kostoglou, M.; Iatrou, H.; Bikiaris, D.N. Chitosan Derivatives with Mucoadhesive and Antimicrobial Properties for Simultaneous Nanoencapsulation and Extended Ocular Release Formulations of Dexamethasone and Chloramphenicol Drugs. *Pharmaceutics* **2020**, *12*, 594. [CrossRef]
12. Huang, H.; Qi, X.; Chen, Y.; Wu, Z. Thermo-Sensitive Hydrogels for Delivering Biotherapeutic Molecules: A Review. *Saudi Pharm. J.* **2019**, *27*, 990–999. [CrossRef]
13. Martín-Illana, A.; Notario-Pérez, F.; Cazorla-Luna, R.; Ruiz-Caro, R.; Veiga, M.D. Smart Freeze-Dried Bigels for the Prevention of the Sexual Transmission of HIV by Accelerating the Vaginal Release of Tenofovir During Intercourse. *Pharmaceutics* **2019**, *11*, 232. [CrossRef]
14. Ilomuanya, M.O.; Hameedat, A.T.; Akang, E.N.; Ekama, S.O.; Silva, B.O.; Akanmu, A.S. Development and Evaluation of Mucoadhesive Bigel Containing Tenofovir and Maraviroc for HIV Prophylaxis. *Futur. J. Pharm. Sci.* **2020**, *6*, 1–12. [CrossRef]
15. Mhlanga, N.; Ray, S.S. Kinetic Models for the Release of the Anticancer Drug Doxorubicin from Biodegradable polylactide/Metal Oxide-Based Hybrids. *Int. J. Biol. Macromol.* **2015**, *72*, 1301–1307. [CrossRef] [PubMed]
16. Shahriari, M.; Zahiri, M.; Abnous, K.; Taghdisi, S.M.; Ramezani, M.; Alibolandi, M. Enzyme Responsive Drug Delivery Systems in Cancer Treatment. *J. Control Release* **2019**, *308*, 172–189. [CrossRef] [PubMed]
17. Huang, R.; Wan, B.; Hultz, M.; Diaz, J.M.; Tang, Y. Phosphatase-Mediated Hydrolysis of Linear Polyphosphates. *Environ. Sci. Technol.* **2018**, *52*, 1183–1190. [CrossRef]
18. Jiang, H.; Shi, X.; Yu, X.; He, X.; An, Y.; Lu, H. Hyaluronidase Enzyme-Responsive Targeted Nanoparticles for Effective Delivery of 5-Fluorouracil in Colon Cancer. *Pharm. Res.* **2018**, *35*, 73. [CrossRef]
19. Kaus, N.H.M.; Shaarani, S.; Hamid, S.S. The Influence of Pluronic F68 and F127 Nanocarrier on Physicochemical Properties, in Vitro Release, and Antiproliferative Activity of Thymoquinone Drug. *Pharmacogn. Res.* **2017**, *9*, 12–20. [CrossRef] [PubMed]
20. Lou, J.; Hu, W.; Tian, R.; Zhang, H.; Jia, Y.; Zhang, J.; Zhang, L. Optimization and Evaluation of a Thermoresponsive Ophthalmic in Situ Gel Containing Curcumin-Loaded Albumin Nanoparticles. *Int. J. Nanomed.* **2014**, *9*, 2517–2525.
21. Russo, E.; Villa, C. Poloxamer Hydrogels for Biomedical Applications. *Pharmaceutics* **2019**, *11*, 671. [CrossRef]
22. Dezzutti, C.S.; Rohan, L.C.; Wang, L.; Uranker, K.; Shetler, C.; Cost, M.; Lynam, J.D.; Friend, D. Reformulated Tenofovir Gel for Use as a Dual Compartment Microbicide. *J. Antimicrob. Chemother.* **2012**, *67*, 2139–2142. [CrossRef] [PubMed]
23. Andonova, V.; Peneva, P.; Georgiev, G.S.; Toncheva, V.T.; Apostolova, E.; Peychev, Z.; Dimitrova, S.; Katsarova, M.; Petrova, N.; Kassarova, M. Ketoprofen-Loaded Polymer Carriers in Bigel Formulation: An Approach to Enhancing Drug Photostability in Topical Application Forms. *Int. J. Nanomed.* **2017**, *12*, 6221–6238. [CrossRef] [PubMed]
24. Cutler, B.; Justman, J. Vaginal Microbicides and the Prevention of HIV Transmission. *Lancet Infect. Dis.* **2008**, *8*, 685–697. [CrossRef]
25. Notario-Pérez, F.; Ruiz-Caro, R.; Veiga-Ochoa, M.-D. Historical Development of Vaginal Microbicides to Prevent Sexual Transmission of HIV in Women: From past Failures to Future Hopes. *Drug Des. Dev. Ther.* **2017**, *11*, 1767–1787. [CrossRef] [PubMed]
26. Ibrahim, M.M.; Hafez, A.S.; Mahdy, M.M. Organogels, Hydrogels and Bigels as Transdermal Delivery Systems for Dilti-Azem Hydrochloride. *Asian J. Pharm. Sci.* **2013**, *8*, 46–54.
27. Lackman-Smith, C.S.; Snyder, B.A.; Marotte, K.M.; Osterling, M.C.; Mankowski, M.K.; Jones, M.; Nieves-Duran, L.; Richardson-Harman, N.; Cummins, J.E., Jr.; Sanders-Beer, B.E. Safety and Anti-HIV Assessments of Natural Vaginal Cleansing Products in an Established Topical Microbi-Cides in Vitro Testing Algorithm. *AIDS Res. Ther.* **2010**, *7*, 22. [CrossRef]
28. Zhang, T.; Zhang, C.; Agrahari, V.; Murowchick, J.B.; Oyler, N.A.; Youan, B.-B.C. Spray Drying Tenofovir Loaded Mucoadhesive and PH-Sensitive Microspheres Intended for HIV Prevention. *Antivir. Res.* **2013**, *97*, 334–346. [CrossRef]
29. Destache, C.J.; Mandal, S.; Yuan, Z.; Kang, G.; Date, A.; Lu, W.; Shibata, A.; Pham, R.; Bruck, P.; Rezich, M.; et al. Topical Tenofovir Disoproxil Fumarate Nanoparticles Prevent HIV-1 Vaginal Transmission in a Humanized Mouse Model. *Antimicrob. Agents Chemother.* **2016**, *60*, 3633–3639. [CrossRef]

30. Ekama, S.; Ilomuanya, M.O.; Azubuike, C.P.; A Bamidele, T.; Fowora, M.A.; Aina, O.O.; Ezechi, O.C.; Igwilo, C.I. Mucoadhesive Microspheres of Maraviroc and Tenofovir Designed for Pre-Exposure Prophylaxis of HIV-1: An in Vitro Assessment of the Effect on Vaginal Lactic Acid Bacteria Microflora. *HIV/AIDS Res. Palliat. Care* **2021**, *13*, 399–413. [CrossRef] [PubMed]
31. Agrahari, V.; Zhang, C.; Zhang, T.; Li, W.; Gounev, T.K.; Oyler, N.A.; Youan, B.-B.C. Hyaluronidase-Sensitive Nanoparticle Templates for Triggered Release of HIV/AIDS Microbicide In Vitro. *AAPS J.* **2013**, *16*, 181–193. [CrossRef]
32. Gouda, R.; Baishaya, H.; Qing, Z. Application of Mathematical Models in Drug Release Kinetics of Carbidopa and Levodopa ER Tablets. *J. Develop. Drugs* **2017**, *6*, 1–8.
33. Date, A.A.; Shibata, A.; Mcmullen, E.; La Bruzzo, K.; Bruck, P.; Belshan, M.; Zhou, Y.; Destache, C.J. Thermosensitive Gel Containing Cellulose Acetate Phthalate-Efavirenz Combination Nanoparticles for Prevention of HIV-1 Infection. *J. Biomed. Nanotechnol.* **2015**, *11*, 416–427. [CrossRef] [PubMed]
34. Hamed, R.; Aburezeq, A.; Tarawneh, O. Development of Hydrogels, Oleogels, and Bigels as Local Drug Delivery Systems for Periodontitis. *Drug Dev. Ind. Pharm.* **2018**, *44*, 1488–1497. [CrossRef]
35. Ilomuanya, M.O.; Elesho, R.F.; Amenaghawon, A.N.; Adetuyi, A.O.; Velusamy, V.; Akanmu, A.S. Development of Trigger Sensitive Hyaluronic acid/Palm Oil-Based Organogel for in Vitro Release of HIV/AIDS Microbicides Using Artificial Neural Networks. *Futur. J. Pharm. Sci.* **2020**, *6*, 1–14. [CrossRef]

gels

MDPI

Article

A Ponceau S Staining-Based Dot Blot Assay for Rapid Protein Quantification of Biological Samples

Dario Lucas Helbing [1,2,3,*,†], Leopold Böhm [1,2,3,†], Nova Oraha [1,2,3], Leonie Karoline Stabenow [2,3] and Yan Cui [1,4,*]

1 Leibniz Institute on Aging—Fritz Lipmann Institute, 07745 Jena, Germany; leopold.boehm@leibniz-fli.de (L.B.); nova.oraha@leibniz-fli.de (N.O.)
2 Faculty of Medicine, Friedrich-Schiller-University Jena, 07743 Jena, Germany; Leonie.Stabenow@med.uni-jena.de
3 Institute of Molecular Cell Biology, University Hospital Jena, Friedrich-Schiller-University, 07745 Jena, Germany
4 International Center for Aging and Cancer, Hainan Medical University, Haikou 571199, China
* Correspondence: dario.helbing@leibniz-fli.de (D.L.H.); yancui.research@gmail.com (Y.C.)
† These authors contributed equally to this work.

Abstract: Despite the availability of a wide range of commercial kits, protein quantification is often unreliable, especially for tissue-derived samples, leading to uneven loading in subsequent experiments. Here we show that the widely used Bicinchoninic Acid (BCA) assay tends to underestimate protein concentrations of tissue samples. We present a Ponceau S staining-based dot-blot assay as an alternative for protein quantification. This method is simple, rapid, more reliable than the BCA assay, compatible with biological samples lysed in RIPA or 2x SDS gel-loading buffer, and also inexpensive.

Keywords: protein quantification; gel electrophoresis; Western blot; dot blot

Citation: Helbing, D.L.; Böhm, L.; Oraha, N.; Stabenow, L.K.; Cui, Y. A Ponceau S Staining-Based Dot Blot Assay for Rapid Protein Quantification of Biological Samples. *Gels* **2022**, *8*, 43. https://doi.org/10.3390/gels8010043

Academic Editor: Hiroyuki Takeno

Received: 28 November 2021
Accepted: 1 January 2022
Published: 7 January 2022

1. Introduction

Many methods exist for quantifying total protein contents in cell or tissue lysates. However, the conventional colorimetric methods, such as the Bicinchoninic Acid (BCA) [1], Lowry [2] and Bradford [3] assays, have several drawbacks: (1) their requirement for large volumes of samples ranging from 10 to 25 μL per replicate could be difficult to meet for samples with low protein contents, such as mouse nerve or lymph node lysates; (2) colorimetric assays tend to saturate if the protein concentration is too high; in this case, the samples must be diluted for repetition, consuming more materials and time; (3) the widely used SDS-based lysis buffers often contain bromophenol blue, a substance that interferes with most colorimetric assays, making colorimetric assays unsuitable for bromophenol blue-containing lysates [1,4]; and (4) some native components in biological samples, such as high concentrations of lipids in nerve or brain lysates, may interfere with the standard BCA assay [5]. Therefore, protein concentrations of biological samples determined by the common assays are often imprecise, resulting in unequal sample loading in subsequent experiments such as Western blots.

Here, we demonstrate a modified Ponceau S-based dot blot (PDB) assay that has distinct advantages over similar methods described elsewhere [6–8] concerning its ability to quantify proteins in cell or tissue lysates, its highly linear standard curve with a wide range from 0.25 to 12 μg/μL, and its overall improved workflow. We further show that protein loading based on the PDB assay was equal in Western blots, essentially eliminating the need for error-prone normalization of immunoblot band intensities to those of a loading control [9]. Therefore, the PDB assay is a reliable method for protein quantification, thereby facilitating subsequent experiments. Our work, in tandem with several other publications, demonstrates the versatility of dot blot in protein quantification and detection [10–12].

2. Materials and Methods

2.1. Reagents

Coomassie Blue solution: 0.05% (*w*/*v*) CBB-R250 (Carl Roth GmbH + Co. KG, Karlsruhe, Germany, #6862), 40% (*v*/*v*) ethanol, 10% (*v*/*v*) acetic acid in water.

Direct Blue solution: 0.008% (*w*/*v*) Direct Blue 71 (Merck KGaA, Darmstadt, Germany, #212407), 40% (*v*/*v*) ethanol and 10% (*v*/*v*) acetic acid in water.

Micro BCA Protein assay kit, Thermo Fisher Scientific Inc., Waltham, MA, USA, #23225.

Pierce Bovine Serum Albumin Standard (BSA) ampules, Thermo Fisher Scientific Inc., Waltham, MA, USA, #23209.

Ponceau S solution: 0.1% (*w*/*v*) Ponceau S (Merck KGaA, Darmstadt, Germany, #P3504-10G) in 5% (*v*/*v*) acetic acid.

RIPA buffer: Pierce RIPA buffer (Thermo Fisher Scientific Inc., Waltham, MA, USA) or self-made RIPA buffer (2.5% (*v*/*v*) 1 M Tris-HCl, 0.88% (*w*/*v*) NaCl, 1% (*w*/*v*) NP-40, 1% (*w*/*v*) sodium deoxycholate, 0.1% (*w*/*v*) SDS in water, adjusted to pH 7.6), 1 tablet/50 mL cOmplete protease inhibitor (Roche Diagnostics GmbH, Mannheim, Germany), 1 tablet/10 mL RIPA phosSTOP phosphatase inhibitor (Roche Diagnostics GmbH, Mannheim, Germany).

2x SDS Gel Loading buffer (2x SDS LB): 0.1 M Tris-HCl, 4% (*v*/*v*) SDS, 20% (*v*/*v*) Glycerol, 0.2% (*v*/*v*) Bromophenol blue (Carl Roth GmbH + Co. KG, Karlsruhe, Germany, #A512.1) in water, adjusted to pH 6.8.

2.2. Ethics Statement

The Thuringian State Office for Consumer Protection (TLV), Department of Health and Technical Consumer Protection has examined and prospectively approved the animal experiment application under the permit number 03-046/16. It is advised by the Thuringian Animal Protection Commission.

2.3. Experimental Animals

Mice (C57BL/6J) were housed under a 12 h dark/light cycle and had free access to food and water. All mice were handled in strict adherence to local governmental and institutional animal care regulations and euthanized by trained personnel using CO_2 inhalation. All efforts were made to minimize suffering.

2.4. Lysate Preparation

Cell lysates were prepared by adding 2x SDS LB to cell culture plates after washing off cell debris and culture media and then collected with cell scrapers. Lysates were then transferred to Eppendorf tubes and vortexed for 10 s.

Sciatic nerves, brains and spleens were harvested from euthanized mice and lysed using ceramic beads in a Precellys 24 homogenizer (Bertin Instruments, Montigny-le-Bretonneux, France) in either RIPA buffer or 2x SDS LB.

2.5. BCA Assay

The microscale BCA assay was performed according to the manufacturer's instruction. Protein concentrations were calculated using the linear equation based on the trend line of the standard curve generated with Microsoft Excel.

2.6. Dot Blot, Ponceau S Staining, Direct Blue Staining and Coomassie Blue Staining

Lysates in 2x SDS LB were boiled for 8 min at 98 °C beforehand. Different volumes of lysates and BSA solution were applied dot-wise to dry nitrocellulose membranes. Membranes were then left to dry for 15 min. Membranes with samples lysed in 2x SDS LB were then washed three times (5 min each) in deionized (DI) water on a shaker before staining with Ponceau S solution for 1 min, and membranes with samples lysed in RIPA buffer were stained directly with Ponceau S solution for 1 min. Membranes were then briefly washed with DI water to remove the Ponceau S solution. Membranes were then placed into a plastic film and scanned with an Epson Perfection V750 Pro scanner.

For re-staining of membranes with Direct Blue solution, membranes previously stained with Ponceau S were washed overnight with TBS-T, stained with Direct Blue staining solution for 5 min, washed with a solution of 40% (*v*/*v*) ethanol and 10% (*v*/*v*) acetic acid in DI water to remove non-specific background staining, and then imaged as described above. For re-staining of membranes with Coomassie Blue, membranes previously stained with Direct Blue were washed overnight with a solution of 50% (*v*/*v*) ethanol and 15% (*w*/*v*) 1 M sodium bicarbonate in water. Membranes were then stained with Coomassie Blue solution for 3 min and imaged as described above.

For experiments using diluted BSA solutions, a BSA solution was diluted 1:1 with either 0.5 M NaCl, PBS, TBS-T or a 0.1% (*v*/*v*) SDS solution, resulting in BSA solutions with a protein concentration of 1 µg/µL. BSA solutions were applied dot-wise onto a nitrocellulose membrane with different volumes to create a standard curve for each solution based on Ponceau S staining.

We included a movie demonstrating the experimental workflow and the analysis with Fiji in the Supplementary Materials (Video S1).

2.7. Protein Quantification with Fiji and Microsoft Excel

After transforming a source image into an 8-bit greyscale image and inverting it in ImageJ/Fiji, the rectangle tool and ROI manager were used to define a series of regions of interest (ROIs), covering each dot with a single ROI of the same size, and the integrated density of each ROI was measured. The integrated density values of technical replicates of the samples and the standards were averaged and, if necessary, corrected for the baseline signal by subtracting the trendlines' y-axis intercept. A standard curve was generated based on the mean integrated density values of the standards and the equation of a corresponding linear trendline used to calculate protein concentrations. The goodness of fit/coefficient of determination for each trendline was calculated in Microsoft Excel and displayed as R^2 in the corresponding figure.

2.8. Cost Calculations

We calculated the costs regarding the PDB assay and the micro BCA assay for measuring 12 biological samples and standards. Both the commercial BCA assay and a self-made variant were compared. The reagents for a self-made BCA assay were as follows:

Reagent A: 1% (*w*/*v*) sodium bicinchoninate, 2% (*w*/*v*) sodium carbonate, 0.16% sodium tartrate (*w*/*v*), 0.4% (*w*/*v*) NaOH, 0.95% (*w*/*v*) sodium bicarbonate in water, pH adjusted to 11.25.

Reagent B: 4% (*w*/*v*) cupric sulfate in water.

A range of total costs was calculated for each method to reflect the market price ranges of the required materials.

We calculated the costs of RIPA buffer and SDS Gel loading buffer per ml and compared the commercial and self-made RIPA buffers with the self-made SDS Gel loading buffer.

We determined the prices of reagents on the websites of established distributors and used the smallest available package size for our calculations. Basic laboratory chemicals such as Tris-HCl or NaCl were not included in our calculations. We included the protease and the phosphatase inhibitor prices in both cost calculations for the commercial and the self-made RIPA buffers.

2.9. Immunoblotting

Immunoblotting was performed as previously described [13]. Antibodies are listed in Table 1. Blots were developed with Pierce ECL Western Blotting Substrate (Thermo Fisher Scientific Inc., Waltham, MA, USA).

Table 1. List of antibodies.

Antibody	Host Species	Supplier	Cat. No.	Dilution
ERK 1/2	Mouse	Cell Signaling	4696	1:2000
GAP-43	Rabbit	Santa Cruz	10,786	1:500
GAPDH	Mouse	Santa Cruz	32,233	1:5000
Histone H1	Mouse	Santa Cruz	8030	1:500
LC3A/B	Rabbit	Cell Signaling	4108	1:1000
MBP	Rat	Novus Biologicals	NB600-717	1:1000
MEK 1/2	Rabbit	Cell Signaling	8727	1:2000
Merlin	Rabbit	Cell Signaling	12,896	1:1000
NF-M	Mouse	Santa Cruz	16,143	1:500
P0	Chicken	Abcam	39,375	1:2000
p-ERK 1/2	Rabbit	Cell Signaling	4370	1:2000
p-MEK1/2	Mouse	Abcam	91,545	1:2000
Anti-chicken HRP	Goat	Abcam	97,135	1:5000
Anti-Rabbit HRP	Goat	Agilent Dako	P0448	1:2000
Anti-Mouse HRP	Goat	Agilent Dako	P0447	1:2000
Anti-rat HRP	Rabbit	Invitrogen	61-9520	1:2000
Anti-Goat HRP	Rabbit	Agilent Dako	P0449	1:1000

2.10. Statistical Procedures and Figure Preparation

The p-values were calculated by Student's t-test with Graphpad Prism 7.0. Statistical significance was accepted at $p \leq 0.05$. All data are presented as mean +/− SEM (Standard error of the mean). Missing error bars are due to technical reasons when the error bar would be shorter than the height of a single datapoint. All figures were made with Graphpad Prism 7.0 and assembled in Adobe Photoshop CS6 or Microsoft PowerPoint.

3. Results and Discussion

To assess the reliability of the PDB assay for protein quantification, we spotted different volumes of a BSA solution (0.25–4 µg of BSA) on a nitrocellulose membrane. Ponceau S staining was performed after air-drying the membrane (Figure 1A). The sizes of the dots did not correlate with the total protein amounts (data not shown) and we therefore measured the integrated density, which is the sum of grey values in the region of interest. It should be noted that a gray value measured by ImageJ is opposite to the actual intensity, so images should be inverted before measurement. The integrated density and protein amount showed a linear trendline, indicated by a coefficient of determination R^2 near the optimal value of 1 (Figure 1B). For comparison, we performed a BCA assay, which also showed a trendline with R^2~1 (Figure 1C), indicating that both assays perform well within their respective detection range. However, according to the manufacturer, beyond the maximum concentration of 40 µg/mL, the BCA assay—like any other colorimetric assay—is no longer reliable and further dilution of the sample is required. In contrast, the PDB assay yielded a linear standard curve up to 12 µg of the protein per dot (Figure S1).

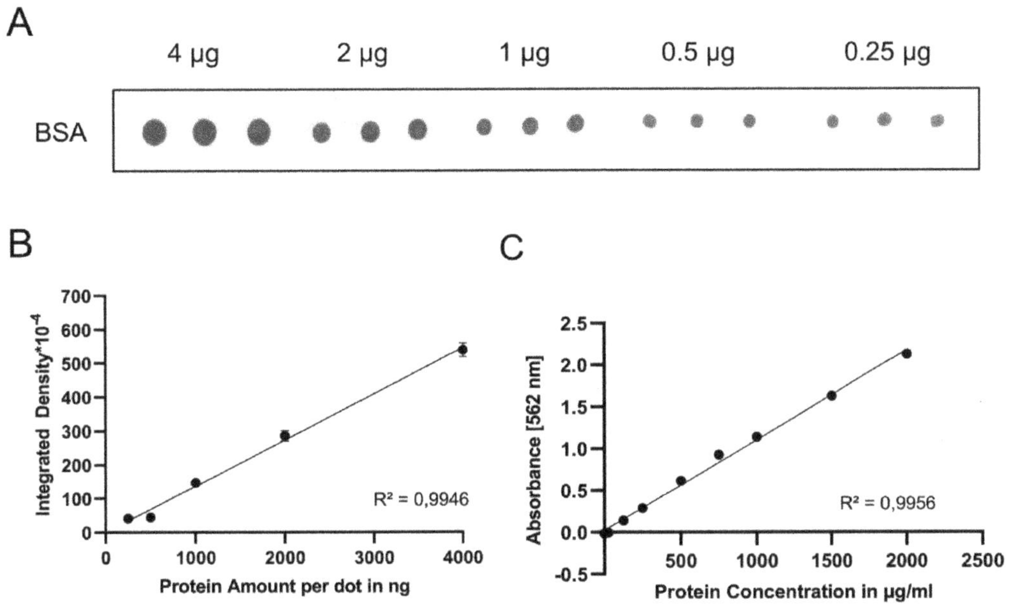

Figure 1. Standard curves of the PDB and the BCA assays. (**A**) Ponceau S-stained dot blot. Variable volumes of a BSA solution (2 µg/µL) containing indicated BSA amounts were spotted in triplicate onto a nitrocellulose membrane. (**B**) Inverted standard curve based on Figure 1A. The experiment was repeated with three batches of BSA solutions. (**C**) Standard curve of three BCA assays performed with different BSA solutions. Error bars indicate SEM. R^2 = Coefficient of determination.

Visible in Figure S1, the bottom range of the PDB assay standard curve may lose fidelity if the total protein amount is below 1 µg and the respective BSA volumes are not spotted perfectly, which can occur if the applied volume per drop is small. Therefore, the BSA standards may be diluted if the samples are suspected of having low protein contents. To identify suitable diluents, we diluted the BSA standards with TBS-T, 0.1% SDS, PBS and 0.5 M NaCl, and compared the corresponding PDB assay standard curves (Figure 2).

As indicated by higher R^2 values and lower error deviations between technical replicates, TBS-T and 0.1% SDS in water outperformed PBS and 0.5 M NaCl (Figure 2B–E). Using diluted standards, samples with very low protein concentrations just above 0.125 µg/µL can be analyzed.

As described above, fixing the protein concentration while varying the volume yielded a linear standard curve in the PDB assay. Similarly, fixing the volume while varying the protein concentration also yielded a linear standard curve (Figure S2), suggesting that the integrated density (if not saturated) is determined by the protein amount and less affected by the dot size.

To test whether the PDB assay can be used to quantify proteins in lysates of biological tissues, spleens from four mice were lysed in 1 mL of RIPA buffer and a PDB assay was performed (Figure 3A–C). For comparison, the same lysates were also measured with a BCA assay (Figure 3C).

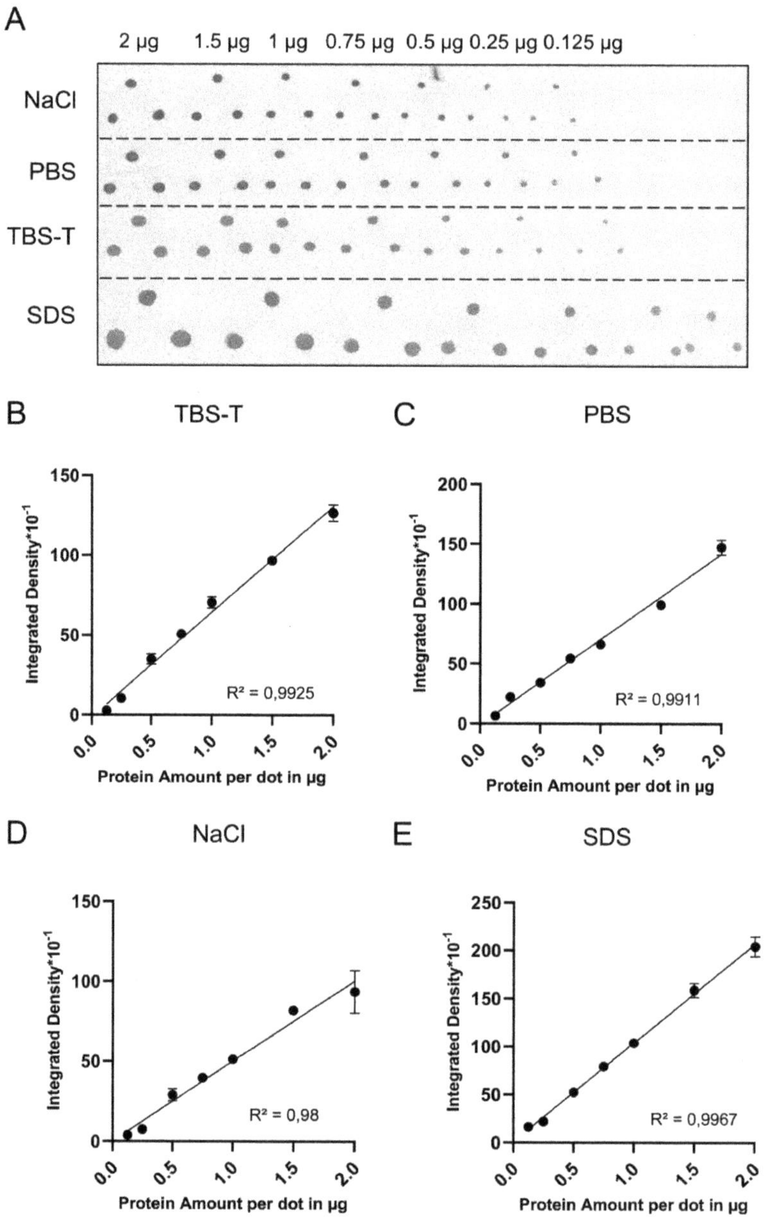

Figure 2. Comparison of diluents for the PDB assay. (**A**) Ponceau S-stained dot blot. BSA was diluted in different solutions to a final concentration of 1 µg/µL. Variable volumes of the solutions containing indicated BSA amounts were spotted in triplicate onto a nitrocellulose membrane. (**B–E**) Inverted standard curves based on Figure 2A. Error bars indicate SEM. R^2 = Coefficient of determination.

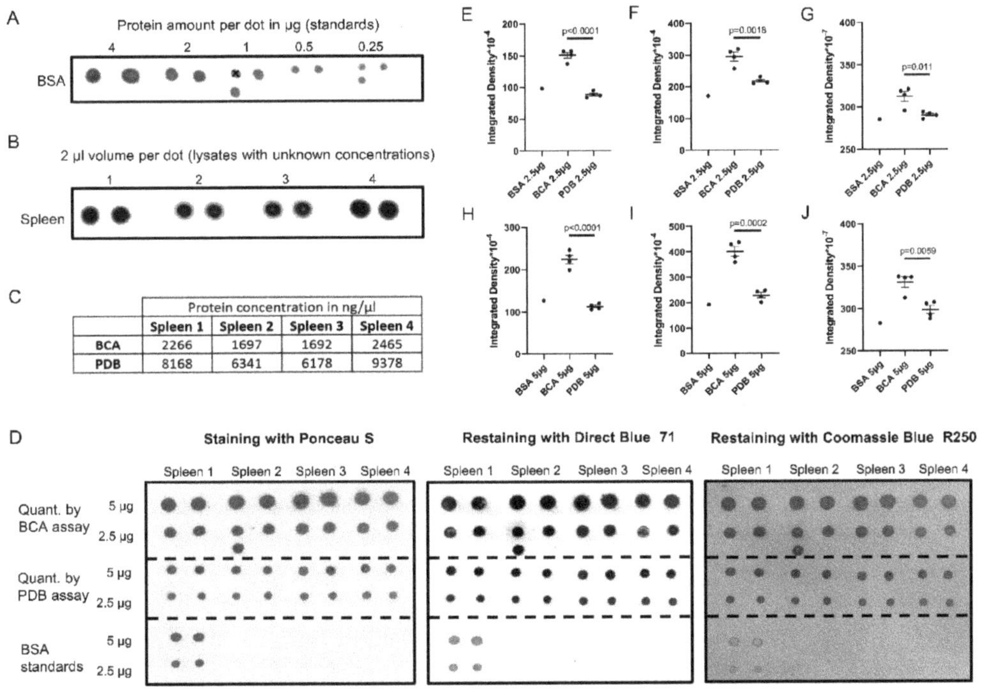

Figure 3. Underestimation of protein concentrations by the BCA assay. (**A**) Ponceau S-stained dot blot. Variable volumes of a BSA solution (2 µg/µL) containing indicated BSA amounts were spotted onto a nitrocellulose membrane. A cross indicates an incorrectly applied sample that was excluded for the analysis. (**B**) Ponceau S-stained dot blot. A quantity of 2 µL of spleen lysates from four mice was spotted in duplicates onto a nitrocellulose membrane. (**C**) Calculated protein concentrations of the spleen samples based on either the PDB or the BCA assay. (**D**) Retest of the protein concentrations of the spleen samples by different staining of the dot blot. Spleen samples containing 5 or 2.5 µg of total proteins (calculation based on either the PDB or the BCA assay), in addition to the BSA standards (2 µg/µL), were spotted in duplicate onto a nitrocellulose membrane. The blot was sequentially stained with Ponceau S, Direct Blue 71 and Coomassie Blue R250, with destaining between stainings. The integrated density values were then plotted: (**E**,**H**) for Ponceau S staining; (**F**,**I**) for Direct Blue 71 staining; (**G**,**J**) for Coomassie Blue R250 staining. Error bars indicate SEM. p-values were calculated using Student's t-test.

Both standard curves had to be extrapolated in order to calculate the protein concentrations of the spleen lysates because of high protein contents, which should generally be avoided. The BCA assay showed sample concentrations above the range of the respective standard curve. It would be necessary to repeat the measurement either with diluted samples or with another serial of BSA standards. In contrast, the calculation of the protein concentrations based on the PDB assay is less problematic because the standard curve for the PDB assay shows a linear correlation until 12 µg total protein amount per dot (Figure S1). It should be mentioned that, in routine experiments where the protein concentrations are estimated to be low, such high standards might not be necessary.

Strikingly, the protein concentrations of the spleen lysates based on the PDB assay were three to four times higher than those determined by the BCA assay (Figure 3C). To validate the results, we performed a PDB assay spotted with 5 and 2.5 µg of total proteins, which was based on the concentrations determined by either the BCA or the PDB assay (Figure 3D, left panel). Comparison of the integrated densities revealed a higher variability of the dots based on the BCA assay (Figure 3E,H). Importantly, the integrated densities of the dots based on the BCA assay were generally much higher than those of the BSA standards. In contrast, the integrated densities of the dots based on the PDB assay were similar to those of the BSA standard (Figure 3E,H). This retest suggests that the PDB assay is more reliable than the BCA assay, whereas the latter may significantly underestimate the total protein contents of unknown samples

Protein inputs on blots can also be stained with other dyes. Different dyes may have different binding modes to proteins. Unequal binding to heterogenous proteins will lead to a bias in total protein quantification. Ponceau S binds to positively charged amino acids and nonpolar regions of proteins [14]; therefore, it stains heterogenous proteins relatively equally. To compare Ponceau S staining with two other staining methods, we destained the same blot and re-stained it with Direct Blue 71 [15] and Coomassie Blue R250 sequentially. The overall staining profiles and integrated densities were comparable between the three dyes, despite the fact that the BSA standards were less stained by Direct Blue 71 or Coomassie Blue R250 (Figure 3D–J). It is unclear whether this difference was due to each dye's binding preference or loss of BSA from the blot during extensive destaining. It is also apparent that Coomassie Blue R250 staining resulted in a high background. Due to its properties of having a simple recipe, fast procedure, relatively equal staining of heterogenous proteins, adequate sensitivity, and low background, Ponceau S staining is an excellent choice for protein quantification.

We have shown that the PDB assay is compatible with biological tissues lysed in RIPA buffer, but some researchers may prefer 2x SDS LB because some proteins are known to be insoluble in RIPA buffer, leading to an overall loss of 10% to 30% of all proteins [9]. We therefore tested the compatibility of the PDB assay with samples lysed in 2x SDS LB. Because an initial experiment with BSA diluted in 2x SDS LB did not yield a linear standard curve (data not shown), we decided to add a washing step to wash out the SDS from the dots, which we suspected interfered with the binding of Ponceau S to the proteins. Washing the membrane three times with DI water (5 min each) after drying it for 15 min allowed us to generate a standard curve similar to that based on undiluted BSA standards (Figure S3), allowing the PDB assay to measure samples lysed in 2x SDS LB. RIPA buffer emerged to be unsuitable to dilute BSA standards because it results in the formation of coffee rings on blots (Figure S4).

To compare the protein extraction efficiency between 2x SDS LB and RIPA buffer, we performed a PDB assay with tissues lysed in these two buffers and conducted a subsequent Western blot. Two sciatic nerves and cerebral hemispheres, each from one mouse (two mice in total), were pooled and lysed. Due to their high lipid content, both tissues were hard to lyse. We measured the total protein contents of the lysates by the PDB assay, and then loaded the lysates containing 50, 25, and 15 μg of total proteins for SDS-PAGE and Western blot (Figure 4C).

Ponceau S staining of the membrane after protein transfer revealed equal loading of the samples (Figure S5), indicating that the PDB assay performs similarly as lysates in RIPA buffer or 2x SDS LB. Interestingly, we observed striking differences in the band intensities between the nerve and the brain lysates. We hypothesize that this is due to a high abundance of albumin (the strong band below 70 kDa) and the IgG heavy (the strong band slightly above 55 kDa) and light chain (the strong band between 25 and 35 kDa) in the peripheral nerves, which are absent in the brains because of the blood–brain barrier [16,17]. This circumstance will lead to an overestimation of the actual protein content of the sciatic nerve and thus biased loading. This could be a major issue when comparing protein expression between the peripheral and the central nervous systems.

Our Western blot results showed that the cytoskeleton-associated protein merlin, as well as the nuclear protein Histone H1, the autophagic vesicle membrane proteins LC3A/B, and the myelin basic protein MBP, were better extracted by 2x SDS LB than by RIPA buffer (Figure 4C). In contrast, the cytoplasmic proteins MEK 1/2, ERK 1/2, GAPDH, and GAP-43 were indistinguishably extracted by the two buffers. These findings confirm the literature reports that 2x SDS lyses biological tissues more effectively than RIPA buffer, which can be attributed to the insolubility of cytoskeleton-associated and extracellular matrix proteins in RIPA [12].

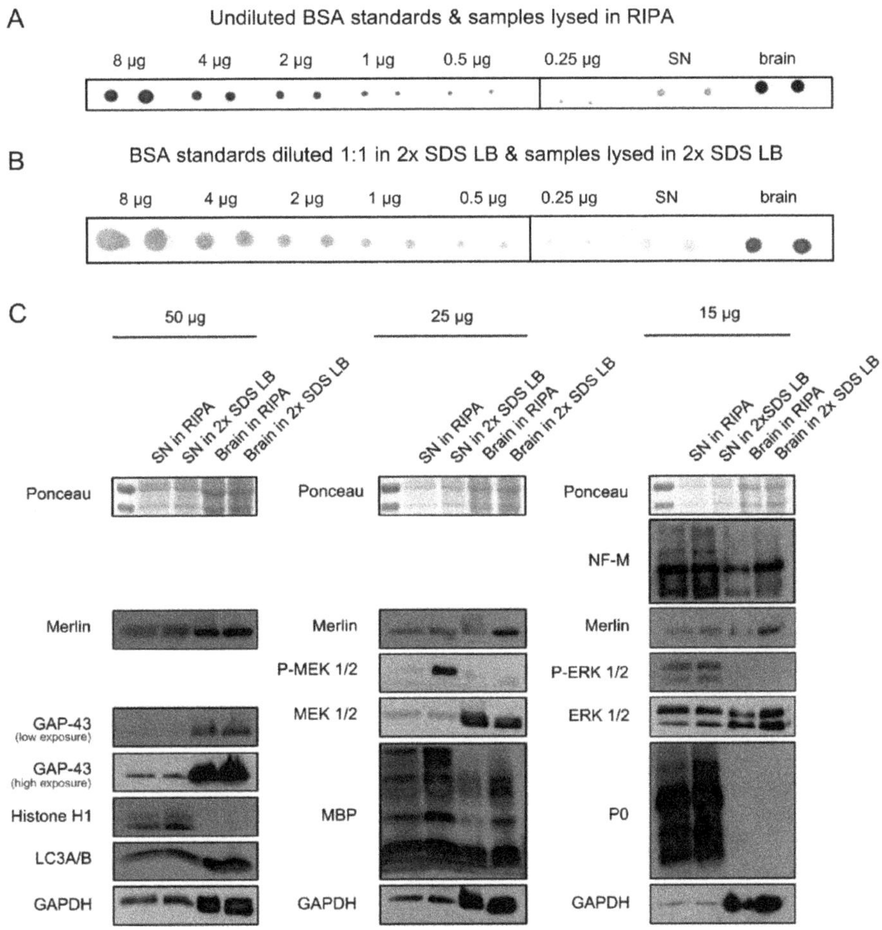

Figure 4. Western blots based on protein quantification with the PDB assay. (**A**,**B**) Ponceau S-stained dot blot. Undiluted BSA standards and BSA standards diluted 1:1 in 2x SDS lysis buffer were spotted in duplicate onto a membrane (fixed concentration, variable volumes). A quantity of 1 µL of sciatic nerve (SN) and brain samples lysed in 2x SDS LB or RIPA buffer was also applied onto the same membrane for quantification with the PDB assay. (**C**) The nerve and brain lysates containing 50, 25 or 15 µg of total proteins (based on the PDB assay) were loaded for SDS-PAGE and Western blot. After protein transfer to a nitrocellulose membrane, the membrane was also stained with Ponceau S.

To our surprise, p-ERK1/2 and p-MEK1/2 levels were higher in 2x SDS LB lysates than in RIPA lysates. We could not exclude the possibility that p-ERK1/2 and p-MEK1/2 in the respective nerves were indeed different. If this is not the case, the result suggests that RIPA buffer, which contains a designated phosphatase buffer, is inferior to 2x SDS LB in terms of phosphatase inhibition, which may be explained by the strong denaturing effect of SDS. This observation may be important for studying rapidly changing signaling processes, such as degeneration and regeneration in the nervous system [18,19], where 2x SDS LB may be a better choice for tissue lysis.

In the course of our experiments, we found that the PDB assay, especially when using 2x SDS LB instead of RIPA buffer to lyse biological tissue, can be much cheaper than the established workflow in our laboratory, which relies on the widely used BCA assay for protein quantification. Therefore, we calculated the costs for quantification of 12 samples with a commercial or a self-made BCA assay, or a PDB assay. In this study, we only used the commercial BCA assay, and the reagents can be self-made. The self-made BCA kit would cost between EUR 15.29 and 24.91, a commercial BCA kit was EUR 13.47, and the PDB assay cost only EUR 2.05 for quantification of 12 samples. From our experience, Ponceau S solution can be reused at least 20 times to stain transfer membranes or dot blots. A laboratory quantifying 1000 samples with the PDB instead of the BCA assay would save about EUR 950 (Figure S6). Because we have shown that direct lysis of biological tissues in 2x SDS LB is compatible with protein quantification by PDB assay and superior to RIPA buffer in terms of protein solubilization, we also checked the overall price difference between RIPA and 2x SDS LB, which amounted to a difference of between EUR 1380 and 2310 for 1000 lysates (Figure S7). One laboratory could save over EUR 2300 per 1000 lysates by switching from tissue lysis with RIPA buffer and protein quantification with the BCA assay to 2x SDS LB and our PDB assay.

The time needed for a PDB assay or a BCA assay is comparable: both assays will take around 45 min from sample preparation to measurement with a plate reader (BCA assay) or to scanning (PDB assay); the subsequent analysis including standard curve preparation, linear regression, and calculation of protein concentrations is almost the same.

Previously, Morcol et al. described a similar method to quantify purified proteins [8]. Here, we demonstrate an improved PDB assay that can be used to quantify tissue lysates and is compatible with 2x SDS LB. The PDB assay has the following advantages: (1) it can be easily adjusted to quantify samples with very high or low protein contents; (2) it requires far fewer materials than the BCA assay, saving valuable samples; (3) it only needs a simple laboratory equipment; and (4) it is significantly cheaper and more reliable than the BCA assay, thereby facilitating downstream experiments relying on correct sample inputs.

4. Conclusions

The PDB assay is a cheap and reliable method for quantifying proteins in tissue lysates.

Supplementary Materials: The following are available online at https://www.mdpi.com/article/10.3390/gels8010043/s1. Figure S1: Extended range for the PDB assay; Figure S2: The integrated density is determined by the protein amount instead of the dot size; Figure S3: Standard curves related to Figure 4A,B; Figure S4: BSA standards diluted with RIPA buffer forms "coffee rings" on dot blots; Figure S5: Ponceau S staining of the membrane after SDS-PAGE and protein transfer, uncropped, related to Figure 4C; Figure S6: Cost comparison between the PDB and the BCA assays; Figure S7: Cost comparison between the PDB assay with samples lysed in 2x SDS LB and the BCA assay with samples lysed in RIPA buffer; Video S1: Workflow of the PDB assay.

Author Contributions: Conceptualization, D.L.H., L.B. and Y.C.; methodology, D.L.H., L.B., Y.C. and N.O.; software D.L.H. and L.B.; validation D.L.H. and L.B.; formal analysis D.L.H. and L.B.; investigation, D.L.H. and L.B.; resources, D.L.H., L.B. and Y.C.; data curation, D.L.H. and L.B.; writing—original draft preparation, D.L.H., L.B., Y.C. and L.K.S.; writing—review and editing, Y.C., D.L.H., L.B. and L.K.S.; visualization, D.L.H., L.B. and N.O.; supervision, D.L.H. and Y.C.; project administration, D.L.H., L.B. and Y.C.; funding acquisition, Y.C. All authors have read and agreed to the published version of the manuscript.

Funding: FLI is a member of the Leibniz Association and is financially supported by the Federal Government of Germany and the State of Thuringia. This work was supported in part by funding from the Deutsche Forschungsgemeinschaft (DFG) granted to YC (MO 1421/5-1).

Institutional Review Board Statement: Not applicable.

Informed Consent Statement: Not applicable.

Data Availability Statement: All data underlying the results are available as part of the article and no additional source data are required.

Acknowledgments: The authors would like to thank Helen Morrison for providing laboratory space and support, and Debra Weih for critical reading and editing of the manuscript.

Conflicts of Interest: The authors declare no conflict of interest.

References

1. Smith, P.K.; Krohn, R.I.; Hermanson, G.T.; Mallia, A.K.; Gartner, F.H.; Provenzano, M.D.; Fujimoto, E.K.; Goeke, N.M.; Olson, B.J.; Klenk, D.C. Measurement of protein using bicinchoninic acid. *Anal. Biochem.* **1985**, *150*, 76–85. [CrossRef]
2. Lowry, O.H.; Rosebrough, N.J.; Farr, A.L.; Randall, R.J. Protein measurement with the Folin phenol reagent. *J. Biol. Chem.* **1951**, *193*, 265–275. [CrossRef]
3. Bradford, M.M. A rapid and sensitive method for the quantitation of microgram quantities of protein utilizing the principle of protein-dye binding. *Anal. Biochem.* **1976**, *72*, 248–254. [CrossRef]
4. Morton, R.E.; Evans, T.A. Modification of the bicinchoninic acid protein assay to eliminate lipid interference in determining lipoprotein protein content. *Anal. Biochem.* **1992**, *204*, 332–334. [CrossRef]
5. Kessler, R.J.; Fanestil, D.D. Interference by lipids in the determination of protein using bicinchoninic acid. *Anal. Biochem.* **1986**, *159*, 138–142. [CrossRef]
6. Bannur, S.V.; Kulgod, S.V.; Metkar, S.S.; Mahajan, S.K.; Sainis, J.K. Protein Determination by Ponceau S Using Digital Color Image Analysis of Protein Spots on Nitrocellulose Membranes. *Anal. Biochem.* **1999**, *267*, 382–389. [CrossRef] [PubMed]
7. Goldman, A.; Harper, S.; Speicher, D.W. Detection of Proteins on Blot Membranes. *Curr. Protoc. Protein Sci.* **2016**, *86*, 10.8.1–10.8.11. [CrossRef] [PubMed]
8. Morçöl, T.; Subramanian, A. A Red-Dot-Blot Protein Assay Technique in the Low Nanogram Range. *Anal. Biochem.* **1999**, *270*, 75–82. [CrossRef] [PubMed]
9. Janes, K.A. An analysis of critical factors for quantitative immunoblotting. *Sci. Signal.* **2015**, *8*, rs2. [CrossRef] [PubMed]
10. Ortega Ibarra, J.M.; Cifuentes-Castro, V.H.; Medina-Ceja, L.; Morales-Villagran, A. Nano dot blot: An alternative technique for protein identification and quantification in a high throughput format. *J. Neurosci. Methods* **2021**, *358*, 109194. [CrossRef] [PubMed]
11. Qi, X.; Zhang, Y.; Zhang, Y.; Ni, T.; Zhang, W.; Yang, C.; Mi, J.; Zhang, J.; Tian, G. High Throughput, Absolute Determination of the Content of a Selected Protein at Tissue Levels Using Quantitative Dot Blot Analysis (QDB). *J. Vis. Exp.* **2018**, e56885. [CrossRef] [PubMed]
12. Tian, G.; Tang, F.; Yang, C.; Zhang, W.; Bergquist, J.; Wang, B.; Mi, J.; Zhang, J. Quantitative dot blot analysis (QDB), a versatile high throughput immunoblot method. *Oncotarget* **2017**, *8*, 58553–58562. [CrossRef] [PubMed]
13. Morrison, H.; Sherman, L.S.; Legg, J.; Banine, F.; Isacke, C.; Haipek, C.A.; Gutmann, D.H.; Ponta, H.; Herrlich, P. The NF2 tumor suppressor gene product, merlin, mediates contact inhibition of growth through interactions with CD44. *Genes Dev.* **2001**, *15*, 968–980. [CrossRef] [PubMed]
14. Moritz, C.P. Tubulin or Not Tubulin: Heading Toward Total Protein Staining as Loading Control in Western Blots. *Proteomics* **2017**, *17*, 1600189. [CrossRef] [PubMed]
15. Hong, H.-Y.; Yoo, G.-S.; Choi, J.-K. Direct Blue 71 staining of proteins bound to blotting membranes. *Electrophoresis* **2000**, *21*, 841–845. [CrossRef]
16. Olsson, Y.; Klatzo, I.; Sourander, P.; Steinwall, O. Blood-brain barrier to albumin in embryonic new born and adult rats. *Acta Neuropathol.* **1968**, *10*, 117–122. [CrossRef] [PubMed]
17. Seitz, R.J.; Heininger, K.; Schwendemann, G.; Toyka, K.V.; Wechsler, W. The mouse blood-brain barrier and blood-nerve barrier for IgG: A tracer study by use of the avidin-biotin system. *Acta Neuropathol.* **1985**, *68*, 15–21. [CrossRef] [PubMed]
18. Kim, E.K.; Choi, E.-J. Pathological roles of MAPK signaling pathways in human diseases. *Biochim Biophys Acta.* **2010**, *1802*, 396–405. [CrossRef]
19. Napoli, I.; Noon, L.; Ribeiro, S.; Kerai, A.P.; Parrinello, S.; Rosenberg, L.H.; Collins, M.; Harrisingh, M.C.; White, I.J.; Woodhoo, A.; et al. A Central Role for the ERK-Signaling Pathway in Controlling Schwann Cell Plasticity and Peripheral Nerve Regeneration In Vivo. *Neuron* **2012**, *73*, 729–742. [CrossRef]

gels

Article

Modulation of the Structure and Stability of Novel Camel Lens Alpha-Crystallin by pH and Thermal Stress

Ajamaluddin Malik [1,*], Javed Masood Khan [2], Abdullah S. Alhomida [1] and Mohammad Shamsul Ola [1]

[1] Department of Biochemistry, College of Science, King Saud University, Riyadh 11451, Saudi Arabia; alhomida@ksu.edu.sa (A.S.A.); mola@ksu.edu.sa (M.S.O.)
[2] Department of Food Science and Nutrition, Faculty of Food and Agricultural Sciences, King Saud University, Riyadh 11451, Saudi Arabia; jmkhan@ksu.edu.sa
* Correspondence: amalik@ksu.edu.sa

Abstract: Alpha-crystallin protein performs structural and chaperone functions in the lens and comprises alphaA and alphaB subunits at a molar ratio of 3:1. The highly complex alpha-crystallin structure challenges structural biologists because of its large dynamic quaternary structure (300–1000 kDa). Camel lens alpha-crystallin is a poorly characterized molecular chaperone, and the alphaB subunit possesses a novel extension at the N-terminal domain. We purified camel lens alpha-crystallin using size exclusion chromatography, and the purity was analyzed by gradient (4–12%) sodium dodecyl sulfate–polyacrylamide gel electrophoresis. Alpha-crystallin was equilibrated in the pH range of 1.0 to 7.5. Subsequently, thermal stress (20–94 °C) was applied to the alpha-crystallin samples, and changes in the conformation and stability were recorded by dynamic multimode spectroscopy and intrinsic and extrinsic fluorescence spectroscopic methods. Camel lens alpha-crystallin formed a random coil-like structure without losing its native-like beta-sheeted structure under two conditions: >50 °C at pH 7.5 and all temperatures at pH 2.0. The calculated enthalpy of denaturation, as determined by dynamic multimode spectroscopy at pH 7.5, 4.0, 2.0, and 1.0 revealed that alpha-crystallin never completely denatures under acidic conditions or thermal denaturation. Alpha-crystallin undergoes a single, reversible thermal transition at pH 7.5. The thermodynamic data (unfolding enthalpy and heat capacity change) and chaperone activities indicated that alpha-crystallin does not completely unfold above the thermal transition. Camels adapted to live in hot desert climates naturally exhibit the abovementioned unique features.

Keywords: alpha-crystallin; dynamic multimode spectroscopy; circular dichroism; fluorescence; thermal stability

Citation: Malik, A.; Khan, J.M.; Alhomida, A.S.; Ola, M.S. Modulation of the Structure and Stability of Novel Camel Lens Alpha-Crystallin by pH and Thermal Stress. *Gels* **2022**, *8*, 273. https://doi.org/10.3390/gels8050273

Academic Editors: Hiroyuki Takeno and Vijay Kumar Thakur

Received: 14 March 2022
Accepted: 22 April 2022
Published: 27 April 2022

1. Introduction

Alpha-crystallin belongs to the small heat shock protein (sHsp) superfamily, is highly expressed in the eye lens, and has at least two known functions. First, alpha-crystallin is a structural protein that maintains an appropriate refractive index (ability to focus light on the retina). Second, as a molecular chaperone, it maintains lens clarity throughout the lifespan of an organism [1]. Eye lens proteins are frequently exposed to environmental stress, including UV-radiation and high temperatures. The mature lens fibers lack a protein folding machinery and all organelles to minimize light scattering. Consequently, no new protein can be synthesized, and damaged proteins cannot be replaced. Therefore, the eye lens in all organisms must maintain damaged proteins such as alpha-crystallin in a soluble state throughout life. The camel has adapted to thrive in extreme desert climates, which includes high temperatures, solar radiation, dryness, and low nutrition. High ambient temperature and UV–Vis radiation may increase the lens temperature and induce protein misfolding and aggregation [2]. Epidemiological studies have shown a positive association between early-onset and a high grade of cataracts and prolonged sunlight exposure [3,4].

Therefore, the role of the camel lens in maintaining lenticular alpha-crystallin in a soluble state throughout its entire life presents challenges. Unfolding and aggregation of lenticular proteins cause lens opacity, resulting in cataract formation.

Lens alpha-crystallin is a large, heterogeneous multimeric protein comprising two subunits (alphaA and alphaB chains), each approximately 175 amino acids and exhibiting 60% homology. In the human eye lens, alpha-crystallin comprises 15–50 subunits of two homologous forms, alphaA and alphaB, each approximately 20 kDa [5,6] and at a 3:1 molar ratio [5]. Camels have evolved uniquely (anatomically, physiologically, and biochemically) to adapt to the scorching climate where most other mammals cannot survive. To our best knowledge, the camel eye lens has at least two novel features: the recruitment of high levels of taxon-specific zeta-crystallin and the presence of an extended N-terminal domain in the alphaB-crystallin protein. Camel alphaA-crystallin (XP_010998042.1) comprises 173 amino acid residues, identical to human alphaA-crystallin. However, camel alphaB-crystallin (XP_010984284.1) contains an additional 44 residues compared with human alphaB-crystallin (219 residues vs. 175 residues, respectively) [7].

The expression of alphaA-crystallin is primarily lens specific; in other tissues, it is expressed in trace amounts. By contrast, alphaB-crystallin expression is stress-inducible and widespread throughout the body, particularly in the heart, muscle, and brain [8,9]. AlphaB-crystallin overexpression is linked to several protein misfolding and neurodegenerative diseases, including myopathies [10,11], Parkinson's disease [12,13], Alzheimer's disease [14,15], Creutzfeldt–Jakob disease [16,17], multiple sclerosis [18,19], and cancer [20,21].

Alpha-crystallin acts as a "holdase" in an ATP-independent manner [1,22]. The size of the hetero-oligomeric quaternary structure of alpha-crystallin is diverse with an average molecular weight of 700 kDa, and its size ranges from 300 to 1000 kDa [5,23]. The size variation is caused by several factors (e.g., pH, ionic strength, temperature, and metal ions). Temperature is a critical factor for alpha-crystallin oligomerization [23,24]. Recently, we reported that the chaperone activity of camel alpha-crystallin is activated in a stepwise manner during heat stress [7]. Moreover, camel alpha-crystallin retains a native beta-sheeted dominant secondary structure up to 50 °C. High thermal stress (above 50 °C) leads to a structural transition in alpha-crystallin with a gain of a random-coiled-like structure without losing beta-sheeted content [7]. In previous studies of alpha-crystallin, temperature was associated with the single thermal transition and activation of its chaperone activity [25,26].

Thermal transition was reported to occur above 50 °C with a transition mid-point (T_m) of approximately 61–63 °C [7,27–29]. Interestingly, alpha-crystallin efficiently retained its function during and above thermal transition [7,30,31]. Surprisingly, the calculated enthalpy of denaturation (ΔH) for alpha-crystallin at pH 7.5 using different techniques [Differential scanning calorimetry (DSC), dynamic multimode spectroscopy (DMS), and Fourier transform infrared spectroscopy (FTIR)] was significantly lower than the theoretically estimated enthalpy of denaturation [7,32]. The experimentally calculated heat capacity change of alpha-crystallin denaturation (ΔCp) was also less than half of the theoretical (ΔCp) value [26,32,33]. Several reports have shown the structural integrity of alpha-crystallin under thermal denaturation at pH 7.5. However, ambiguities exist regarding the folding species of alpha-crystallin above the transition. Whether alpha-crystallin is fully unfolded, partially unfolded, or retains a native-like structure remains unclear [28,29,34,35]. Alpha-crystallin protein presents challenges for structural determination, and its crystal structure is unavailable. In particular, camel lens alpha-crystallin comprises an extended N-terminal domain and is poorly characterized. In the present study, we used multi-spectroscopic techniques to characterize the folding and thermodynamic characteristics of alpha-crystallin in the pH range of 1.0–7.5 and temperature range of 20–94 °C. Many proteins are unfolded at low pH due to loos of electrostatic interactions. Several types of forces such as ionic, hydrophobic, H-bond, and covalent interactions are responsible to maintain the structure–function relationship of the proteins. The change in medium may perturb these interactions and result in protein unfolding. In this study, we have evaluated the role of pH and temperature on the stability of alpha-crystallin.

2. Materials and Methods

Superdex 200 and Superdex 75 prepacked columns were obtained from GE Healthcare Life Sciences, Chicago, USA. The 4–12% gradient SDS-PAGE gels were purchased from Invitrogen and Bradford's reagent was obtained from Pierce. All other chemicals were of analytical grade.

2.1. Extraction and Purification of Alpha-Crystallin from Camel Lens

Fresh camel eye lenses were obtained from a local slaughterhouse and transported under chilled conditions. Two lenses were gently stirred in 50 mL of extraction buffer [20 mM sodium phosphate buffer (pH 7.8) containing 0.2 mM EDTA] for 30 min to extract the soluble lens protein. The supernatant was collected after centrifugation at 13,000 rpm for 15 min. Alpha-crystallin was purified using two different size exclusion chromatography columns (Superdex 200 and Superdex 75 gel filtration columns). The Superdex 200 and Superdex 75 gel filtration columns were equilibrated with 20 mM sodium phosphate buffer (pH 7.8) containing 0.2 mM EDTA. The clear soluble lens extract was passed through a Superdex 200 column, and the purity of the eluted fractions was evaluated by 4–12% gradient sodium dodecyl sulfate–polyacrylamide gel electrophoresis. Subsequently, fractions containing relatively pure alpha-crystallin were pooled and passed through a Superdex 75 gel filtration column. The fraction purity was re-analyzed by 4–12% sodium dodecyl sulfate–polyacrylamide gel electrophoresis [7]. The pure fractions were concentrated to 8 mg/mL and stored at −20 °C. The protein was quantified using the Bradford assay.

2.2. Equilibration of Alpha-Crystallin at Different pH Values

Camel lens alpha-crystallin (0.3 mg mL^{-1}) was equilibrated overnight with a 20 mM buffer (pH 1.0–7.5) at room temperature. To obtain the desired pH, the following buffers were used: KCl-HCl (pH 1.0 and 1.5), Gly-HCl (pH 2.0–3.0), acetate buffer (pH 4.0–5.0), and phosphate buffer (pH 6.0–7.5).

2.3. Far-UV CD Spectroscopy of Alpha-Crystallin at Different pH Values

Far-UV CD measurements of alpha-crystallin equilibrated at different pH values were obtained using a Chirascan^{Plus} spectropolarimeter (Applied Photophysics Ltd., London, UK) and coupled with a Peltier temperature controller. The far-UV CD spectra of alpha-crystallin were measured at a concentration of 0.3 mg/mL in a 1-mm-pathlength cuvette at 22 °C. Three spectra for each sample were scanned from 200 to 250 nm with a 1-nm bandwidth, and the data were collected at 0.5 s per point. The air baseline and buffer background were subtracted from each spectrum of alpha-crystallin.

2.4. Intrinsic Fluorescence Spectroscopy of Alpha-Crystallin at Different pH Values

The tryptophan fluorescence emission spectra of alpha-crystallin at different pH values were measured at room temperature using a Cary Eclipse Fluorescence Spectrophotometer (Agilent Technologies, Santa Clara, CA, USA) coupled with a Peltier temperature controller [36]. Alpha-crystallin (0.1 mg/mL) at different pH values (1.0 to 7.5) in a 10-mm-pathlength cuvette was excited at 295 nm (bandwidth, 5 nm each) to record tryptophan fluorescence emission spectra.

2.5. Dynamic Multimode Spectroscopy of Alpha-Crystallin at Different pH Values

DMS was performed using a Chirascan^{Plus} spectrophotometer [37]. Based on the observation of major secondary and tertiary structural transitions in camel lens alpha-crystallin with respect to pH values, four different pH values (1.0, 2.0, 4.0, and 7.5) were selected for detailed spectroscopic and thermodynamic studies. Camel lens alpha-crystallin (0.2 mg/mL) was dissolved in 20 mM buffer at pH 1.0, 2.0, 4.0, and 7.5, and temperature-dependent conformational changes were measured in 1-mm-pathlength cuvettes using internal temperature probes. Alpha-crystallin was gradually heated from 20 °C to 94 °C at a rate of 1 °C/min, and far-UV CD spectra were recorded between 200 and 250 nm. The thermal transition data were processed using Chirascan Global 3 software provided by the manufacturer.

2.6. ANS (8-Anilino-1-naphthalene sulfonate) Fluorescence Measurements of Alpha-Crystallin at Different Temperatures and pH Values

ANS fluorescence of alpha-crystallin (0.2 mg mL^{-1}) at pH 1.0, 2.0, 4.0, and 7.5, respectively, were recorded at different temperatures (5 °C increments at each step) from 20 °C to 90 °C using a Peltier-controlled Cary Eclipse Fluorescence Spectrophotometer. ANS (50 μM) was added to the alpha-crystallin samples at pH 1.0, 2.0, 4.0, and 7.5, respectively, and the samples were equilibrated for 1 h at room temperature. The solution temperature was monitored using an internal temperature probe. ANS-treated alpha-crystallin was gradually heated and allowed to equilibrate for 5 min at each temperature step. The ANS fluorescence emission spectra were recorded between 400 and 600 nm (5.0-nm slit) by exciting the samples at 375 nm (2.5-nm slit).

3. Results

3.1. Effect of pH on the Secondary Structure of Camel Lens Alpha-Crystallin

Pure alpha-crystallin was obtained as previously described [7]. Far-UV CD (200–250 nm) was used to characterize the effect of acidic pH on the secondary structure of alpha-crystallin (0.3 mg mL^{-1}) (Figure 1). The far-UV CD spectra of alpha-crystallin at pH 7.5 revealed a single negative minimum at 217, which is a characteristic feature of beta-sheeted proteins. Changes in the negative minima of alpha-crystallin were insignificant, between pH 4.0 and 7.5 (Figure 1, inset), but the ellipticity at 217 nm gradually decreased as the pH was reduced from 7.5 to 5.0, indicating a loss of secondary structure (beta-sheeted structure) (Supplementary Figure S1). The maximum loss of secondary structure was observed at pH 5.0. The alpha-crystallin quickly regained a beta-sheeted structure below pH 5.0, particularly at pH 4.5, 4.0, and 3.0, respectively, which was close to the native secondary structure. Interestingly, at pH 3.0, 2.5, and 2.0, respectively, the alpha-crystallin secondary structure transformed into a random coil structure without affecting its beta-sheeted core structure (Figure 1). When alpha-crystallin was further incubated at pH 1.0 and 1.5, negative ellipticity was regained to that of native alpha-crystallin (Figure 1, inset). The far-UV CD data indicated that the secondary structure of alpha-crystallin was unstable with respect to pH changes. Figure 1 shows the different folding states of alpha-crystallin in which the far-UV CD minima varied with pH: native state (pH 7.5), beginning of the random-coil-like structure (pH 4.0), random-coiled structure (pH 2.0), and gain of the native-like secondary structure (pH 1.0).

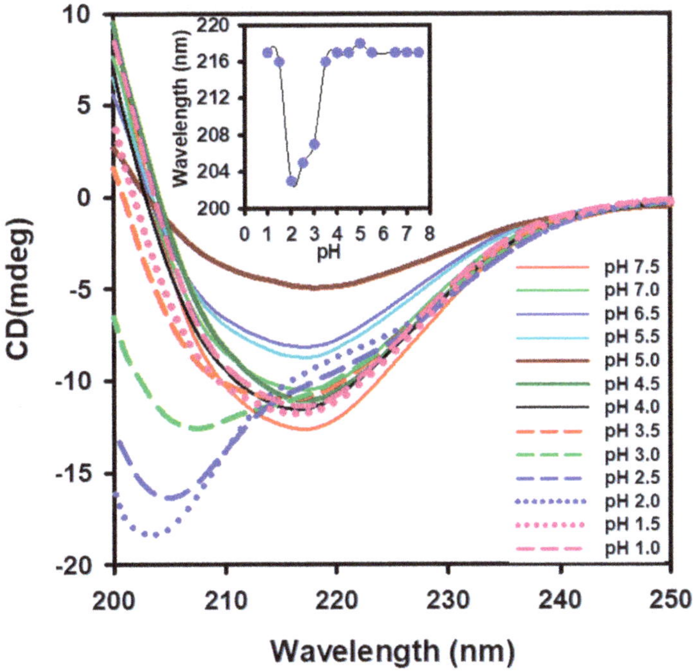

Figure 1. Far-UV circular dichroism (CD) spectra of 0.3 mg mL^{-1} of alpha-crystallin at pH 1.0 to 7.5. Changes in the ellipticity minima at different pH values are plotted in the inset figure. Below pH 3.5, alpha-crystallin gained a random coil-like structure but was restored to a native-like structure at pH 1.0.

3.2. Effect of pH on the Tertiary Structure of Alpha-Crystallin

Intrinsic fluorescence spectroscopy was used to investigate changes in the alpha-crystallin tertiary structure with respect to pH changes. Measurements of intrinsic fluorescence are useful readouts of the microenvironment surrounding aromatic residues and provide information regarding even subtle changes in the tertiary structure of proteins [38–40]. Figure 2 shows the tryptophan fluorescence spectra of alpha-crystallin at pH 1.0 to 7.5, revealing that alpha-crystallin at pH 7.5 exhibited a maximum fluorescence intensity at 336 nm. This finding confirmed that alpha-crystallin existed in a well-folded form. As the pH was reduced, the fluorescence emission maximum (λ_{max}) of alpha-crystallin was unchanged up to pH 5.5. However, below pH 5.5 down to pH 2.0, a gradual redshift in the λ_{max} was observed, indicating exposure of tryptophan residues to the polar environment (Figure 2, inset). The redshift in the wavelength maximum occurs when the microenvironment surrounding tryptophan residues becomes polar (aqueous), indicating protein unfolding or a loss of protein tertiary structure. The maximum redshift of alpha-crystallin was found at pH 2.0, indicating that the alpha-crystallin tertiary structure was maximally lost. Interestingly, at pH values below 2.0, the λ_{max} of the alpha-crystallin returned to 336 nm (i.e., the native-like structure), indicating that alpha-crystallin was again refolded at pH 2.0. The λ_{max} of alpha-crystallin was 352 nm; in the completely unfolded state (in 6 M GdnHCl), alpha-crystallin showed a λ_{max} of 363 nm, indicating a partially unfolded state of alpha-crystallin at pH 2.0.

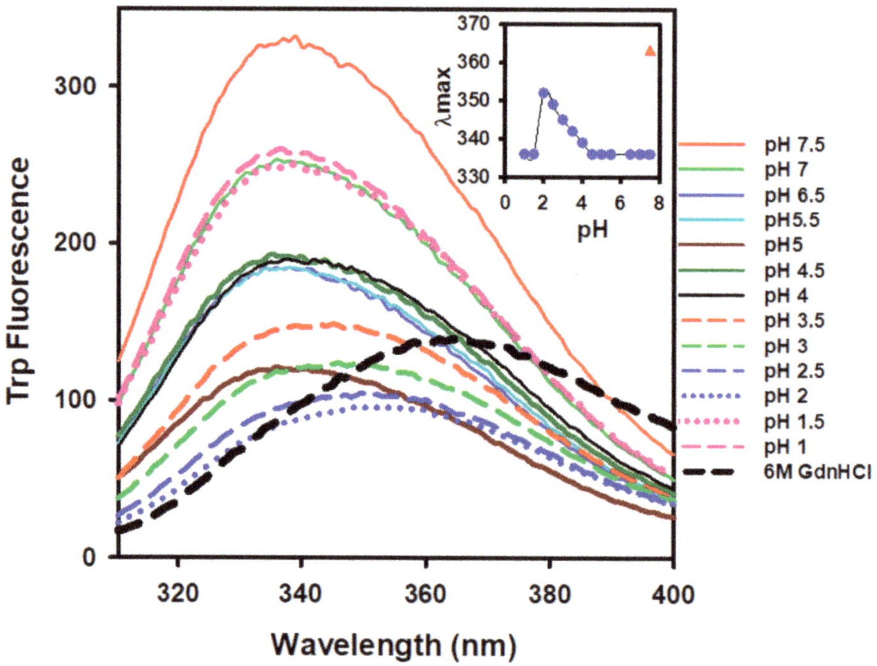

Figure 2. Tryptophan fluorescence spectra of camel lens alpha-crystallin at pH 1.0–7.5. Alpha-crystallin (0.1 mg mL^{-1}) equilibrated at different pH values was excited at 295 nm. Emission spectra were collected from 300 to 400 nm at room temperature (bandwidth 5 nm each). Each spectrum recorded at a different pH is color-coded. The inset figure shows the λmax plot with respect to pH.

3.3. Changes in Surface Hydrophobicity at Selected pH Values

The partially unfolded states or molten globule states of proteins are frequently characterized by measuring changes in ANS fluorescence. ANS has a significantly lower binding affinity with native and fully denatured proteins because the appropriate hydrophobic patches are unavailable for ANS binding. However, the partially unfolded or molten globule state of the protein exposes hydrophobic patches and provides a suitable environment for ANS binding and producing high ANS fluorescence intensity [41,42]. The exposure of hydrophobicity of alpha-crystallin (0.2 mg mL^{-1}) at four different pH values (1.0, 2.0, 4.0, and 7.5, respectively) was measured at room temperature (Figure 3). Alpha-crystallin in the native state (pH 7.5) exhibited a poor ANS binding signal, indicating that alpha-crystallin in the native state has low surface hydrophobicity (Figure 3). This finding indicates that alpha-crystallin is well-folded. However, at pH 4.0, 2.0, and 1.0, respectively, the fluorescence intensity of ANS increased in response to a change in pH, confirming that the surface hydrophobicity of alpha-crystallin was increased with respect to the change in pH. The increased surface hydrophobicity with respect to pH resulted from the formation of the molten, globule-like state of alpha-crystallin.

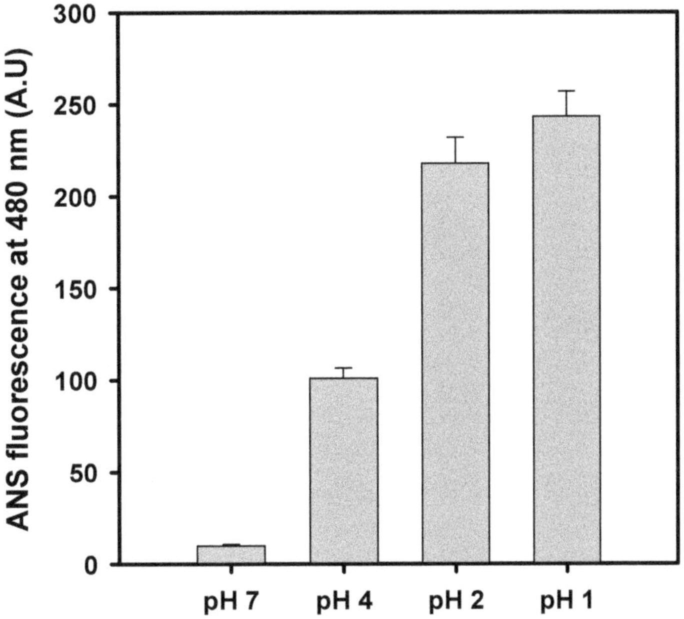

Figure 3. Surface hydrophobicity measurements of alpha-crystallin at four selected pH values. Alpha-crystallins (0.2 mg mL^{-1}) equilibrated at pH 1.0, 2.0, 4.0, and 7.5, respectively, were treated with 50 μM ANS. The samples were excited at 375 nm (2.5 nm slit) and the emission spectra were recorded between 400 and 600 nm (5.0 nm slit).

3.4. Thermodynamic and Spectroscopic Properties of Alpha-Crystallin at Different pH Values as Determined by Dynamic Multimode Spectroscopy

The spectroscopic and thermodynamic properties of alpha-crystallin at four different pH values (1.0, 2.0, 4.0, and 7.5) were examined by DMS [43]. Alpha-crystallin revealed distinct secondary and tertiary structures at pH 1.0, 2.0, 4.0, and 7.5, respectively. Therefore, we selected these pH values to evaluate the thermodynamic and folding characteristics of alpha-crystallin. Alpha-crystallin samples (0.2 mg mL^{-1}) at pH 1.0, 2.0, 4.0, and 7.5, respectively, were heat-stressed from 20 °C to 94 °C at a rate of 1 °C/min under identical conditions. The far-UV CD spectra (200 to 250 nm) were recorded as a function of temperature. Figure 4A–D shows the changes in the secondary structure conformation of alpha-crystallin at different temperatures and pH values. Alpha-crystallin in the native state (pH 7.5 and room temperature) exhibited a single minimum at 217 nm, representing a characteristic feature of a beta-sheeted rich protein (Figure 4A). The peaks at 217 nm were unchanged during heat stress (20–94 °C), indicating that the alpha-crystallin beta-sheeted core structure was preserved during heat stress. Moreover, the ellipticity at 217 nm was unchanged between 20 °C and 50 °C, indicating that the secondary structure was intact over this temperature range. Interestingly, above 50 °C, a sharp increase in the ellipticity minima at 203 nm was observed without altering the 217 nm ellipticity, indicating the formation of a random coil-like structure while maintaining the original beta-sheeted structure (Figure 4A, inset).

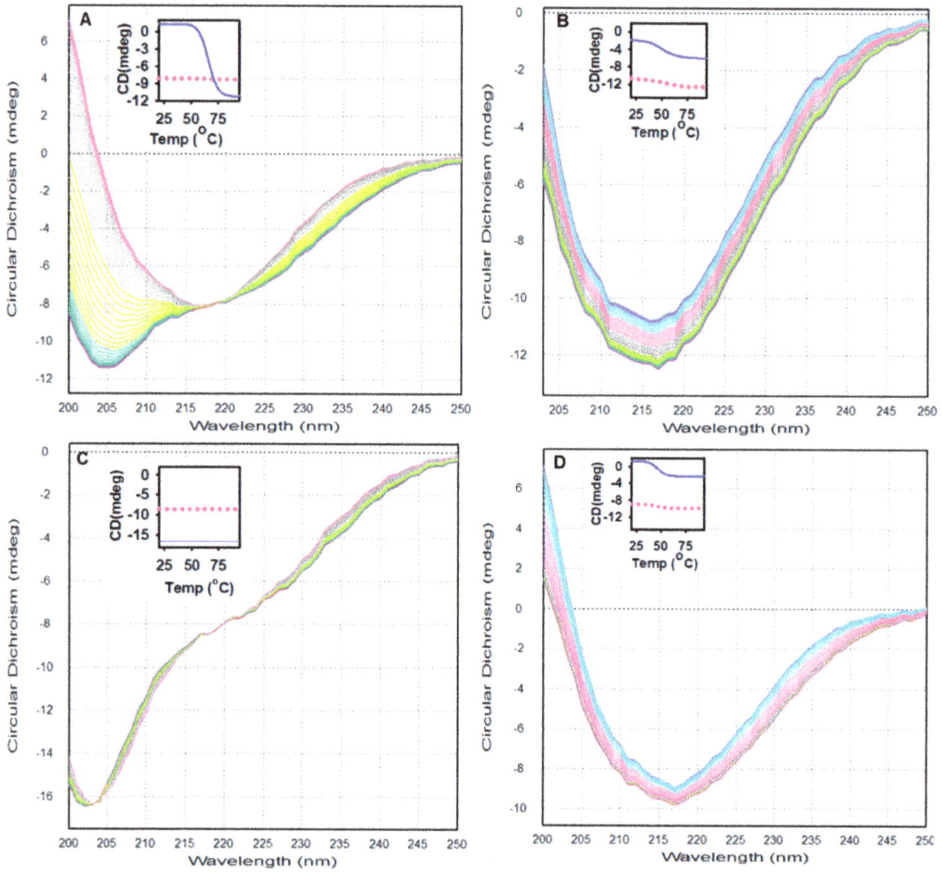

Figure 4. Far-UV circular dichroism (CD) spectra of camel lens alpha-crystallin at different temperatures and pH values. Alpha-crystallin (0.2 mg mL^{-1}) was heat-denatured at a constant rate (1 °C min^{-1}) at pH 7.5 (**A**), pH 4.0 (**B**), pH 2.0 (**C**), and pH 1.0 (**D**). Far-UV CD spectra were collected from 200 to 250 nm at intervals of 1 °C from 20 °C to 94 °C. In the inset figure, the blue line shows the changes at 203 nm and the pink dots at 217 nm, with respect to temperature.

Our results also demonstrated that far-UV CD ellipticity at 217 nm was unchanged at pH 7.5 and 2.0 (Figure 4A,C). By contrast, a slight increase in the ellipticity minima at 217 nm at pH 4.0 and 1.0 (Figure 4B,D) was detected during thermal denaturation. These results indicated that the core beta-sheeted structure of alpha-crystallin remained intact during thermal denaturation (Figure 4). Moreover, the far-UV CD spectra of thermally stressed (>80 °C) alpha-crystallin at pH 7.5 were similar to those of alpha-crystallin at pH 2.0 at all temperatures (20 °C–94 °C). These data showed the presence of random-coiled and beta-sheeted structures under two conditions: alpha-crystallin above 50 °C at pH 7.5 and alpha-crystallin at all temperatures at pH 2.0 (Figure 4A,C). The far-UV CD spectra of alpha-crystallin at pH 4.0 and 1.0 were similar to native-like alpha-crystallin, and these conformations did not undergo any major structural transitions during heat stress, except a slight gain of ellipticity minima at 217 nm (Figure 4B,D). Moreover, thermal stress at pH 1.0, 2.0, 4.0, and 7.5, respectively was reversible, and no aggregation was detected.

The thermal transition midpoints (T_m) and enthalpy of alpha-crystallin at pH 1.0, 2.0, 4.0, and 7.5, respectively, were determined (Table 1) using Global 3 analysis software provided by Applied Photophysics Ltd., UK. The three-dimensional model of the thermal transitions in alpha-crystallin at pH 1.0, 2.0, 4.0, and 7.5, respectively, was generated using Global 3 analysis software (Figure 5).

Table 1. Thermal transition midpoints TM and enthalpies of alpha-crystallin at pH 1.0, 2.0, 4.0, and 7.5, respectively.

pH	Van't Hoff Enthalpy (kJ/mol)	Transition Temperature (°C)
7.5	237.0 ± 1.9	60.9 ± 0.1
4.0	108.9 ± 4.7	48.1 ± 0.4
2.0	177.3 ± 9.2	59.1 ± 0.4
1.0	211.3 ± 10.2	43.5 ± 0.3

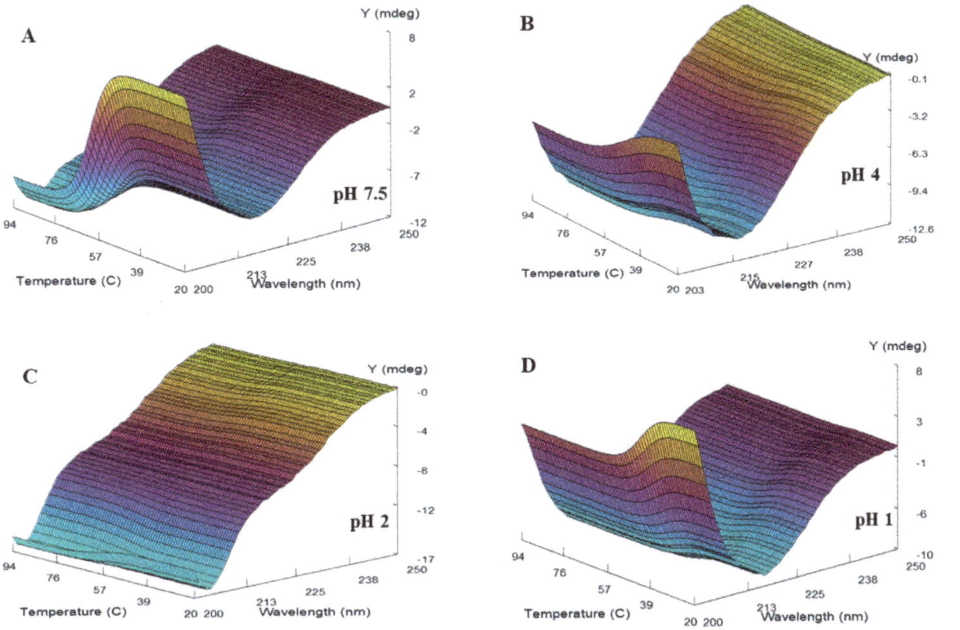

Figure 5. Calculated temperature, wavelength, and far-UV CD signal of alpha-crystallin at different pHs. The three-dimensional model of alpha-crystallin at pH 7.5 (**A**), 4.0 (**B**), 2.0 (**C**), and 1.0 (**D**), respectively, was calculated using Global 3 software and the far-UV CD signal obtained during temperature ramping (1 °C min^{-1}).

3.5. Changes in the Surface Hydrophobicity of Alpha-Crystallin at Different Temperatures and pH Values

Extrinsic fluorophore ANS was used to monitor the exposure of hydrophobic patches in alpha-crystallin in response to thermal stress. ANS fluorescence of alpha-crystallin at pH 1.0, 2.0, 4.0, and 7.5, respectively, was evaluated at different temperatures, from 20 °C to 90 °C (Figure 6). When alpha-crystallin at pH 7.5 was heat stressed from 20 °C, a slight redshift in the wavelength maxima was observed above 35 °C, and a sharp redshift was observed above 55 °C (Figure 6A), indicating exposure of hydrophobic residues at the surface of alpha-crystallin in response to heat stress. Alpha-crystallin at pH 7.5 exhibited low ANS fluorescence intensity and displayed poor ANS binding with native state alpha-crystallin at pH 7.5 (Figure 6A). Because ANS is a temperature-sensitive probe, a gradual decrease in ANS fluorescence intensity was observed at all pH values. At acidic pH (4.0, 2.0, and 1.0), the ANS fluorescence of alpha-crystallin exhibited increased fluorescence intensity resulting from the exposure of hydrophobic patches (Figure 6B–D). A slight redshift in emission was observed only above 65 °C when alpha-crystallin was at pH 1.0, 2.0, and 4.0, respectively (Figure 6E).

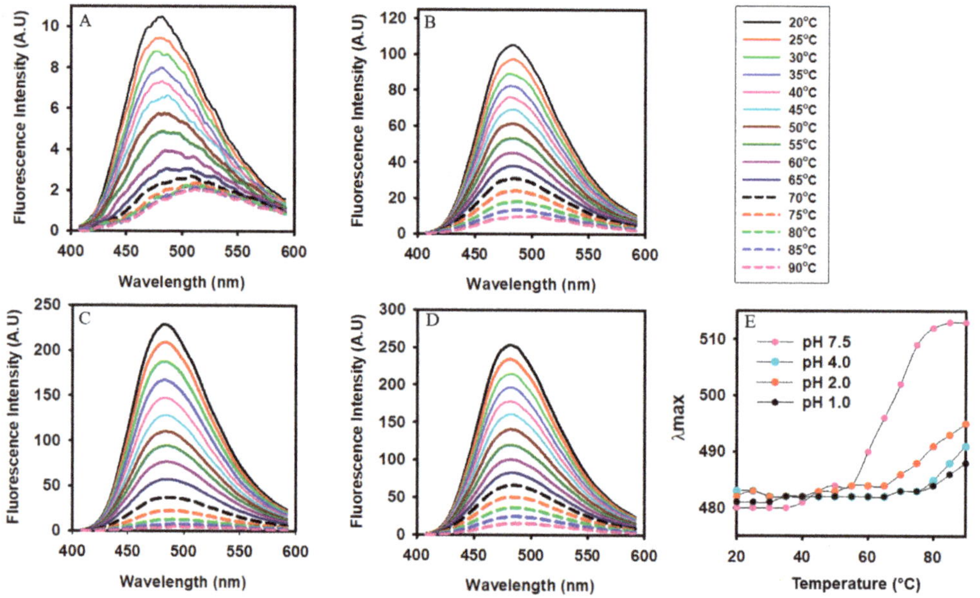

Figure 6. Differential scanning fluorometry of alpha-crystallin using an extrinsic fluorophore at four different pH values. ANS fluorescence was monitored by incubating alpha-crystallin at pH 7.5, 4.0, 2.0, and 1.0, respectively, from 20 °C to 90 °C by stepwise temperature increases of 5 °C. (**A**) pH 7.5, (**B**) pH 4.0, (**C**) pH 2.0, and (**D**) pH 1.0. (**E**) Effect of temperature on the λ_{max} of alpha-crystallin at pH 7.5, 4.0, 2.0, and 1.0, respectively.

4. Discussion

Ocular proteins are exposed to environmental stress (solar radiation and ambient temperature), making them susceptible to unfolding and aggregation. Alpha-crystallin naturally maintains ocular proteins in a soluble state. The Arabian camel has evolved to live in a stressful desert climate of intense heat, solar radiation, and dryness. The camel possesses several unique anatomical, physiological, and biochemical features to survive and thrive in the extreme desert climate [44–48]. The camel eye lens has two modifications with respect to crystallin proteins: it contains levels of zeta-crystallin (a taxon-specific crystallin) [49] and an extended N-terminal domain in the alphaB-crystallin protein [7]. Camel alphaA-crystallin (NCBI Reference Sequence: XP_010998042.1) is identical in length and shares 93% homology with human alphaA-crystallin, whereas camel alphaB-crystallin (NCBI Reference Sequence: XP_010984284.1) contains 44 additional residues at its N-terminus and shows an overall 78% similarity with human alphaB-crystallin [7].

The alpha-crystallin architecture is highly complex, and its quaternary structure changes (forms larger functional oligomers) as the temperature increases. Alpha-crystallin presents challenges for structural biologists to study. Despite several attempts, the crystal structure of the alpha-crystallin has not been solved. Its three-dimensional structure has been reconstituted using multiple techniques, including cryo-electron microscopy, NMR, dynamic light scattering, analytical ultracentrifugation, and structural modeling [50,51]. Because the hetero-oligomeric quaternary structure of alpha-crystallin changes in response to temperature, interpreting the structural changes in alpha-crystallin using spectroscopic techniques precisely is challenging. We used multiple spectroscopic methods (DMS, intrinsic, and ANS fluorescence) to characterize the changes in the secondary and tertiary structure and surface hydrophobicity of camel lens alpha-crystallin in response to changes in pH and temperature.

At pH 7.5, camel lens alpha-crystallin exhibited a single minimum at 217 nm, indicating the presence of a beta-sheeted dominant structure. Earlier studies reported a beta-sheeted dominant structure in bovine alpha-crystallin at pH 7.5 [32,34]. The effect of pH from 1.0 to 7.5 on the secondary structure of camel lens alpha-crystallin revealed that the single negative minimum remained at 217 nm between pH 7.5 and 4.0. During the pH scanning experiment, multiple conformational changes were detected in the secondary structure of alpha-crystallin. Initially, loss of the beta-sheeted structure occurred up to pH 5.0. A further reduction of pH to 4.0 induced a beta-sheeted structure, resulting in the formation of native-like alpha-crystallin. A subsequent reduction of pH resulted in major conformational changes, as shown in Figure 1. Below pH 4.0, alpha-crystallin contained native-like beta-sheeted and random-coiled structures. Interestingly, at pH 1.5 and 1.0, the far-UV CD spectra nearly overlapped with alpha-crystallin at pH 7.5, indicating a restoration of the native-like beta-sheeted structure and a loss of the random-coiled structure.

For a detailed investigation of the spectroscopic and thermodynamic properties of alpha-crystallin, four different pH values were selected: pH 7.5, native alpha-crystallin; pH 4.0, native-like alpha-crystallin; pH 2.0, alpha-crystallin with random coils; and pH 1.0, native-like alpha-crystallin. DMS based on far-UV CD spectroscopy was used to obtain thermodynamic and spectroscopic data. DMS is an information-rich technique that precisely determines changes in the secondary structure of proteins under different conditions over the entire temperature range [43,52]. Figure 4A–D and Table 1 show the thermodynamic and spectroscopic data obtained by DMS. Below 50 °C (pre-transition) and at pH 7.5, alpha-crystallin exists as a beta-sheeted dominant protein (Figure 4A). It retained a native beta-sheeted secondary structure up to 50 °C at pH 7.5. However, above 50 °C, the minima shifted toward 203 nm and maintained a shoulder at 217 nm. The ellipticity at 217 nm in the pre- and post-transition spectra remained the same. Chemical denaturants (urea and single minimum) induce a decrease in ellipticity at 217 nm (loss of beta-sheeted secondary structure) and subunit dissociation in the bovine alpha-crystallin [53]. Our study and earlier studies showed that the core of the beta-sheeted structures in alpha-crystallin at pH 7.5 remained intact under thermal denaturation temperatures up to 94 °C [28,32,34,35]. However, thermal stress above 50 °C induced a random-coil-like structure at pH 7.5.

Large conformational changes in alpha-crystallin occurred over the pH range of 1.0 to 7.5 (Figure 1). Initially, a loss of ellipticity at 217 nm occurred after shifting the pH from 7.5 to 5.0. Further reduction of the pH resulted in the formation of a beta-sheeted core structure in alpha-crystallin. Interestingly, the temperature did not affect the beta-sheeted secondary structure at all the pH values (7.5, 4.0, 2.0, and 1.0) tested; a slight gain of a beta-sheeted structure occurred at pH 4.0 and 1.0 (Figure 4). The formation of a random-coiled structure in alpha-crystallin occurred under two conditions: (i) >50 °C at pH 7.5 and (ii) 20 °C at pH 2.0.

Camel lens alpha-crystallin at pH undergoes a single thermal transition between 50 °C and 80 °C. After the thermal transition, whether alpha-crystallin was fully folded, partially folded, or fully denatured was unclear. However, the folding species was reversible. In the present study, the mid-point of the thermal transition (T_m) was 60.9 ± 0.1 °C, and the enthalpy of denaturation was 237.0 ± 1.9 kJ/mol. In earlier studies, the thermal stability and structural changes of eye lens alpha-crystallin from other sources were determined [28,35,54]. Alpha-crystallin underwent a single thermal transition (T_m) at approximately 61–64 °C [27–29]. The minor ambiguity in the T_m may be due to the difference in the detection methods (far-UV CD, DSC, or FTIR), buffer pH, ionic strength, or experimental conditions. The enthalpy of bovine lens alpha-crystallin was 235 kJ/mol by DSC [32]. Usually, the unfolding enthalpy of monomeric globular proteins of similar sizes is approximately 2.92 kJ/mol per residue [55]. Therefore, the calculated unfolding enthalpy of the alpha-crystallin subunits was more than 500 kJ/mol. Moreover, oligomerization increased the unfolding enthalpy of the proteins [56,57]. Although the crystal or NMR structure of alpha-crystallin is unavailable, data from multi-technique investigations have led to a consensus that subunit assembly of alpha-crystallin is controlled by the hydrophobicity of the N-terminal domains [50,51,58]. Therefore, dissociation of the oligomeric structure would result in the exposure of the buried hydrophobic patches, leading to an endothermic effect, which increases the overall unfolding enthalpy. Therefore, the calculated unfolding enthalpy of oligomeric alpha-crystallin at pH 7.5 may be much higher than 500 kJ/mol. Less than half of the enthalpic changes were determined in this study and earlier studies [32]. Accordingly, the DMS data presented in this work and an earlier study revealed that alpha-crystallin retained a secondary structure above the thermal transition (T_m). These data also indicate that the thermal transition of alpha-crystallin does not dissociate its subunits. Denaturation without subunit dissociation was also observed in the Arc repressor [59]. The presence of a secondary structure in the alpha-crystallin or Arc repressor enabled a persistent interaction among subunits, maintaining the oligomeric structure during thermal stress.

Alpha-crystallin retained a native-like beta-sheeted secondary structure under thermal stress at pH 7.5, 4.0, 2.0, and 1.0, respectively. The unfolding enthalpies of alpha-crystallin at acidic pH values (4.0, 2.0, and 1.0, respectively) were less marked compared with those at pH 7.5 (Table 1). Thus, camel alpha-crystallin cannot be fully denatured at high temperatures (94 °C) and low pH values (e.g., pH 1.0).

Camel lens alpha-crystallin loses tertiary structure below pH 4.0 at room temperature (Figure 2). The tryptophan fluorescence spectra exhibited a 16 nm redshift at pH 2.0 and 20 °C. AlphaA-crystallin contains one tryptophan residue, whereas alphaB-crystallin contains two tryptophan residues [7]. These tryptophan residues are partially buried and located at the N-terminal domains (Supplementary Figure S2). Acid denaturation at pH 2.0 leads to partial unfolding of the N-terminal domain of alpha-crystallin. The complete unfolding of alpha-crystallin by chemical denaturants (6 M GdnHCl) resulted in a 27 nm redshift in tryptophan fluorescence (Figure 2). The far-UV CD data showed that camel alpha-crystallin retained a native-like beta-sheeted secondary structure at pH 2.0 but lost tertiary structure at this pH. Thus, alpha-crystallin forms a molten-like, globular structure at pH 2.0. Changes in the tertiary structure below pH 4.0 caused an increase in hydrophobicity (Figure 3). An increase in protonation below pH 2.0 caused charge–charge repulsion and forced alpha-crystallin to attain native-like secondary and tertiary structures (Figures 1 and 2). Moreover, the unfolding enthalpy at pH 1.0 was closer to that of alpha-crystallin at pH 7.5 (Table 1).

Alpha-crystallin at pH 7.5 exhibited a slight increase in hydrophobicity above 35 °C and a large exposure of hydrophobic patches above 55 °C. Alpha-crystallin at acidic pH values (4.0, 2.0, and 1.0, respectively) resulted in little exposure of hydrophobic patches above 65 °C (Figure 6). We recently reported that the chaperoning effect of alpha-crystallin was activated in a stepwise manner and correlated with the biphasic changes in the tertiary structure and surface hydrophobicity of alpha-crystallin during thermal stress [7]. Alpha-crystallin even retained its chaperone activity above the post-transition temperature (89 °C) at pH 7.5 [7]. These data suggest that above the thermal transition state (denatured state), alpha-crystallin is not denatured. It retains a core beta-sheeted structure, maintains an oligomeric state, and performs chaperone activity efficiently (~90% protection at 89 °C when the alpha-crystallin to substrate w/w ratio was 0.87:1) [7]. This state is possible only when each monomeric subunit of alpha-crystallin retains a substrate-binding site (i.e., remains in the functional state). Therefore, a higher temperature does not denature alpha-crystallin at pH 7.5 but results in conformational changes required to activate its chaperone activities.

5. Conclusions

In this study, the thermodynamic parameters obtained during the thermal transition of camel lens alpha-crystallin and those determined in earlier studies for other alpha-crystallins represent less than half of the theoretically calculated values for complete protein denaturation. Even at acidic pH values, the enthalpies were much lower than those at pH 7.5. This finding showed that alpha-crystallin was never completely denatured at an extreme pH or temperature or both. Folding species formed post-transition were neither dissociated nor aggregated and were reversible. Interestingly, the folding species formed post-transition at pH 7.5 remained fully active (i.e., they retained almost a 1:1 substrate binding site). Maintaining a functional state post-thermal transition is a feature that is incongruent with the unfolded state. This phenomenon may be due to the natural selection of alpha-crystallin to suppress aggregation in the lens under stress and maintain clarity in the lens throughout life. These tasks may be more challenging to perform in the camel eye lens, which is exposed to high temperatures, solar radiation, and dryness from the desert climate. To overcome these larger tasks, camel lens alpha-crystallin has the natural ability to retain secondary and oligomeric structures and maintain solubility and activity at extreme temperatures.

Supplementary Materials: The following supporting information can be downloaded at: https://www.mdpi.com/article/10.3390/gels8050273/s1, Figure S1. Changes in far-UV CD signal at 217 nm at different pH values are plotted. With a de-crease in pH, the far-UV CD ellipticities initially decreased between pH 7.5 and 5.0. The far-UV CD ellipticities at 217 nm rapidly increased to a native-like structure below pH 5.0; Figure S2. Modeled 3D structure of camel lens alpha-crystallin. The tryptophan residue in the A- and B-chain of alpha-crystallin is shown in blue: (A) alphaA-crystallin and (B) alphaB-crystallin.

Author Contributions: Conceptualization, A.M.; methodology, A.M. and J.M.K.; software, A.M.; formal analysis, A.M. and M.S.O.; resources, A.S.A. and M.S.O.; writing—original draft preparation, A.M. and J.M.K.; writing—review and editing, A.M., A.S.A. and J.M.K.; funding acquisition, J.M.K. All authors have read and agreed to the published version of the manuscript.

Funding: This research was funded by [King Saud University] grant number [RSP-2021/360] and The APC was funded by [RSP-2021/360].

Informed Consent Statement: Not applicable.

Data Availability Statement: Not applicable.

Acknowledgments: The authors are grateful to the Researchers Supporting Project Number (RSP-2021/360), King Saud University, Riyadh, Saudi Arabia.

Conflicts of Interest: The authors declare no conflict of interest.

References

1. Horwitz, J. Alpha-crystallin can function as a molecular chaperone. *Proc. Natl. Acad. Sci. USA* **1992**, *89*, 10449–10453. [CrossRef] [PubMed]
2. Al-Ghadyan, A.A.; Cotlier, E. Rise in lens temperature on exposure to sunlight or high ambient temperature. *Br. J. Ophthalmol.* **1986**, *70*, 421–426. [CrossRef] [PubMed]
3. Neale, R.E.; Purdie, J.L.; Hirst, L.W.; Green, A.C. Sun exposure as a risk factor for nuclear cataract. *Epidemiology* **2003**, *14*, 707–712. [CrossRef] [PubMed]
4. Heys, K.R.; Friedrich, M.G.; Truscott, R.J. Presbyopia and heat: Changes associated with aging of the human lens suggest a functional role for the small heat shock protein, alpha-crystallin, in maintaining lens flexibility. *Aging Cell* **2007**, *6*, 807–815. [CrossRef] [PubMed]
5. Horwitz, J. Alpha-crystallin. *Exp. Eye Res.* **2003**, *76*, 145–153. [CrossRef]
6. Augusteyn, R.C. Alpha-crystallin: A review of its structure and function. *Clin. Exp. Optom.* **2004**, *87*, 356–366. [CrossRef] [PubMed]
7. Malik, A.; Almaharfi, H.A.; Khan, J.M.; Hisamuddin, M.; Alamery, S.F.; Haq, S.H.; Ahmed, M.Z. Protection of zeta-crystallin by alpha-crystallin under thermal stress. *Int. J. Biol. Macromol.* **2021**, *167*, 289–298. [CrossRef]

8. Srinivasan, A.N.; Nagineni, C.N.; Bhat, S.P. Alpha A-crystallin is expressed in non-ocular tissues. *J. Biol. Chem.* **1992**, *267*, 23337–23341. [CrossRef]
9. Bhat, S.P.; Nagineni, C.N. Alpha B subunit of lens-specific protein alpha-crystallin is present in other ocular and non-ocular tissues. *Biochem. Biophys. Res. Commun.* **1989**, *158*, 319–325. [CrossRef]
10. Selcen, D.; Engel, A.G. Myofibrillar myopathy caused by novel dominant negative alpha B-crystallin mutations. *Ann. Neurol.* **2003**, *54*, 804–810. [CrossRef]
11. Fischer, D.; Matten, J.; Reimann, J.; Bonnemann, C.; Schroder, R. Expression, localization and functional divergence of alphaB-crystallin and heat shock protein 27 in core myopathies and neurogenic atrophy. *Acta Neuropathol.* **2002**, *104*, 297–304. [CrossRef] [PubMed]
12. Liu, Y.; Zhou, Q.; Tang, M.; Fu, N.; Shao, W.; Zhang, S.; Yin, Y.; Zeng, R.; Wang, X.; Hu, G.; et al. Upregulation of alphaB-crystallin expression in the substantia nigra of patients with Parkinson's disease. *Neurobiol. Aging* **2015**, *36*, 1686–1691. [CrossRef] [PubMed]
13. Klettner, A.; Richert, E.; Kuhlenbaumer, G.; Nolle, B.; Bhatia, K.P.; Deuschl, G.; Roider, J.; Schneider, S.A. Alpha synuclein and crystallin expression in human lens in Parkinson's disease. *Mov. Disord.* **2016**, *31*, 600–601. [CrossRef] [PubMed]
14. Narayanan, S.; Kamps, B.; Boelens, W.C.; Reif, B. AlphaB-crystallin competes with Alzheimer's disease beta-amyloid peptide for peptide-peptide interactions and induces oxidation of Abeta-Met35. *FEBS Lett.* **2006**, *580*, 5941–5946. [CrossRef] [PubMed]
15. Mao, J.J.; Katayama, S.; Watanabe, C.; Harada, Y.; Noda, K.; Yamamura, Y.; Nakamura, S. The relationship between alphaB-crystallin and neurofibrillary tangles in Alzheimer's disease. *Neuropathol. Appl. Neurobiol.* **2001**, *27*, 180–188. [CrossRef] [PubMed]
16. Wang, K.; Zhang, J.; Xu, Y.; Ren, K.; Xie, W.L.; Yan, Y.E.; Zhang, B.Y.; Shi, Q.; Liu, Y.; Dong, X.P. Abnormally upregulated alphaB-crystallin was highly coincidental with the astrogliosis in the brains of scrapie-infected hamsters and human patients with prion diseases. *J. Mol. Neurosci.* **2013**, *51*, 734–748. [CrossRef]
17. Renkawek, K.; de Jong, W.W.; Merck, K.B.; Frenken, C.W.; van Workum, F.P.; Bosman, G.J. Alpha B-crystallin is present in reactive glia in Creutzfeldt-Jakob disease. *Acta Neuropathol.* **1992**, *83*, 324–327. [CrossRef]
18. van Noort, J.M.; Bsibsi, M.; Nacken, P.J.; Verbeek, R.; Venneker, E.H. Therapeutic Intervention in Multiple Sclerosis with Alpha B-Crystallin: A Randomized Controlled Phase IIa Trial. *PLoS ONE* **2015**, *10*, e0143366. [CrossRef]
19. Stoevring, B.; Vang, O.; Christiansen, M. (Alpha)B-crystallin in cerebrospinal fluid of patients with multiple sclerosis. *Clin. Chim. Acta* **2005**, *356*, 95–101. [CrossRef]
20. Malin, D.; Petrovic, V.; Strekalova, E.; Sharma, B.; Cryns, V.L. AlphaB-crystallin: Portrait of a malignant chaperone as a cancer therapeutic target. *Pharmacol. Ther.* **2016**, *160*, 1–10. [CrossRef]
21. Malin, D.; Strekalova, E.; Petrovic, V.; Deal, A.M.; Al Ahmad, A.; Adamo, B.; Miller, C.R.; Ugolkov, A.; Livasy, C.; Fritchie, K.; et al. AlphaB-crystallin: A novel regulator of breast cancer metastasis to the brain. *Clin. Cancer Res.* **2014**, *20*, 56–67. [CrossRef] [PubMed]
22. Marini, I.; Moschini, R.; Del Corso, A.; Mura, U. Alpha-crystallin: An ATP-independent complete molecular chaperone toward sorbitol dehydrogenase. *Cell Mol. Life Sci.* **2005**, *62*, 599–605. [CrossRef] [PubMed]
23. Siezen, R.J.; Bindels, J.G.; Hoenders, H.J. The quaternary structure of bovine alpha-crystallin. Size and charge microheterogeneity: More than 1000 different hybrids? *Eur. J. Biochem.* **1978**, *91*, 387–396. [CrossRef] [PubMed]
24. Siezen, R.J.; Bindels, J.G.; Hoenders, H.J. The quaternary structure of bovine alpha-crystallin. Effects of variation in alkaline pH, ionic strength, temperature and calcium ion concentration. *Eur. J. Biochem.* **1980**, *111*, 435–444. [CrossRef]
25. Das, K.P.; Surewicz, W.K. Temperature-induced exposure of hydrophobic surfaces and its effect on the chaperone activity of alpha-crystallin. *FEBS Lett.* **1995**, *369*, 321–325. [CrossRef]
26. Raman, B.; Ramakrishna, T.; Rao, C.M. Temperature dependent chaperone-like activity of alpha-crystallin. *FEBS Lett.* **1995**, *365*, 133–136. [CrossRef]
27. Gesierich, U.; Pfeil, W. The conformational stability of alpha-crystallin is rather low: Calorimetric results. *FEBS Lett.* **1996**, *393*, 151–154. [CrossRef]
28. Surewicz, W.K.; Olesen, P.R. On the thermal stability of alpha-crystallin: A new insight from infrared spectroscopy. *Biochemistry* **1995**, *34*, 9655–9660. [CrossRef]
29. Walsh, M.T.; Sen, A.C.; Chakrabarti, B. Micellar subunit assembly in a three-layer model of oligomeric alpha-crystallin. *J. Biol. Chem.* **1991**, *266*, 20079–20084. [CrossRef]
30. Lee, J.S.; Satoh, T.; Shinoda, H.; Samejima, T.; Wu, S.H.; Chiou, S.H. Effect of heat-induced structural perturbation of secondary and tertiary structures on the chaperone activity of alpha-crystallin. *Biochem. Biophys. Res. Commun.* **1997**, *237*, 277–282. [CrossRef]
31. Raman, B.; Rao, C.M. Chaperone-like activity and temperature-induced structural changes of alpha-crystallin. *J. Biol. Chem.* **1997**, *272*, 23559–23564. [CrossRef] [PubMed]
32. Rasmussen, T.; van de Weert, M.; Jiskoot, W.; Kasimova, M.R. Thermal and acid denaturation of bovine lens alpha-crystallin. *Proteins* **2011**, *79*, 1747–1758. [CrossRef] [PubMed]
33. Privalov, P.L.; Makhatadze, G.I. Heat capacity of proteins. II. Partial molar heat capacity of the unfolded polypeptide chain of proteins: Protein unfolding effects. *J. Mol. Biol.* **1990**, *213*, 385–391. [CrossRef]
34. Farnsworth, P.N.; Groth-Vasselli, B.; Greenfield, N.J.; Singh, K. Effects of temperature and concentration on bovine lens alpha-crystallin secondary structure: A circular dichroism spectroscopic study. *Int. J. Biol. Macromol.* **1997**, *20*, 283–291. [CrossRef]

35. Maiti, M.; Kono, M.; Chakrabarti, B. Heat-induced changes in the conformation of alpha- and beta-crystallins: Unique thermal stability of alpha-crystallin. *FEBS Lett.* **1988**, *236*, 109–114. [CrossRef]
36. Khan, J.M.; Malik, A.; Ahmed, A.; Rehman, M.T.; AlAjmi, M.F.; Khan, R.H.; Fatima, S.; Alamery, S.F.; Abdullah, E.M. Effect of cetyltrimethylammonium bromide (CTAB) on the conformation of a hen egg white lysozyme: A spectroscopic and molecular docking study. *Spectrochim. Acta A Mol. Biomol. Spectrosc.* **2019**, *219*, 313–318. [CrossRef]
37. Malik, A.; Albogami, S.; Alsenaidy, A.M.; Aldbass, A.M.; Alsenaidy, M.A.; Khan, S.T. Spectral and thermal properties of novel eye lens zeta-crystallin. *Int. J. Biol. Macromol.* **2017**, *102*, 1052–1058. [CrossRef]
38. Zhu, Y.; Lu, Y.; Ye, T.; Jiang, S.; Lin, L.; Lu, J. The Effect of Salt on the Gelling Properties and Protein Phosphorylation of Surimi-Crabmeat Mixed Gels. *Gels* **2021**, *8*, 10. [CrossRef]
39. Al-Shabib, N.A.; Khan, J.M.; Malik, A.; Sen, P.; Alsenaidy, M.A.; Husain, F.M.; Alsenaidy, A.M.; Khan, R.H.; Choudhry, H.; Zamzami, M.A.; et al. A quercetin-based flavanoid (rutin) reverses amyloid fibrillation in beta-lactoglobulin at pH 2.0 and 358 K. *Spectrochim. Acta A* **2019**, *214*, 40–48. [CrossRef]
40. Khan, J.M.; Ahmed, A.; Alamery, S.F.; Farah, M.A.; Hussain, T.; Khan, M.I.; Khan, R.H.; Malik, A.; Fatima, S.; Sen, P. Millimolar concentration of sodium dodecyl sulfate inhibit thermal aggregation in hen egg white lysozyme via increased alpha-helicity. *Colloid Surf. A* **2019**, *572*, 167–173. [CrossRef]
41. Khan, J.M.; Khan, M.R.; Sen, P.; Malik, A.; Irfan, M.; Khan, R.H. An intermittent amyloid phase found in gemini (G5 and G6) surfactant induced beta-sheet to alpha-helix transition in concanavalin A protein. *J. Mol. Liq.* **2018**, *269*, 796–804. [CrossRef]
42. Park, S.J.; Borin, B.N.; Martinez-Yamout, M.A.; Dyson, H.J. The client protein p53 adopts a molten globule-like state in the presence of Hsp90. *Nat. Struct. Mol. Biol.* **2011**, *18*, 537–541. [CrossRef] [PubMed]
43. Malik, A.; Haroon, A.; Jagirdar, H.; Alsenaidy, A.M.; Elrobh, M.; Khan, W.; Alanazi, M.S.; Bazzi, M.D. Spectroscopic and thermodynamic properties of recombinant heat shock protein A6 from Camelus dromedarius. *Eur. Biophys. J.* **2015**, *44*, 17–26. [CrossRef] [PubMed]
44. Ouajd, S.; Kamel, B. Physiological Particularities of Dromedary (Camelus dromedarius) and Experimental Implications. *Scand. J. Lab. Anim. Sci.* **2009**, *36*, 19–29.
45. Warda, M.; Prince, A.; Kim, H.K.; Khafaga, N.; Scholkamy, T.; Linhardt, R.J.; Jin, H. Proteomics of old world camelid (Camelus dromedarius): Better understanding the interplay between homeostasis and desert environment. *J. Adv. Res.* **2014**, *5*, 219–242. [CrossRef] [PubMed]
46. Kadim, I.T.; Mahgoub, O.; Purchas, R.W. A review of the growth, and of the carcass and meat quality characteristics of the one-humped camel (Camelus dromedaries). *Meat Sci.* **2008**, *80*, 555–569. [CrossRef]
47. Wu, H.; Guang, X.; Al-Fageeh, M.B.; Cao, J.; Pan, S.; Zhou, H.; Zhang, L.; Abutarboush, M.H.; Xing, Y.; Xie, Z.; et al. Camelid genomes reveal evolution and adaptation to desert environments. *Nat. Commun.* **2014**, *5*, 5188. [CrossRef]
48. Khalkhali-Evrigh, R.; Hafezian, S.H.; Hedayat-Evrigh, N.; Farhadi, A.; Bakhtiarizadeh, M.R. Genetic variants analysis of three dromedary camels using whole genome sequencing data. *PLoS ONE* **2018**, *13*, e0204028. [CrossRef]
49. Duhaiman, A.S.; Rabbani, N.; AlJafari, A.A.; Alhomida, A.S. Purification and characterization of zeta-crystallin from the camel lens. *Biochem. Biophys. Res. Commun.* **1995**, *215*, 632–640. [CrossRef]
50. Braun, N.; Zacharias, M.; Peschek, J.; Kastenmuller, A.; Zou, J.; Hanzlik, M.; Haslbeck, M.; Rappsilber, J.; Buchner, J.; Weinkauf, S. Multiple molecular architectures of the eye lens chaperone alphaB-crystallin elucidated by a triple hybrid approach. *Proc. Natl. Acad. Sci. USA* **2011**, *108*, 20491–20496. [CrossRef]
51. Ryazantsev, S.N.; Poliansky, N.B.; Chebotareva, N.A.; Muranov, K.O. 3D structure of the native alpha-crystallin from bovine eye lens. *Int. J. Biol. Macromol.* **2018**, *117*, 1289–1298. [CrossRef] [PubMed]
52. Malik, A.; Fouad, D.; Labrou, N.E.; Al-Senaidy, A.M.; Ismael, M.A.; Saeed, H.M.; Ataya, F.S. Structural and thermodynamic properties of kappa class glutathione transferase from Camelus dromedarius. *Int. J. Biol. Macromol.* **2016**, *88*, 313–319. [CrossRef] [PubMed]
53. Siezen, R.J.; Bindels, J.G. Stepwise Dissociation Denaturation and Reassociation Renaturation of Bovine Alpha-Crystallin in Urea and Guanidine-Hydrochloride—Sedimentation, Fluorescence, near-Ultraviolet and Far Ultraviolet Circular-Dichroism Studies. *Exp. Eye Res.* **1982**, *34*, 969–983. [CrossRef]
54. Das, B.K.; Liang, J.J.; Chakrabarti, B. Heat-induced conformational change and increased chaperone activity of lens alpha-crystallin. *Curr. Eye Res.* **1997**, *16*, 303–309. [CrossRef]
55. Robertson, A.D.; Murphy, K.P. Protein Structure and the Energetics of Protein Stability. *Chem. Rev.* **1997**, *97*, 1251–1268. [CrossRef]
56. Neet, K.E.; Timm, D.E. Conformational stability of dimeric proteins: Quantitative studies by equilibrium denaturation. *Protein Sci.* **1994**, *3*, 2167–2174. [CrossRef]
57. Steif, C.; Weber, P.; Hinz, H.J.; Flossdorf, J.; Cesareni, G.; Kokkinidis, M. Subunit interactions provide a significant contribution to the stability of the dimeric four-alpha-helical-bundle protein ROP. *Biochemistry* **1993**, *32*, 3867–3876. [CrossRef]
58. Augusteyn, R.C. Alpha-Crystallin polymers and polymerization: The view from down under. *Int. J. Biol. Macromol.* **1998**, *22*, 253–262. [CrossRef]
59. Robinson, C.R.; Rentzeperis, D.; Silva, J.L.; Sauer, R.T. Formation of a denatured dimer limits the thermal stability of Arc repressor. *J. Mol. Biol.* **1997**, *273*, 692–700. [CrossRef]

gels

MDPI

Article

The Effect of Different Additives on the Hydration and Gelation Properties of Composite Dental Gypsum

Liang Ma [1], Qianting Xie [1], Amutenya Evelina [1], Wenjun Long [1], Cunfa Ma [1], Fengshan Zhou [1,*] and Ruitao Cha [2]

1 Beijing Key Laboratory of Materials Utilization of Nonmetallic Minerals and Solid Wastes, National Laboratory of Mineral Materials, School of Materials Science and Technology, China University of Geosciences Beijing, Haidian District, Beijing 100083, China; Maxliang@cugb.edu.cn (L.M.); cugbxqt@163.com (Q.X.); 9103190001@cugb.edu.cn (A.E.); longwenjun@cugb.edu.cn (W.L.); 2103190084@cugb.edu.cn (C.M.)

2 CAS Key Laboratory for Biomedical Effects of Nanomaterials and Nanosafety, National Center for NanoScience and Technology, No. 11, Haidian District, Beijing 100190, China; chart@nanoctr.cn

* Correspondence: zhoufs@cugb.edu.cn

Abstract: Dental mold gypsum materials require fine powder, appropriate liquidity, fast curing, and easy-to-perform clinical operations. They require low linear expansion coefficient and high strength, reflecting the master model and facilitating demolding. In this article, the suitable accelerators and reinforcing agents were selected as additives to modify dental gypsum. The main experimental methods used were to compare the trends of linear expansion coefficients of several commercially available dental gypsum products over 72 h and to observe the cross-sectional microstructure of cured bodies before and after dental gypsum modification using scanning electron microscopy. By adjusting the application of additives, the linear expansion coefficient of dental gypsum decreased from 0.26% to 0.06%, while the flexural strength increased from 6.7 MPa to 7.4 MPa at 2 h. Formulated samples showed good stability and gelation properties with linear expansion completed within 12 h. It is indicated that the performance of dental gypsum materials can be improved by adding additives and nanomaterials, which provided a good reference for clinical preparation of high-precision dental prosthesis.

Keywords: dental gypsum; linear expansion coefficient; 2 h flexural strength; stability; gelation properties

Citation: Ma, L.; Xie, Q.; Evelina, A.; Long, W.; Ma, C.; Zhou, F.; Cha, R. The Effect of Different Additives on the Hydration and Gelation Properties of Composite Dental Gypsum. *Gels* **2021**, *7*, 117. https://doi.org/10.3390/gels7030117

Academic Editor: Hiroyuki Takeno

Received: 15 July 2021
Accepted: 6 August 2021
Published: 11 August 2021

1. Introduction

Models of oral tissues are used in dentistry to assess, treat, and manufacture indirect restorations [1]. As an essential auxiliary material, dental gypsum has been used for simulating oral cavity models. Put a certain amount of dental gypsum powder and water in a small rubber bowl to knead the gypsum slurry and then cast the slurry into the oral mold to prepare dental restoration, such as an inlay, a crown, a bridge, a partial denture, and a complete denture [2–4]. The manufacture of denture model is the basis of denture processing, and the dimensional stability of gypsum is fundamental to achieving a precise fit between dental structure and restorative material [5,6].

The main component of dental gypsum is α-hemihydrate gypsum dissolved and recrystallized in saturated steam medium or liquid water solution. Doctors usually choose the corresponding gypsum products to make prostheses by referring to various performance parameters of gypsum materials. For example, an ordinary dental stone should be used when making ordinary elastic dentures or movable dentures, and a dental stone with high strength and low expansion should be chosen when making precision prostheses such as fixed dentures and attachments [7,8]. According to the specification of the American Dental Association (ADA) [9], and the International Organization for Standardization (ISO 6873), dental gypsum can be classified into five types [10]:

Type I—Impression plaster. The linear expansion value is: ADA: 0–0.15%, ISO: 0–0.15%; Type II—Model plaster. The linear expansion value is: ADA: 0–0.30%, ISO: 0–0.30%; Type III—Dental stone. The linear expansion value is: ADA: 0–0.20%, ISO: 0–0.20%; Type IV—Dental stone (low expansion, high strength). The linear expansion value is: ADA: 0–0.10%, ISO: 0–0.15%; Type V—Dental stone (high expansion, high strength). The linear expansion value is: ADA: 0.10–0.30%, ISO: 0.16–0.30%.

After dental impression filling, the dental gypsum material begins to solidify, and the volume shrinkage of the gypsum material occurs at the early stage of solidification. When the mixture of a model material is rigid, the expansion in all directions will affect the size of the model [11,12]. On the one hand, the gypsum material will form calcium sulfate dihydrate crystal during hydration with continuous growth and re-formation of crystal, resulting in the expansion of gypsum volume; on the other hand, the dissolution of calcium sulfate dihydrate requires excessive water, which will form stomata after evaporation of water except for crystalline water, thereby significantly increasing the volume of gypsum [13–15].

Clinical studies show that the resin denture made of plaster model material with low expansion coefficient has the highest accuracy. To ensure the consistency between the models and the oral cavity conditions as well as improving patient satisfaction, dental gypsum requires an accuracy capable of reproducing details of an oral cavity, with excellent strength and high stability, which prevents the reproduced details from being broken or damaged [16,17]. Currently, modification methods of dental gypsum are divided into the following three types, according to additive addition ways.

The first way is to coat some inorganic and organic solutions on the surface of the gypsum model after solidification to solve problems of surface hardness and wear resistance of the gypsum model [18,19]. The second way is adding some salts, alkalis, and organic substances into water to form a new solution for mixing gypsum to help improve mechanical properties of the model during the mixing process of gypsum materials. However, the current application of this method is still limited, and the research hotspots focus on the application of disinfection and antimicrobial additives [20,21]. The third way employs the addition of some salt, alkali, or some organic substances grounded into a powder to the natural gypsum powder for improving the rheological and gelation properties of gypsum to produce a high-quality dental gypsum product [22,23]. This type of addition is currently the most widely used and has the most significant impact.

However, the research methods for the expansion properties of dental gypsum at home and abroad are still immature. In this study, the effects of additives on the linear expansion properties of dental gypsum were discussed by looking for effective solid additives and appropriate dosages. Since meeting the rheological properties and strength requirements of gypsum materials in clinical operations, the linear expansion coefficient is significantly reduced. It solves the problems of easy deformation, unstable size, and easy damage of dental gypsum materials, and its comprehensive performance is obviously better than existing dental gypsum products at home and abroad.

2. Results

In this study, single factor and orthogonal experiments were used to study the effects of various additives on the gelation properties of dental gypsum. The water–powder ratio was 23%, 0.1% BR was very beneficial to the adjustment of the weak-gel state, and the fixed dosage was 0.1%. Under the action of BR (borax cross-linking agent), the polymer network formed by the cross-linking of linear polymer additives enables the gypsum gelling system to effectively prevent the gypsum gel from sticking in the blending stage when it is regulated to the gel state; it also has a certain swelling-reducing effect, which can greatly improve the comprehensive blending properties and mechanical properties of dental gypsum.

2.1. *Effect of Accelerant on the Gelation Characteristics of Dental Gypsum*

Commonly used accelerants were acids (HCl, HNO_3, H_2SO_4) and their salts. A systematic study on the accelerating effect of various anions and calcium indicated that K^+ and SO_4^{2+} constituted the best accelerating anion-calcium pair. The effects of calcium sulfate dihydrate, potassium sulfate, and aluminum stearate on the gelation properties of dental gypsum were studied with 23% water powder ratio, 0.30% CPS (Copolymer water reducer), 0.03% SG (Sodium gluconate retarder), 0.10% BR (Borax cross-linker agent), and 2.00% Nano-SiO_2 as experimental control group.

2.2. *Linear Expansion Coefficient*

Three kinds of accelerants were selected in this experiment, and their effects on linear expansion coefficient of dental gypsum were shown in Figure 1.

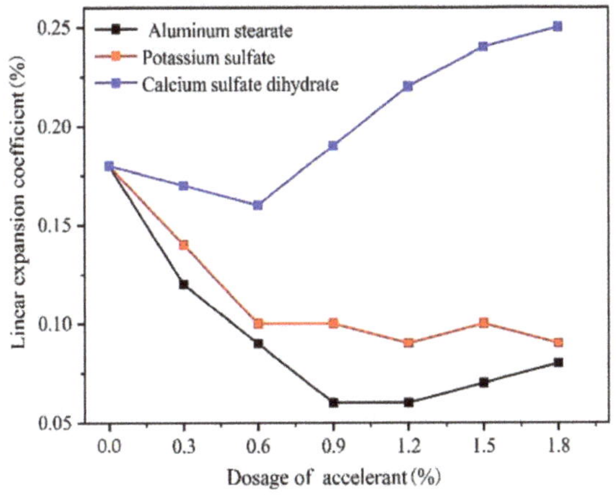

Figure 1. The effect of accelerant on the linear expansion coefficient of dental gypsum at different dosages (ww^{-1}).

When the dosage of calcium sulfate dihydrate exceeded 0.60%, the linear expansion value increased significantly. A small amount of potassium sulfate and aluminum stearate could significantly reduce the linear expansion coefficient of dental gypsum, among which the inhibitory effect on linear expansion of aluminum stearate was most obvious. When the dosage of aluminum stearate reached 0.90%, the 2 h linear expansion coefficient of dental gypsum was as low as 0.06%, which obviously weakened the formation of setting expansion. As the dosage of potassium sulfate and aluminum stearate continued to increase, the linear expansion coefficient showed a trend of slowly rising or basically unchanged.

When the crystal of hemihydrate gypsum was covered with an adsorbed layer of water molecules, the accelerant stabilized the layer and arranged the neighboring water molecules, resulting in a multilayer water structure. As the area and depth of coverage of the adsorbed water layer increased, the dissolution rate increased, thereby accelerating the reaction rate and shortening the setting time [24]. Given the influence of aluminum stearate on the setting time, when the final setting time was 6–12 min, the adjustable range of aluminum stearate was 0.30–1.20%.

Figure 2a,b shows cross-sectional morphologies of a blank dental gypsum sample and dental gypsum sample with 0.80% aluminum stearate observed under the electron scanning microscope, respectively. The water–powder ratio was 27% of standard consistency water demand. Compared with the blank group, after adding aluminum stearate, calcium sulfate dihydrate showed a larger crystal size, fewer inter-crystalline bonds, and a looser structure. According to the theory of expansion energy (crystal with unit mass, the expansion energy of crystal with coarse crystal is smaller), the 2 h linear expansion coefficient of dental gypsum was significantly reduced, which was consistent with the experimental results.

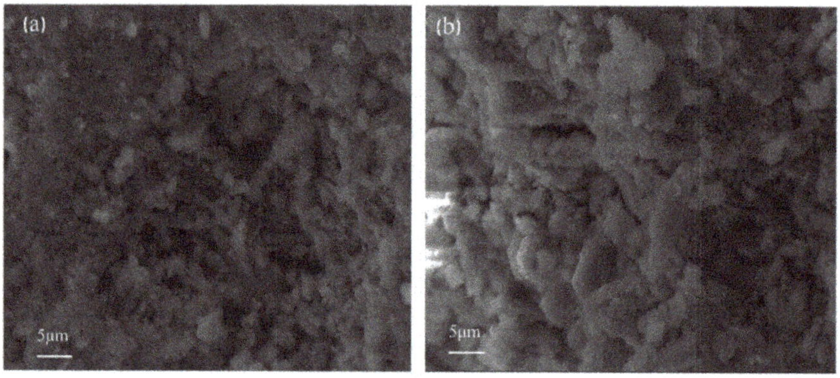

Figure 2. SEM images of gypsum (multiple = 3.5 k, scale = 5.00 μm) (**a**) blank dental gypsum, (**b**) 0.80% aluminum stearate.

2.3. Mechanical Properties

Figure 3 shows that the effect of the accelerant on mechanical properties was mainly reflected in the 2 h flexural strength. After adding 0.9% aluminum stearate and potassium sulfate, 2 h flexural strength loss was 23.90% and 22.40%, respectively, and wet compressive strength loss was 9.90% and 12.10%, respectively. The results indicated that potassium sulfate was suitable for a shorter setting time, and aluminum stearate was suitable for a small amount of use to prevent a sharp decrease in strength.

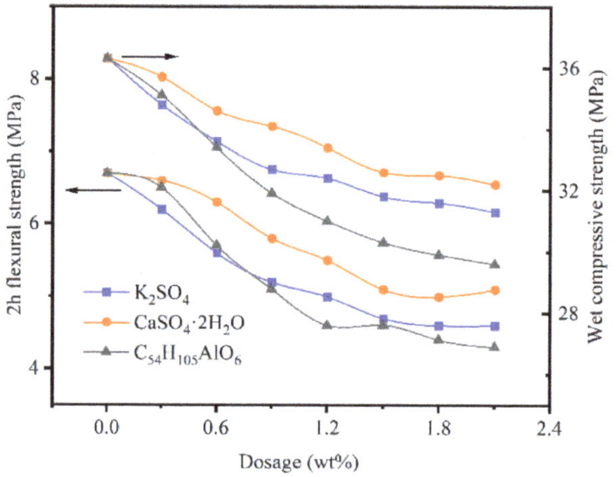

Figure 3. The effect of accelerant on the mechanical strength of dental gypsum.

It is worth noting that, when the dosage of potassium sulfate is too large, on the surface of the gypsum model appears a blooming phenomenon, which would affect the appearance of the model and destroy the compactness of the internal structure, leading to a sharp decrease in the strength and durability of the gypsum.

2.4. Effect of the Retarder on Linear Expansion Coefficient

In the application process of dental gypsum, a retarder was used as one of the necessary admixtures to adjust the setting time and linear expansion coefficient of the gypsum slurry to meet the operating requirements. At present, the commonly used gypsum retarders could be roughly divided into three categories: inorganic salts, organic acids, and organic macromolecules.

The effects of JR (methionine retarder), SG (sodium gluconate retarder) and GR (bone glue retarder) on the expansion properties of dental gypsum were studied with 23.00% water–powder ratio, 0.30% CPS (copolymer water reducer), 0.10% BR (borax cross-linker agent), and 2.00% Nano-SiO_2 and 0.90% aluminum stearate as experimental control group.

Figure 4 shows that the retarding effect on dental gypsum of three kinds of retarders was significant when the dosage was at a ratio of 10 thousandths, belonging to a high-efficiency retarder. At the same dosage, the retardation effect of GR was the strongest, followed by SG and JR.

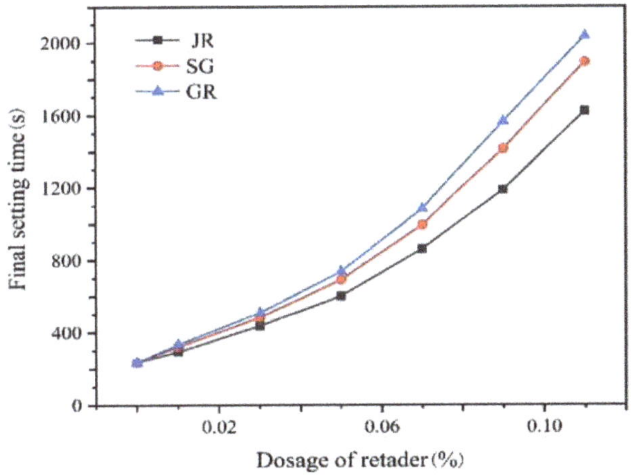

Figure 4. Effects of different kinds of retarders on the final setting time of dental gypsum at different dosages (ww^{-1}).

Figure 5 indicates that three retarders reduced the linear expansion coefficients of gypsum in varying degrees. When the dosage was more than 0.09%, the linear expansion coefficient tended to be 0.02–0.03%, which belonged to zero expansion. When the dosage was 0.09%, the setting time was about 20 min, which was not suitable for clinical use. Considering the strength loss and the actual cost of the solidified body, SG was selected as the best retarder. The final setting time was about 8 min when the dosage was 0.03%, and the expansion coefficient was 0.06%. According to the actual requirements of the final setting time range (6–10 min), the dosage shall not be higher than 0.05%.

2.5. *Effect of Water Reducer on Linear Expansion Coefficient*

It is well known that a water reducer is a substance that facilitates the kneading of powder and liquid to decrease the amount of water for kneading, thereby enhancing the strength of the composition after setting; however, its effect on the thermal expansion of gypsum has been rarely studied.

Concerning the experience of building gypsum, four representative water reducers were selected to study the modification effect of the linear expansion coefficient of dental gypsum. The control group was 23% water–powder ratio gypsum with 0.03% SG, 0.10% BR, 0.90% aluminum stearate, and 2.00% Nano-SiO_2.

Figure 6 shows that the expansion coefficient of gypsum increased first and then decreased after adding PAC-HR-01 (polycarboxylate water reducing agent). When the dosage reached 0.30%, the expansion coefficient was as low as 0.02%, belonging to zero expansion, which proved that the increase in dosage of PAC-HR-01 could effectively inhibit the formation of setting expansion. However, the slump flow at dosage of 0.30% was as high as 89 mm, and the slurry was easy to leak into the mold and increase the loss rate of the mold. Considering the influence of PAC-HR-01 on fluidity and setting expansion of dental gypsum, the dosage should not exceed 0.15%.

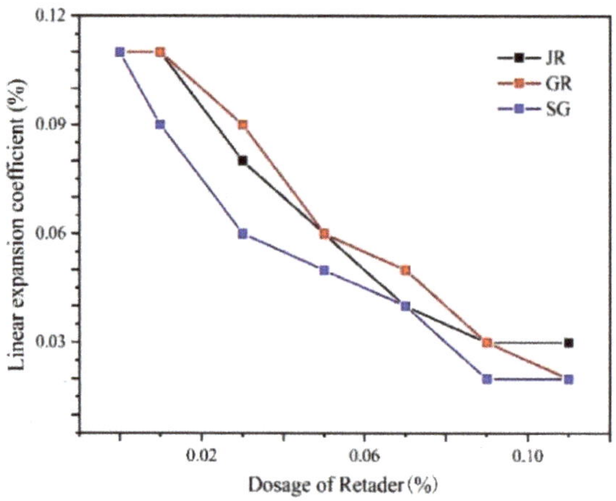

Figure 5. Effects of different kinds of retarders on the linear expansion coefficient of dental gypsum at different dosages (ww^{-1}).

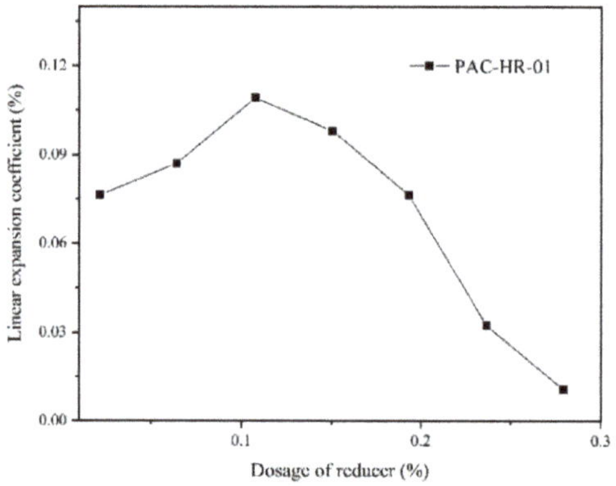

Figure 6. The effect of PAC-HR-01 on the linear expansion coefficient of dental gypsum at different dosages (ww^{-1}).

It can be seen from Figure 7 that, with the increase in AF-JFL-1 (anthracene water reducing agent), FDN-C (naphthalene water reducing agent), and SM-F10 (sulfonated melamine water reducing agent) dosage, the linear expansion coefficient of dental gypsum increased continuously. It can also be seen that the influence of CPS (copolymer water reducer) on the linear expansion coefficient was optimum. The linear expansion coefficient was minimized at a dosage of 0.30% CPS. Meanwhile, CPS reduced the viscosity while improving the fluidity of the slurry, leading to minute independent bubbles being dispersed on the solidified material, thereby providing better kneading performance and operability.

Analysis of the reasons for the different trends in the two stages after adding CPS: When the dosage of CPS was less than 0.30%, the calcium sulfate hemihydrate was fully hydrated; the number of crystals of calcium sulfate dihydrate increased while the length-diameter ratio decreased, thereby optimizing the microstructure of the hardened body of the gypsum. Therefore, an appropriate amount of water reducer could reduce the expansion value of gypsum. With further increase in CPS, a large number of acidic molecules would be adsorbed on the crystal surface in the form of chemisorption, which reduced the free energy of the crystal surface, leading to coarsening of the crystal. Therefore, the linear expansion of gypsum became more obvious when the dosage of CPS exceeded 0.30%.

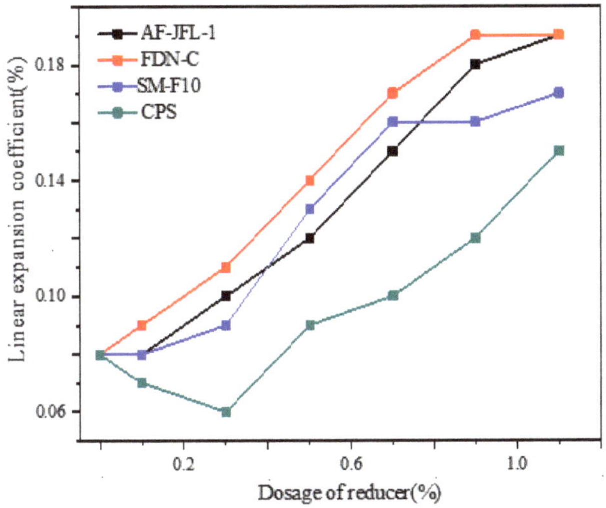

Figure 7. The effect of AF-JFL-1, FDN-C, SM-F10, and CPS on the linear expansion coefficient of dental gypsum at different dosages (ww^{-1}).

2.6. Effect of Reinforcer on the Gelation Characteristics of Dental Gypsum

Due to the high solubility of hydrated products in the hardened gypsum paste, the gypsum mold in the saturated water state would have a large degree of strength loss, and the introduction of accelerants and retarders would also cause a certain degree of strength loss. In order to reduce the solubility of the dental gypsum hydration product and to increase the bonding force between the hydration products, an appropriate amount of reinforcer should be added there to make up for this defect.

2.6.1. Linear Expansion Coefficient

In this experiment, MCC (Microcrystalline cellulose), NCC (Nanocellulose), GA (Gum Arabic), and three inorganic nanomaterials were used to improve the strength of dental gypsum, and the expansion performance of the mold gypsum sample was accordingly improved. The control group was 23.00% water–powder ratio, 0.30% CPS, 0.03% SG, 0.10% BR, and 0.90% aluminum stearate.

Figure 8 indicates that the linear expansion coefficient of dental gypsum increased with Nano-$CaCO_3$, Nano-TiO_2, MCC, NCC, and GA. GA was the most conducive for the formation of linear expansion, and the expansion coefficient increased slowly when the dosage of GA exceeded 2.00%, followed by NCC. The effects of MCC, Nano-$CaCO_3$, and Nano-TiO_2 were similar. When the dosage of Nano-SiO_2 was less than 2.00%, the linear expansion coefficient would not decrease, obviously, whereas the overall growth rate was the smallest.

2.6.2. Mechanical Properties

Figure 9 shows that the flexural strength and wet compressive strength of the dental gypsum increased first and then remained unchanged or decreased after adding inorganic nanomaterials. The enhancement effect of Nano-SiO_2 was most obvious, followed by Nano-TiO_2 and Nano-$CaCO_3$.

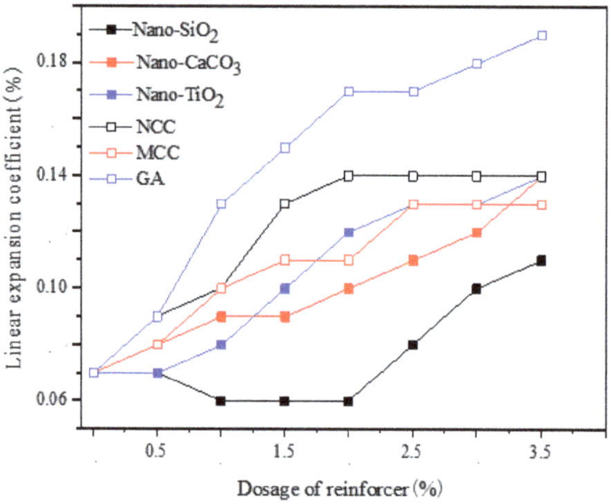

Figure 8. The effect of reinforcer on the linear expansion coefficient of α-hemihydrate gypsum at different dosages (ww^{-1}).

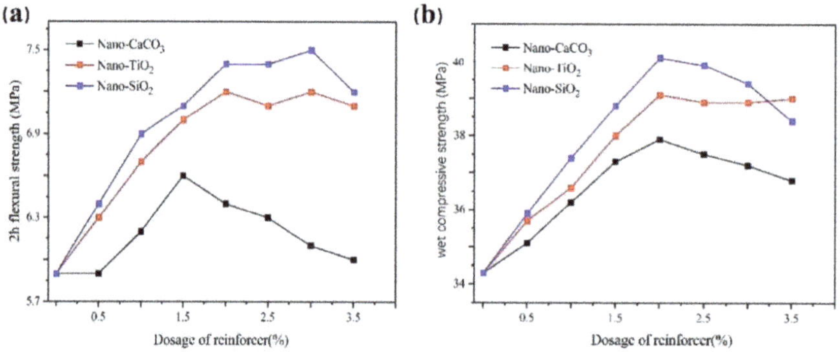

Figure 9. The effect of inorganic nano-materials on the mechanical strength of dental gypsum at different dosages (ww^{-1}) (**a**) 2 h flexural strength, (**b**) wet compressive strength.

When the dosage of Nano-SiO_2 was 2.00%, the 2 h flexural strength of dental gypsum was 7.4 MPa, increasing by 25.40%. Moreover, the wet compressive strength was increased to the maximum, up to 40.1 MPa, by 16.90%. At this time, the slump flow of slurry was 67 mm, whose operating state has met clinical requirement. With the increase in Nano-SiO_2, 2 h flexural strength increased slightly and then decreased slowly, and wet compressive strength decreased directly. This was because the fluidity of the slurry was damaged after the increase in dosage, and small holes appeared in the cross section of the molded specimen, resulting in a gradual decrease in mechanical strength.

MCC, NCC, and GA as reinforcers were also used to regulate the mechanical properties of dental gypsum, and the influence results were shown in Figure 10.

The reinforcement effect of NCC was the best. With the increase in NCC, MCC, and GA, 2 h flexural strength and wet compressive strength of gypsum model increased first and then fluctuated or decreased. When the dosage of MCC was 1.50%, the mechanical properties of gypsum were optimum, and 2 h flexural strength and compressive strength were 6.9 MPa and 39.4 MPa, respectively. After adding 2.00% GA, 2 h flexural strength and wet compressive strength were 7.1 MPa and 37.9 MPa, respectively, which rose to the highest level. When the dosage of NCC was 1.50%, 2 h flexural strength increased to 8.4 MPa, which was 42.40% higher than the contrast sample, and 42.4 MPa in wet compression. In particular, the improved 2 h flexural strength was obviously much better than most of the commercial dental gypsum products.

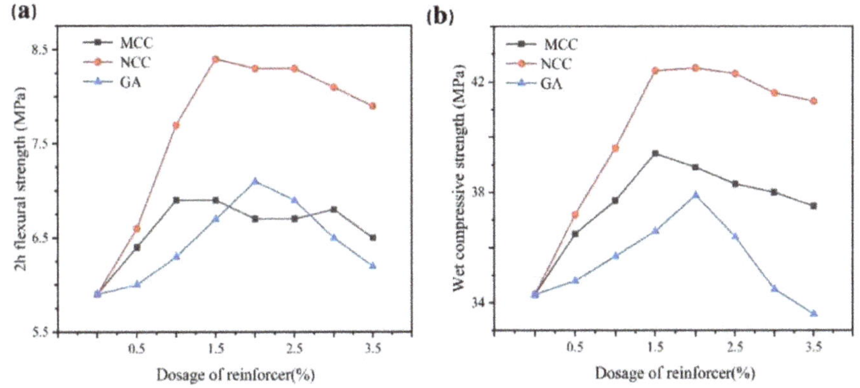

Figure 10. The effect of NCC, MCC, and GA on the mechanical strength of dental gypsum at different dosages (ww^{-1}) (**a**) 2 h flexural strength, (**b**) wet compressive strength.

Figure 11 shows the micro morphology of the cross section of dental gypsum after hydration for 2 h with different reinforcement materials under the electron scanning microscope. Without reinforcer, the calcium sulfate dihydrate crystal was a lamellar and layered structure with loose crystal network and large pores between grains. When 1.00% Nano-SiO_2 was added, the structure became dense, and internal defects were reduced. After adding 1.00% NCC, the array of crystal particles in the solidified body was more regular, uniform, and orderly; the crystal spacing was further reduced, and the structure was more complete, indicating that the reinforcement effect of NCC was more obvious, which was consistent with the experimental results.

Figure 11. SEM images of gypsum (**a**) 0%, (**b**): 1% Nano-SiO_2, (**c**): 1% NCC (multiple = 2.7 k, scale = 5.00 μm).

3. Discussion

Before designing any composite structure, three interdependent factors must be considered: (1) a selection of the suitable matrix and dispersed materials; (2) a choice of appropriate fabrication and processing methods; and (3) both the internal and external structural design of the device itself [25,26]. As an essential index to evaluate the internal structure of dental gypsum, the linear expansion coefficient is a key element [27,28]. It determines whether a certain gypsum material can be used as a high-precision dental restoration material [29]. Taking the linear expansion coefficient of dental gypsum as an index, the water–powder ratio was 23.00%, and the dosage of BR was 0.10%. SG, CPS, Nano-SiO_2, and aluminum stearate were selected as four influencing factors to carry out the orthogonal experiment of four factors and three levels. The factors and horizontal distribution results of the experimental scheme were shown in Table 1. Take the average value of the three experiments as the final result, accurate to 0.001%.

Table 1. Orthogonal scheme factor and horizontal distribution table.

Run	A (SG) %	B (CPS) %	C (Nano-SiO_2) %	D ($C_{54}H_{105}AlO_6$) %	Expansion Value %
1	1 (0.01)	1 (0.2)	1 (1.0)	1 (0.3)	0.131
2	1 (0.01)	2 (0.3)	2 (1.5)	2 (0.6)	0.078
3	1 (0.01)	3 (0.4)	3 (2.0)	3 (0.9)	0.064
4	2 (0.02)	1 (0.2)	2 (1.5)	3 (0.9)	0.082
5	2 (0.02)	2 (0.3)	3 (2.0)	1 (0.3)	0.114
6	2 (0.02)	3 (0.4)	1 (1.0)	2(0.6)	0.102
7	3 (0.03)	1 (0.2)	3 (2.0)	2 (0.6)	0.076
8	3 (0.03)	2 (0.3)	1 (1.0)	3 (0.9)	0.058
9	3 (0.03)	3 (0.4)	2 (1.5)	1 (0.3)	0.125
K_1	0.273	0.289	0.291	0.370	
K_2	0.298	0.250	0.285	0.256	
K_3	0.259	0.291	0.254	0.204	
k_1	0.091	0.096	0.097	0.123	
k_2	0.099	0.083	0.095	0.085	
k_3	0.086	0.097	0.085	0.068	
ΔR	0.013	0.014	0.012	0.055	

It can be seen from the range of R in Table 2 that, among the factors affecting the setting expansion, the addition amount of aluminum stearate was most significant, followed by CPS, SG, and Nano-SiO_2. It can be concluded from the value of k that the best process solution for the low-expansion and high-strength dental gypsum was D3C3B2A3. It means that the optimum ingredient ratio for the formula sample is 0.90% aluminum stearate, 0.30% CPS, 0.03% SG, 2.00% Nano-SiO_2, and 0.10% BR. The test was repeated three times, and the average linear expansion coefficient of the dental gypsum was 0.06%, which was within the error tolerance range.

Table 2. Linear expansion coefficient within 72 h (%).

Sample		CK	Heraeus	Dentona CAD/CAM	Bowin KKK	Formula Sample
Time interval (h)	2	0.26	0.26	0.13	0.22	0.06
	6	0.26	0.26	0.15	0.24	0.06
	12	0.27	0.27	0.15	0.24	0.07
	24	0.27	0.27	0.18	0.25	0.07
	36	0.29	0.28	0.18	0.26	0.07
	48	0.30	0.28	0.19	0.27	0.07
	60	0.31	0.28	0.20	0.27	0.07
	72	0.31	0.28	0.20	0.27	0.07
growth rate(%)		19.23	7.69	53.85	22.73	16.67

To further prove the advantages of the selected scheme, the linear expansion of each sample was measured continuously for 72 h, and 8 time intervals were selected for analysis. We compared the blank control group (CK) with three kinds of dental gypsum products with good general performance in order to study the differences of setting expansion and their stability. These three dental gypsum products with good performance and the best commercial quality are Dentona, Heraeus, and Bowin. We purchased all three products, formulated them into gypsum materials with a 23% water–powder ratio, and tested their linear expansion coefficients. The results indicate that, among these three dental gypsum products, Dentona has the best performance, and its expansion value is only reduced to 0.13–0.20% when it meets the high compressive strength of 39.2 MPa. The linear expansion coefficients of Heraeus and Bowin are above 0.2% in 2 h, the size changes are greater, and the retardation time is longer.

It can be seen from Table 2 that the linear expansion coefficients of the blank control group, Heraeus, and Bowin were more than 0.20%, which were not suitable for the production of high-precision dental models [1]. The linear expansion value of Dentona increased from 0.13% to 0.20% within 72 h, which could not meet the occlusal requirements of precision dental castings for clinical applications because it was entirely higher than the 0.06% expansion value required by the application level. The linear expansion of formula sample was basically completed within 12 h, and its linear expansion coefficient is as low as only 0.07%. In the subsequent time frame, the formula sample performance is very stable, no expansion phenomenon occurs, and the linear expansion rate is only 16.70%, which can be used as a high-precision clinical dental gypsum material.

Therefore, the dental gypsum prepared in this study with low linear expansion coefficient, high strength, and stable performance, after being optimized by compound admixtures, was obviously superior to most representatives of commercial dental gypsum products.

4. Conclusions

The dental gypsum prepared in this study has the best comprehensive properties of low expansion and high strength. The linear expansion coefficient was reduced from 0.26% to 0.06%, inclined to zero expansion with a stable expansion property. Meanwhile, 2 h flexural strength increased from 6.7 to 7.4 MPa, resulting in compression resistance, increased to 40.1 MPa. It was suitable for the manufacture of high-precision dental prostheses. After using Nano-SiO_2, the linear expansion coefficient may be slightly reduced, but 2 h flexural strength and wet compressive strength are increased significantly.

The dental gypsum material prepared in this study has excellent properties. It will help many patients have a more comfortable treatment experience if it can be applied in the dental field.

5. Materials and Methods

5.1. Materials

Dental gypsum was provided by Tangshan Xinghua Gypsum Co., Ltd. (Hebei, China), technical grade. The physical properties of the gypsum powder are shown in Table 3. The chemical composition and phase composition of dental gypsum were analyzed by X-ray fluorescence spectrometry (XRF) and X-ray diffraction (XRD), as shown in Figure 12.

Table 3. Physical property of α-dental gypsum used in this paper.

Tested Parameters	Properties
Standard consistency water demand (%)	27
Initial setting time (s)	634
Final setting time (s)	930
Linear expansion coefficient (%)	0.26
2 h flexural strength (MPa)	6.70
Wet compressive strength (MPa)	36.30

According to the XRD test results, the main phase component in the sample was hemihydrate gypsum. It can be seen from the three strong peaks and the corresponding normalized intensity values that the crystallinity of dental gypsum sample was good.

Polycarboxylate water reducing agent (PAC-HR-01) was purchased from Nantong Runfeng Petrochemical Co., Ltd., Nantong, China, technical grade. Anthracene water reducing agent (AF-JFL-1) was purchased from Tianjin Feilong Concrete Admixture Co., Ltd., Tianjin, China, technical grade. Sulfonated melamine water reducing agent (SM-F10) was purchased from Shanghai Chenqi Chemical Co., Ltd., Shanghai, China, technical grade. Naphthalene water reducing agent (FDN-C) and Borax cross-linker agent (BR) were purchased from Yousuo Chemical Technology Co., Ltd., Beijing, China, technical grade. Multi-component copolymer water reducer (CPS) was purchased from Shijiazhuang Chenxiang Nonmetallic Mineral Research Institute.

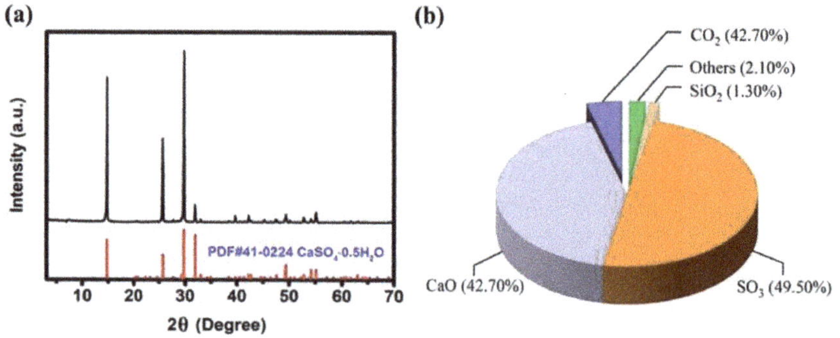

Figure 12. (**a**) XRD experimental results of α-dental gypsum (35 kV (30 mA)$^{-1}$, 8° min^{-1}, 0.02° step^{-1}), (**b**) main chemical composition.

Bone glue retarder (GR) was purchased from Suzhou Rongguang Chemical Co., Ltd., Suzhou, China, technical grade. Sodium gluconate retarder (SG) was purchased from Pudong Xingbang Chemical Development Co., Ltd., Shanghai, China, technical grade. Methionine retarder (JR) was purchased from Shanghai Yanyu New Building Materials Co., Ltd., Shanghai, China, technical grade.

Potassium sulfate and aluminum stearate were purchased from Sinopharm Chemical Reagent Beijing Co., Ltd., Beijing, China, analytical reagent; calcium sulfate dihydrate was purchased from Beihua Kaiyuan Chemical Co., Ltd., Beijing, China, analytical reagent.

Nano-TiO_2, Nano-SiO_2, and Nano-$CaCO_3$ were purchased from Jinan Texing Chemical Co., Ltd., Jinan, China, technical grade. Nanocellulose (NCC) was provided by National Center for Nanoscience technology, analytical reagent. Microcrystalline cellulose (MCC) was purchased from Beijing Coupling Technology Co., Ltd., Beijing, China, analytical reagent. Gum Arabic (GA) was purchased from Tianjin Lichang Chemical Co., Tianjin, China, Ltd., technical grade.

Dentona, Heraeus, and Bowin dental gypsum products were purchased from Beijing Xin Kang Venture Trading Co., Ltd., Beijing, China, technical grade.

5.2. Instruments

Vicat apparatus, model SN09; high-strength gypsum deformer, model BX-100a. Both were provided by Shanghai Rongjida Instrument Technology Co., Ltd., Shanghai, China. Electric bending test machine was provided by Shanghai Shenrui Test Equipment Manufacturing Co., Ltd., Shanghai, China, model SD-75. Universal testing machine was provided by Shimadzu Co., Ltd., Beijing, China, model AGS-10KNG. The X-ray diffractometer (XRD) was provided by Rigaku Co., Ltd., Beijing, China, model D(max-rB). The X-ray fluorescence spectrum analyzer (XRF) was provided by Rigaku Co., Ltd., Beijing, China, model ZSX Primus II. The scanning electron microscope (SEM) was provided by JEOL Co., Ltd., Beijing, China, model JSM-IT300.

5.3. Methods

5.3.1. Preparation of Low-Expansion Gypsum Powder for Dentistry

First, the water reducer, retarder, accelerant, and reinforcer were pulverized to a 100 mesh particle size in a tank pulverizer. The above additives were weighed according to a specific ratio, mixed with 50 g dental gypsum in the mixer for 60 s, and the mixer was then shut down for 5 min. An addition of 100 g α-dental gypsum powder was added to the mixer followed by 60 s of mixing, and the mixer was shut down for 5 min. A 300 g dental gypsum powder was again added to the mixer followed by 60 s of mixing and shut down for 5 min, then mixed up with 550 g of dental gypsum for 60 s, and shut down for 5 min once again; finally, the low-expansion and high-strength gypsum powder for dental model was prepared.

5.3.2. Mixing and Molding

An amount of distilled water was added appropriately to the mixing bowl according to different water–powder ratio (23–27%), the low-expansion high-strength gypsum powder was then slowly added into the mixing bowl to avoid large air retention. The powder was first placed in water for 30–60 s to ensure that the solvent water, gypsum powder, and additives have a penetration and surface pre-wetting effect of preventing agglomeration. Through preliminary performance tests, we found that the difference in performance between the 23–27% water–powder ratio gypsum materials was small, so we chose to use the smallest water–powder ratio for the experiments in order to save cost and maximize clinical application. The mixture was manually stirred and a homogeneous mixture with a weak-gel state was obtained after 60–90 s. The prepared slurry was slowly poured into a Vicat apparatus, high-strength gypsum deformer, and triple mortar mold coated by mineral oil, subsequently removing the surface bubbles by shaking. After scraping off the overflow slurry with a spatula, cover the surface of the test piece with a layer of PTFE film and place it at room temperature (20 ± 5 °C) for airtight curing and start molding.

Author Contributions: Conceptualization, L.M., Q.X., and F.Z.; methodology, L.M., Q.X., and R.C.; software, L.M.; validation, L.M., F.Z., and R.C.; formal analysis, Q.X.; investigation, L.M. and Q.X.; resources, L.M. and Q.X.; data curation, L.M. and Q.X.; writing—original draft preparation, L.M. and Q.X.; writing—review and editing, L.M., Q.X., F.Z., A.E., R.C., W.L., and C.M.; supervision, L.M. and Q.X.; project administration, F.Z.; funding acquisition, F.Z. All authors have read and agreed to the published version of the manuscript.

Funding: This research was funded by the Fundamental Research Funds for the Central Universities, grant number Nos. 2-9-2019-141.

Institutional Review Board Statement: Not applicable.

Informed Consent Statement: Not applicable.

Acknowledgments: We would like to acknowledge Mingzheng Wang for providing language help and writing assistance for this article.

Conflicts of Interest: The authors declare no conflict of interest.

References

1. Nejatian, T.; Firouzmanesh, P.; Yaqin, A.U. Dental gypsum and investments. *Adv. Dent. Biomater.* **2019**, *3*, 37–39.
2. Abdelaziz, K.M.; Combe, E.C.; Hodges, J.S. The effect of disinfectants on the properties of dental gypsum, part 2: Surface properties. *J. Prosthodont.* **2002**, *11*, 234–240. [CrossRef]
3. Oancea, L.; Bilinschi, L.G.; Burlibasa, M.; Petre, A.; Sandu, M.; Costela, S. Effects of disinfectant solutions incorporated in dental stone on setting expansion, compression and flexural strength of dental models. *Rom. Biotechnol. Lett.* **2020**, *25*, 2095–2102. [CrossRef]
4. Rudolph, H.; Salmen, H.; Moldan, M.; Kuhn, K.; Sichwardt, V.; Wöstmann, B. Accuracy of intraoral and extraoral digital data acquisition for dental restorations. *J. Appl. Oral Sci.* **2008**, *24*, 85–94. [CrossRef]
5. Michalakis, K.X.; Stratos, A.; Hirayama, H.; Pissiotis, A.L.; Touloumi, F. Delayed setting and hygroscopic linear expansion of three gypsum products used for cast articulation. *J. Prosthet. Dent.* **2009**, *102*, 313–318. [CrossRef]
6. Silva, M.A.B.; Vitti, R.P.; Consani, S.; Sinhoreti, M.A.C.; Mesquita, M.F.; Consani, R.L.X. Linear dimensional change, compressive strength and detail reproduction in type IV dental stone dried at room temperature and in a microwave oven. *J. Appl. Oral Sci.* **2012**, *20*, 588–593. [CrossRef] [PubMed]
7. Sakaguchi, R.; Ferracane, J.; Powers, J.M. Craig's Restorative Dental Materials. *Br. Dent. J.* **2019**, *226*, 293–314.
8. Hashedi, A.A.; Laurenti, M.; Mezour, M.A.; Basiri, T.; Touazine, H.; Jahazi, M. Advanced inorganic nanocomposite for decontaminating titanium dental implants. *J. Biomed. Mater.* **2019**, *107*, 761–772. [CrossRef]
9. American Dental Association. The dentist, the forensic pathologist, and the identification of human remains. *J. Am. Dent. Assoc.* **1972**, *85*, 1324–1329. [CrossRef] [PubMed]
10. Duke, P.; Moore, B.K.; Haug, S.P.; Andres, C.J. Study of the physical properties of type IV gypsum, resin-containing, and epoxy die materials. *J. Prosthet. Dent.* **2000**, *83*, 466–473. [CrossRef]
11. Mahler, D.B.; Asgarzadeh, K. The Volumetric Contraction of Dental Gypsum Materials on Setting. *J. Dent. Res.* **1953**, *32*, 354–361. [CrossRef] [PubMed]
12. Lautenschlager, E.P.; Corbin, F. Investigation on the expansion of dental stone. *J. Dent. Res.* **1969**, *48*, 206–210. [CrossRef]

13. Asaoka, K.; Bae, J.Y.; Lee, H.H. Porosity of dental gypsum-bonded investments in setting and heating process. *Dent. Mater. J.* **2012**, *31*, 120–124. [CrossRef] [PubMed]
14. Lodovici, E.; Meira, J.B.C.; Filho, L.E.; Ballester, R.Y. Expansion of high flow mixtures of gypsum-bonded in investments in contact with absorbent liners. *Dent. Mater.* **2005**, *21*, 573–579. [CrossRef]
15. Heshmati, R.H.; Nagy, W.W.; Wirth, G.G.; Dhuru, V.B. Delayed linear expansion of improved dental stone. *J. Prosthet. Dent.* **2002**, *88*, 26–31. [CrossRef]
16. Carvalho, M.A.; Calil, C.C.; Savastano, H.; Tubino, R.; Carvalho, M.T. Microstructure and Mechanical Properties of Gypsum Composites Reinforced with Recycled Cellulose Pulp. *Am. J. Mater.* **2008**, *11*, 391–397. [CrossRef]
17. Michalakis, K.X.; Asar, N.V.; Kapsampeli, V.; Trikka, P.M.; Pissiotis, A.L.; Hirayama, H. Delayed linear dimensional changes of five high strength gypsum products used for the fabrication of definitive casts. *J. Prosthet. Dent.* **2012**, *108*, 189–195. [CrossRef]
18. Kumar, S.R.; Patnaik, A.; Bhat, I.K. Optimum selection of nano- and microsized filler for the best combination of physical, mechanical, and wear properties of dental composites. *Proc. Inst. Mech. Eng. Part L J. Mater. Des. Appl.* **2016**, *232*, 416–428. [CrossRef]
19. He, L.H.; Vuuren, L.J.; Planitz, N.; Swain, M. A micro-mechanical evaluation of the effects of die hardener on die stone. *Dent. Mater. J.* **2010**, *29*, 433–437. [CrossRef]
20. Kalahasti, D.; Hegde, V.; Kosaraju, K.; Baliga, X.; Reddy, N.K.; Sujatha, B.K. Evaluation of Efficacy of Microwave Irradiation in Disinfecting Dental Gypsum Casts: An Ex Vivo Study. *J. Indian Prosthodont. Soc.* **2014**, *14*, 381–392. [CrossRef]
21. Paula, P.R.; Lucas, M.G.; Spolidorio, D.M.P. Antimicrobial activity of disinfectant agents incorporated into type IV dental stone. *Gerodontology* **2012**, *29*, 267–274.
22. Dalmay, P.; Smith, A.; Chotard, T.; Sahay, T.P.; Gloaguen, V.; Krausz, P. Properties of cellulosic fibre reinforced plaster: Influence of hemp or flax fibres on the properties of set gypsum. *J. Mater. Sci.* **2010**, *45*, 793–803. [CrossRef]
23. Yan, M.; Takahashi, H.; Nishimura, F. Dimensional accuracy and surface property of titanium casting using gypsum-bonded alumina investment. *Dent. Mater. J.* **2004**, *23*, 539–544. [CrossRef]
24. Lewry, A.J.; Williamson, J. The setting of gypsum plaster.3. The effect of additives and impurities. *J. Mater. Sci.* **1994**, *29*, 6085–6090. [CrossRef]
25. Touraj, N.; Zohaib, K.; Muhammad, Z.; Shariq, N.; Sana, Z.; Masoud, M. Dental biocomposites. *Biomater. Oral Dent. Tissue Eng.* **2017**, *5*, 65–84.
26. Romanec, C.; Rosu, S.; Macovei, G.; Scutariu, M.M.; Dragomir, B.; Olteanu, N.D. Morphofunctional Features in Angle Second Class Malocclusion on Dental Gypsum Models. *Mater. Plast.* **2018**, *55*, 686–690. [CrossRef]
27. Choi, J.W.; Ahn, J.J.; Son, K.; Huh, J.B. Three-Dimensional Evaluation on Accuracy of Conventional and Milled Gypsum Models and 3D Printed Photopolymer Models. *Materials* **2019**, *12*, 3499. [CrossRef]
28. Prombonas, A.E.; Paralika, M.A.; Sotiriou, M.P.; Vlissidis, D.S. The peak-amplitude method of vibration analysis for nondestructively studying the structural integrity of dental gypsum. *J. Biomed. Mater. Res.* **2002**, *63*, 605–609. [CrossRef] [PubMed]
29. Akasaka, T.; Miyaji, H.; Imamura, T.; Kaga, N.; Yokoyama, A.; Yoshida, Y. Submicro-patterning of curable dental materials by molding methods: A screening trail. *Dig. J. Nanomater. Biostruct.* **2017**, *12*, 281–292.

gels

MDPI

Article

Genipin-Based Crosslinking of Jellyfish Collagen 3D Hydrogels

Laura Riacci [1,2,*,†], Angela Sorriento [1,2,*,†] and Leonardo Ricotti [1,2]

1 The BioRobotics Institute, Scuola Superiore Sant'Anna, 56127 Pisa, Italy; leonardo.ricotti@santannapisa.it
2 Department of Excellence in Robotics & AI, Scuola Superiore Sant'Anna, 56127 Pisa, Italy
* Correspondence: laura.riacci@santannapisa.it (L.R.); angela.sorriento@santannapisa.it (A.S.)
† These authors share equal contribution.

Abstract: Collagen-based hydrogels are an attractive option in the field of cartilage regeneration with features of high biocompatibility and low immunogenic response. Crosslinking treatments are often employed to create stable 3D gels that can support and facilitate cell embodiment. In this study, we explored the properties of JellaGel™, a novel jellyfish material extracted from *Rhizostoma pulmo*. In particular, we analyzed the influence of genipin, a natural crosslinker, on the formation of 3D stable JellaGel™ hydrogels embedding human chondrocytes. Three concentrations of genipin were used for this purpose (1 mM, 2.5 mM, and 5 mM). Morphological, thermal, and mechanical properties were investigated for the crosslinked materials. The metabolic activity of embedded chondrocytes was also evaluated at different time points (3, 7, and 14 days). Non-crosslinked hydrogels resulted in an unstable matrix, while genipin-crosslinked hydrogels resulted in a stable matrix, without significant changes in their properties; their collagen network revealed characteristic dimensions in the order of 20 μm, while their denaturation temperature was 57 °C. After 7 and 14 days of culture, chondrocytes showed a significantly higher metabolic activity within the hydrogels crosslinked with 1 mM genipin, compared to those crosslinked with 5 mM genipin.

Keywords: jellyfish collagen; hydrogels; crosslinking; cell-laden biomaterials; chondrocytes

Citation: Riacci, L.; Sorriento, A.; Ricotti, L. Genipin-Based Crosslinking of Jellyfish Collagen 3D Hydrogels. *Gels* **2021**, *7*, 238. https://doi.org/10.3390/gels7040238

Academic Editor: Hiroyuki Takeno

Received: 26 October 2021
Accepted: 25 November 2021
Published: 27 November 2021

1. Introduction

Collagen is one of the main components of the connective tissue extracellular matrix; it is a natural substrate for the support and growth of a variety of cells and tissues in the body and acts as a structure in conjunction with other extracellular molecules, such as glycosaminoglycans and fibronectin. Due to its properties (excellent biocompatibility, low antigenicity, and biodegradability) it has been broadly explored in the field of tissue engineering [1,2]. Collagen has been classified into two main categories, based on their supramolecular structures: non-fibrillar collagens and fibrillar collagens. The non-fibrillar collagens belong to a family of structurally related short-chain proteins that do not form large fibril bundles. Instead, the fibrillar collagens form highly organized fibers and fibrils, providing structural support for the body in the skeleton, skin, hollow organs, and capsules from different organs. Fibrils are mainly organized into bundles or lamellae; the size and higher-order arrangement of fibrils give rise to tissue-specific biomechanical and other biological properties. The fibrillar collagens include type I, II, III, V, and XI. Type I collagen is produced by fibroblasts and other cells, such as osteoblasts; as a key component of the extracellular matrix, it is abundant in bone, tendon, skin, ligaments, and cornea, and comprises between 80% and 99% of the total body collagen. Type II collagen is the major collagen type in cartilaginous tissues, although it is also present in other connective tissues, such as the *nucleus pulposus* of the intervertebral disk and the vitreous humor. Type III collagen is a normal constituent of the skin (10–20% of the total collagen) and it is found in many other connective tissues; it provides resistance to forces and stretching. Type V collagen is abundant in vascular tissues [3], while type XI collagen is found in cartilaginous tissue [3].

Collagen as a biomaterial is typically extracted from mammalians (especially from dermis, tendons, and bones that are rich in fibrillar collagen). Although purified collagen can be isolated from human peripheral tissues or human placenta, animal species, such as rats, bovines, pigs, and sheep are often the preferred source [4]. In particular, primary sources of animal-derived collagen are bovine skin and tendons and porcine skin. However, animal-derived collagen presents many disadvantages related to possible immunogenicity and transmission of diseases, such as bovine spongiform encephalopathy [5]. Moreover, some patients refuse to receive components derived from those animals, for ethical and religious reasons [6].

Some marine organisms (e.g., the jellyfish) have a high content of fibrillar collagen (more than 60% of their weight), representing an intriguing alternative for collagen extraction. Jellyfish-derived collagen shows high biocompatibility and low immunogenic response [7]. These properties, together with ease of handling [6] and large bioavailability, make jellyfish collagen an excellent candidate for replacing mammalian one [8]. Jellyfish collagen also shows promising features for cartilage regeneration; indeed, it has been proven that chondrocytes preserve their phenotype when included in three-dimensional (3D) matrices made of jellyfish collagen [6,9]. Collagen from different jellyfish species shows similarities to the collagen of different vertebrates. In fact, it has been shown that some jellyfish collagens are comparable to vertebrate collagen IV or V, while others resemble vertebrate collagen I [10]. Collagen derived from *Rhizostoma pulmo* has shown a high degree of similarity with mammalian type I collagen, but also showing collagen type II-"like" properties.

In this paper, we focused on JellaGel™, a new type of jellyfish collagen, extracted from *Rhizostoma pulmo* and belonging to the category of fibrillar collagen. JellaGel™ was recently marketed and specifically formulated to form three-dimensional gels, thus seeming interesting for tissue engineering applications. Since JellaGel™ is a new and rather unexplored material, almost no works on it are available in the literature, except from a preliminary study reporting the expression of cellular filopodia within the hydrogel matrix, although it was not crosslinked [11].

Crosslinking treatments are frequently used to create stable 3D scaffolds (including collagen ones) supporting cell encapsulation. Indeed, crosslinking is an important aspect that would turn otherwise weak gels into more robust materials, enhancing their stability over time. Genipin is a hydrolytic product of geniposide isolated from the fruits of *Gardenia jasminoides* and it is classified as a natural crosslinking agent [12]. It can spontaneously react with the amine groups of amino acids (including those constituting collagen) to form dark blue pigments [13]. The reaction between genipin and collagen induces the formation of cyclic structures, which enable intramolecular and intermolecular crosslinks [14]. The cytotoxicity of genipin is considerably lower than that of other chemical crosslinking reagents, such as glutaraldehyde (that is 5000–10,000 times more cytotoxic). Accordingly, the biocompatibility of materials crosslinked through genipin is superior to those crosslinked by glutaraldehyde or epoxy compounds [13]. In the literature, genipin-crosslinked collagen matrices (of animal origin) have shown good viability of chondrocytes or stem cells seeded on them [12,15].

To the best of our knowledge, no works concerning genipin-based crosslinking of JellaGel™ are available in the literature. The aim of this paper is to investigate the influence of genipin in the formation of 3D crosslinked JellaGel™ hydrogels and on the embodiment of human chondrocytes. Three genipin concentrations were tested and for each concentration material properties were assessed. In particular, Scanning Electron Microscope (SEM) and Differential Scanning Calorimetry (DSC) were used to analyze the morphological and thermal properties of the hydrogels, respectively. Rheometric measurements were used to characterize the matrix mechanical properties. Finally, the influence of different genipin concentrations on the metabolic activity of chondrocytes embedded within the JellaGel™ hydrogels was assessed at different time points (3, 7, and 14 days).

2. Materials and Methods

2.1. Hydrogel Preparation

JellaGel™ solution was purchased by the company Jellagen Marine Biotechnologies (Cardiff, UK) and prepared according to the manufacturer's instructions, as reported in the following. A solution of 10× Phosphate Buffered Saline (PBS, P4417, Sigma–Aldrich, Darmstadt, Germany) was first prepared dissolving one tablet in 20 mL of deionized water. Then, 25 mg/L of Phenol red (P3532, Sigma–Aldrich) were added. For each sample, 222 µL of such solution were mixed to 2 mL of JellaGel™ solution in a glass vial, thus reaching a ratio of 9:1 between JellaGel™ and PBS. The pH of the solution was adjusted until the pH range was 7.5–8.5 using sodium hydroxide solution (S5881, Sigma–Aldrich) at 2M and 0.2M.

To improve the stability of collagen hydrogels, genipin was used as a crosslinker at different concentrations. First, 125 mg of genipin powder (G4796, Sigma–Aldrich) were dissolved in 12.5 mL of Dulbecco's Phosphate Buffered Saline modified without calcium and magnesium chloride (DPBS, D8537, Sigma–Aldrich) to achieve a final genipin concentration of 10 mg/mL. After adjusting the pH of the JellaGel™ solution as described above, the genipin solution was added in an appropriate quantity to reach the final concentrations of 1 mM, 2.5 mM, and 5 mM, under constant stirring for 3 min. The solution was left at room temperature for 15 min and then incubated at 37 °C overnight to allow gel formation. After the overnight incubation, the samples (without cells embedded) were used for SEM, DSC, and rheometric analyses.

To investigate cell metabolic activity, chondrocytes were encapsulated in the 3D crosslinked JellaGel™ hydrogels, as follows. Human Chondrocytes (HC), derived from normal human articular cartilage, were purchased from Cell Applications (Cat. number 402-05a) and expanded in a human chondrocyte Growth Medium (Cell Applications, San Diego, California, Cat. number 411-500) at 37 °C and 5% CO_2 atmosphere. To obtain cell-laden hydrogels, the pelleted cells were first resuspended in 50 µL of medium and then added in 2 mL of JellaGel™ solution with a final density of 400,000 cells/mL. Before starting the hydrogel preparation procedure described above, all the solutions were filtered using a filter with a nominal pore diameter of 0.22 µm (16532-K, Sartorius) and the vials and the stir bars were sterilized in the autoclave for 90 min at a temperature of 121 °C and a pressure of 1 atm. The cells were evenly mixed in the solution through a magnetic stirrer at a velocity of 400 rpm. After 3 min, the stirring was stopped and the solution containing the cells was left at room temperature for 15 min. Then, the compound was gently transferred (without shaking it) into an incubator (37 °C and 5% CO_2). After 2 h, each sample was provided with 4 mL of cell medium and then kept in the incubator overnight. The samples were entirely immersed in the medium but the bottom side detached from the culture well, allowing the scaffold to float just below the medium level. This guaranteed a supply of oxygen and nutrients on all hydrogel sides. The medium was changed every two days. The sample preparation procedure is depicted in Figure 1.

2.2. SEM Imaging

SEM imaging was used to analyze the morphology of one representative sample for each genipin concentration. After the preparation described above, the sample (without cells) was dried overnight and gold-sputtered before SEM acquisition. SEM scans were carried out using an EVO MA10 SEM microscope (Zeiss), setting a beam voltage of 5 kV, a current of 90 pA at a working distance of around 10 mm. The images were obtained using a magnification of 1600x. A morphometric analysis was performed on SEM images to evaluate the diameter of the formed collagen fibers. For the measurement of the fiber diameters, we used the open source tool DiameterJ (https://imagej.net/plugins/diameterj, accessed on 10 November 2021), already validated for this purpose [16]. DiameterJ follows a two-step process: (i) first, several image segmentations in a binary image are provided starting from the original image; (ii) then, the analysis of the segmented images is performed. For each image, four segmented images were analyzed, and the final histogram was derived from the sum of frequencies of the four segmented images. A weighted mean and weighted standard deviation were also calculated for each experimental condition.

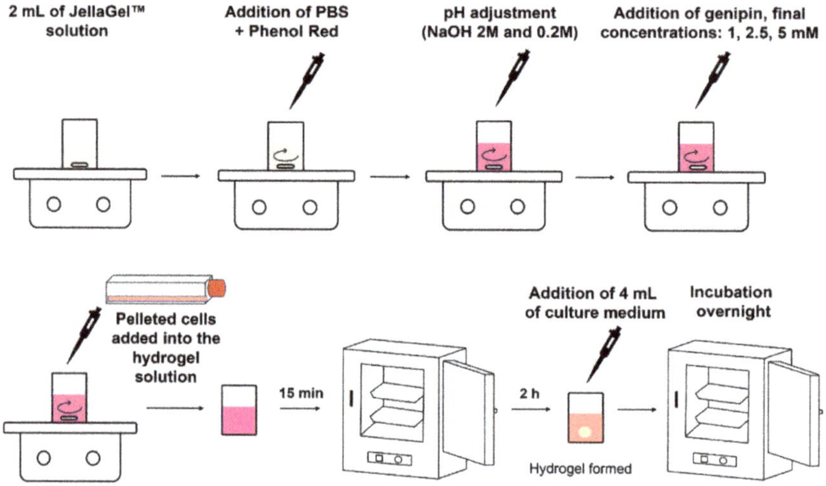

Figure 1. Graphical representation of the hydrogel preparation procedure: the different phases needed for preparing the hydrogels starting from JellaGel™ solution and the key parameters for each step are depicted.

2.3. DSC Analysis

DSC was used to investigate the effect of chemical crosslinking on the thermal characteristics of jellyfish collagen. Three samples (without cells) for each genipin concentration were used for this analysis. The samples were freeze-dried and analyzed through a DSC system 1 STAR (Mettler Toledo), with a heating rate of 2 °C min^{-1} and a temperature range from 25 °C to 75 °C. An empty melting pan was used as the reference sample.

2.4. Rheometric Measurements

Rheometric analyses were performed on three independent samples for each genipin concentration, using a rheometer (Anton Paar MCR-302) at a temperature of 25 °C in a plate–plate geometry, with a diameter of 25 mm and a gap between the two plates of 1 mm. After the preparation, the samples (without cells) were gently transferred on the rheometer plate to start the measurements.

The storage modulus (G'), the loss modulus (G") and the shear stress (τ) were measured in oscillation mode from 0.01 to 1000% strain at a frequency of 1 Hz and the results were plotted in a log-linear scale graph by using GraphPad 8.0.2.

2.5. Assessment of Chondrocyte Metabolic Activity

Cell metabolic activity was assessed after 3, 7, and 14 days on JellaGel™ samples with embedded chondrocytes, by using the PrestoBlue™ Cell Viability Reagent (A13261, Invitrogen). Three concentrations of genipin (1, 2.5 and 5 mM) and three samples for each concentration were tested to investigate the effect of the crosslinking level on metabolic activity. The same samples were analyzed at different time points (3, 7, and 14 days). Before each metabolic analysis, the samples were moved to a 6-well plate to avoid any possible contribution to the metabolic activity outcome due to cells not embedded in the hydrogel, but rather attached at the bottom of the vial or in the supernatant. PrestoBlue™ is a resazurin-based solution that uses the reducing ability of live cells to quantitatively evaluate metabolic activity. When cells are alive and healthy, they maintain a reduced environment within their cytosol. Upon entering a living cell, the PrestoBlue™ reagent is reduced to resorufin, which is red in color and highly fluorescent. Metabolically active cells continuously convert the PrestoBlue™ reagent and they can be monitored by measuring the change in fluorescence. Non metabolically active cells cannot reduce the dye and thus they do not generate a change in the signal [17].

At the desired time-points, PrestoBlue™ was first diluted with culture medium in a ratio 1:10. Then, 2 mL of such solution were added to each sample. After incubation for 2 h at 37 °C, the solution was split into 4 wells (300 µL for each sample), obtaining a total of 12 measurements (three independent samples and four measurements for each sample) for each genipin concentration and each time-point. A VICTOR Multilabel plate reader (PerkinElmer, Waltham, MA, USA) was used to read the fluorescence signal, setting an excitation wavelength of 560 nm and an emission wavelength of 590 nm.

2.6. Statistical Analyses

For DSC, rheometric and metabolic tests, three independent samples for each genipin concentration were prepared and characterized. A Kolmogorov–Smirnov test allowed assessing the non-normality of data distribution. The statistical comparison between the samples crosslinked with different genipin concentrations, at each time point, was performed using a non-parametric Kruskal–Wallis test, whereas a post-hoc test was performed using a non-parametric Mann–Whitney test for unpaired data. Statistical significance was corrected for multiple comparisons according to the Bonferroni–Holm rule and a *p*-value of 0.05 was set as the significance threshold.

3. Results

3.1. Preparation of JellaGel™ Hydrogels

Figure 2 shows representative images for the control (JellaGel™ without the addition of genipin) and samples crosslinked with different genipin concentrations. The samples featured by higher concentrations of genipin (2.5 mM and 5 mM) showed better-formed collagen networks than the ones obtained with a genipin concentration of 1 mM. Samples not provided with genipin (control) were not able to form a *stable* gel. For such a reason, we excluded control samples from the subsequent analyses.

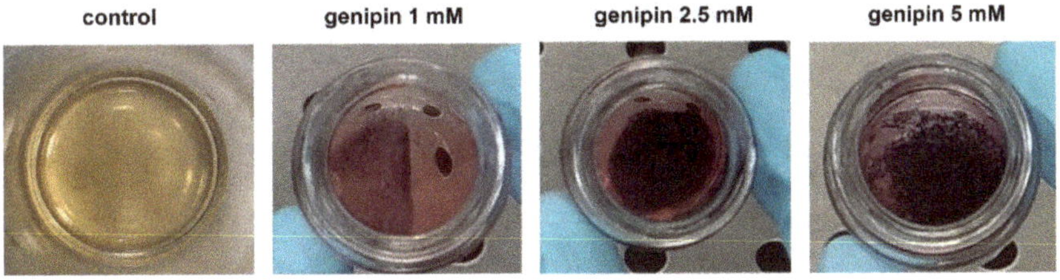

Figure 2. JellaGel™ hydrogels crosslinked with genipin at different concentrations. One representative sample for the control (no genipin) and for each genipin concentration is shown. The dark structures visible in the images correspond to the collagen gel networks, formed due to the bonds established between the collagen chains.

A schematic representation of the crosslinking mechanism between genipin and collagen is depicted in Figure 3. The extraction of genipin and its chemical reaction with natural biomaterials (such as collagen) has been extensively studied in the past, thanks to the promising biosafety and specific crosslinking performances of genipin [18]. Genipin is able to crosslink materials that contain primary amine groups (Figure 3a). Although genipin is colorless, the release of dark blue pigments occurs when it reacts with the primary amines. The crosslinking mechanism starts with a ring-opening reaction caused by the amino group via a nucleophilic attack on the olefinic carbon atom of the genipin (Figure 3b). Consequently, the genipin forms a covalent bond with the amino group of the polymer (Figure 3b). An unstable intermediate aldehyde group is formed, and it is again attacked by another amine group from another polymer (Figure 3c), forming a new covalent bond, which leads to the formation of the crosslink (Figure 3d) [19].

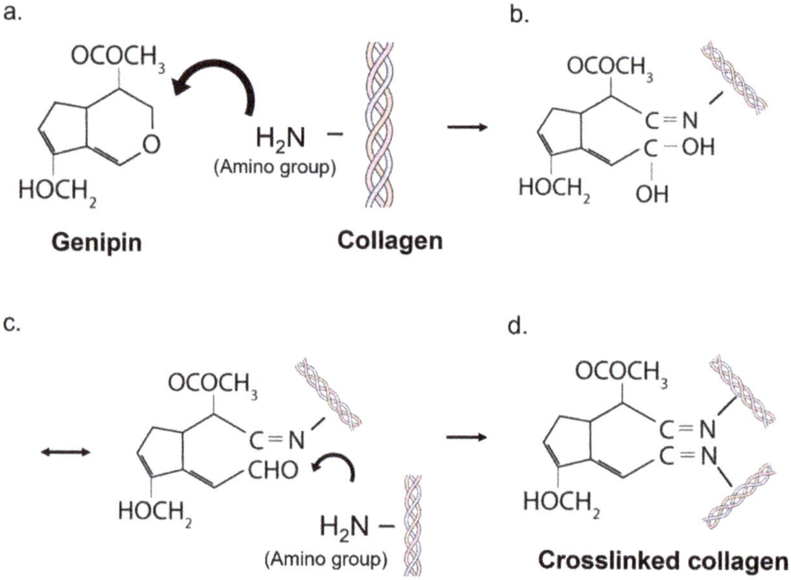

Figure 3. Schematic representation of the crosslinking mechanism of genipin. (**a**) genipin interaction with primary amine groups; (**b**) ring-opening reaction in genipin and covalent bond with the amino group of collagen; (**c**) formation of an unstable intermediate aldehyde group; (**d**) formation of a new covalent bond with another polimer, which leads to the formation of the crosslink.

3.2. SEM Imaging

In Figure 4, representative SEM images are reported for the different sample types crosslinked with genipin.

The diameters of the collagen fibers formed in 3D JellaGel™ hydrogels resulted in the order of hundreds of nanometers. In particular, the weighted mean values of the fiber diameters were 0.58 µm, 0.62 µm, and 0.66 µm for 1 mM, 2.5 mM and 5 mM genipin concentrations, respectively. No considerable differences in the collagen fiber microstructure were observed for the different genipin concentrations; the dimensions and organization of the network of fibers were similar for all the experimental conditions, as confirmed by the corresponding histograms.

3.3. DSC Analysis

The DSC analysis was performed on three dried samples for each genipin concentration. The results are shown in Figure 5.

The mean temperature peak was similar for all the genipin concentrations: 57.0 °C for 1 mM, 57.7 °C for 2.5 mM, and 57.3 °C for 5 mM. No significant differences in the temperature peak values were found, thus indicating no influence of the crosslinking degree on the denaturation temperature. The enthalpy change was also calculated by integrating the area under the thermogram, to determine the energy of the bonds keeping the protein in the folded conformation. Also, enthalpy values did not show significant differences for the different genipin concentrations. A relatively high variability of the enthalpy values was observed, especially for the 5 mM genipin concentration (-47.7 ± 32 J/g).

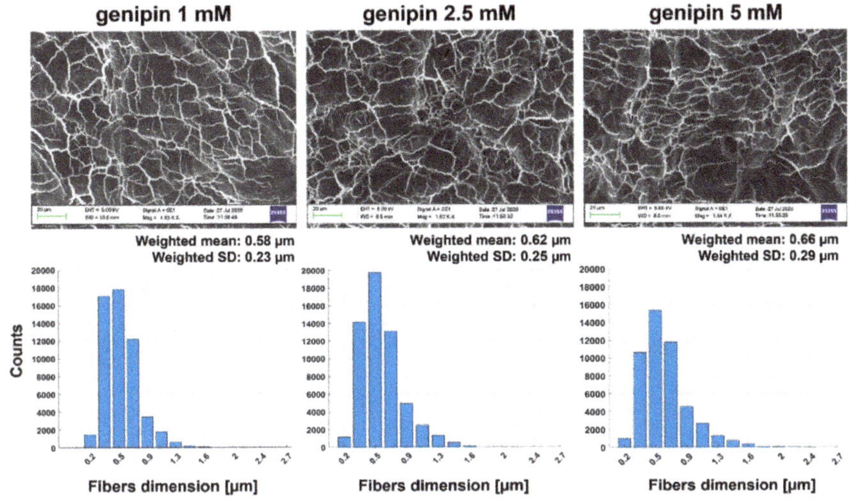

Figure 4. SEM images of JellaGel™ hydrogels crosslinked with genipin at different concentrations. The images are reported at a magnification of 1600×. They show the internal gel microarchitecture with the domains created by the collagen network. For each concentration, a histogram showing the diameter of the collagen fibers is also reported, highlighting the weighted mean and standard deviation for each concentration. Scale bars = 20 μm.

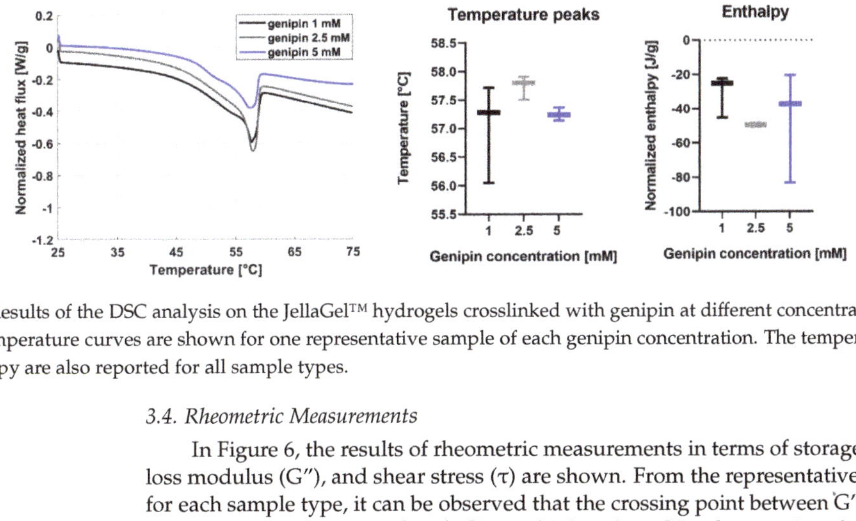

Figure 5. Results of the DSC analysis on the JellaGel™ hydrogels crosslinked with genipin at different concentrations. Heat flux vs. temperature curves are shown for one representative sample of each genipin concentration. The temperature peak and enthalpy are also reported for all sample types.

3.4. Rheometric Measurements

In Figure 6, the results of rheometric measurements in terms of storage modulus (G′), loss modulus (G″), and shear stress (τ) are shown. From the representative curves shown for each sample type, it can be observed that the crossing point between G′ and G″ occurs at ~100% strain. More in detail, G′ was higher than G″ at low strains, thus indicating a more elastic behavior at strain values smaller than 100%, for all samples. On the other hand, G′ was lower than G″ at high strains, turning into a more viscous behavior at strains over 100%. The mean values of G′ and G″ were comparable for all the genipin concentrations; indeed, no significant differences were found among the three tested concentrations for G′ and G″. However, the G′ and G″ values resulted rather variable among independent samples crosslinked with the same genipin concentration, especially in case of G′ for 1 mM (23.97 ± 19 Pa).

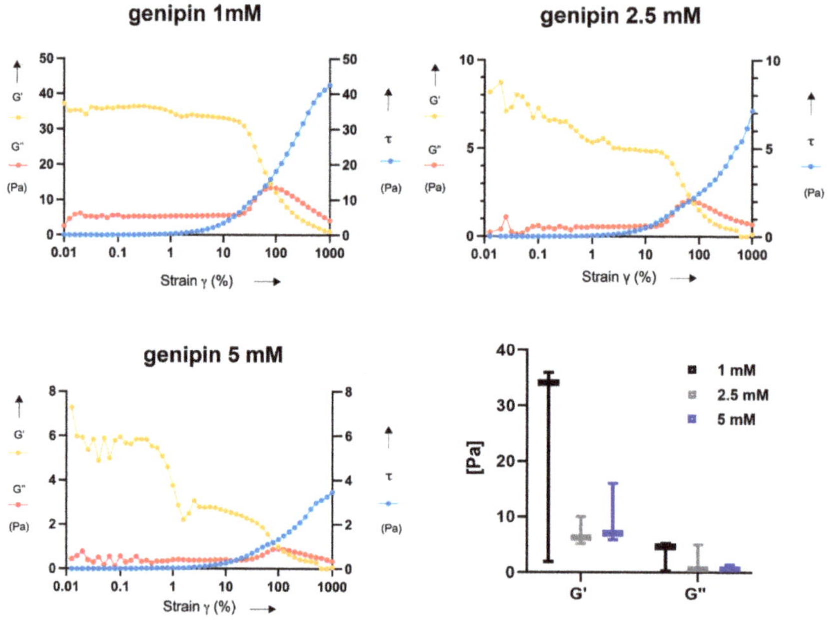

Figure 6. Results of rheometric measurements. Representative graphs of the trend of storage modulus (G′), loss modulus (G″), and shear stress (τ) versus the shear strain (γ) are reported for each genipin concentration. Boxplots with G′ and G″ values, calculated in the linear region of the curves, are also reported for all samples at the three genipin concentrations.

3.5. Cell Metabolic Activity

Figure 7 shows the results obtained in terms of metabolic activity of chondrocytes embedded within the JellaGel™ hydrogels crosslinked with the three genipin concentrations, for different culture time-points (3, 7, and 14 days).

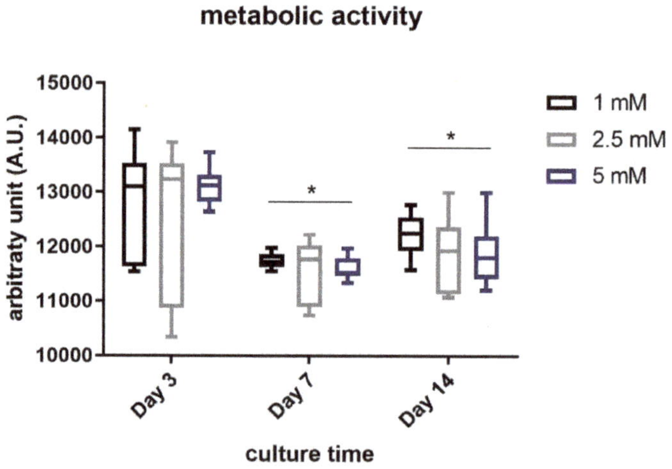

Figure 7. Metabolic activity of human chondrocytes embedded in JellaGel™ hydrogels crosslinked with genipin at different concentrations, after 3, 7, and 14 days of culture. * = $p < 0.05$.

After 3 days, no significant differences were found among the three concentrations. The metabolic activity of the cells embedded within the hydrogels crosslinked with 1 mM genipin was significantly (even if slightly) higher than the one observed in the hydrogels crosslinked using 5 mM genipin both at day 7 ($p = 0.02$) and day 14 ($p = 0.04$).

4. Discussion

In this work, the influence of three different genipin concentrations (1, 2.5, and 5 mM) on the formation of 3D crosslinked JellaGel™ hydrogels was explored for the first time. One representative gel for each genipin concentration is shown in Figure 2. As mentioned in the Results section, the gels prepared with higher concentrations of genipin (2.5 mM and 5 mM) appeared thicker and better formed than those with a genipin concentration of 1 mM. Genipin results were crucial for obtaining stable gels, which were instead impossible to achieve without any crosslinking agent. This made it impossible to include a non-crosslinked control in the subsequent characterization tests.

The SEM images shown in Figure 3. demonstrated that the different genipin concentrations did not affect the microstructure of collagen fibers within the hydrogels. This finding is in agreement with the state-of-the-art: it has been demonstrated in fact that a crosslinking procedure using different ratios of EDC/NHS did not significantly influence the microstructure of jellyfish collagen samples [20]. Moreover, Figure 3 suggested that the characteristic dimensions of the collagen network structure formed in the JellaGel™ hydrogels were in the order of ~0.6 μm. A direct comparison with previous works is difficult to perform. Indeed, the different collagen sources and the different crosslinking agent used can influence the dimension of the fibers. Moreover, previous similar studies did not perform a quantitative analysis of the diameters of the collagen fibers [21,22]. However, collagen fibers generally present a cord or tape shape 1–20 μm wide. These fibers consist of closely packed thin collagen fibrils (30–100 nm thick), which are organized to form a three-dimensional network as a whole [23]. In our case, we found fiber diameters of hundreds of nanometers, in agreement with the general fibers' diameters reported in previous literature [23,24]. The results of the DSC analysis are reported in Figure 4. They suggest that the denaturation temperature was not influenced by the crosslinking degree. The temperature peak was found at around 57 °C for all concentrations. The denaturation temperature of pure collagen is about 37 °C [25]. Hence, the crosslinking process, for all genipin concentrations, guaranteed achieving a higher denaturation temperature, allowing gel stability at body temperature. Our findings are in agreement with results reported in the literature. Hoyer et al. [10] found an increase of about 12 K in the denaturation temperature of jellyfish collagen crosslinked with 1% EDC with respect to pure jellyfish collagen. The enthalpy results suggest that the energy needed to keep the proteins in their original state did not depend on the genipin concentration used to crosslink the gel. It is worth mentioning that the enthalpy change depends on the percentage of native proteins in the original solution. Hence, the great variability observed in the enthalpy values may be associated with the variability in the gel formation.

In rheological analyses, the correct loading of the testing plate is an important aspect, which influences the rheological results [26]. Probably, the procedure for sample preparation required for this test negatively influenced the results, increasing their variability; indeed, it was difficult to properly place the samples on the rheometer testing plate without almost destroying them. This aspect can explain the relatively high variability found in the rheological results, presented in Figure 5. Besides rheometric analyses, the encountered manipulability issues are an important aspect in view of possible future pre-clinical and clinical translation of this material.

Finally, we evaluated the metabolic activity of chondrocytes embedded in JellaGel™ hydrogels, as shown in Figure 6. This test could assess the suitability of the JellaGel™ matrix as a possible 3D environment hosting chondrocytes. Results revealed that the cell metabolic activity did not decrease over the culture time, until 14 days. This suggests that the crosslinked matrices can safely host chondrocytes for several days, without hampering nutrient diffusion, not interfering with cell metabolic processes. Eun Song et al. [20] demonstrated that the viability of human chondrocytes seeded on jellyfish collagen scaffolds crosslinked with EDC was stable over time. In our work, we evaluated for the first time the metabolic activity of chondrocytes embedded in JellaGel™ hydrogels crosslinked with different genipin concentrations. Wang et al. [27] investigated cytotoxic effects on chondrocytes incubated with a culture medium containing different genipin concentrations. After 24 h of incubation, the death of chondrocytes exposed to 5 mM genipin was ~ 65% higher than the control. Indeed, 5 mM was identified as the minimum concentration of genipin that can induce toxicity when placed in direct contact with cells. Zhou et al. [28] also demonstrated that the cytotoxicity on adipose stem cells (ASCs) embedded in genipin-crosslinked hydrogels was affected by genipin concentration. In their case, cytotoxicity and proliferation assays demonstrated that the best genipin concentration was 0.02% *w/v* (0.2 mg/mL). In addition, cell proliferation decreased on day 7 and day 14. These results are in agreement with those found in our work: the cells embedded in hydrogels crosslinked with 1 mM genipin (0.226 mg/mL) showed a slight better metabolic activity compared to the other genipin concentrations (especially compared to 5 mM) at 7 (p = 0.02) and 14 days (p = 0.04).

In previous reports, the potential of jellyfish collagen has been demonstrated in the field of cartilage regeneration [6,29]. JellaGel™ was formulated to create 3D hydrogels providing and maintaining a realistic, near-native microenvironment for cells [11]. However, natural materials are often affected by some limitations related to their usability and reproducibility. JellaGel™ being a natural material, it revealed a rather poor reproducibility in the material preparation procedure. This issue typically affects natural materials that, since they derive from animal sources, often imply a high batch-to-batch variability and also challenging processability [30]. We demonstrated that genipin-based crosslinking allows to obtain stable gels able to embed cells, although with a relatively high variability of the hydrogel properties.

Overall, our results assessed a good biological activity, confirming JellaGel™ as a biocompatible matrix for chondrocytes. On the other hand, results revealed some limitations of this material concerning poor reproducibility in the material processing, high variability, and manipulability issues. These issues should be solved to enable future in vivo applications of this material.

In this work, the authors mainly focused on the physical characterization of crosslinked JellaGel™ hydrogels without chondrocytes embedded, and investigated the influence of the genipin crosslinker on the basic properties of JellaGel™. However, the authors are aware that the inclusion of cells in a 3D hydrogel may alter its properties. Hence, future tests will be needed to perform a more refined characterization of JellaGel™ with embedded chondrocytes, to also evaluate how the cell production of extracellular matrix and the formation of new cartilaginous tissue over time may affect the physical properties of the material.

5. Conclusions

In this study, three different genipin concentrations (1 mM, 2.5 mM, and 5 mM) were used to obtain 3D crosslinked JellaGel™ hydrogels. Scanning electron microscopy, differential scanning calorimetry, and rheometry allowed for the analysis of the morphological, thermal, and mechanical properties of the materials, respectively. The hydrogels treated with the highest concentrations (2.5 mM and 5 mM) of genipin appeared by visual inspection thicker and more well-formed than the one crosslinked with 1 mM genipin. Morphological results revealed no changes in the microstructure of the hydrogels increasing the level of genipin concentration. The average diameters of the collagen fibers were 0.58 μm, 0.62 μm, and 0.66 μm for 1 mM, 2.5 mM, and 5 mM genipin concentrations, respectively. Similarly, the denaturation temperature and the enthalpy change of the hydrogels were not affected by the degree of crosslinking. The denaturation temperature result of 57 °C for all the genipin concentrations made genipin-crosslinked JellaGel™ hydrogels stable at body temperature. Rheometric analyses showed no significant differences in the mechanical properties of gels crosslinked with different genipin concentrations, with a crossing point between G′ and G″ occurring at ~100% strain. Finally, genipin-crosslinked JellaGel™ hydrogels proved to constitute a suitable matrix for human chondrocyte embodiment. The metabolic activity of the embedded cells did not decrease from day 3 until day 14 of culture. The cells encapsulated in samples crosslinked with 1 mM genipin showed a slightly higher metabolic activity than the others. The material analyzed in this study also revealed some negative aspects, such as a relatively high variability (typical of natural materials), low reproducibility, and low manipulability. New material processing strategies or different crosslinking methods should be devised in the future, to overcome the mentioned limitations.

Author Contributions: All the authors conceived and designed the study. L.R. (Laura Riacci). and A.S. were the main operators of the experiments and data analysis. All authors contributed to the interpretation of the results. L.R. (Laura Riacci). and A.S. wrote the original manuscript. All authors reviewed the manuscript. L.R. (Leonardo Ricotti). secured funding for this study. All authors have read and agreed to the published version of the manuscript.

Funding: This research was funded by the European Union's Horizon 2020 research and innovation program, Grant agreement no. 814413, project ADMAIORA (Advanced nanocomposite MAterIals fOr in situ treatment and ultrasound-mediated management of osteoarthritis).

Data Availability Statement: The datasets generated during and/or analyzed during the current study are available from the corresponding author on reasonable request.

Conflicts of Interest: The authors declare no conflict of interest.

References

1. Glowacki, J.; Mizuno, S. Collagen scaffolds for tissue engineering. *Biopolymers* **2008**, *89*, 338–344. [CrossRef]
2. Lee, C.H.; Singla, A.; Lee, Y. Biomedical applications of collagen. *Int. J. Pharm.* **2001**, *221*, 1–22. [CrossRef]
3. Patino, M.G.; Neiders, M.E.; Andreana, S.; Noble, B.; Cohen, R.E. Collagen: An Overview. *Implant. Dent.* **2002**, *11*, 280–285. [CrossRef] [PubMed]
4. Dong, C.; Lv, Y. Application of Collagen Scaffold in Tissue Engineering: Recent Advances and New Perspectives. *Polymers* **2016**, *8*, 42. [CrossRef]
5. Gómez-Guillén, M.; Giménez, B.; López-Caballero, M.; Montero, M. Functional and bioactive properties of collagen and gelatin from alternative sources: A review. *Food Hydrocoll.* **2011**, *25*, 1813–1827. [CrossRef]
6. Sewing, J.; Klinger, M.; Notbohm, H. Jellyfish collagen matrices conserve the chondrogenic phenotype in two- and three-dimensional collagen matrices. *J. Tissue Eng. Regen. Med.* **2015**, *11*, 916–925. [CrossRef]
7. Widdowson, J.P.; Picton, A.J.; Vince, V.; Wright, C.; Mearns-Spragg, A. In vivocomparison of jellyfish and bovine collagen sponges as prototype medical devices. *J. Biomed. Mater. Res. Part B Appl. Biomater.* **2017**, *106*, 1524–1533. [CrossRef]
8. Addad, S.; Exposito, J.-Y.; Faye, C.; Ricard-Blum, S.; Lethias, C. Isolation, Characterization and Biological Evaluation of Jellyfish Collagen for Use in Biomedical Applications. *Mar. Drugs* **2011**, *9*, 967–983. [CrossRef]
9. Rigogliuso, S.; Salamone, M.; Barbarino, E.; Barbarino, M.; Nicosia, A.; Ghersi, G. Production of Injectable Marine Collagen-Based Hydrogel for the Maintenance of Differentiated Chondrocytes in Tissue Engineering Applications. *Int. J. Mol. Sci.* **2020**, *21*, 5798. [CrossRef]
10. Hoyer, B.; Bernhardt, A.; Lode, A.; Heinemann, S.; Sewing, J.; Klinger, M.; Notbohm, H.; Gelinsky, M. Jellyfish collagen scaffolds for cartilage tissue engineering. *Acta Biomater.* **2014**, *10*, 883–892. [CrossRef]
11. Karahan, O.A. 3rd general meeting and working group meetings of the COST Action 16203: Stem cells of marine/aquatic invertebrates: From basic research to applications (maristem), December 3, 2019, METU-Culture and Convention Center (METU-CCC), METU Ankara, T. *Intervertebrate Surviv. J.* **2020**, *17*, 32–35.
12. Zhou, X.; Tao, Y.; Chen, E.; Wang, J.; Fang, W.; Zhao, T.; Liang, C.; Li, F.; Chen, Q. Genipin-cross-linked type II collagen scaffold promotes the differentiation of adipose-derived stem cells into nucleus pulposus-like cells. *J. Biomed. Mater. Res. Part A* **2018**, *106*, 1258–1268. [CrossRef]
13. Bi, L.; Cao, Z.; Hu, Y.; Song, Y.; Yu, L.; Yang, B.; Mu, J.; Huang, Z.; Han, Y. Effects of different cross-linking conditions on the properties of genipin-cross-linked chitosan/collagen scaffolds for cartilage tissue engineering. *J. Mater. Sci. Mater. Med.* **2010**, *22*, 51–62. [CrossRef] [PubMed]
14. Hwang, Y.-J.; Larsen, J.; Krasieva, T.B.; Lyubovitsky, J.G. Effect of Genipin Crosslinking on the Optical Spectral Properties and Structures of Collagen Hydrogels. *ACS Appl. Mater. Interfaces* **2011**, *3*, 2579–2584. [CrossRef]

15. Yan, L.; Wang, Y.-J.; Ren, L.; Wu, G.; Caridade, S.; Fan, J.-B.; Wang, L.-Y.; Ji, P.-H.; Oliveira, J.M.; Oliveira, J.T.; et al. Genipin-cross-linked collagen/chitosan biomimetic scaffolds for articular cartilage tissue engineering applications. *J. Biomed. Mater. Res. Part A* **2010**, *95*, 465–475. [CrossRef]
16. Hotaling, N.; Bharti, K.; Kriel, H.; Simon, C. Dataset for the validation and use of DiameterJ an open source nanofiber diameter measurement tool. *Data Brief* **2015**, *5*, 13–22. [CrossRef] [PubMed]
17. Hanson, B.J.; Hancock, M.; Kopp, L.; Lakshmipathy, U. The PrestoBlue^{TM} Cell Viability Reagent: Reliable Cell Viability Data in as Few as 10 Minutes. *Mol. Probes Life Technol.* **2011**, 1. Available online: https://assets.thermofisher.cn/TFS-Assets/LSG/manuals/MAN0018371-PrestoBlueHS-CellViabilityReagent-PI.pdf (accessed on 15 October 2021).
18. Wang, Z.; Liu, H.; Luo, W.; Cai, T.; Li, Z.; Liu, Y.; Gao, W.; Wan, Q.; Wang, X.; Wang, J.; et al. Regeneration of skeletal system with genipin crosslinked biomaterials. *J. Tissue Eng.* **2020**, *11*. [CrossRef]
19. Pal, K.; Paulson, A.T.; Rousseau, D. Biopolymers in Controlled-Release Delivery Systems. In *Handbook of Biopolymers and Biodegradable Plastics*; Elsevier: Amsterdam, The Netherlands, 2013; pp. 329–363.
20. Song, E.; Kim, S.Y.; Chun, T.; Byun, H.-J.; Lee, Y.M. Collagen scaffolds derived from a marine source and their biocompatibility. *Biomaterials* **2006**, *27*, 2951–2961. [CrossRef]
21. Mercado, K.P.; Helguera, M.; Hocking, D.C.; Dalecki, D. Noninvasive Quantitative Imaging of Collagen Microstructure in Three-Dimensional Hydrogels Using High-Frequency Ultrasound. *Tissue Eng. Part C: Methods* **2015**, *21*, 671–682. [CrossRef] [PubMed]
22. Ulrich, T.A.; Jain, A.; Tanner, K.; MacKay, J.L.; Kumar, S. Probing cellular mechanobiology in three-dimensional culture with collagen–agarose matrices. *Biomaterials* **2010**, *31*, 1875–1884. [CrossRef]
23. Ushiki, T. Collagen Fibers, Reticular Fibers and Elastic Fibers. A Comprehensive Understanding from a Morphological Viewpoint. *Arch. Histol. Cytol.* **2002**, *65*, 109–126. [CrossRef] [PubMed]
24. Fratzl, P. *Collagen: Structure and Mechanics, an Introduction in Collagen*; Springer: Boston, MA, USA, 1998; pp. 1–13.
25. Zhang, Z.; Li, G.; Shi, B. Physicochemical properties of collagen, gelatin and collagen hydrolysate derived from bovine limed split wastes. *J. Soc. Leather Technol. Chem.* **2006**, *90*, 23–28.
26. Chen, M.H.; Wang, L.L.; Chung, J.J.; Kim, Y.-H.; Atluri, P.; Burdick, J.A. Methods To Assess Shear-Thinning Hydrogels for Application As Injectable Biomaterials. *ACS Biomater. Sci. Eng.* **2017**, *3*, 3146–3160. [CrossRef]
27. Wang, C.; Lau, T.T.; Loh, W.L.; Su, K.; Wang, D.-A. Cytocompatibility study of a natural biomaterial crosslinker-Genipin with therapeutic model cells. *J. Biomed. Mater. Res. Part B Appl. Biomater.* **2011**, *97B*, 58–65. [CrossRef] [PubMed]
28. Zhou, X.; Wang, J.; Fang, W.; Tao, Y.; Zhao, T.; Xia, K.; Liang, C.; Hua, J.; Li, F.; Chen, Q. Genipin cross-linked type II collagen/chondroitin sulfate composite hydrogel-like cell delivery system induces differentiation of adipose-derived stem cells and regenerates degenerated nucleus pulposus. *Acta Biomater.* **2018**, *71*, 496–509. [CrossRef]
29. Keller, L. Combined Jellyfish Collagen Type II, Human Stem Cells and Tgf-β3 as a Therapeutic Implant for Cartilage Repair. *J. Stem Cell Res. Ther.* **2017**, *7*, 4. [CrossRef]
30. Coenen, A.M.; Bernaerts, K.; Harings, J.A.; Jockenhoevel, S.; Ghazanfari, S. Elastic materials for tissue engineering applications: Natural, synthetic, and hybrid polymers. *Acta Biomater.* **2018**, *79*, 60–82. [CrossRef]

gels

Article

Robust and Highly Stretchable Chitosan Nanofiber/Alumina-Coated Silica/Carboxylated Poly (Vinyl Alcohol)/Borax Composite Hydrogels Constructed by Multiple Crosslinking

Hiroyuki Takeno [1,2,*] and Nagisa Suto [1]

[1] Division of Molecular Science, Graduate School of Science and Technology, Gunma University, Kiryu 376-8515, Japan; t191a047@gunma-u.ac.jp

[2] Center for Food Science and Wellness, Gunma University, 4-2 Aramaki, Maebashi 371-8510, Japan

* Correspondence: takeno@gunma-u.ac.jp

Abstract: We investigated the mechanical and structural properties of composite hydrogels composed of chitosan nanofiber (ChsNF), positively charged alumina-coated silica (ac-SiO_2) nanoparticles, carboxylated poly (vinyl alcohol) (cPVA), and borax. ChsNF/cPVA/borax hydrogels without ac-SiO_2 exhibited high Young's modulus but poor elongation, whereas cPVA/ac-SiO_2/borax hydrogels without ChsNF had moderate Young's modulus but high elongation. ChsNF/ac-SiO_2/cPVA/borax hydrogels using both ChsNF and ac-SiO_2 as reinforcement agents exhibited high extensibility (930%) and high Young's modulus beyond 1 MPa at a high ac-SiO_2 concentration. The network was formed by multiple crosslinking such as the complexation between borate and cPVA, the ionic complexation between ac-SiO_2 and cPVA, and the hydrogen bond between ChsNF and cPVA. Structural analysis by synchrotron small-angle X-ray scattering revealed that the nanostructural inhomogeneity in ChsNF/ac-SiO_2/cPVA/borax hydrogel was suppressed compared to those of the ChsNF/cPVA/borax and cPVA/ac-SiO_2/borax hydrogels.

Keywords: chitosan nanofiber; composite hydrogel; nanoparticles; tough hydrogels; multiple crosslinking

Citation: Takeno, H.; Suto, N. Robust and Highly Stretchable Chitosan Nanofiber/Alumina-Coated Silica/Carboxylated Poly (Vinyl Alcohol)/Borax Composite Hydrogels Constructed by Multiple Crosslinking. *Gels* **2022**, *8*, 6. https://doi.org/10.3390/gels8010006

Academic Editor: Qiang Chen

Received: 30 November 2021
Accepted: 20 December 2021
Published: 22 December 2021

1. Introduction

Hydrogels composed of biocompatible polymers have attracted numerous researchers and engineers because of potential applications in the biomedical field. Chitosan (Chs), which is known as one of the biocompatible polymers, is produced by alkali deacetylation of chitin, the major component in the exoskeleton of crustaceans [1]. Chs is soluble in acidic aqueous solutions, where it has positive charges due to the protonation of amine groups on the backbone. Accordingly, Chs can form a polyelectrolyte complex with the oppositely (negatively) charged polyelectrolyte such as κ-carrageenan or xanthan gum [2,3]. However, generally, the mechanical strength of polyelectrolyte complex gels is comparatively weak. As one of the methods to enhance the mechanical strength of polymer hydrogels or produce mechanically tough polymer hydrogels, the addition of reinforcing agents such as inorganic nanoparticles [4,5] or clay nanoparticles [6–8] is effective. To acquire effective reinforcement effects, it is necessary to disperse the nanoparticles and to connect between the nanoparticles and the polymer [9,10]. In these composite hydrogels, the nanoparticles act as a multiple-crosslinker, i.e., a lot of polymer chains are attached to one nanoparticle, i.e., multiple crosslinking points are formed, so that mechanically tough hydrogels are formed. Additionally, our previous studies clarified that the molecular mass of the constituent polymer is a key factor in the enhancement of mechanical performance [11,12]. Besides, we recently reported that composite hydrogels using two kinds of reinforcing agents such as clay and silica nanoparticles were robust and highly stretchable [13,14]; two identically charged nanoparticles formed multiple crosslinking and suppressed inhomogeneities in the composite hydrogels [13,14]. Thus, the construction of the gel network by multiple crosslinking was effective in fabricating tough polymer hydrogels.

In the case of biocompatible polymer hydrogels, nanofibers have been used to enhance the mechanical performance, e.g., mechanically tough composite hydrogels using cellulose nanofiber have been developed [9,15,16]. Similarly, nanofibers of chitosan can reinforce polymer hydrogels [17,18]. Nitta et al. used a chitosan nanofiber (ChsNF) to reinforce the mechanical strength of polyethylene glycol (PEG) hydrogels so that the compressive modulus and fracture stress of the ChsNF/PEG composite hydrogels attained the values of ~10 kPa and ~15 kPa, respectively [17]. Zhou and Wu showed that the compressive stress of a ChsNF/poly (acrylamide) (PAM) composite hydrogel at 95% strain attained the value of ~50 kPa, which was 7.7 times higher than that of PAM hydrogel without ChsNF [18].

In this study, we tried to enhance the mechanical performance of the hydrogels composed of chitosan and carboxylated poly (vinyl alcohol) (cPVA). For this purpose, we used ChsNF and positively charged alumina-coated silica nanoparticles (ac-SiO_2) as reinforcing agents. As a result, we found that the ChsNF/ac-SiO_2/cPVA/borax composite hydrogels were highly stretchable and robust.

2. Results and Discussion

2.1. Mechanical Properties of the Composite Hydrogels

First, we examined the effect of the addition of ChsNF on the mechanical properties of cPVA/borax hydrogels. Figure 1 depicts representative tensile stress–strain curves (a), the Young's modulus E (b), the fracture stress σ_f (c), and the fracture strain ε_f (d) for ChsNF/cPVA/borax hydrogels at different ChsNF concentrations. Both E and σ_f largely increased with increasing ChsNF concentrations but ε_f dramatically decreased. This behavior has often been seen for many composite hydrogels; although the increase in the content of the reinforcing agent leads to enhancement of the mechanical strength, the degree of elongation lowers because of the difficulty in the dispersion of the reinforcing agent.

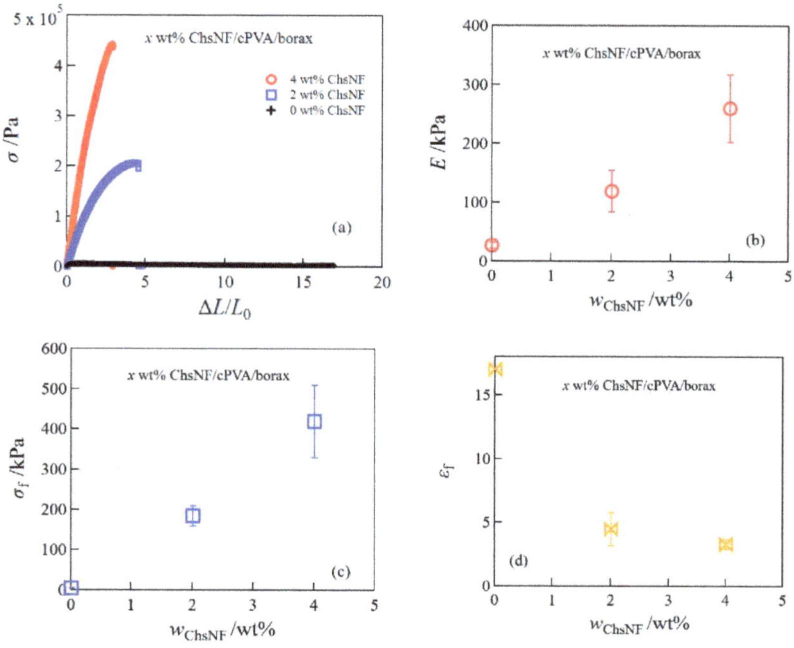

Figure 1. Tensile stress–strain curves (**a**), the Young's modulus (**b**), fracture stress (**c**), and fracture strain (**d**) for ChsNF/cPVA/borax hydrogels.

Next, we investigated the effect of the addition of ac-SiO_2 nanoparticles on the mechanical properties of cPVA/borax hydrogels (Figure 2). Similarly, both E and σ_f, for the ac-SiO_2/cPVA/borax hydrogels, significantly increased with the increase in ac-SiO_2. The increase in E was remarkably larger at higher ac-SiO_2 concentrations. At high concentrations of nanoparticles, the inter-particle distance became closer, so that many polymer chains could be attached to one nanoparticle, i.e., the number of crosslinking points increased. ε_f showed a gradual decrease with the increase in the ac-SiO_2 concentration; the degree of elongation exhibited a value of ~1000% even at high concentrations of ac-SiO_2. Although the tensile stress of ac-SiO_2/cPVA/borax hydrogels increased with the addition of ac-SiO_2, the reinforcement effect was lower compared to ChsNF/cPVA/borax hydrogels.

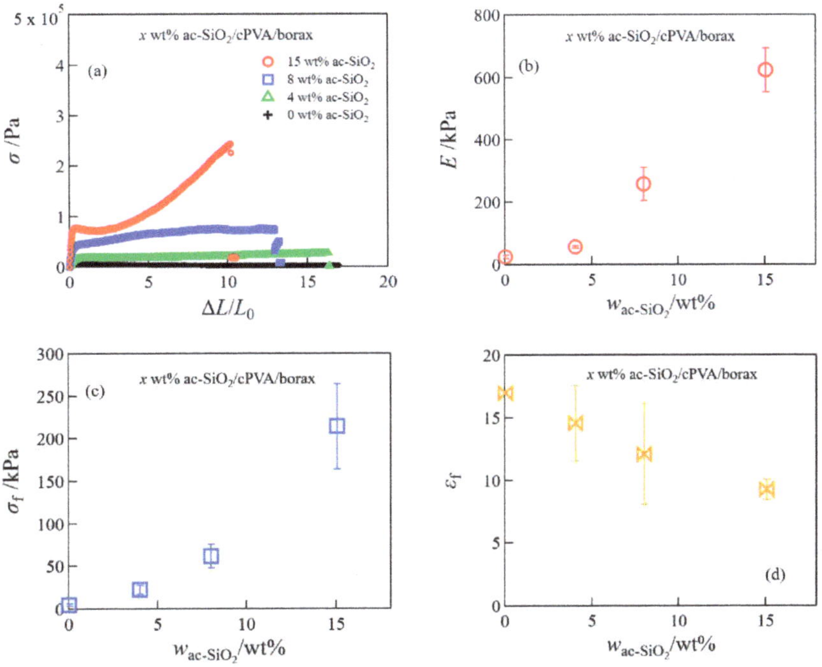

Figure 2. Tensile stress–strain curves (**a**), the Young's modulus (**b**), fracture stress (**c**), and fracture strain (**d**) for ac-SiO_2/cPVA/borax hydrogels.

For further improvement of the mechanical performance, i.e., expecting the composite hydrogels with both high mechanical strength and high extensibility, we prepared ChsNF/ac-SiO_2/cPVA/borax hydrogels using both ChsNF and ac-SiO_2 nanoparticles as reinforcing agents. Figure 3 depicts the representative stress–strain curves (a), the Young's modulus (b), the fracture stress (c), and the fracture strain for 2 wt% ChsNF/ac-SiO_2/cPVA/borax hydrogels at different ac-SiO_2 concentrations. The ChsNF/ac-SiO_2/cPVA/borax hydrogels exhibited excellent mechanical properties with high mechanical strength and high elongation. Especially, the composite hydrogel at 15 wt% ac-SiO_2 obtained the Young's modulus of 1.3 MPa and an elongation of 930%.

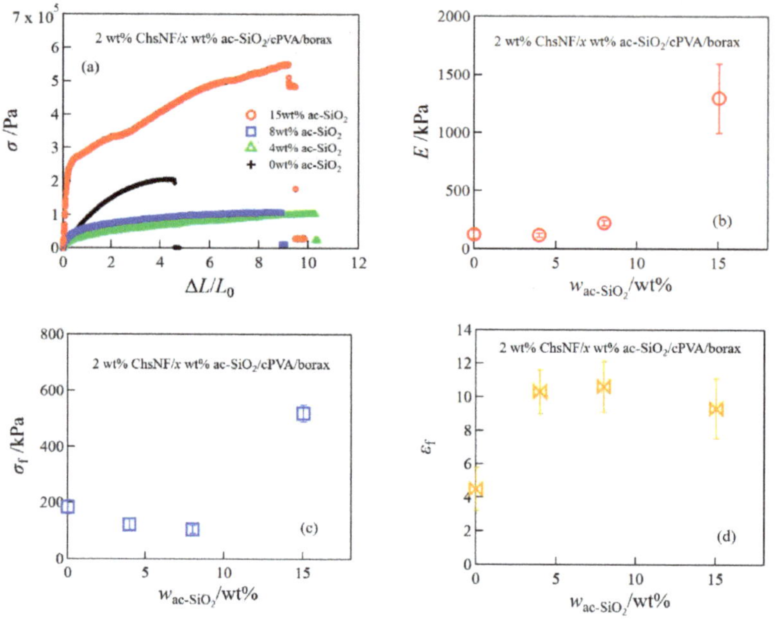

Figure 3. Tensile stress–strain curves (**a**), the Young's modulus (**b**), fracture stress (**c**), and fracture strain (**d**) for ChsNF/ac-SiO_2/cPVA/borax hydrogels.

2.2. Fourier-Transform Infrared (FT-IR) Spectroscopy

We performed FT-IR measurements to examine the interactions between different components. Figure 4a,b depict the FT-IR spectra for the cPVA/borax and ChsNF/borax systems. A characteristic peak was observed at 1339 cm^{-1} for the cPVA/borax systems; the intensity became larger as the borax content increased. This band was assigned to the asymmetric stretching vibration of B-O-C, indicating the tetrahedral complexation between PVA and borate [19–21]. For the ChsNF/borax system, the characteristic band was observed at 1316 cm^{-1}, which was at the same position as for pure borax. This result suggested that tetrahedral complexation was, to a significant extent, not formed between ChsNF and borate. Figure 4c,d show the FT-IR spectra for the cPVA/ac-SiO_2 and ChsNF/cPVA systems. The characteristic bands observed at 1077 and 791 cm^{-1} for pure ac-SiO_2 were ascribed to antisymmetric and symmetric Si-O-Si (or Si-O-Al) stretching vibrations [13,22–24]. The characteristic band at 3314 cm^{-1} for pure cPVA was assigned to the stretching vibration of hydrogen-bonded hydroxyl groups, whereas the band at 1586 cm^{-1} was assigned to the COO^- antisymmetric stretching vibration [25]. The characteristic band of the COO^- antisymmetric stretching vibration was shifted to a higher wavenumber (1590 cm^{-1}) for the cPVA/ac-SiO_2 system, whereas it was observed at the same wavenumber for the ChsNF/cPVA system. These results suggested that ion complexation between cPVA and ac-SiO_2 was formed, whereas it was not formed between cPVA and ChsNF. The band at 3314 cm^{-1} arising from the OH stretching vibration for pure cPVA was shifted to a higher wavenumber (3331 cm^{-1}) for ChsNF/cPVA. Besides, the band observed at 1089 cm^{-1} for pure cPVA, which was assigned to the stretching vibration of C-O [26,27], was shifted to a lower value (1079 cm^{-1}) for ChsNF/cPVA. This result suggested hydrogen bonding between ChsNF and cPVA.

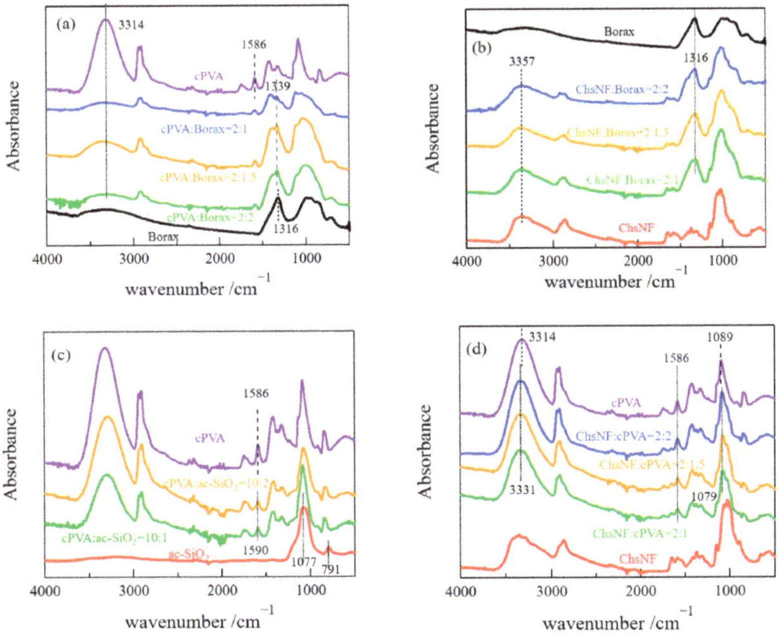

Figure 4. FT-IR spectra for cPVA/borax (**a**), ChsNF/borax (**b**), cPVA/ac-SiO_2 (**c**), and ChsNF/cPVA (**d**).

2.3. Synchrotron Small-Angle X-ray Scattering and Wide-Angle X-ray Scattering

Synchrotron SAXS and WAXS measurements were performed to examine the structures of the composite hydrogels. Firstly, we conducted the XRD measurement for a freeze-dried sample of ChsNF dispersion (Figure 5a). The XRD curve had sharp diffraction peaks at $2\theta \approx 10.6°$ and $20.0°$, which were assigned to the (020) and (110) reflections, respectively [28,29]. We estimated the crystallinity of ChsNF from the XRD curve and obtained a value of 35%.

Figure 5b depicts the WAXS curves for the ac-SiO_2/cPVA/borax hydrogels at different ac-SiO_2 concentrations. For comparison, the WAXS curve for water is appended in the figure. The figure does not show any peak except for the amorphous peak of water, suggesting that cPVA and ac-SiO_2 were amorphous in the hydrogels. Figure 5c shows the WAXS curves for the ChsNF/ac-SiO_2/cPVA/borax hydrogels at different ac-SiO_2 concentrations. All the curves were found to have a small peak at q = 1.4 Å^{-1}, which corresponded to the (110) reflection of the chitosan crystal. The peak intensity slightly increased with the increase in the ac-SiO_2 concentrations. This result suggests that the addition of ac-SiO_2 seemingly induced the crystallization of ChsNF. However, the situation may have been unexpected; the pH values only showed a slight decrease with the addition of ac-SiO_2 (pH = 8.5 for 0 wt% ac-SiO_2 and 8.2 for 8 wt% ac-SiO_2). As another possibility, a preferential orientation of ChsNF may have influenced the intensity of the Bragg reflection, although we could not estimate the orientation from the 2D SAXS images of the ChsNF/ac-SiO_2/cPVA/borax hydrogels; this was because the X-ray pattern of the composite hydrogels was dominated by the scattering intensity of ac-SiO_2, as mentioned below.

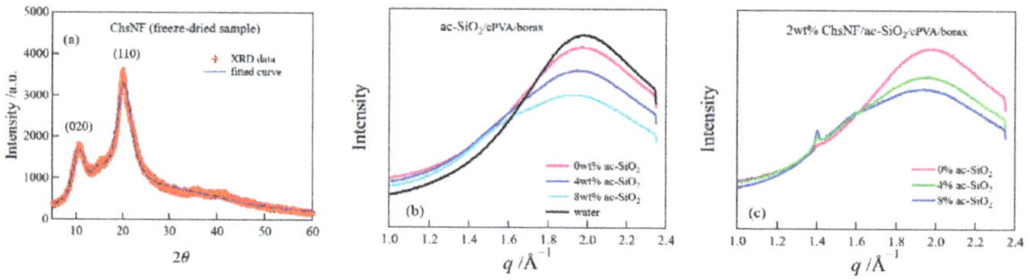

Figure 5. XRD curve for ChsNF powder (**a**), WAXS curves for ac-SiO_2/cPVA/borax hydrogels (**b**), and ChsNF/ac-SiO_2/cPVA/borax (**c**) hydrogels.

Figure 6a depicts the SAXS profiles of the cPVA/borax and ChsNF/cPVA/borax hydrogels. First, the analysis of the SAXS data for cPVA/borax hydrogel was conducted. Shibayama et al. analyzed the small-angle neutron scattering data for PVA/borax hydrogels using a generalized Zimm model for fractals with the fractal dimension of D [30]. According to the model, the scattering function is described as follows:

$$I(q) = \frac{I(0)}{\left\{1 + \frac{(D+1)}{3}\xi^2 q^2\right\}^{D/2}} \quad (1)$$

where $I(0)$ and q are the scattering intensity at $q = 0$ and the magnitude of the wavevector defined by $q = 4\pi \sin(\theta/2)/\lambda$. Here, θ and λ are the scattering angle and the wavelength, respectively. ξ is the correlation length, which represents the spatial length of concentration fluctuations. We carried out the fitting analysis for the SAXS data of cPVA/borax hydrogel using Equation (1); the fitted curve represented the scattering data well. As a result, the values of $D = 1.0$ and $\xi = 101$ Å were obtained. The latter value was almost the same as that of the PVA/borax gel [30]. The SAXS intensity for the ChsNF/cPVA/borax hydrogel largely increased compared to that of the cPVA/borax hydrogel; in particular, the scattering intensity at small q increased upward, reflecting the inhomogeneous distribution of ChsNF in the gel. Accordingly, we analyzed the SAXS data with the Debye–Buche function that had been used for the analysis of the inhomogeneous structure [31]

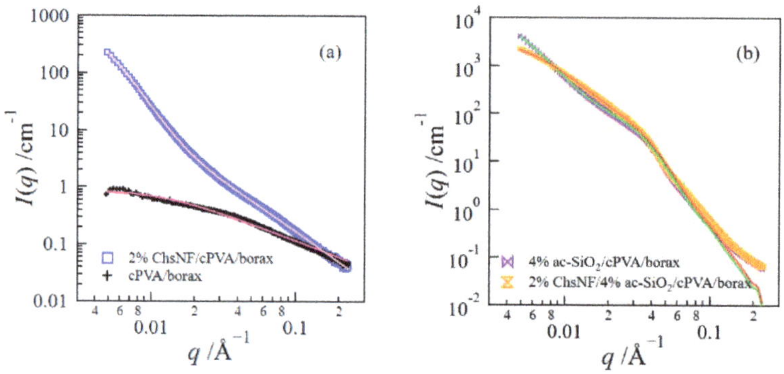

Figure 6. SAXS curves for cPVA/borax and ChSNF/cPVA/borax hydrogels (**a**), and for ac-SiO_2/cPVA/borax hydrogels and ChsNF/ac-SiO_2/cPVA/borax hydrogels (**b**).

$$I_{DB}(q) = \frac{I_1}{\left(1 + \xi_{DB}{}^2 q^2\right)^2} \quad (2)$$

where I_1 is the prefactor of the scattering intensity and ξ_{DB}—denotes a parameter that characterizes the spatial length of the inhomogeneous structure in the model. The fitting analysis using Equations (1) and (2) was conducted for the SAXS curve of the ChsNF/cPVA/borax hydrogel, so that we obtained the values of $D = 1.8$, $\xi = 47$ Å and $\xi_{DB} = 275$ Å. Thus, the SAXS analysis revealed that the inhomogeneous structure, with a size of tens of nanometers, was formed in the composite gel, which have may have caused poor elongation for this composite hydrogel; the lowering of elongation due to an inhomogeneous distribution of reinforcing agents was also reported in previous studies [12,32]. The correlation length of Equation (1) for the ChsNF/cPVA/borax hydrogel was smaller than that of the cPVA/borax hydrogel. The hydrogen bond between ChsNF and cPVA shown by the FT-IR measurements may have suppressed the concentration fluctuations in the gel network. Figure 6b shows the scattering curves for the ac-SiO_2/cPVA/borax and ChsNF/ac-SiO_2/cPVA/borax hydrogels. The scattering profiles for both hydrogels were similar except for the scattering behavior at small q. This result suggests that their scattering curves were significantly dominated by the scattering from ac-SiO_2 nanoparticles constituted of heavy atoms; heavier atoms have a larger scattering length in X-ray scattering [33].

The scattering function of spherical particles with a radius of R can be expressed by

$$I(q) = I_1 P_{sphere}(q) S(q) \quad (3)$$

with

$$P_{sphere}(q) = \left[\frac{3\{\sin(qR) - qR\cos(qR)\}}{(qR)^3} \right]^2 \quad (4)$$

where I_1 is a prefactor of the scattering intensity. $P_{sphere}(q)$ and $S(q)$ are the form factor and the structure factor of spherical particles; the former and the latter correspond to the scattering from the intra-particle interference and the inter-particle interference, respectively. We considered the size distribution of the spherical particles using a Gaussian distribution with the mean radius of R_0 and the standard deviation σ. Furthermore, we adopted the Percus–Yevick (PY) hard-sphere model [34] to calculate the structure factor $S(q)$. The detailed representation of the function was described elsewhere [13,14]; in short, the function can be expressed using parameters of an interaction radius R_{HS} and the volume fraction of spheres ϕ.

The scattering curve for the ac-SiO_2/cPVA/borax hydrogel was analyzed with a combination of the PY and DB models, i.e., Equations (2)–(4). The fitted curve is shown in Figure 6b, and the obtained parameters are summarized in Table 1. The mean radius of the ac-SiO_2 nanoparticles obtained in the fitting analysis agreed well with the value (12 nm) of the particle diameter shown in the product catalog. Subsequently, we analyzed the scattering curve for the ChsNF/ac-SiO_2/cPVA/borax hydrogel using Equations (2)–(4). In the analysis, we fixed the parameters in Equations (3) and (4) using the values obtained in the analysis of the ac-SiO_2/cPVA/borax hydrogel, because the scattering curves for both composite hydrogels were almost the same except for the scattering behavior at small q arising from the inhomogeneous structure that could be expressed by the DB model. Consequently, the spatial length of the inhomogeneous structure obtained in the fitting analysis for the ChsNF/ac-SiO_2/cPVA/borax hydrogel was much smaller, which suggested that the addition of ChsNF to the ac-SiO_2/cPVA/borax hydrogel suppressed the inhomogeneity in the composite hydrogel. Thus, the ChsNF/ac-SiO_2/cPVA/borax hydrogel possessed excellent mechanical performance, having both robustness and a high degree of elongation, as a result of the lowering of the inhomogeneity in the gel.

Table 1. The result of the fitting analysis.

Sample	R_0/Å	σ/Å	R_{HS}/Å	ϕ	ξ_{DB}/Å
ac-SiO_2/cPVA/borax	59	0.44	59	0.15	252
ChsNF/ac-SiO_2/cPVA/borax	59 (fix)	0.44 (fix)	59 (fix)	0.15 (fix)	126

3. Summary

We investigated the mechanical and structural properties of composite hydrogels using chitosan nanofibers and alumina–coated silica nanoparticles as reinforcing agents and borax as a crosslinker. The composite hydrogels exhibited mechanically robust and highly stretchable properties. This study showed that the combined use of bio-nanofiber and inorganic nanoparticles as reinforcing agents is effective in the fabrication of robust and highly stretchable composite hydrogels. The composite hydrogels were constructed by the multiple crosslinking composed of ion complexation between cPVA and ac-SiO_2, the hydrogen bond between ChsNF and cPVA, and the complexation between cPVA and borate. Synchrotron SAXS analysis revealed that the inhomogeneity in the ChsNF/ac-SiO_2/cPVA/borax hydrogel was significantly suppressed so that the composite hydrogel exhibited excellent mechanical performance with both high mechanical strength and high degrees of elongation.

4. Experimental

4.1. Materials

In this study, we used chitosan nanofiber with a diameter of 20–50 nm (ChsNF) purchased from Sugino Machine Ltd. Carboxylated poly (vinyl alcohol) (cPVA) with a saponification value larger than 99% (Gohsenol T-330H) and alumina-coated silica nanoparticles (ac-SiO_2) with particle sizes of 12 nm (SNOWTEX ST-AK) were kindly supplied from Mitsubishi Chemical Corp. and Nissan Chemical Corp., respectively. Sodium tetraborate decahydrate (borax) was purchased from Kanto Chemical Co., Inc. (Tokyo, Japan).

4.2. Gel Preparation

After a ChsNF suspension and an ac-SiO_2 suspension were added into a vial, ChsNF and ac-SiO_2 nanoparticles were dispersed using an ultrasonic homogenizer (QSONICA Model Q55) for 30 min. Afterward, cPVA was added to the ChsNF/ac-SiO_2 suspension and dissolved at 90 °C. After the suspension was condensed in a vacuum oven to reach the desired concentration, borax was added. After the mixture was thoroughly mixed using a glass rod, it was placed in a mold with 1 mm thickness and was pressed at 70 °C. The final concentrations of cPVA and borax were 10 wt% and 3 wt%, respectively. The concentrations of ChsNF and ac-SiO_2 are shown in the text. The pH values for the ChsNF/ac-SiO_2/cPVA/borax hydrogels were measured with a PH mater (F-71, Horiba, Kyoto, Japan) and a pH electrode (ISFET 0040-10D, Horiba).

4.3. Tensile Tests

We performed tensile tests for the composite hydrogels using TENSILE TESTER STM-20 (ORIENTEC). The measurements were conducted at a stretching speed of 10 mm/min for the specimens with 1 mm thickness, 10 mm length, and 15 mm width. The tensile stress σ and strain ε were calculated from the relations of $\sigma = F/S_0$ and $\varepsilon = \Delta L/L_0$, where F and ΔL are the tensile force and the deformation, respectively. S_0 and L_0 denote the initial area and initial length of the test specimen, respectively. The Young's modulus E was estimated from the slope of the stress–strain curve at small strains. The average values of E, σ_f, and ε_f were obtained from three tests.

4.4. Fourier-Transform Infrared (FT-IR) Measurements

FT-IR spectroscopy (JASCO FT/IR 4700) was used to examine the interactions between different components using the attenuated total reflection (ATR) method. The FT-IR spectra were recorded in the wavenumber range of 500–4000 cm^{-1}. Freeze-dried samples were used for FT-IR measurements.

4.5. X-ray Diffraction Measurements

X-ray diffraction (XRD) measurement was conducted to explore the structure of ChsNF using an X-ray diffractometer (RIGAKU, RINT2200VF) at the Center for Instrumental

Analysis of Gunma University. CuK α radiation was used in this measurement, and the diffracted intensity was detected at the diffraction angles of 5°–60°. The sample for XRD measurement was prepared as follows: a ChsNF suspension was freeze-dried, and the freeze-dried sample was filled in an aluminum spacer for XRD measurements. The crystallinity of ChsNF was estimated from the ratio of the area of crystalline peaks to the whole area.

4.6. Synchrotron Small-Angle X-ray Scattering/Wide-Angle X-ray Scattering

Synchrotron SAXS and WAXS measurements were performed to investigate the structure of the composite hydrogels. The experiments were conducted at the beamline 6A at the photon factory of the High Energy Accelerator Research Organization (KEK) in Tsukuba, Japan. An X-ray beam with a wavelength of 1.5 Å was used for the measurements, and the scattered intensity was detected using two-dimensional detectors—PILATUS 1M for SAXS and PILATUS 100K for WAXS. The detected X-ray images were circularly averaged to obtain the scattering curves as a function of q [35]. Moreover, the scattering intensity was corrected by the beam intensity, transmittance, and background scattering, and was reduced to the absolute units [36].

Author Contributions: H.T. analyzed the experimental data and wrote the paper. N.S. performed the experiments and analyzed the data. All authors have read and agreed to the published version of the manuscript.

Funding: This research was funded by JSPS KAKENHI, Grant Number JP19K05594.

Institutional Review Board Statement: Not Applicable.

Informed Consent Statement: Not Applicable.

Data Availability Statement: Data are contained within the article.

Acknowledgments: This work was supported by JSPS KAKENHI, Grant Number JP19K05594. The synchrotron SAXS measurements were conducted under the approval of the Photon Factory Program Advisory Committee. The XRD measurements were performed at the Center for Instrumental Analysis of Gunma University.

Conflicts of Interest: The authors declare no conflict of interest.

References

1. Nilsen-Nygaard, J.; Strand, S.; Vårum, K.; Draget, K.; Nordgård, C. Chitosan: Gels and interfacial properties. *Polymers* **2015**, *7*, 552–579. [CrossRef]
2. Shumilina, E.V.; Shchipunov, Y.A. Chitosan-carrageenan gels. *Colloid J.* **2002**, *64*, 372–378. [CrossRef]
3. Luo, Y.; Wang, Q. Recent development of chitosan-based polyelectrolyte complexes with natural polysaccharides for drug delivery. *Int. J. Biol. Macromol.* **2014**, *64*, 353–367. [CrossRef]
4. Xia, S.; Song, S.; Ren, X.; Gao, G. Highly tough, anti-fatigue and rapidly self-recoverable hydrogels reinforced with core-shell inorganic-organic hybrid latex particles. *Soft Matter* **2017**, *13*, 6059–6067. [CrossRef]
5. Zhong, M.; Liu, X.Y.; Shi, F.K.; Zhang, L.Q.; Wang, X.P.; Cheetham, A.G.; Cui, H.G.; Xie, X.M. Self-healable, tough and highly stretchable ionic nanocomposite physical hydrogels. *Soft Matter* **2015**, *11*, 4235–4241. [CrossRef]
6. Haraguchi, K. Synthesis and properties of soft nanocomposite materials with novel organic/inorganic network structures. *Polym. J.* **2011**, *43*, 223–241. [CrossRef]
7. Takeno, H.; Nakamura, W. Structural and mechanical properties of composite hydrogels composed of clay and a polyelectrolyte prepared by mixing. *Colloid Polym. Sci.* **2013**, *291*, 1393–1399. [CrossRef]
8. Huang, B.; Liu, M.; Zhou, C. Chitosan composite hydrogels reinforced with natural clay nanotubes. *Carbohydr. Polym.* **2017**, *175*, 689–698. [CrossRef]
9. Takeno, H.; Inoguchi, H.; Hsieh, W.-C. Mechanical and structural properties of cellulose nanofiber/poly(vinyl alcohol) hydrogels cross-linked by a freezing/thawing method and borax. *Cellulose* **2020**, *27*, 4373–4387. [CrossRef]
10. Takeno, H.; Sato, C. Effects of molecular mass of polymer and composition on the compressive properties of hydrogels composed of laponite and sodium polyacrylate. *Appl. Clay Sci.* **2016**, *123*, 141–147. [CrossRef]
11. Takeno, H.; Kimura, Y.; Nakamura, W. Mechanical, swelling, and structural properties of mechanically tough clay-sodium polyacrylate blend hydrogels. *Gels* **2017**, *3*, 10. [CrossRef] [PubMed]

12. Takeno, H.; Kimura, Y. Molecularweight effects on tensile properties of blend hydrogels composed of clay and polymers. *Polymer* **2016**, *85*, 47–54. [CrossRef]
13. Takeno, H.; Aoki, Y.; Kimura, K. Effects of addition of silica nanospheres on mechanical properties of clay/sodium polyacrylate hydrogels. *Mater. Today Commun.* **2021**, *28*, 102710. [CrossRef]
14. Takeno, H.; Aoki, Y.; Kimura, K. Effects of silica and clay nanoparticles on the mechanical properties of poly(vinyl alcohol) nanocomposite hydrogels. *Colloids Surf. A Physicochem. Eng. Asp.* **2021**, *630*, 127592. [CrossRef]
15. De France, K.J.; Hoare, T.; Cranston, E.D. Review of hydrogels and aerogels containing nanocellulose. *Chem. Mater.* **2017**, *29*, 4609–4631. [CrossRef]
16. Yuan, N.X.; Xu, L.; Zhang, L.; Ye, H.W.; Zhao, J.H.; Liu, Z.; Rong, J.H. Superior hybrid hydrogels of polyacrylamide enhanced by bacterial cellulose nanofiber clusters. *Mater. Sci. Eng. C* **2016**, *67*, 221–230. [CrossRef]
17. Nitta, S.; Kaketani, S.; Iwamoto, H. Development of chitosan-nanofiber-based hydrogels exhibiting high mechanical strength and ph-responsive controlled release. *Eur. Polym. J.* **2015**, *67*, 50–56. [CrossRef]
18. Zhou, C.; Wu, Q. A novel polyacrylamide nanocomposite hydrogel reinforced with natural chitosan nanofibers. *Colloids Surf. B Biointerfaces* **2011**, *84*, 155–162. [CrossRef]
19. Dixit, A.; Bag, D.S.; Kalra, S.J.S. Synthesis of strong and stretchable double network (dn) hydrogels of pva-borax and p(am-co-hema) and study of their swelling kinetics and mechanical properties. *Polymer* **2017**, *119*, 263–273. [CrossRef]
20. Spoljaric, S.; Salminen, A.; Luong, N.D.; Seppälä, J. Stable, self-healing hydrogels from nanofibrillated cellulose, poly(vinyl alcohol) and borax via reversible crosslinking. *Eur. Polym. J.* **2014**, *56*, 105–117. [CrossRef]
21. Kobayashi, M.; Kitaoka, Y. Complex formation of boric acids with di- and tricarboxylic acids and poly(vinyl alcohol) in aqueous solutions. *Macromol. Symp.* **1997**, *114*, 303–308. [CrossRef]
22. Pauly, C.S.; Genix, A.C.; Alauzun, J.G.; Sztucki, M.; Oberdisse, J.; Mutin, P.H. Surface modification of alumina-coated silica nanoparticles in aqueous sols with phosphonic acids and impact on nanoparticle interactions. *Phys. Chem. Chem. Phys.* **2015**, *17*, 19173–19182. [CrossRef]
23. Tang, S.; Zou, P.; Xiong, H.; Tang, H. Effect of nano-sio2 on the performance of starch/polyvinyl alcohol blend films. *Carbohydr. Polym.* **2008**, *72*, 521–526. [CrossRef]
24. Zhang, W.; Dehghani-Sanij, A.A.; Blackburn, R.S. Ir study on hydrogen bonding in epoxy resin-silica nanocomposites. *Prog. Nat. Sci.* **2008**, *18*, 801–805. [CrossRef]
25. Nara, M.; Torii, H.; Tasumi, M. Correlation between the vibrational frequencies of the carboxylate group and the types of its coordination to a metal ion: An ab initio molecular orbital study. *J. Phys. Chem.* **1996**, *100*, 19812–19817. [CrossRef]
26. Ghaderi, J.; Hosseini, S.F.; Keyvani, N.; Gómez-Guillén, M.C. Polymer blending effects on the physicochemical and structural features of the chitosan/poly(vinyl alcohol)/fish gelatin ternary biodegradable films. *Food Hydrocoll.* **2019**, *95*, 122–132. [CrossRef]
27. Wu, Y.; Guan, Y.; Gao, H.; Zhou, L.; Peng, F. Novel high-strength montmorillonite/polyvinyl alcohol composite film enhanced by chitin nanowhiskers. *J. Appl. Polym. Sci.* **2020**, *138*, app50344. [CrossRef]
28. Hai, T.A.P.; Sugimoto, R. Fluorescence control of chitin and chitosan fabricated via surface functionalization using direct oxidative polymerization. *RSC Adv.* **2018**, *8*, 7005–7013.
29. Zhang, Y.Q.; Xue, C.H.; Xue, Y.; Gao, R.C.; Zhang, X.L. Determination of the degree of deacetylation of chitin and chitosan by X-ray powder diffraction. *Carbohydr. Res.* **2005**, *340*, 1914–1917. [CrossRef] [PubMed]
30. Shibayama, M.; Kurokawa, H.; Nomura, S.; Muthukumar, M.; Stein, R.S.; Roy, S. Small-angle neutron-scattering from poly(vinyl alcohol)-borate gels. *Polymer* **1992**, *33*, 2883–2890. [CrossRef]
31. Debye, P.; Bueche, A.M. Scattering by an inhomogeneous solid. *J. Appl. Phys.* **1949**, *20*, 518–525. [CrossRef]
32. Takeno, H.; Nagai, S. Mechanical properties and structures of clay-polyelectrolyte blend hydrogels. *Gels* **2018**, *4*, 71. [CrossRef] [PubMed]
33. Takeno, H. Synchrotron small-angle X-ray scattering and small-angle neutron scattering studies of nanomaterials. In *X-ray and Neutron Techniques for Nanomaterials Characterization*; Kumar, C.S.S.R., Ed.; Springer: Berlin/Heidelberg, Germany, 2016; pp. 717–760.
34. Percus, J.K.; Yevick, G.J. Analysis of classical statistical mechanics by means of collective coordinates. *Phys. Rev.* **1958**, *110*, 1–13. [CrossRef]
35. Shimizu, N.; Yatabe, K.; Nagatani, Y.; Saijyo, S.; Kosuge, T.; Igarashi, N. Software development for analysis of small-angle X-ray scattering data. *AIP Conf. Proc.* **2016**, *1741*, 050017.
36. Zhang, F.; Ilavsky, J.; Long, G.G.; Quintana, J.P.G.; Allen, A.J.; Jemian, P.R. Glassy carbon as an absolute intensity calibration standard for small-angle scattering. *Metall. Mater. Trans. A* **2010**, *41A*, 1151–1158. [CrossRef]

gels

MDPI

Article

Effective Carbon/TiO_2 Gel for Enhanced Adsorption and Demonstrable Visible Light Driven Photocatalytic Performance

Anam Safri and Ashleigh Jane Fletcher *

Department of Chemical and Process Engineering, University of Strathclyde, Glasgow G1 1XJ, UK; anam.safri@strath.ac.uk
* Correspondence: ashleigh.fletcher@strath.ac.uk; Tel.: +44-141-5482-431

Abstract: A new strategy to synthesise carbon/TiO_2 gel by a sol–gel method is proposed. Textural, morphological, and chemical properties were characterised in detail and the synthesised material was proven to be an active adsorbent, as well as a visible light photocatalyst. Homogenously distributed TiO_2 is mesoporous with high surface area and, hence, exhibited a high adsorption capacity. The adsorption equilibrium experimental data were well explained by the Sips isotherm model. Kinetic experiments demonstrated that experimental data fitted a pseudo second order model. The modification in electronic structure of TiO_2 resulted in a reduced bandgap compared to commercial P25. The absorption edge studied through UV-Vis shifted to the visible region, hence, daylight photocatalytic activity was efficient against degradation of MB dye, as an example pollutant molecule. The material was easily removed post treatment, demonstrating potential for employment in industrial water treatment processes.

Keywords: adsorption; carbon/TiO_2 gels; resorcinol formaldehyde RF/TiO_2 gels; photocatalysis; adsorption kinetics; methylene blue dye degradation

Citation: Safri, A.; Fletcher, A.J. Effective Carbon/TiO_2 Gel for Enhanced Adsorption and Demonstrable Visible Light Driven Photocatalytic Performance. *Gels* **2022**, *8*, 215. https://doi.org/10.3390/gels8040215

Academic Editors: Hiroyuki Takeno and Avinash J. Patil

Received: 11 February 2022
Accepted: 29 March 2022
Published: 1 April 2022

1. Introduction

Adsorption of carbon is perhaps the most widely used water treatment technique. However, there is an ongoing effort to develop efficient adsorbents with reduced regeneration costs. Currently, the combination of carbon and titanium dioxide (TiO_2) appears to offer a promising route to obtain an adsorbent with self-regeneration properties. Additionally, the synergistic effect of both carbon and TiO_2 enhances the degradation process due to respective adsorptive and photocatalytic properties. Literature reports several studies to address the synergy of adsorption and photodegradation by experimental demonstration of various carbon/TiO_2 composite materials [1,2]. However, there is still a need to better understand the phenomenon of pollutant-adsorbent interactions in order to have a good knowledge to design an efficient water treatment process. Additionally, the improvement in design involves the type of materials and synthesis process employed to attain maximum efficiency of the system.

Previously, carbon has been combined with TiO_2 through various approaches, in different forms, such as carbon nanotubes [3–5], graphene [6–8], and activated carbon [9,10]. Lately, focus has been shifted to highly porous carbon materials as support matrix for industrial applications, due to the high surface area and tuneable porosity. Ideally, well-developed mesoporous structures with large pore volumes and uniform pore size distributions are preferred, due to enhanced accessible surface sites contributing to superior adsorption capacity of pollutants from the aqueous phase. However, the preparation process of these mesoporous carbons is costly and complicated, usually resulting in materials with moderate or low surface area. The efficiency of the material is also limited, since most TiO_2 nanoparticles incorporated in the pores of the carbon are unavailable for photocatalysis [11].

Amongst mesoporous carbon materials, carbon gels are a new type of nanocarbon with potential applications in photocatalysis [2,12,13]. Carbon gels produced by polycondensation of resorcinol (R) with formaldehyde (F) are highly porous and have flexible properties. A comprehensive review of sol–gel synthesis of RF gel reveals that the material can be easily tailored to attain desired properties, mainly tuneable porosity, and acts as a support for metals [14]. Hence, RF gels can be promising materials for water treatment applications, mainly due to their stability, owing to aromatic resorcinol rings and their overall interconnected mesoporous carbon structure. For industrial applications where a continuous process system is often required, carbon derived from RF gels can be more efficient and cost-effective than commercial adsorbents, which are in the form of granules or powders and are unsuitable for use in continuous systems.

The aim of this study is to synthesize an adsorbent with visible light driven photocatalytic activity by incorporating TiO_2 nanoparticles into RF gels. A typical synthesis route of an RF gel [15] is modulated in this study to integrate TiO_2 nanoparticles by formulating a twostep synthesis scheme. In addition to enhancement in textural properties of this newly synthesised adsorbent, improvement in photocatalytic properties is expected by (i) modification in electronic structure of TiO_2, due to the presence of RF gel as a carbon source, shifting the absorption edge to the visible light region, hence, enabling TiO_2 to activate under visible light irradiation; (ii) the carbon phase can entrap the photogenerated electron and hole pairs, which would otherwise recombine and dissipate heat energy; and (iii) the porous RF gel helps facilitate dispersion of TiO_2 and easy post treatment removal of the adsorbent/photocatalyst.

Here, we report a study of the textural and optical characteristics of the adsorbent/photocatalyst. Detailed adsorption experiments were carried out to study the effect of several parameters on adsorption capacity. Additionally, the interaction behaviour between potential pollutants and the material were investigated, using methylene blue (MB) as a model adsorptive. Equilibrium sorption data were modelled using Langmuir, Freundlich, Sips, and Toth isotherm models. Kinetic analyses were carried out by comparing the experimental data with pseudo first order and pseudo second order expressions, as well as a diffusion model to better understand the transfer behaviour of the adsorbate species. Further, photocatalytic application tests were performed under visible light irradiation and the data were modelled to study the kinetics of photocatalysis.

2. Results and Discussion

2.1. Morphology

The morphology of sample, studied using FESEM, is shown in Figure 1. Figure 1a shows a heterogenous nature of synthesised RF/TiO_2 with homogenously distributed TiO_2, as represented in Figure 1b. The overall structure shows the nanospheres connected to form a three-dimensional porous network, as represented in Figure 1c [14]. The heterogenous surface is more evident in Figure 1d where organic and inorganic phases can be differentiated. The diameter of microspheres ranged around 0.76–1.66 μm, indicating that the size of the primary particles was slightly larger than pristine RF, which generally is in the nanometre range [16]. Energy dispersive X-ray (EDX) spectra of the microspheres is shown in Figure 1e, (EDX zone shown in supplementary information, Figure S1) which evidently corresponds to the recorded spectra.

2.2. FTIR Analysis

The IR absorption bands of RF/TiO_2 overall resembled those of the pristine RF gel, as also observed through FESEM images with clear uniform spheres illustrating a porous network and the retention of the gel structure even after addition of TiO_2. Typical characteristic peaks, such as the previously reported C=C stretching, CH_2, and C-O-C of aromatic rings, methylene bridges, and methylene ether bridges [17,18], were observed. The broad peak at 3300 cm^{-1} is characteristic of stretching vibrations associated with phenolic OH groups. Weak vibrations in the range of 2000–1700 cm^{-1} are attributed to CH bending of aromatic compounds. The absorption bands at 1605 and 1473 cm^{-1} correspond to aromatic ether bridges, attributed to condensation of resorcinol to form the RF gel network. A strong IR peak, expected in the range 1740–1700 cm^{-1}, associated with C=O stretching of aldehyde, was not observed, which confirms that the sol–gel reaction was complete. In comparison to a spectrum of pristine RF, a few additional peaks were observed that verify the chemical linkages between RF and TiO_2, as marked in Figure 2. It has been established that the oxygenated surface groups of carbon materials support the attachment of TiO_2. [19]. Here, crosslinking of TiO_2 with RF, via the hydroxyl groups, can be observed through the peaks in the vicinity of 1400 cm^{-1}, attributed to OH groups of RF, which appeared weak in the spectrum of RF/TiO_2, signifying the reaction of OH and TiO_2. Meanwhile, new signals observed at 1200 and 1084 cm^{-1} suggest formation of Ti-O-C functionalities. Similar crosslinking has previously been reported in TiO_2/phenol resol hybrid structures, where chemical interactions between TiO_2 and phenol resol form Ti-O-C complexes. This heterojunction is responsive to visible light due to formation of a charge complex between the interface of TiO_2 and mesoporous phenol resol producing new electronic interactions [20]. Hence, it can be concluded that the interactions between RF and TiO_2 are chemical in nature. Additional signals below 1000 cm^{-1}, such as bands at 963 and 880 cm^{-1}, are associated with titanium ethoxide functional groups. Additionally, the broad band observed in the range of 600 cm^{-1} corresponds to the vibration of Ti-O-Ti bonds [21].

2.3. Surface Area Analysis

A nitrogen sorption isotherm was measured to determine the specific surface area and pore volume of RF/TiO_2. Figure 3 shows N_2 sorption isotherm and pore size distribution (inset Figure 3). As can be seen, the isotherm of RF/TiO_2 is of Type IV classification [22] with a sharp capillary condensation at P/P_o = 0.4–0.9 and a well-defined hysteresis loop of Type H1, associated with open ended pores whilst suggesting a mesoporous structure [16]. Pore filling occurs at low relative pressure and the calculated mesoporosity in the structure was ~94%. The S_{BET}, corresponding pore size and total pore volume of as prepared RF/TiO_2 is 439 $m^2\ g^{-1}$, 9.4 nm and 0.71 $cm^3\ g^{-1}$, respectively. The S_{BET} value of pristine RF gel obtained in this study is 588 $m^2\ g^{-1}$. The reason in reduced S_{BET} value for RF/TiO_2 is attributed to blockage of pores of RF gel matrix with inclusion of TiO_2 nanoparticles. Meanwhile, in comparison with pristine TiO_2, the S_{BET} value is significantly higher for the synthesised RF/TiO_2. Additionally, noteworthy S_{BET} value for pristine TiO_2 (i.e., 111 $m^2\ g^{-1}$) is obtained in this study, contrary to commercial P25 with S_{BET} value of 57 $m^2\ g^{-1}$.

2.4. Effect of pH

The influence of MB sorption was studied by varying the solution pH from 2–12 (25 mL, 100 mg L^{-1}, 0.01 g of adsorbent). The adsorption capacities at different pH values are shown in Figure 4. The efficiency of uptake increases from 47.24 to 65.96 mg g^{-1} when the pH increases from 2–5. Thereafter, a sharp increase in adsorption capacity is observed at pH $\geq$ 6. The variation in adsorption behaviour of MB on RF/TiO_2 can be explained by considering the structure of MB and evaluated point of zero charge (pzc). The pHpzc value for RF/TiO_2 is determined to be 7.2 (Figure S2).

RF/TiO_2 can be amphoteric having both positively and negatively charged surface sites in aqueous solution due to the varying amount and nature of surface oxygen [23]. At pH lower than the pHpzc, the surface of RF/TiO_2 is positively charged, which repels the cationic dye (MB), and resultant interactions are hindered in acidic media due to electrostatic repulsion between the competing H^+ ions on the surface of adsorbent and MB dye molecules. As the pH increases, the surface of RF/TiO_2 becomes deprotonated and the adsorption sites available for interaction with cationic species increase, therefore, increased adsorption capacity is observed. This suggests that the electrostatic forces of attraction between MB and the surface of RF/TiO_2 increases due to increased ion density and positive charges on the surface. Further, the OH groups on the surface of RF/TiO_2 can also attract MB dye molecules under higher pH conditions.

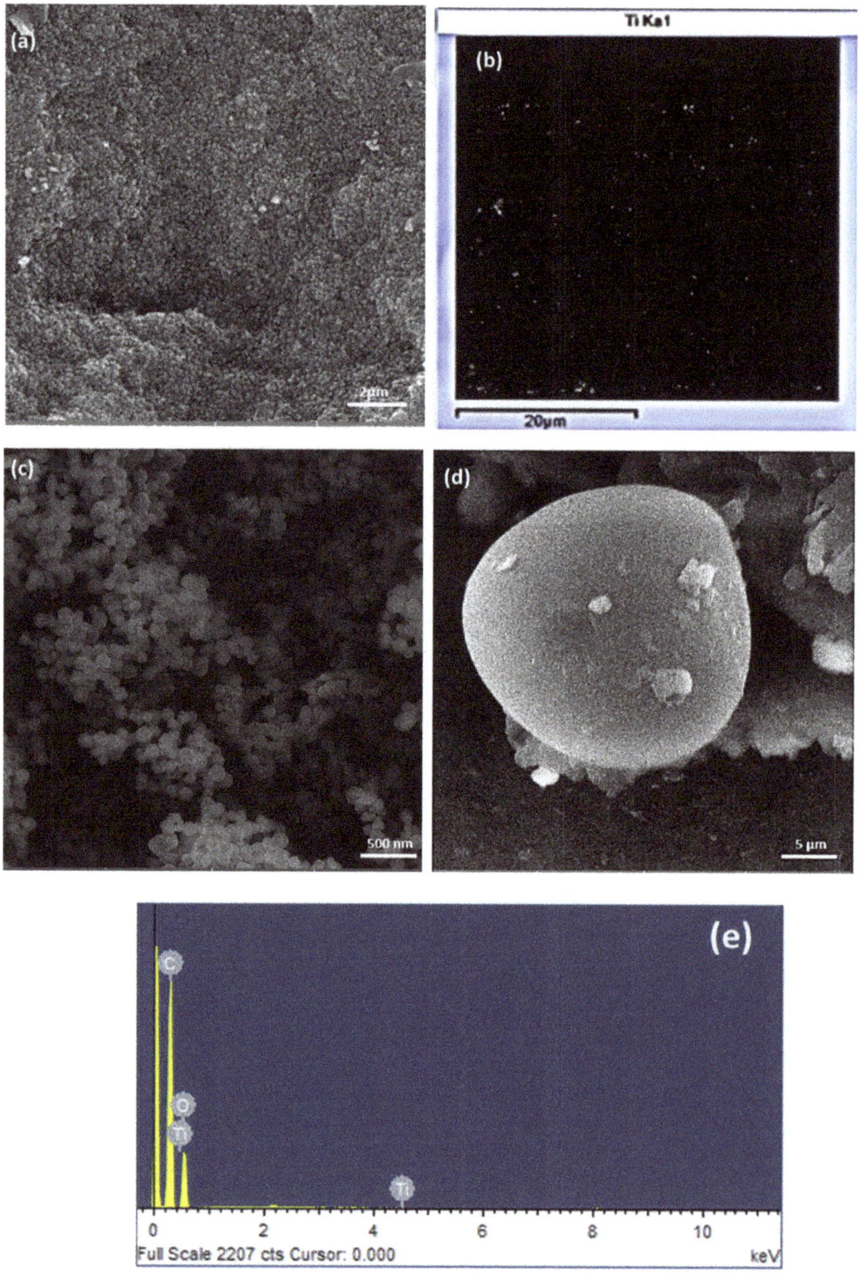

Figure 1. Morphology of RF/TiO_2 sample (**a**) FESEM image of RF/TiO_2, (**b**) TiO_2 distribution determined by EDX on the sample, (**c**) distinct appearance of micro/nanospheres, (**d**) isolated microsphere with differentiation between organic–inorganic phase, and (**e**) corresponding EDX spectra.

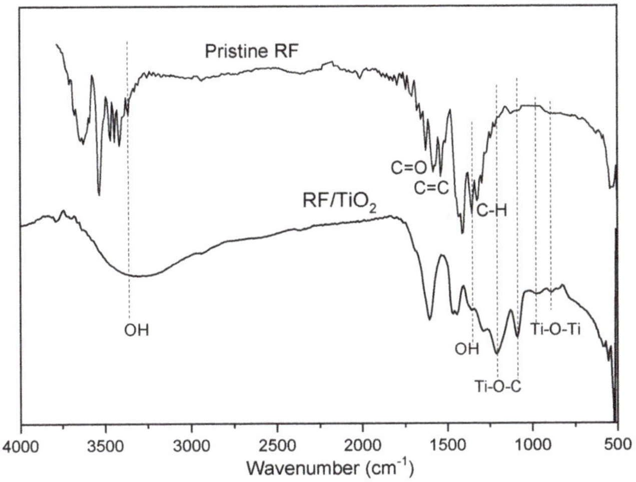

Figure 2. FTIR spectra of synthesised RF/TiO_2 gel compared to pristine RF gel.

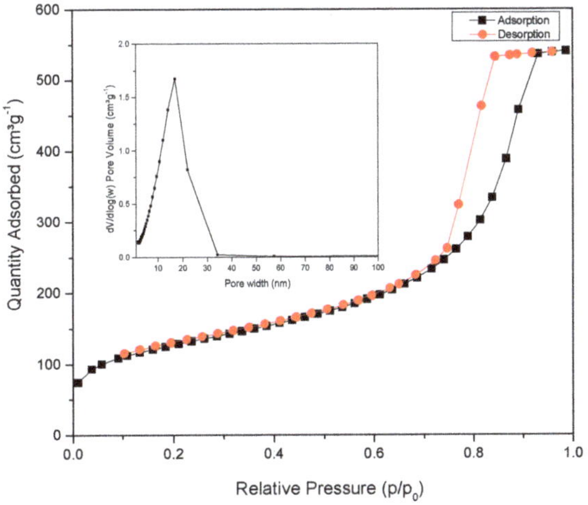

Figure 3. Nitrogen sorption isotherms and pore size distribution of synthesised RF/TiO_2 gel.

Overall, a good adsorption capacity for MB is observed at pH higher than the pHpzc due to an increased number of negative sites in the higher pH range. This is in good agreement with the fact that, due to the presence of COO^- and OH- functional groups, MB dye adsorption is favoured at pH > pHpzc [24]. The same trend has been observed in previous studies with activated carbon and TiO_2 composites where reduced activity was observed at acidic pH and maximum activity was observed in the pH range 6–10 [25–27].

2.5. *Effect of Contact Time*

Figure 5 shows the effect of contact time on the amount of MB molecules adsorbed by RF/TiO_2 gel under different initial MB concentrations. As shown, the adsorption capacity increases with increase in initial concentration. The equilibrium adsorption capacity increases from 102 mg g^{-1} to 207 mg g^{-1} by increasing the initial concentration of MB from 50 mg L^{-1} to 200 mg L^{-1}. Initially, the adsorption capacity overall is rapid for timeframes up to 30 min. This trend is expected, due to the greater driving force of MB dye molecules and immediate availability of vacant adsorption sites, hence resulting in increased in frequency of collisions between MB dye molecules and the RF/TiO_2 gel. Additionally, mesoporosity throughout the RF/TiO_2 gel structure provides a high surface area for greater adsorption of MB molecules. It is noteworthy that at higher MB concentration the adsorption rate is greater and adsorption capacity attains equilibrium faster than at low concentration. The reason is attributed to immediate occupancy of available active sites by a large amount of adsorbate molecules. This rapid occurrence of sorption is due to the presence of mesoporosity within the RF/TiO_2 gel, which corresponds to a large portion of the adsorption sites. In this case, the mesoporous structure provides a large surface area to solution volume within the porous network of the adsorbent gel. Additionally, within the mesopores, MB dye molecules are confined to be in close proximity to the surface. Such observations have been reported in previous research, particularly for activated carbons [28]. Over time, saturation of active sites occurs, and adsorption becomes difficult on the fewer available active sites due to repulsive forces between the MB molecules and the RF/TiO_2 gel surface. Additionally, the blockage of pores and charge repulsion of MB dye species may decelerate the adsorption progress. Similar phenomena have been explained for porous TiO_2 and other carbon/TiO_2 porous composite materials, where it may have taken longer for the adsorbate to diffuse deeper in the fine pores [29]. Thereafter, the adsorption capacity increases gradually until 90 min, and equilibrium is attained for the entire concentration range. Thus, equilibrium time was considered as 90 min which was considered sufficient for removal of MB ions by RF/TiO_2 gel. Hence, the contact time was set to 90 min in the remaining experiments to ensure equilibrium was achieved.

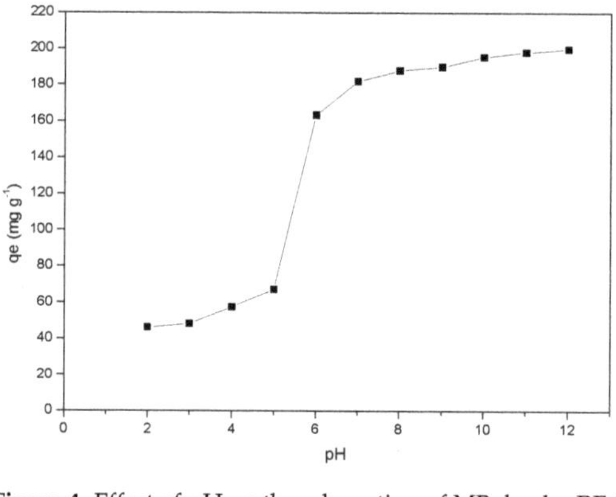

Figure 4. Effect of pH on the adsorption of MB dye by RF/TiO_2 gel.

2.6. *Effect of Sorbent Dose*

The percentage removal of MB dye increased with increase in the adsorbent dose from 0.005 to 0.01 g but remained almost constant with further increase in the dose range 0.01 to 0.1 g, as represented in Figure 6. Percentage removal was calculated using Equation (2), and showed an increase with increase in adsorbent dose, due to greater availability of vacant active sites, a large surface area, and a greater number of adsorptive sites present on the surface of RF/TiO_2. With further increase in adsorbent dose (>0.01 g), the rate of MB removal becomes low, as the concentrations at the surface and solution reach equilibrium. The resultant reduction in adsorption rate is attributed to unoccupied adsorbent sites, as well as overcrowding or aggregation of adsorbent particles [30]. Hence, the surface area available for MB adsorption per unit mass of the adsorbent reduces, whereby percentage removal was not significantly enhanced with further increase adsorbent dose.

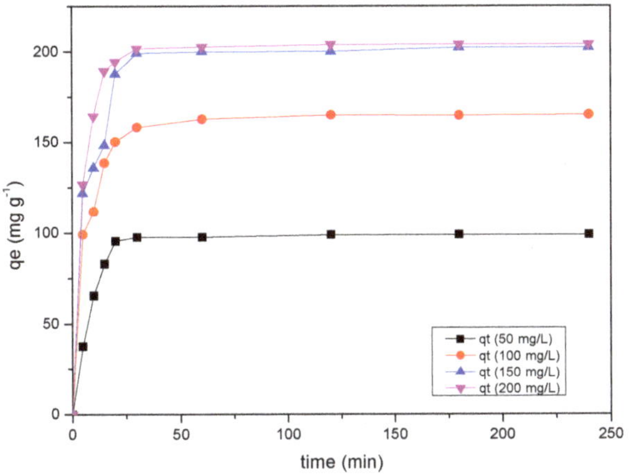

Figure 5. Effect of adsorption on contact time and initial concentration of MB dye by RF/TiO_2 gel.

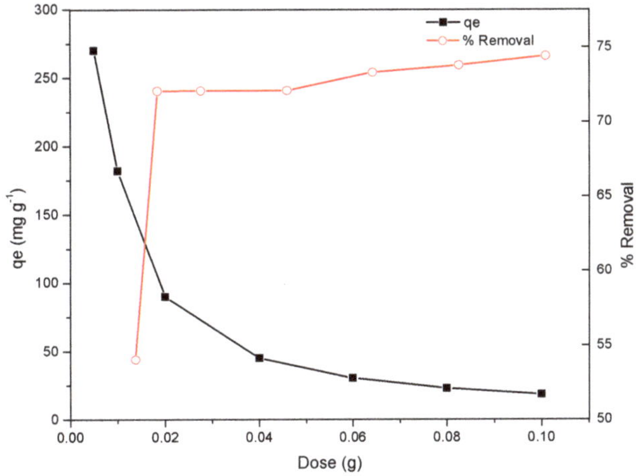

Figure 6. Effect of adsorbent dose on the removal and adsorption of MB dye by RF/TiO_2 gel.

2.7. Adsorption Kinetics

The adsorption kinetics were studied using a contact time of 240 min in the concentration range 50–200 mg L^{-1}. The experimental data obtained for MB dye adsorption capacity vs. time (t) were fitted with PFO and PSO, as presented in Figure 7a–d. The parameters determined, including measured equilibrium adsorption capacity qe (experimental), theoretical equilibrium adsorption capacity qe (calculated), first order rate constant K_1, second order rate constant K_2, and regression coefficient R^2, are presented in Table 1.

As observed from the data, the correlation factor R^2 deviates significantly from 1 for PFO and, therefore, pseudo first order model does not exhibit good compliance with the experimental data for the entire concentration range. This implies that the adsorption reaction is not inclined towards physisorption, and the MB dye molecules adsorb to specific sites on the surface of RF/TiO_2 gel. The argument regarding the failure of the pseudo first order model suggests that several other interactions are responsible for the sorption mechanism. Hence, the correlation coefficients R^2 of the pseudo second order model were compared with pseudo first order parameters. R^2 values for pseudo second order behaviour are approximately 0.99 for the entire concentration range, indicating that the system is more appropriately described by the pseudo second order equation. The dependence on initial concentration of MB dye is verified by good compliance of the experimental data with the pseudo second order equation, where the adsorption capacity is affected by the initial MB dye concentration, subsequent surface-active sites, and adsorption rate (Other error analyses are represented in Table S1).

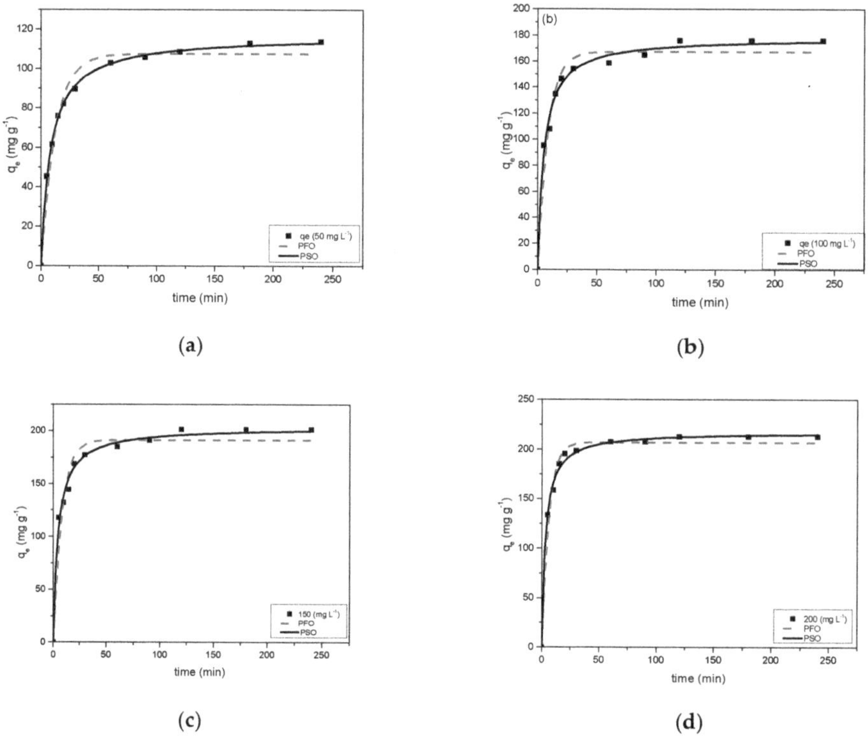

Figure 7. MB uptakes on RF/TiO_2 gel at (**a**) 50 mg L^{-1}, (**b**) 100 mg L^{-1}, (**c**) 150 mg L^{-1}, (**d**) 200 mg L^{-1}, and fitted data for pseudo first order and pseudo second order kinetic models.

Table 1. Kinetic parameters obtained by fitting kinetic data for MB adsorption to RF/TiO_2.

Model	50 mg L^{-1}	100 mg L^{-1}	150 mg L^{-1}	200 mg L^{-1}
q_e experimental (mg g^{-1})	112.75	175.98	201.46	212.56
Pseudo first order				
q_e, mg g^{-1}	107.65	167.45	183.10	206.99
K_1 (min^{-1})	0.08087	0.11886	0.1261	0.17078
R^2	0.9758	0.9632	0.948	0.985
Pseudo second order				
q_e, mg g^{-1}	116.97	178.54	203.58	217.59
K_2 ($\times 10^{-3}$ g mg^{-1} min^{-1})	1.01	1.10	1.06	1.48
R^2	0.998	0.989	0.987	0.996

The equilibrium sorption capacity increased from 116.97 to 217.59 mg g^{-1} when initial dye concentration was increased from 50 to 200 mg g^{-1} confirming that MB dye removal is dependent on initial concentration, where the rate limiting step is determined by both adsorbate (MB) and adsorbent (RF/TiO_2) concentration. This signifies that the sorption mechanism is chemisorption. Previous studies have explained theoretically that if diffusion is not the rate limiting factor, then higher adsorbate concentrations would give a good pseudo first order fit whereas, for low concentrations, pseudo second order better represents the kinetics of sorption, analogous to the observations made here [31]. Previously, the adsorption processes of MB on TiO_2/carbon composites have also exhibited strong dependencies of pseudo second order fitting parameters on initial concentrations [27].

Figure 8 shows a plot of MB dye uptake (qe) on synthesised RF/TiO_2 against $(time)^{0.5}$. The plots exhibit multi-linearity, rather than two straight lines, indicating that the adsorption process is influenced by several steps. The initial segment of the plots shows that diffusion across the boundary of the adsorbent only lasts for a short time in comparison to the whole adsorption process. This second section is attributed to diffusion into the mesopores of the adsorbent, i.e., the MB dye molecules enter less accessible pore sites. Resultantly, the diffusion resistance increases, and the diffusion rate decreases. This stage is a slow and gradual stage of the adsorption process. The third segment represents the final equilibrium stage where intra-particle diffusion slows down to an extremely low rate due to the remaining concentration of the MB dye molecules in the solution. This implies a slow transport rate of MB dye molecules from the solution (through the gel–dye solution interface) to available sites. Here, the surface of the RF/TiO_2 gel, and micropores, may be responsible for the uptake of MB dye molecules.

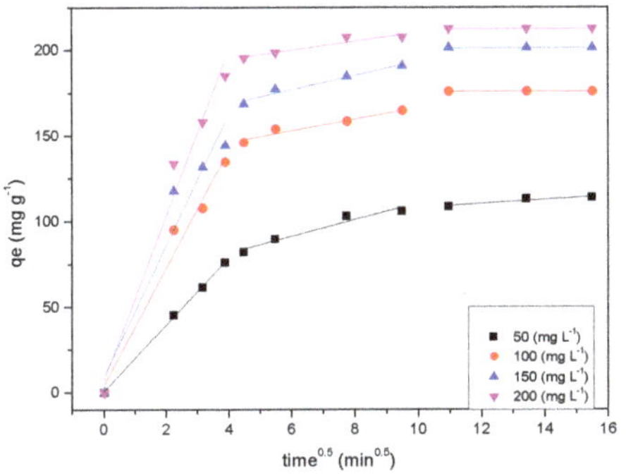

Figure 8. Intra-particle diffusion kinetics of MB dye adsorption on RF/TiO_2.

2.8. Adsorption Isotherms

The equilibrium data were analysed using Langmuir, Freundlich, Sips, and Toth isotherm equations to obtain the best fit. The isotherm data plots, and fitting model parameters are shown in Figure 9 and Table 2, respectively. Comparison of the correlation factor R^2 indicates that qe,exp fitted well to the Sips model with the lowest χ^2 value. The qe,cal value, calculated using the Sips model, is closest to qe,exp with R^2 closest to 1. The Sips model is a combination of the Langmuir and Freundlich adsorption isotherms, hence, the model suggests both monolayer and multilayer adsorption. At low MB dye concentrations, the model predicts Freundlich adsorption isotherms as a heterogenous adsorption system and localised adsorption without adsorbate–adsorbate interactions, whereas at high concentrations the model predicts monolayer adsorption as in Langmuir isotherm [32,33]. In the present study, the value of constant ns from Equation (11), the heterogeneity factor, is greater than 1 (i.e., n_s = 1.91), hence, the adsorption system is predicted to be heterogenous [33]. Further, the Toth isotherm model validates multilayer and heterogeneous adsorption, where the factor n_T determines heterogeneity. Here, again the value of n_T is greater than 1, and, therefore, the system confirms heterogeneity. It is evident that the equilibrium uptakes follow the Sips model according to the correlation factor R^2 (other error analyses are represented in Table S2) and the isotherm models fit the data in the order Sips > Toth > Langmuir > Freundlich.

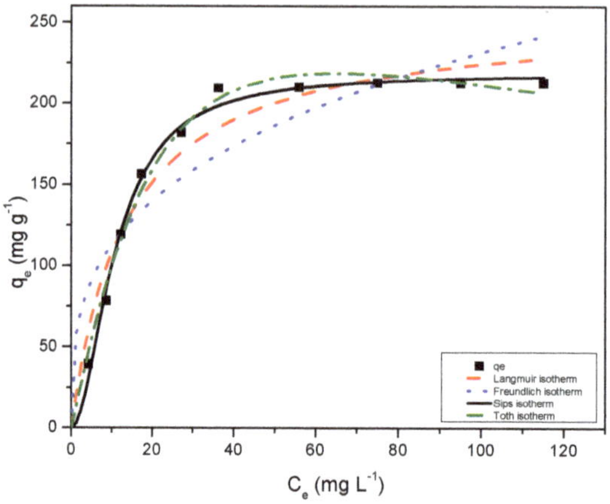

Figure 9. Adsorption data for RF/TiO_2 onto MB dye corresponding fits to Langmuir, Freundlich, Sips, and Toth equation.

Table 2. Isotherm parameters obtained by fitting MB adsorption data for RF/TiO_2 to the Langmuir, Freundlich, Sips, and Toth equations.

Langmuir	q_m (mg g^{-1})	254.65
	K_L (L mg^{-1})	0.0732
	R^2	0.960
Freundlich	K_F mg g^{-1} (L mg^{-1})$^{1/n}$	54.85
	n_F	3.1993
	R^2	0.865
Sips	q_s (mg g^{-1})	218.71
	K_S	0.010
	n_s	1.913
	R^2	0.994
Toth	q_m (mg g^{-1})	558.47
	K_T	0.0295
	n_T	1.403
	R^2	0.991

2.9. Thermodynamic Study

Thermodynamic parameters for the adsorption system are recorded in Table 3. Negative values of free energy changes are evident from the data, which signifies the spontaneous adsorption of MB dye molecules on the sample for the studied temperature range. Adsorption capacity increases with an increase in temperature and a positive ΔH^0 (Table 3) suggests that the adsorption is endothermic in nature. Positive ΔS^0 indicates some structural changes in the MB dye and RF/TiO_2 gel causing an increase in the degree of freedom of the MB dye species and consequently increased randomness at the adsorbent–adsorbate interface. At high temperature, the release of high-energy desolvated water molecules from the MB dye molecules and/or aggregates arise after adsorption on RF/TiO_2 gel, which relates to a positive ΔS^0 [34]. Before sorption begins, the MB ions are surrounded by highly ordered water clusters strongly bound via hydrogen bonding. Once MB ions come in close contact with the surface of RF/TiO_2, the interaction results in agitation of the ordered water molecules, subsequently increasing the randomness of the system. Although, the adsorption of MB dye onto RF/TiO_2 gel may reduce the freedom of the system, the entropy increase in water molecules is much higher than the entropy decrease in MB ions. Therefore, the driving force for the adsorption of MB on RF/TiO_2 is controlled by an entropic effect rather than an enthalpic change. Similar phenomena have previously been reported in order to explain the fact that thermodynamic parameters are not only related to the properties of the adsorbate but also to the properties of other solid particles [35,36].

Table 3. Thermodynamic data for MB adsorption onto RF/TiO_2 at various temperatures.

T (K)	lnk	ΔG^0 (KJ/mol)	ΔS^0 (J/mol)	112
281	1.29	−3.01	ΔH^0 (KJ/mol)	28.2
296	2.20	−5.41		
305	2.40	−6.09		
313	2.50	−6.51		

3. Photocatalytic Tests

Photocatalytic activity was determined by testing the efficiency of RF/TiO_2 against degradation of methylene blue (MB) under visible light irradiation. The maximum absorbance vs. wavelength spectra (in the range of 550–700 nm) were collected and subsequent activity, after 30 min, intervals was recorded, as shown in Figure 10.

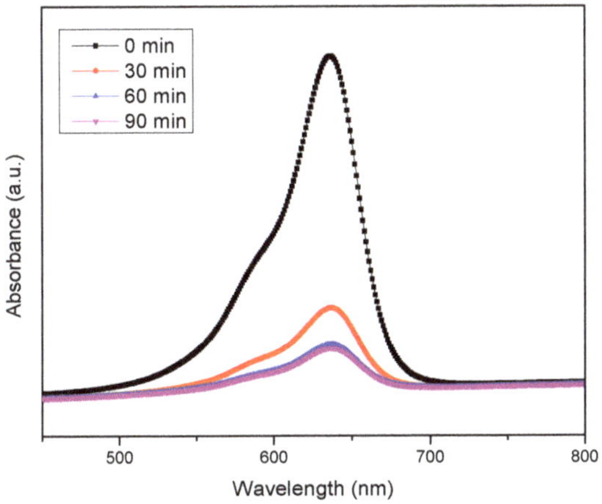

Figure 10. UV-Vis spectra of MB dye degradation using RF/TiO_2 gel.

Within the studied systems, no photodegradation activity (reduction in concentration and decolourisation of MB dye) was observed in the absence of adsorbent/catalyst, as well as in the presence of pristine RF, indicating that the properties of MB are more stable. Additionally, RF solely may not be recommended for photocatalysis due to slow charge transfer properties, which has also been proven by the study carried out by Zang, Ni, and Liu, where the researchers employed pristine RF resins for visible light photocatalysis [37]. Slight photodegradation is observed in the presence of pristine TiO_2, which may be attributed to the potential absorbance of UV-Vis light from the surroundings confirming that the process of MB degradation is light driven. Although the TiO_2 obtained for use in this study has a high surface area, which may possess good adsorption properties to exhibit efficient adsorption of MB dye, since TiO_2 only activates upon UV light irradiation (~280 nm), it does not produce enough reactive oxide species (ROS) to be an effective photodegradation system [38].

The dye degradation data obtained after treatment with RF/TiO_2 showed ~73% MB dye removal after 90 min. This is attributed to the synergy of RF and TiO_2, enabling an absorption shift to a higher wavelength, as λmax is detected at 410 nm (Figure 11). Further analysis indicates modification in the electronic structure and a subsequent reduction in bandgap occurs due to doping of TiO_2 similar to when combined with carbon [39]. The calculated band gap energy is 2.97 eV, as shown in Figure 11 (inset). The value achieved is significantly lower than pristine TiO_2 (i.e., 3.2 eV [21]), indicating photodegradation of MB dye under visible light irradiation. The RF matrix enables entrapment of a photogenerated electron and hole pairs and, therefore, rapid generation of ROS is possible for efficient degradation of the MB dye. These findings are comparable to other carbon/TiO_2 systems where synergistic effects have substantially enhanced the performance of the system due to improved optical properties of the material [1,40,41].

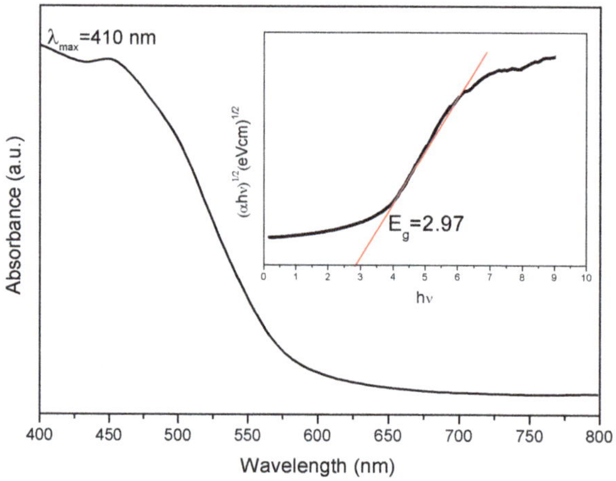

Figure 11. Absorption vs. wavelength spectrum of RF/TiO_2 dispersed in ethanol measured through UV-Vis spectrophotometer, inset shows calculated band gap of synthesised RF/TiO_2.

The photodegradation of MB dye, that is, the reduction in concentration with time is recorded in Figure 12 and the recorded data is modelled using pseudo first order kinetics, shown in Figure 12a,b. The data are fitted to the first order kinetic equation ($\ln (C_o/Ce) = kt$) to evaluate the value of the rate constant by slope of plot $\ln(C_o/Ce)$ vs. time (t) in minutes, where C_o and Ce is the initial at t = 0 and final concentration at given time of MB concentration, respectively. The value of k here is the measure of photocatalytic performance, as it defines the concentration of reacting substances, that is, photogenerated reactive oxide species, therefore, a higher value of k signifies higher photocatalytic efficiency. As compared to no catalyst ($k = 2.43 \times 10^{-6}$ min^{-1}) pristine TiO_2 ($k = 1.74 \times 10^{-3}$ min^{-1}) and pristine RF ($k = 6.89 \times 10^{-4}$ min^{-1}), the rate of RF/TiO_2 was the highest ($k = 1.27 \times 10^{-2}$ min^{-1}). Clearly, the rate constant obtained for photodegradation of MB using RF/TiO_2 was the highest. Mainly, improved optical property was the most important advancement in forming RF/TiO_2 gel which is photocatalytically active under visible light (410 nm) irradiation.

The RF/TiO_2 material created in this study exhibits excellent photoactivity under visible light, which can further be explained by the mechanism of MB photodegradation represented in Equations (1)–(8). The system activates when RF/TiO_2 absorbs light with photon energy (hν) and generates conduction band (CB) electron (e^-) and valence band (VB) hole (h^+) pairs upon under visible light irradiation. The holes interact with moisture on the surface of the adsorbent gel yielding hydroxy free radicals or reactive oxide species (H^+ or $OH^\bullet$), which are oxidation agents that can mineralise a wide range of organic pollutants, ultimately producing CO_2 and H_2O as end products. The reaction sequence below represents the photodegradation of MB, showing a simplified mechanism of photoactivation by a photocatalyst (Equations (1)–(4)) [19]. For the mechanism of photoinactivation of MB in the presence of RF/TiO_2, hydroxy free radicals or reactive oxide species (H^+ or $OH^\bullet$) attack the aromatic ring of the MB structure, degrading it into a single ring structure product, which then finally degrades to CO_2 and H_2O (Equations (5)–(8)) [42,43].

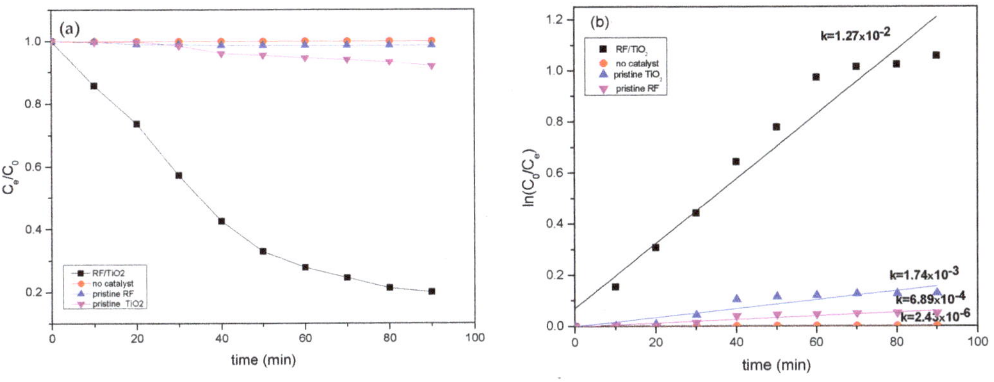

Figure 12. (**a**) Photocatalytic performance regarding MB dye degradation without catalyst, and with pristine RF, pristine TiO_2 and RF/TiO_2 gel (**b**) First-order kinetics of photoactivity without catalyst, and with pristine RF, pristine TiO_2, and RF/TiO_2 gel.

$$RF/TiO_2 + h\nu = e^-_{CB} + h^+_{VB} \tag{1}$$

$$h^-_{CB} + H_2O = H^+ + OH^{\cdot} \tag{2}$$

$$e^-_{CB} + O_2 = O_2^{\cdot -} \tag{3}$$

$$O_2^{\cdot -} + H^+ = HO_2^{\cdot} \tag{4}$$

$$MB + RF/TiO_2 = MB^{\cdot +} + e^-_{CB}(RF/TiO_2) \tag{5}$$

$$O_2 + e^- = O_2^{\cdot -} \tag{6}$$

$$MB^{\cdot +} + OH^- = MB + OH^{\cdot} \tag{7}$$

$$MB^{\cdot +} + OH^{\cdot} = H_2O + CO_2 + \text{other products} \tag{8}$$

4. Conclusions

An RF/TiO_2 gel was successfully synthesised using sol–gel techniques via a straightforward route. The synergy of RF and TiO_2 exhibited excellent adsorption–photodegradation activity due to the corresponding characteristics, mainly mesoporosity and photocatalysis. The synergy of contributing materials allowed modification in the electronic structure of TiO_2 by formation of Ti-O-C chemical linkages, responsible for a reduction in the band gap of TiO_2 for photodegradation upon visible light irradiation. Kinetic studies revealed a pseudo second order reaction, signifying chemisorption phenomenon is involved in the adsorption mechanism. The adsorption isotherm study showed that the system was heterogeneous following the Sips model equation. The spontaneity of the process was validated via thermodynamic studies, which signified an entropically driven adsorption mechanism. Effective photodegradation results were observed due to the high adsorption capacity and improved optical properties of the material, enabling significant MB dye degradation within 90 min. Overall, the material possesses properties that have potential to effectively reduce/eliminate a wide range of pollutants and, therefore, can be employed as a low-cost photocatalytic adsorbent for water treatment Especially in the industrial applications where post treatment separation and recovery of the adsorbent is difficult, employing this material can reduce the costs since in this case the adsorbent precipitates and easy separation is possible just by filtration or even decantation.

5. Material and Methods

5.1. Synthesis

Synthesis of RF and TiO_2 precursors was carried out in two separate systems, which were integrated and processed further in order to obtain the final gel.

System 1: Preparation of Titania Sol

For preparation of the titania sol, 1.78 g of titanium precursor: titanium isopropoxide (TTIP) (98+%, ACROS Organics™, Geel, Belgium) was dissolved in ethanol and stirred for 30 min. A mixture of water and HCl was added dropwise to the titania/EtOH solution under constant stirring, at room temperature, to begin hydrolysis. After 2 h, a homogenous solution was obtained. The molar ratios for these parameters were 1 TTIP:10 EtOH:0.3 HCl:0.1 H_2O.

System 2: Preparation of RF sol

In total, 7.74 g of resorcinol (SigmaAldrich, ReagentPlus, 99%, Poole, UK) was added to 50 mL of deionised water until completely dissolved. 0.0249 g of sodium carbonate (Na_2CO_3, Sigma-Aldrich, anhydrous, ≥99.5%), as a catalyst, and 4.23 g of formaldehyde (37wt%) were added to the dissolved resorcinol under continuous stirring, at room temperature.

Finally, the prepared titania sol (system 1) was gradually transferred to the RF sol (from system 2) under constant stirring, at room temperature. The resulting sol was stirred at room temperature for 2 h after which the sol mixture was aged at 85 °C for 72 h.

After aging, the process of solvent exchange and drying the RF/TiO_2 gel, first involved cutting the gel into smaller pieces. These pieces were then immersed in acetone for 72 h to facilitate solvent exchange prior to drying, followed by vacuum drying at 110 °C for 48 h to obtain the final RF/TiO_2 adsorbent gel. In this way, the final gel corresponded to 10 wt% TiO_2 (theoretical percentage) incorporated in the RF gel matrix.

5.2. Adsorbent Characterisation

Morphology of the synthesised sample was studied by field emission electron scanning microscope (FESEM) TESCAN-MIRA. The functional groups on the surface of synthesised RF/TiO_2, and the chemical linkages between the constituent RF and TiO_2 components, were verified using Fourier Transform Infrared Spectroscopy (FTIR) (MB3000 series, scan range 4000–400 nm). BET surface area measurements were carried out using a Micromeritics ASAP 2420 to obtain N_2 adsorption isotherm at 77 K and pore size was determined via BJH theory [22]. A UV-Vis Spectrophotometer (Varian Cary 5000 UV-Vis NIR Spectrophotometer Hellma Analytics) was used to collect absorption spectra and the data used to interpret the change in electronic structure of RF/TiO_2 [44]. The data were manipulated to calculate the band gap energy values through the Tauc method described in previous studies [44].

6. Adsorption Experiments

6.1. Effect of pH

The effect of pH on the sorption of methylene blue (MB) dye was investigated with 0.01 g of sample by adjusting the pH of solution (25 mL, 100 mg L^{-1} MB) between 2 and 12, at 23 °C. The pH was adjusted using 0.01 M HCl and 0.01 M NaOH. After 2 h of agitation, the solution was centrifuged for 15 min and the supernatant solution was collected via syringe. The initial and final concentrations were measured using a UV-Vis spectrophotometer (Varian Cary 5000 UV-Vis NIR Spectrophotometer, Agilent, UK) and onward calculations were performed.

6.2. Effect of Sorbent Dose

The amount of sorbent dose was gradually increased from 0.005 to 0.01 g to study the effect of sorbent dose on the adsorption capacity. pH and temperature were maintained at 7.0 and 23 °C, respectively, against 25 mL of 100 mg L^{-1} MB concentrated solution. The pH was adjusted using 0.01 M HCl and 0.01 M NaOH. After 2 h of agitation, the solution was centrifuged for 15 min and the supernatant solution was collected via syringe. The initial and final concentrations were measured using a UV-Vis spectrophotometer (Varian

Cary 5000 UV-Vis NIR Spectrophotometer Hellma Analytics) and onward calculations were performed.

6.3. Effect of Initial Concentration

All adsorption experiments were performed at 23 °C in 125 mL conical flasks, using a shaker (VWR 3500 Analog orbital shaker) set to 125 rpm. The first experiment was conducted to study the isothermal equilibrium and the effect of initial MB concentration. Standard solutions of MB were prepared using distilled water, with initial concentrations in the range of 20–200 mg L^{-1}. Then, 25 mL aliquots were distributed into each flask, and 0.01 g of the adsorbent gel was added individually to each flask. The pH values of all solutions were recorded and adjusted to 7.0, if required, using 1 M HCl and 1 M NaOH. After 2 h of agitation, the solution was centrifuged for 15 min and the supernatant solution was collected via syringe. The initial and final concentrations were measured using UV-Vis spectrophotometer (Varian Cary 5000 UV-Vis NIR Spectrophotometer Hellma Analytics).

The equilibrium adsorption capacity, q_e (mg g^{-1}), was calculated using:

$$q_e = \frac{(C_o - C_e) \times V(l)}{W} \tag{9}$$

while the respective percentage removal of MB was calculated as:

$$\text{Removal }\% = \frac{C_o - C_e}{C_o} \times 100\% \tag{10}$$

where C_o and C_e are the initial MB and final concentration, respectively. W is the weight (g) of the adsorbent and V is the volume (L) of MB solution.

6.4. Effect of Contact Time

The effect of contact time was studied by adding MB solution (pH 7.0, 25 mL, 100 mg L^{-1}) and 0.01 g adsorbent gel into a flask, which was then agitated for a pre-determined contact time between 5 min and 4 h. The samples were prepared and treated as described in Section 2.5 and the amount of adsorption was calculated using Equation (11):

$$q_t = \frac{(C_o - C_t) \times V}{W} \tag{11}$$

where C_t is the equilibrium MB concentration at a given time, and C_o, V, and W are as previously defined. Equilibrium concentration was determined by plotting q_t versus time of aliquots collected at different time intervals. Adsorption-photodegradation (absorption) changes of MB dye with time were also recorded via UV-Vis spectrophotometry.

6.5. Effect of Temperature

The effect of temperature on the removal of MB (pH 7.0, 25 mL, 20–200 mg L^{-1}) was investigated by adding a known concentration MB solution and 0.01 g adsorbent gel to a flask. A hot plate with a stirrer (120 rpm) was used to maintain a constant temperature of 8, 23, 32, and 40 °C, under stirring, for 120 min after which the absorbance versus wavelength spectra were recorded to measure the final concentration, and subsequent adsorption was calculated using Equation (9).

6.6. Kinetic Models

The kinetic-based models: pseudo first order (PFO) and pseudo second order (PSO) were applied to study the adsorption kinetics and to explain the mode of sorption of MB onto the synthesised RF/TiO_2. The PFO model [33] has been frequently used to describe kinetic processes under non-equilibrium conditions. PFO is based on the assumption that the rate of adsorption is proportional to the driving force, that is, the difference between the equilibrium concentration and solid phase concentration, presented as a differential Equation (12):

$$\frac{dq_t}{dt} = k_1(q_e - q_t) \quad (12)$$

Integrating Equation (13) with the initial condition of $q_t = 0$ at t = 0, the PFO model can be rewritten, in a linear form, as:

$$q_{t=}q_e\left(1 - e^{-k_1 t}\right) \quad (13)$$

Several studies have also reported the use of PSO [45] to interpret data obtained for the sorption of contaminants from water, including dyes, organic molecules, and metal ions. The PSO model assumes that the overall adsorption rate is proportional to the square of the driving force and can be expressed as Equation (14):

$$\frac{dq_t}{dt} = k_1(q_e - q_t)^2 \quad (14)$$

Integrating Equation (14) with the initial condition of $q_t = 0$ at t = 0, and q_t = t at t = t, the PSO model can be rewritten as:

$$q_t = \frac{k_2 t q_e^2}{1 + k_2 t q_e} \quad (15)$$

In Equations (13)–(15), qt (mg g^{-1}) and q_e (mg g^{-1}) are the adsorption capacities of MB dye molecules at time t and at equilibrium, respectively. k_1 (mg g^{-1} min^{-1}) and k_2 (mg g^{-1} min^{-1}) are the PFO and PSO rate constants, respectively.

6.7. Sorption Isotherm Models

The equilibrium data for the sorption of MB on RF/TiO_2 adsorbent gel as a function of equilibrium concentration (C_e mg L^{-1}) was analysed in terms of Langmuir, Freundlich, Sips, and Toth isotherm models [2]. The nonlinear form of Langmuir's isotherm model is represented as:

$$q_e = \frac{q_m K_L C_e}{1 + C_e K_L} \quad (16)$$

where q_e (mg g^{-1}) is the MB uptake at equilibrium, C_e (mg L^{-1}) is the equilibrium concentration, q_m (mg g^{-1}) is the amount of adsorbate at complete monolayer coverage, and K_L is the Langmuir constant.

The Freundlich equation can be expressed as follows:

$$q_e = K_F C_e^{1/n} \quad (17)$$

where q_e and C_e are as defined in the Langmuir equation, adsorption affinity is related to the adsorption constant K_F, and n indicates the magnitude of the adsorption driving force and the distribution of energy sites on the adsorbent surface, if n < 1, then adsorption is a chemical process, whereas if n > 1, then adsorption maybe dependent on distribution of the surface sites. Generally, n values within 1–10 represents good adsorption [46].

The Sips isotherm model is a combination of the Langmuir and Freundlich isotherms and is represented as:

$$q_e = \frac{q_s K_s C_e^{ns}}{1 + K_s C_e^{ns}} \quad (18)$$

where q_e and C_e are as defined for Equation (16), K_s is the Sips isotherm model constant (L g^{-1}), and ns; is the Sips isotherm exponent.

The Toth model also describes heterogeneous systems, considering both low- and high-end concentrations. The Toth expression is as follows:

$$q_e = \frac{q_m K_T C_e}{[1 + (K_T C_e)^{\,t}]^{1/t}} \quad (19)$$

where q_e and C_e are as defined for Equation (16), q_m is the maximum adsorption capacity, t is the surface heterogeneity, and K_T is the surface affinity.

6.8. Photodegradation Procedure

Photocatalytic performance of as prepared RF/TiO_2 was investigated through MB dye degradation, by recording the dye degradation spectra with time using UV-Vis Spectrophotometry. 0.01 g of the adsorbent dose were used against 25 mL of 100 mg L^{-1} dye concentration at pH ~7 and a temperature of 23 °C and light intensity of 111 Wm^{-2}. For comparison, the measurements were also recorded in the absence of catalyst, as well as pristine RF and TiO_2. All suspensions were stirred in the dark for 60 min to establish sorption equilibrium before exposure to visible light.

Supplementary Materials: The following supporting information can be downloaded at: https://www.mdpi.com/article/10.3390/gels8040215/s1, Figure S1: Zone of Energy dispersive x-ray (EDX) spectra; Figure S2: Point of zero charge (pHpzc) on the surface of RF/TiO_2; Table S1: Kinetic parameters obtained by fitting kinetic data for MB adsorption to RF/TiO_2.; Table S2: Isotherm parameters obtained by fitting MB adsorption data for RF/TiO_2 to the Langmuir, Freundlich, SIPS and Toth equations.

Author Contributions: Methodology, A.S.; formal analysis, A.S. and A.J.F.; resources, A.J.F.; writing—original draft preparation, A.S.; writing—review and editing, A.J.F.; supervision, A.J.F.; project administration, A.J.F.; funding acquisition, A.J.F. All authors have read and agreed to the published version of the manuscript.

Funding: This research received no external funding.

Acknowledgments: Anam Safri thanks Ashleigh Fletcher and the Chemical and Process Engineering Department at the University of Strathclyde for funding this work. The authors gratefully acknowledge the Materials Science and Engineering Department at Institute of Space Technology, Islamabad, for providing support and facilities to conduct the morphological analysis.

Conflicts of Interest: The authors declare no conflict of interest.

References

1. Xue, G.; Liu, H.; Chen, Q.; Hills, C.; Tyrer, M.; Innocent, F. Synergy between surface adsorption and photocatalysis during degradation of humic acid on TiO_2/activated carbon composites. *J. Hazard. Mater.* **2011**, *186*, 765–772. [CrossRef] [PubMed]
2. Ajiboye, T.O.; Oyewo, O.A.; Onwudiwe, D.C. Adsorption and photocatalytic removal of Rhodamine B from wastewater using carbon-based materials. *FlatChem* **2021**, *29*, 100277. [CrossRef]
3. Quyen, N.D.V.; Khieu, D.Q.; Tuyen, T.N.; Tin, D.X.; Diem, B.T.H.; Dung, H.T.T. Highly effective photocatalyst of TiO_2 nanoparticles dispersed on carbon nanotubes for methylene blue degradation in aqueous solution. *Vietnam. J. Chem.* **2021**, *59*, 167–178.
4. Sampaio, M.J.; Silva, C.G.; Marques, R.R.; Silva, A.M.; Faria, J.L. Carbon nanotube—TiO_2 thin films for photocatalytic applications. *Catal. Today* **2011**, *161*, 91–96. [CrossRef]
5. Murgolo, S.; Petronella, F.; Ciannarella, R.; Comparelli, R.; Agostiano, A.; Curri, M.L.; Mascolo, G. UV and solar-based photocatalytic degradation of organic pollutants by nano-sized TiO_2 grown on carbon nanotubes. *Catal. Today* **2015**, *240*, 114–124. [CrossRef]
6. Morawski, A.W.; Kusiak-Nejman, E.; Wanag, A.; Narkiewicz, U.; Edelmannová, M.; Reli, M.; Kočí, K. Influence of the calcination of TiO_2-reduced graphite hybrid for the photocatalytic reduction of carbon dioxide. *Catal. Today* **2021**, *380*, 32–40. [CrossRef]
7. Minella, M.; Sordello, F.; Minero, C. Photocatalytic process in TiO_2/graphene hybrid materials. Evidence of charge separation by electron transfer from reduced graphene oxide to TiO_2. *Catal. Today* **2017**, *281*, 29–37. [CrossRef]
8. Faraldos, M.; Bahamonde, A. Environmental applications of titania-graphene photocatalysts. *Catal. Today* **2017**, *285*, 13–28. [CrossRef]

9. Zeng, G.; You, H.; Du, M.; Zhang, Y.; Ding, Y.; Xu, C.; Liu, B.; Chen, B.; Pan, X. Enhancement of photocatalytic activity of TiO_2 by immobilization on activated carbon for degradation of aquatic naphthalene under sunlight irradiation. *Chem. Eng. J.* **2021**, *412*, 128498. [CrossRef]
10. Paušová, M.; Riva, M.; Baudys, M.; Krýsa, J.; Barbieriková, Z.; Brezová, V. Composite materials based on active carbon/TiO_2 for photocatalytic water purification. *Catal. Today* **2019**, *328*, 178–182. [CrossRef]
11. Khalid, N.; Majid, A.; Tahir, M.B.; Niaz, N.; Khalid, S. Carbonaceous-TiO_2 nanomaterials for photocatalytic degradation of pollutants: A review. *Ceram. Int.* **2017**, *43*, 14552–14571. [CrossRef]
12. Biener, J.; Stadermann, M.; Suss, M.; Worsley, M.A.; Biener, M.M.; Rose, K.A.; Baumann, T.F. Advanced carbon aerogels for energy applications. *Energy Environ. Sci.* **2011**, *4*, 656–667. [CrossRef]
13. Faria, J.L.; Wang, W. *13 Carbon Materials in Photocatalysis*; John Wiley & Sons: Hoboken, NJ, USA, 2009; p. 481.
14. Al-Muhtaseb, S.; Ritter, J. Preparation and Properties of Resorcinol-Formaldehyde Organic and Carbon Gels. *Adv. Mater.* **2003**, *15*, 101–114. [CrossRef]
15. Prostredný, M.; Abduljalil, M.G.; Mulheran, P.A.; Fletcher, A.J. Process variable optimization in the manufacture of resorcinol–formaldehyde gel materials. *Gels* **2018**, *4*, 36. [CrossRef] [PubMed]
16. Awadallah-F, A.; Al-Muhtaseb, S.A. Nanofeatures of resorcinol-formaldehyde carbon microspheres. *Mater. Lett.* **2012**, *87*, 31–34. [CrossRef]
17. Awadallah-F, A.; Elkhatat, A.M.; Al-Muhtaseb, S.A. Impact of synthesis conditions on meso- and macropore structures of resorcinol-formaldehyde xerogels. *J. Mater. Sci.* **2011**, *46*, 7760–7769. [CrossRef]
18. Principe, I.A.; Fletcher, A.J. Parametric study of factors affecting melamine-resorcinol-formaldehyde xerogels properties. *Mater. Today Chem.* **2018**, *7*, 5–14. [CrossRef]
19. Shevlin, S.A.; Woodley, S.M. Electronic and Optical Properties of Doped and Undoped $(TiO_2)n$ Nanoparticles. *J. Phys. Chem. C* **2010**, *114*, 17333–17343. [CrossRef]
20. Jiang, Y.; Meng, L.; Mu, X.; Li, X.; Wang, H.; Chen, X.; Wang, X.; Wang, W.; Wu, F.; Wang, X. Effective TiO_2 hybrid heterostructure fabricated on nano mesoporous phenolic resol for visible-light photocatalysis. *J. Mater. Chem.* **2012**, *22*, 23642–23649. [CrossRef]
21. Zaleska, A. Doped-TiO_2: A review. *Recent Pat. Eng.* **2008**, *2*, 157–164. [CrossRef]
22. Aranovich, G.L.; Donohue, M.D. Adsorption isotherms for microporous adsorbents. *Carbon* **1995**, *33*, 1369–1375. [CrossRef]
23. Wang, S.; Zhu, Z.H.; Coomes, A.; Haghseresht, F.; Lu, G.Q. The physical and surface chemical characteristics of activated carbons and the adsorption of methylene blue from wastewater. *J. Colloid Interface Sci.* **2005**, *284*, 440–446. [CrossRef] [PubMed]
24. Chham, A.; Khouya, E.; Oumam, M.; Abourriche, A.; Gmouh, S.; Mansouri, S.; Elhammoudi, N.; Hanafi, N.; Hannache, H. The use of insoluble mater of Moroccan oil shale for removal of dyes from aqueous solution. *Chem. Int.* **2018**, *4*, 67–77.
25. Atout, H.; Bouguettoucha, A.; Chebli, D.; Gatica, J.M.; Vidal, H.; Yeste, M.P.; Amrane, A. Integration of Adsorption and Photocatalytic Degradation of Methylene Blue Using TiO_2 Supported on Granular Activated Carbon. *Arab. J. Sci. Eng.* **2017**, *42*, 1475–1486. [CrossRef]
26. Matos, J. *Hybrid TiO_2-C Composites for the Photodegradation of Methylene Blue Under Visible Light*; Bol Grupo Español Carbón: Zaragoza, Spain, 2013.
27. Natarajan, T.S.; Bajaj, H.C.; Tayade, R.J. Preferential adsorption behavior of methylene blue dye onto surface hydroxyl group enriched TiO_2 nanotube and its photocatalytic regeneration. *J. Colloid Interface Sci.* **2014**, *433*, 104–114. [CrossRef] [PubMed]
28. Baker, F.S.; Miller, C.E.; Repik, A.J.; Tolles, E.D. Activated carbon. In *Kirk-Othmer Encyclopedia of Chemical Technology*; Interscience Publishers: Geneva, Switzerland, 2000.
29. Zhang, X.; Zhang, F.; Chan, K.-Y. Synthesis of titania-silica mixed oxide mesoporous materials, characterization and photocatalytic properties. *Appl. Catal. A Gen.* **2005**, *284*, 193–198. [CrossRef]
30. Ashraf, M.A.; Peng, W.; Zare, Y.; Rhee, K.Y. Effects of Size and Aggregation/Agglomeration of Nanoparticles on the Interfacial/Interphase Properties and Tensile Strength of Polymer Nanocomposites. *Nanoscale Res. Lett.* **2018**, *13*, 1–7. [CrossRef] [PubMed]
31. Azizian, S. Kinetic models of sorption: A theoretical analysis. *J. Colloid Interface Sci.* **2004**, *276*, 47–52. [CrossRef]
32. Foo, K.Y.; Hameed, B.H. Insights into the modeling of adsorption isotherm systems. *Chem. Eng. J.* **2010**, *156*, 2–10. [CrossRef]
33. Ayawei, N.; Ebelegi, A.N.; Wankasi, D. Modelling and Interpretation of Adsorption Isotherms. *J. Chem.* **2017**, *2017*, 3039817. [CrossRef]
34. Tan, Y.; Kilduff, J.E. Factors affecting selectivity during dissolved organic matter removal by anion-exchange resins. *Water Res.* **2007**, *41*, 4211–4221. [CrossRef] [PubMed]
35. Li, J.; Zhang, S.; Chen, C.; Zhao, G.; Yang, X.; Li, J.; Wang, X. Removal of Cu(II) and Fulvic Acid by Graphene Oxide Nanosheets Decorated with Fe_3O_4 Nanoparticles. *ACS Appl. Mater. Interfaces* **2012**, *4*, 4991–5000. [CrossRef] [PubMed]
36. Shao, D.D.; Fan, Q.H.; Li, J.X.; Niu, Z.W.; Wu, W.S.; Chen, Y.X.; Wang, X.K. Removal of Eu(III) from aqueous solution using ZSM-5 zeolite. *Microporous Mesoporous Mater.* **2009**, *123*, 1–9. [CrossRef]
37. Zhang, G.; Ni, C.; Liu, L.; Zhao, G.; Fina, F.; Irvine, J.T.S. Macro-mesoporous resorcinol–formaldehyde polymer resins as amorphous metal-free visible light photocatalysts. *J. Mater. Chem. A* **2015**, *3*, 15413–15419. [CrossRef]
38. Chen, X.; Mao, S.S. Titanium dioxide nanomaterials: Synthesis, properties, modifications, and applications. *Chem. Rev.* **2007**, *107*, 2891–2959. [CrossRef]

39. Huang, X.; Yang, W.; Zhang, G.; Yan, L.; Zhang, Y.; Jiang, A.; Xu, H.; Zhou, M.; Liu, Z.; Tang, H.; et al. Alternative synthesis of nitrogen and carbon co-doped TiO_2 for removing fluoroquinolone antibiotics in water under visible light. *Catal. Today* **2019**, *361*, 11–16. [CrossRef]
40. Simonetti, E.A.N.; de Simone Cividanes, L.; Campos, T.M.B.; de Menezes, B.R.C.; Brito, F.S.; Thim, G.P. Carbon and TiO_2 synergistic effect on methylene blue adsorption. *Mater. Chem. Phys.* **2016**, *177*, 330–338. [CrossRef]
41. Wu, C.H.; Kuo, C.Y.; Chen, S.T. Synergistic effects between TiO_2 and carbon nanotubes (CNTs) in a TiO_2/CNTs system under visible light irradiation. *Environ. Technol.* **2013**, *34*, 2513–2519. [CrossRef]
42. Houas, A.; Lachheb, H.; Ksibi, M.; Elaloui, E.; Guillard, C.; Herrmann, J.M. Photocatalytic degradation pathway of methylene blue in water. *Appl. Catal. B: Environ.* **2001**, *31*, 145–157. [CrossRef]
43. Lakshmi, S.; Renganathan, R.; Fujita, S. Study on TiO_2-mediated photocatalytic degradation of methylene blue. *J. Photochem. Photobiol. A Chem.* **1995**, *88*, 163–167. [CrossRef]
44. Makuła, P.; Pacia, M.; Macyk, W. How To Correctly Determine the Band Gap Energy of Modified Semiconductor Photocatalysts Based on UV–Vis Spectra. *J. Phys. Chem. Lett.* **2018**, *9*, 6814–6817. [CrossRef] [PubMed]
45. Ho, Y.S.; McKay, G. Pseudo-second order model for sorption processes. *Process Biochem.* **1999**, *34*, 451–465. [CrossRef]
46. Sahoo, T.R.; Prelot, B. Adsorption processes for the removal of contaminants from wastewater: The perspective role of nanomaterials and nanotechnology. In *Nanomaterials for the Detection and Removal of Wastewater Pollutants*; Elsevier: Amsterdam, The Netherlands, 2020; pp. 161–222.

gels

Article

Luminescent Hydrogel Based on Silver Nanocluster/Malic Acid and Its Composite Film for Highly Sensitive Detection of Fe^{3+}

Xiangkai Liu [1,2], Chunhui Li [1,2], Zhi Wang [1], Na Zhang [1,*], Ning Feng [1], Wenjuan Wang [1] and Xia Xin [1,*]

1 Key Laboratory of Colloid and Interface Chemistry (Ministry of Education), School of Chemistry and Chemical Engineering, Shandong University, Jinan 250100, China; lxk575730669@163.com (X.L.); li_chunhui0411@163.com (C.L.); zwang@mail.sdu.edu.cn (Z.W.); 202020302@mail.sdu.edu.cn (N.F.); wenjuanwang@mail.sdu.edu.cn (W.W.)
2 China Research Institute of Daily Chemistry Co., Ltd., Taiyuan 030001, China
* Correspondence: nzhang@sdu.edu.cn (N.Z.); xinx@sdu.edu.cn (X.X.)

Abstract: Metal nanoclusters (NCs) with excellent photoluminescence properties are an emerging functional material that have rich physical and chemical properties and broad application prospects. However, it is a challenging problem to construct such materials into complex ordered aggregates and cause aggregation-induced emission (AIE). In this article, we use the supramolecular self-assembly strategy to regulate a water-soluble, atomically precise Ag NCs $(NH_4)_9[Ag_9(C_7H_4SO_2)_9]$ (Ag_9-NCs, $[Ag_9(mba)_9]$, H_2mba = 2-mercaptobenzoic acid) and L-malic acid (L–MA) to form a phosphorescent hydrogel with stable and bright luminescence, which is ascribed to AIE phenomenon. In this process, the AIE of Ag_9-NCs could be attributed to the non-covalent interactions between L–MA and Ag_9-NCs, which restrict the intramolecular vibration and rotation of ligands on the periphery of Ag_9-NCs, thus inhibiting the ligand-related, non-radiative excited state relaxation and promoting radiation energy transfer. In addition, the fluorescent Ag_9-NCs/L–MA xerogel was introduced into polymethylmethacrylate (PMMA) to form an excellently fluorescent film for sensing of Fe^{3+}. Ag_9-NCs/L–MA/PMMA film exhibits an excellent ability to recognize Fe^{3+} ion with high selectivity and a low detection limit of 0.3 μM. This research enriches self-assembly system for enhancing the AIE of metal NCs, and the prepared hybrid films will become good candidates for optical materials.

Keywords: silver nanoclusters; malic acid; self-assembly; AIE; sensor

Citation: Liu, X.; Li, C.; Wang, Z.; Zhang, N.; Feng, N.; Wang, W.; Xin, X. Luminescent Hydrogel Based on Silver Nanocluster/Malic Acid and Its Composite Film for Highly Sensitive Detection of Fe^{3+}. *Gels* **2021**, 7, 192. https://doi.org/10.3390/gels7040192

Academic Editor: Hiroyuki Takeno

Received: 28 September 2021
Accepted: 26 October 2021
Published: 31 October 2021

Publisher's Note: MDPI stays neutral with regard to jurisdictional claims in published maps and institutional affiliations.

1. Introduction

Metal nanoclusters (NCs), such as gold, silver, and copper NCs, represent a class of multifunctional materials with attractive optoelectronic and photoluminescence properties [1–5]. It consists of a metal core, composed of several to hundreds of metal atoms and a peripheral organic ligand, forming a unique core-shell structure [6–10]. Due to their large Stokes shift, low toxicity, good biocompatibility, and other excellent characteristics, metal NCs can be used as environmentally friendly and biocompatible color conversion materials, fluorescent probes, and excellent biological probe for protein expression [11–19].

Recently, the aggregation-induced emission (AIE) strategy has been used to enhance the photoluminescence of metal NCs, thereby enhancing its application in fluorescent sensing, light-emitting diodes, and other optoelectronic devices [20–22]. At present, the commonly used methods for AIE are solvent and cation-induced aggregation. However, these two methods cannot obtain ordered aggregates, resulting in instability and poor uniformity of nanocluster aggregation [23–26]. Therefore, an effective strategy for manipulating the spatial arrangement of nanostructured units to form a specific structure: self-assembly, was introduced to solve this problem. By regulating the non-covalent forces between multiple molecules (van der Waals forces, hydrogen bonds, electrostatic interactions, and π-π stacking), metal NCs, and other molecules form ordered aggregates with specific functions, which further improves the photoluminescence performance of the metal NCs [27–29].

For example, Shen et al. used the supramolecular self-assembly strategy to obtain stable colloidal aggregates (nanospheres and nanovesicles) of Ag_6-NCs/PEI, through multiple electrostatic interactions between Ag_6-NCs and polyethyleneimine (PEI) [30]. During the formation of the order structure, the Ag_6 NCs luminescence in the dilute aqueous solution was turned on. Zhang et al. demonstrated the enhancement of the luminescence of Cu NCs by forming compact and ordered self-assembly architectures by changing the annealing temperature in the formation of Cu NCs [31]. Therefore, it can be seen that the self-assembly of metal NCs is very attractive and worth studying.

Fluorescent film sensors are widely considered, due to their advantages, such as convenient portability, real-time detection, and no pollution to the system, for testing [32–37]. Compared with the solution and powder detection forms, the adjustable shape, size, and flexibility of the fluorescent film makes it have wider application prospects. For example, Li et al. optimized the concentration of octadecylamine with amino groups in n-hexane to make the Tris-base modified silver NCs self-assemble into a single-layer silver nanocomposite film at the n-hexane-water interface and use it as a surface enhancement Raman scattering (SERS) active substrate for ultra-sensitive detection of Hg^{2+} ions [38]. The SERS sensing monolayer film also has good reproducibility and recovery rate. Katowah et al. used an in-situ method to prepare a ternary nanocomposite film containing copper oxide/polymethylmethacrylate (PMMA)/various carbon-based nanofillers, as a selective Hg^{2+} ion sensor and the mixed nanofiller significantly improved performance of PMMA film [39].

Herein, a water-soluble, atomically precise Ag NCs $(NH_4)_9[Ag_9(C_7H_4SO_2)_9]$ (Ag_9-NCs, $[Ag_9(mba)_9]$, H_2mba = 2-mercaptobenzoic acid, and the molecular structure of Ag_9-NCs is shown in Figure S1) were selected to self-assembled with L-malic acid (L–MA) to construct phosphorescent hydrogels (Scheme 1). The Ag_9-NCs/L–MA hydrogel has a hollow-tube structure which is regulated by non-covalent interactions, based on hydrogen bonds, which further promotes the AIE phenomenon. In order to further develop its application prospects as fluorescent sensors, Ag_9-NCs/L–MA xerogels are doped into PMMA films to construct fluorescent film sensors for detecting Fe^{3+} ions. The value of the corresponding quenching coefficient, K_{SV}, is 5.2×10^4 M^{-1} in the low concentration range, and the detection limit of Fe^{3+} is down to 0.3 μM.

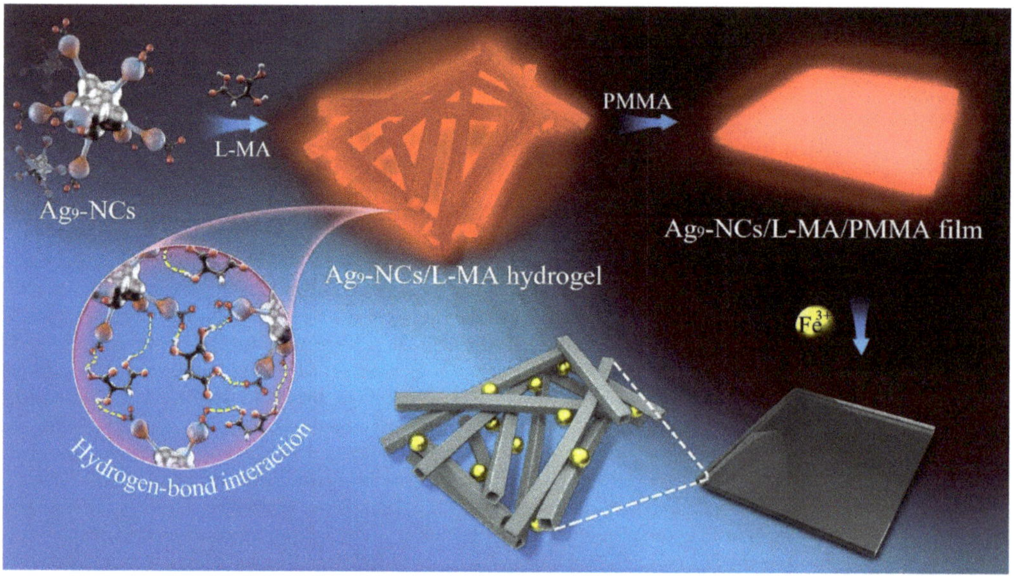

Scheme 1. Schematic illustration of the hollow-tube structure of Ag_9-NCs/L–MA hydrogel and its composite film for highly sensitive detection of Fe^{3+}.

2. Results and Discussion

2.1. Self-Assembly of Ag_9-NCs/L–MA Hydrogels

The Ag_9-NCs in an aqueous solution is in a nonfluorescent state. In order to regulate the aggregation behavior and AIE of Ag_9-NCs, though the non-covalent interaction force dominated by hydrogen bonds, L–MA with hydroxyl and carboxyl functional groups is selected as a hydrogen bond donors, and Ag_9-NCs, with carboxylate on the periphery, act as hydrogen bond acceptors. Firstly, phase behavior study on Ag_9-NCs/L–MA mixed system was performed. Based on our previous study [40,41], $c_{Ag9\text{-}NCs}$ was fixed at 8 mM, while $c_{L\text{-}MA}$ was changed. It can be seen that different phase behaviors of the samples were obtained (Figure 1a). Compared with the solution without fluorescence (0~0.1 M L–MA) and the sample in the two-phase with weaker fluorescence intensity (0.1~0.25 M L–MA), the hydrogel with higher luminous intensity (0.25~0.5 M L–MA) was selected as the main research object. Once the hydrogels formed, the UV absorption exhibits a broad peak from 200 to 550 nm (Figure 1d) and the electrons of the benzene ring system and C=O undergo π-π^* and n-π^* transitions, resulting in ultraviolet absorption bands. Moreover, orange–red fluorescence at 628 nm was excited by blue light at 470 nm (Figure 1d) and it can be observed that as $c_{L\text{-}MA}$ increases, the fluorescence intensity of the hydrogel first increases and then decreases (Figure 1e,f). This is attributed to the fact that when $c_{Ag9\text{-}NCs}$ = 8 mM and $c_{L\text{-}MA}$ = 0.3 M, Ag_9-NCs and L–MA self-assemble to form highly ordered aggregates, and L–MA sufficiently limits the ligands of Ag_9-NCs, while when $c_{L\text{-}MA}$ is too high or too low, the uniformity of the formed aggregates is poor, and the self-assembly strategy cannot be effectively implemented, which leads to the weakening of fluorescence intensity. Thus, 8 mM Ag_9-NCs/0.3 M L–MA hydrogel, which has the highest fluorescence intensity was selected as the typical sample.

Then, the dynamic change of fluorescence intensity with the hydrogel formation was also studied. Once Ag_9-NCs and L–MA mixed, under the drive of non-covalent interaction forces, the fluorescent intensity of 8 mM Ag_9-NCs/0.3 M L–MA hydrogel increased immediately within 30 min, achieving AIE effect. After about 1 h, the self-assembly process is almost complete, the fluorescence intensity stabilized and no longer changed (Figure 1g,h). Moreover, the peak position of the fluorescence spectrum has a slight blue shift with time. It is speculated that Ag_9-NCs, L–MA and water molecules gradually formed the hydrogel through the hydrogen bonds, the vibration of the peripheral ligand of Ag_9-NCs is restricted, and ligand-to-metal charge transfer occurred, resulting in a blue shift of the peak position. Thus, the AIE phenomenon of Ag_9-NCs/L–MA hydrogel can be attributed to the limited intramolecular vibration and rotation of the ligand of Ag_9-NCs, which inhibits the ligand-related non-radiative excited state relaxation and promotes radiation energy transfer. Furthermore, the average of lifetime of Ag_9-NCs solution is 3.277 ns, while the average of the lifetime of 8 mM Ag_9-NCs/0.3 M L–MA hydrogel increased to 7.4383 µs (Figure 1i, Table S1), which is approximately 2270 times that of Ag_9-NCs solution. Therefore, the large Stokes shift (158 nm) and microsecond lifetime (7.4383 µs) indicate that it is essentially phosphors. Besides, the quantum yield of 8 mM Ag_9-NCs/0.3 M L–MA hydrogel, measured using the integrating sphere, is 11.20%, which is higher than that of aggregates formed by the assembly of Ag_9-NCs with other substances [42]. Longer fluorescence lifetime and higher quantum yield of Ag_9-NCs/L–MA hydrogels make it excellent candidates for luminescent sensing materials.

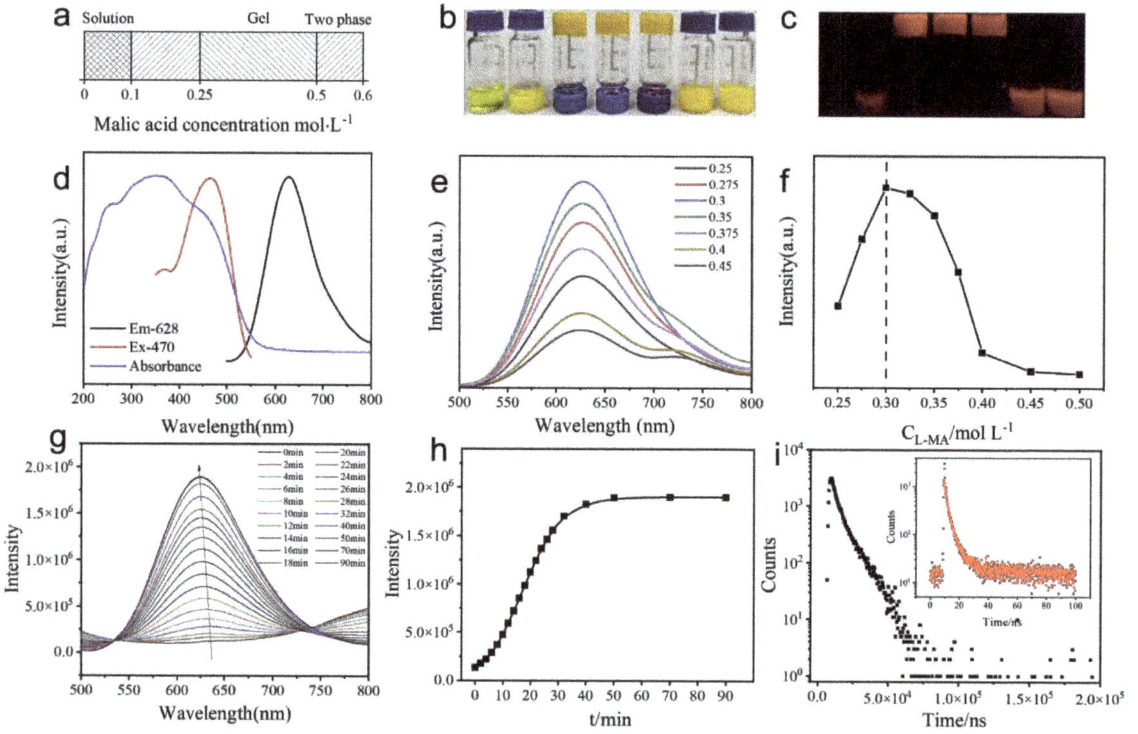

Figure 1. (**a**) Phase transition with different concentration of L–MA. (**b**) Photographs of the hydrogels with different L–MA concentrations. (**c**) Fluorescence photographs of the hydrogels with different L–MA concentrations. (**d**) Photoexcitation (red line, emission = 628 nm), photoemission (black line, excitation = 470 nm) spectra of Ag_9-NCs hydrogels, and UV–vis absorption (blue line). (**e**) Photoluminescence (PL) spectra of hydrogels with different concentration of L–MA for excitation at 470 nm. (**f**) Comparison of the luminescence intensity of hydrogels with different concentration of L–MA at 628 nm. (**g**) PL spectra of Ag_9-NCs hydrogels with different incubation times for excitation at 470 nm. (**h**) Comparison of the luminescence intensity of hydrogels with different incubation times. (**i**) PL decay profiles of Ag_9-NCs hydrogel. Inset: the PL decay profiles of the powder of lyophilized Ag_9-NCs solution.

Scanning electron microscopy (SEM) and transmission electron microscope (TEM) microscopic observations show that the hydrogel is composed of tangled hollow tubes with a very high aspect ratio (20:1), and most of the tubes are between 1–2 μm in length and about 50–100 nm in diameter (Figure 2a–f and Figure S2). The confocal laser scan microscopy (CLSM) image shows that the tubes structure is accompanied by strong fluorescent properties. Further observation by TEM shows that these hollow tubes are composed of a large number of smaller diameter fibers (10–20 nm) (Figure S3). Moreover, it can be seen in SEM that a small amount of hollow tubes are entangled to form a spiral (Figure 2d). It is speculated that the appearance of AIE can be attributed to the appearance of ordered hollow tube structure. The detailed formation mechanism of the hollow-tube structure of Ag_9-NCs/L–MA hydrogel can be shown in Scheme 1.

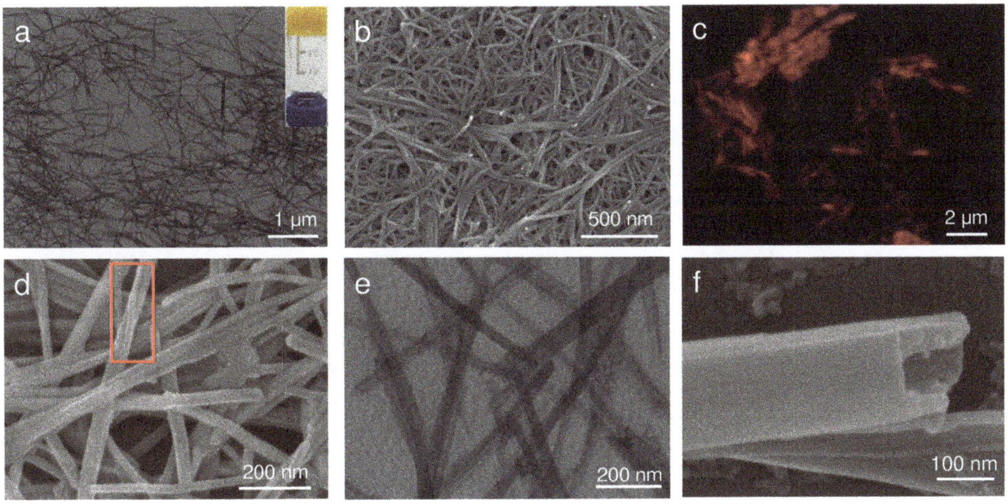

Figure 2. (**a**,**e**) TEM, (**c**) CLSM, and (**b**,**d**,**f**) SEM images of the hollow tubes of 8 mM Ag_9-NCs/0.3 M L–MA hydrogel with different magnifications (the inset of a is a photograph of the hydrogels).

2.2. Structural Analysis of the Hydrogel

In order to further analyze the mechanism of supramolecular self-assembly, a series of characterizations were carried out. Thermogravimetric analyses (TGA) can be used to measure the thermal stability of the studied substances (Figure 3a). The decomposition temperature of L–MA is about 145 °C and the decomposition temperature of the H_2mba ligand of Ag_9-NCs is about 200 °C. After assembly, the decomposition temperature of the H_2mba ligand of Ag_9-NCs for Ag_9-NCs/L–MA xerogel is about 260 °C, indicating that the stability of Ag_9-NCs is further improved by supramolecular self-assembly.

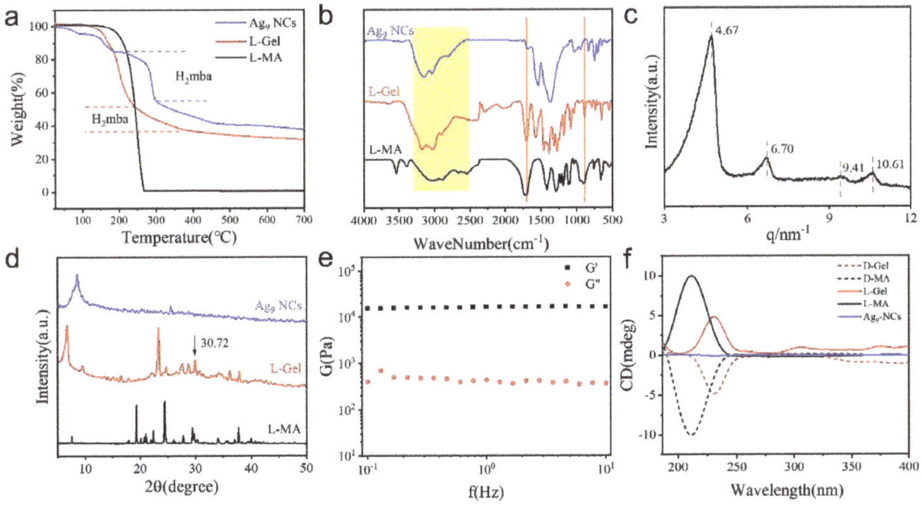

Figure 3. (**a**) TGA and (**b**) FT–IR spectra of lyophilized Ag_9-NCs solution, L–MA and Ag_9-NCs/L–MA xerogel. (**c**) SAXS results of Ag_9-NCs/L–MA xerogel. (**d**) XRD results of lyophilized Ag_9–NCs solution, L–MA and Ag_9-NCs/L–MA xerogel. (**e**) Variation of elastic modulus (G′) and viscous modulus (G″) as a function of frequency (τ = 10 Pa). (**f**) CD spectra of Ag_9–NCs solution, L–MA, Ag_9-NCs/L–MA hydrogels, D–MA and Ag_9-NCs/D–MA hydrogels.

Fourier transform infrared (FT-IR) is an effective tool for analyzing the supramolecular forces (Figure 3b). The broad peak of Ag_9-NCs/L–MA xerogel at 2500 cm^{-1} to 3200 cm^{-1} represents the stretching vibration of the hydroxyl group, which is stronger and wider than that of Ag_9-NCs, indicating the formation of hydrogen bonds in the system. The peak at 1700 cm^{-1} representing the stretching vibration of the C=O in the carboxyl group becomes weaker than that of L–MA after the gel is formed, and slightly shifts to a lower wave number, which represents the formation of hydrogen bonds and confirms that L–MA participates in the construction of the hydrogel. The peak at 3540 cm^{-1} represents that the hydroxyl groups in L–MA undergo intermolecular association through hydrogen bonds to form L–MA dimers, which disappear after gelation, indicating that L–MA forms hydrogen bonds with the peripheral ligands of Ag_9-NCs. The peaks at 1537 cm^{-1} and 1377 cm^{-1} belong to the antisymmetric and symmetric stretching vibration of C=O in carboxylate from ligand of Ag_9-NCs. For Ag_9-NCs/L–MA xerogel, the peak of C=O in carboxylate moves, which further confirms the formation of hydrogen bonds between ligand of Ag_9-NCs and L–MA. Moreover, it can be seen that the absorption peak at 900 cm^{-1}, representing O–H in the carboxyl group of L–MA is strong and after the hydrogel is formed, this peak becomes weaker and narrower, confirming that most of the carboxyl groups of L–MA are involved in the formation of hydrogen bonds, limiting the bending vibration of the O–H bonds in the carboxyl groups.

Small- angle X-ray spectroscopy (SAXS) and X-ray diffraction (XRD) can effectively characterize the deposition pattern and spatial structure of the xerogels. SAXS shows that Ag_9-NCs/L–MA xerogel exhibits four scattering peaks, and the scattering factor q ratio is $1 : \sqrt{2} : 2 : \sqrt{5}$, which is a tetragonal stack (Figure 3c). The smallest repeating unit of the aggregate d = 1.36 nm ($d = 2\pi/q$), which is equivalent to the size of Ag_9-NCs. From the XRD results, it can be observed that L–MA is a triclinic crystal system, while the peak of Ag_9-NCs/L–MA xerogel at 2θ = 20–40° represents various possible regular arrangements of atomic layers in assembled nanostructures, which is very different from L–MA and Ag_9-NCs (Figure 3d), indicating Ag_9-NCs and L–MA formed a multi-complex during the self-assembly. According to the Bragg equation, the interplanar distances of the atomic layer are in the range of 2.37–3.81 Å. Diffraction peaks at 2θ = 30.72 correspond to Ag-Ag, indicating that the d10-d10 argentophilic interaction may exist in the Ag_9-NCs/L–MA xerogel [40,41,43].

The rheological characteristics are of great significance for supramolecular hydrogels, and the mechanical strength of the Ag_9-NCs/L–MA hydrogels can be evaluated by the rheological measurement. In the stress scanning (Figure 3e), the storage modulus (G' = 16,900 Pa) is much larger than the loss modulus (G'' = 614.6 Pa), indicating that the hydrogels exhibit a solid-like nature. The Ag_9-NCs/L–MA hydrogels exhibit good viscoelasticity, a wider linear viscoelastic region, and a higher yield stress (892.8 Pa), indicating that the hydrogels constructed by the Ag_9-NCs are more rigid and exhibit strong damage resistance. In the frequency sweep test (Figure S4), G' is larger than G'' and the moduli of the hydrogels are almost independent of the applied frequency. The above results indicate that the Ag_9-NCs/L–MA hydrogels have high mechanical strength.

The circular dichroism (CD) spectrum has further confirmed that the Ag_9-NCs/L–MA hydrogel possesses supramolecular chirality. L–MA has a positive Cotton effect at 212 nm and no obvious CD signal was detected in the Ag_9-NCs aqueous solution. But the Ag_9-NCs/L–MA hydrogel has a positive Cotton effect at 230 nm, and the original CD signal of L–MA disappears. Moreover, we also used D-MA to construct Ag_9-NCs/D-MA hydrogel and it is interesting to find that the CD spectrum of the Ag_9-NCs/D-MA hydrogel is exactly opposite of the Ag_9-NCs/L–MA hydrogel. It can be concluded that through supramolecular self-assembly, the molecular chirality of L–MA (or D-MA), was successfully transferred to the supramolecular level, which induced the Ag_9-NCs/L–MA hydrogel (Ag_9-NCs/D-MA hydrogel) has supramolecular chirality (Figure 3f).

2.3. Ag_9-NCs/L–MA Xerogel/PMMA Film for Sensing

The composite material obtained by doping the xerogel into the polymer matrix has the advantages of easy processing and good flexibility and can enhance the practical application of the gel [32,33]. PMMA has good UV resistance, chemical durability, and good mechanical properties. Therefore, based on the excellent sensing performance of metal NCs and the advantages of simple preparation of organic films as a matrix material, Ag_9-NCs/L–MA xerogel are introduced into the organic glass matrix to form a new type of organic film. Herein, the 8 mM Ag_9-NCs/0.3 M L–MA xerogel is introduced into the PMMA, and through the solvent volatilization method to get a hybrid film. Firstly, several films were prepared with different doping concentration of 8 mM Ag_9-NCs/0.3 M L–MA xerogel and it can be observed that as the doping concentration increases (from 0.1% to 0.4%), the fluorescence intensity increases but the transmittance decreases (Figure 4a–c). Integrating transmittance and fluorescence spectra, a composite Ag_9-NCs/L–MA/PMMA film with a mass concentration of 0.3% xerogel was selected to study its sensing performance.

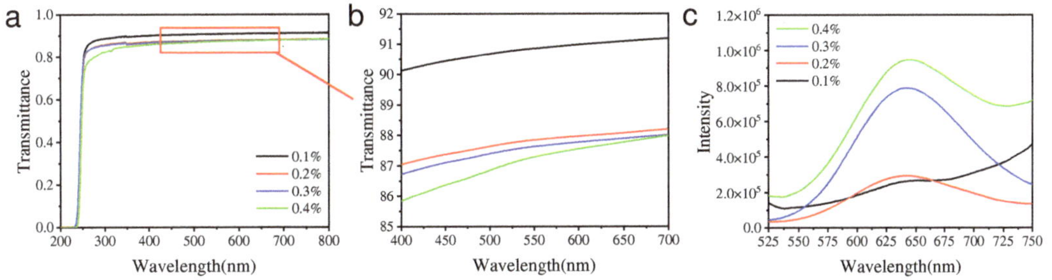

Figure 4. (**a**) transmittance of films with different doping concentrations. (**b**) transmittance of films with different doping concentrations (400–700 nm). (**c**) FL of films with different doping concentrations (from 0.1% to 0.4%).

Then, 0.3% Ag_9-NCs/L–MA xerogel/PMMA films were placed into different metal cation aqueous solutions for 10 min and then, the luminescence spectra of these films were studied (Figures S5 and S6). It is obviously that the composite film shows an obvious quenching effect only toward Fe^{3+} (Figure 5a,b). Furthermore, the 0.3% Ag_9-NCs/L–MA xerogel/PMMA films were placed into aqueous solutions containing different concentrations of Fe^{3+} ions for 10 min to study the corresponding luminescence intensity (Figure 5c,d). With the gradual increase of c_{Fe3+}, the luminescence intensities at 628 nm are significantly weakened. In order to investigate the luminescence quenching efficiency, the Stern–Volmer (S-V) equation was used to quantitatively analyze the quenching curve [32]:

$$I_0/I = 1 + K_{SV}[M] \tag{1}$$

where I_0 and I are the luminescence intensity at 640 nm when c = 0 and c = M, respectively, K_{SV} is the quenching coefficient of Fe^{3+} ions, and $[M]$ is the molar concentration of Fe^{3+} ions. The quenching curve shows a well-fitted linear relationship at low c_{Fe3+} (1–100 μM). The K_{SV} value was calculated to be 2.3×10^4 M^{-1}. According to the detection limit expression defined by IUPAC, the detection limit of the composite membrane is calculated to be 0.3 μM, which indicates that the as-prepared 0.3% Ag_9-NCs/L–MA xerogel/PMMA film material exhibited good sensing performance. The limit of detection of the Ag_9-NCs/L–MA xerogel/PMMA film far less than the maximum concentration of Fe(III) in drinking water of 5.36 μM (specified by the Minister of Health of the People's Republic of China).

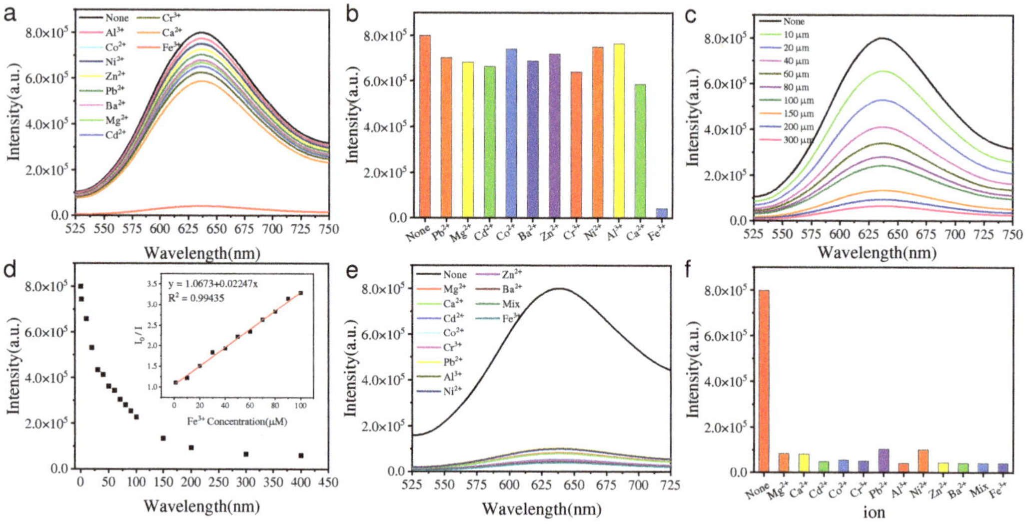

Figure 5. (**a**) Luminescence emission spectra and (**b**) luminescence intensities at 640 nm for 0.3% Ag_9-NCs/L–MA xerogel/PMMA film in different cations. (**c**) Luminescence emission spectra and (**d**) luminescence intensities at 640 nm for 0.3% Ag_9-NCs/L–MA xerogel/PMMA film in Fe^{3+} with different concentrations. The insert of d shows Stern–Volmer quenching curve of the luminescence intensity of 0.3% Ag_9-NCs/L–MA xerogel/PMMA film at 640 nm against Fe^{3+} concentration; (**e**) luminescent emission spectra and (**f**) luminescence intensities at 640 nm for 0.3% Ag_9-NCs/L–MA xerogel/PMMA film in solutions of different cations with Fe^{3+}.

High sensitivity and high selectivity are the basic requirements of fluorescence sensors. Various mixed metal cation solutions containing Fe^{3+} (1.0 equivalent) and different competing cations (1.0 equivalent) were prepared for competition experiments. By comparing the fluorescence intensity of 0.3% Ag_9-NCs/L–MA xerogel/PMMA film in different mixed metal cation solutions (Figure 5e,f), it can be seen that the fluorescence is still quenched in the presence of other competing metal cations, which indicates that the composite film has satisfactory selectivity for Fe^{3+} detection. In addition, the used films were picked out and washed with distilled water several times, and the corresponding luminescence intensities cannot be recovered. Therefore, it would be used as a disposable test strip for detection Fe^{3+} ions in drinking water in the future.

The possible sensing mechanism of Fe^{3+} ions quenching the fluorescence of 0.3% Ag_9-NCs/L–MA xerogel/PMMA film was further analyzed. It can be seen from UV-Vis (Figure 6a) that the absorption peaks of Fe^{3+} and the Ag_9-NCs/L–MA hydrogel overlap in the absorption range, indicating that the competitive absorption of Fe^{3+} reduces the energy transfer efficiency [42]. Moreover, structural collapse or change maybe also cause the luminescence to be quenched [32]. The XRD test was performed on the Ag_9-NCs/L–MA xerogel before and after immersion in Fe^{3+} ion aqueous solution (Figure 6b). The XRD spectra of the samples were not consistent, which means that the structure of the Ag_9-NCs/L–MA xerogel was destroyed after treated by Fe^{3+} ion aqueous solution. Therefore, the collapse of the crystal structure is also one of the reasons for the quenching of luminescence. Thus, it can be concluded that the fluorescence quenching of Ag_9-NCs/L–MA xerogel/PMMA film is caused by the competitive absorption of Fe^{3+} and the destruction of the crystal structure. The fluorescence quenching phenomenon of the Ag_9-NCs/L–MA xerogel/PMMA film is shown in Scheme 1.

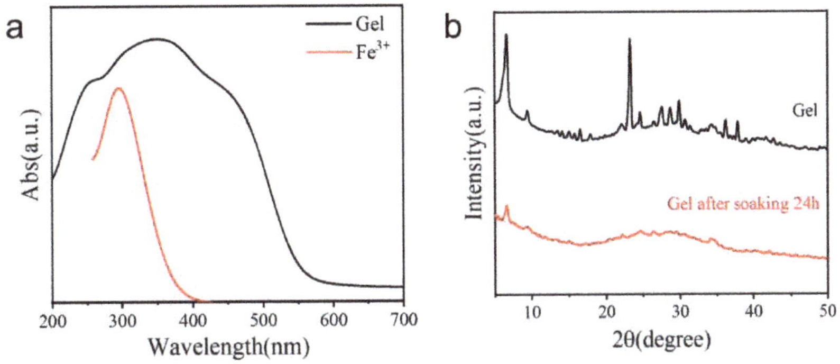

Figure 6. (**a**) UV-Vis absorption of hydrogel and Fe^{3+}. (**b**) XRD of xerogel and xerogel after Fe^{3+} soaking 24 h.

3. Conclusions

In summary, through self-assembly, Ag_9-NCs, and L–MA formed a supramolecular hydrogel, with highly ordered aggregates, by regulating a variety of non-covalent interactions. The nanostructure (tangled hollow tubes) in the hydrogel indicated that L–MA restricts the intramolecular vibration and rotation of the ligand of the Ag_9-NCs, so that it can emit a stable and bright orange–red phosphorescent emission. The excellent photoluminescence properties of Ag_9-NCs/L–MA xerogel make it likely to be used as highly sensitive probes for Fe^{3+}. The Ag_9-NCs/L–MA xerogel/PMMA composite film can selectively identify Fe^{3+}, the quenching coefficient K_{SV} is 2.3×10^4 M^{-1}, and the detection limit is 0.3 μM. The composite film also possesses good recyclability and anti-interference ability, with respect to another ions. The current research aims to achieve AIE, by precisely regulating the formation of ordered aggregates of metal NCs, broadening the research field of metal NCs, and enriching the practical applications of these luminescent materials.

4. Experiment Section

4.1. Materials

Ag_9-NCs was synthesized and purified, according to our previous work, which have a crystal structure [44]. L–MA and D-MA were purchased from Sinopharm Chemical Reagent Co (Shanghai, China) and used without further purification. Ultrapure water used in the experiments, with a resistivity of 18.25 MΩ cm^{-1}, was obtained using a UPH-IV ultrapure water purifier (Sichuan, China). Polymethylmethacrylate (PMMA, average Mw: ~350,000) was purchased from Sigma-Aldrich (Shanghai, China). Dichloromethane (CH_2Cl_2) were obtained from local supplier with the quality of analytical grade and used without further purification.

4.2. Synthesis of Ag_9-NCs/L–MA Hybrid Nanostructures

In this experiment, 0.5 mL of Ag_9-NCs solution (15.87 mM) was added to 0.5 mL L–MA solution (0.6 M) with stirring. The hydrogel was successfully prepared after 8 h of constant temperature (20 °C) in a thermostat. The hydrogels was lyophilized in a vacuum extractor at 60 °C for 5 day to collect the orange–yellow powder.

4.3. Fabrication of Ag_9-NCs/L–MA/PMMA Composite Thin Film

The PMMA powder (200 mg) was dissolved in dichloromethane (8 mL), then followed by addition of the corresponding required of orange–yellow Ag_9-NCs/L–MA xerogels. After being evenly dispersed, place it in a petri dish with a diameter of 6 cm at room temperature for 5 h. Each set of data has been measured 3 times using different batches of film to reduce the error.

4.4. Characterization

A copper mesh was inserted into the gel to obtain a sample and, after drying under an IR lamp for 45 min, TEM images were observed under a JCR-100CX II (JEOL) microscope. The gel was placed on a silica wafer, dried for 45 min under an IR lamp, and observed by field-emission SEM and AFM, respectively. UV−vis data were recorded on a Shimadzu UV-2600 spectrophotometer. Fluorescence data were tested on an LS-55 spectrofluorometer (PerkinElmer, Waltham, MA, USA) and an Edinburgh Instruments FLS920 luminescence spectrometer (xenon lamp, 450 W), respectively. SAXS measurements were performed using an Anton-Paar SAX Sess mc^2 system with nickel-filtered Cu Kα radiation (1.54 Å) operating at 50 kV and 40 mA. XRD patterns were taken on a D8 ADVANCE (Germany Bruker) diffractometer, equipped with Cu Kα radiation and a graphite monochromator. FT-IR spectra in KBr wafer were recorded on a VERTEX-70/70v spectrophotometer. CLSM observations were performed using an inverted microscope (model IX81, Olympus, Tokyo, Japan), equipped with a high-numerical-aperture 60 oil-immersed objective lens. The rheological measurements were carried out on an Anton-Paar Physica MCR302 rheometer with a cone–plate system. Before the frequency sweep, an amplitude sweep at a fixed frequency of 1 Hz was carried out to ensure that the selected stress was in the linear viscoelastic region. The frequency sweep was carried out from 0.01 to 100 Hz at a fixed stress of 10 Pa. TGA was performed under a nitrogen atmosphere at 25–700 °C, with a heating speed of 10 °C min^{-1} on a TA SDT Q600 thermal analyzer. CD spectra were obtained using a JASCO J-810 spectropolarimeter, which was flushed with nitrogen during operation. The absolute fluorescence quantum yields were measured with a spectrofluorometer (FLSP920, Edinburgh Instruments Ltd., Livingston, UK), equipped with an integrating sphere.

Supplementary Materials: The following are available online at https://www.mdpi.com/article/10.3390/gels7040192/s1, Figure S1: The molecular structure of Ag_9-NCs, Figure S2: AFM image of fibers, Figure S3: (a–b) TEM image of the fibers. (c–d) SEM image of fibers, Figure S4: Elastic modulus (G′) and viscous modulus (G″) as a function of the applied stress at a constant frequency (1.0 Hz), Figure S5: Photographs of Ag_9-NCs/L–MA xerogel/PMMA film under different conditions, Figure S6: Luminescence emission spectra of 0.3% Ag_9-NCs/L–MA xerogel/PMMA film in 100 μM Fe^{3+} aqueous solution with different interaction time, Table S1: The average lifetime of Ag_9-NCs and hydrogel.

Author Contributions: X.L.: Investigation, methodology, writing—original draft, formal analysis; C.L.: investigation, methodology, formal analysis; Z.W.: investigation, methodology; N.Z.: investigation, methodology, formal analysis; N.F.: investigation, formal analysis; W.W.: investigation, formal analysis; X.X.: conceptualization, resources, writing—review & editing, funding acquisition, supervision. All authors have read and agreed to the published version of the manuscript.

Funding: This research received no external funding.

Acknowledgments: We gratefully acknowledge the financial support from the National Natural Science Foundation of China (21972077), and Key Technology Research and Development Program of Shandong (2019GGX102019).

Conflicts of Interest: The authors declare no conflict of interest.

References

1. Wu, Z.; Jin, R. On the ligand's role in the fluorescence of gold nanoclusters. *Nano Lett.* **2010**, *10*, 2568–2573. [CrossRef] [PubMed]
2. Cuaran-Acosta, D.; Londono-Larrea, P.; Zaballos-Garcia, E.; Perez-Prieto, J. Reversible pH-induced fluorescence colour change of gold nanoclusters based on pH-regulated surface interactions. *Chem. Commun.* **2019**, *55*, 1604–1606. [CrossRef] [PubMed]
3. Qiao, Y.; Xu, T.; Zhang, Y.; Zhang, C.; Shi, L.; Zhang, G.; Shuang, S.; Dong, C. Green synthesis of fluorescent copper nanoclusters for reversible pH-sensors. *Sens. Actuators B* **2015**, *220*, 1064–1069. [CrossRef]
4. Wang, W.; Zhan, L.; Du, Y.Q.; Leng, F.; Chang, Y.; Gao, M.X.; Huang, C.Z. Label-free DNA detection on the basis of fluorescence resonance energy transfer from oligonucleotide-templated silver nanoclusters to multi-walled carbon nanotubes. *Anal. Methods* **2013**, *5*, 5555. [CrossRef]
5. Burratti, L.; De Matteis, F.; Francini, R.; Lim, J.; Scheu, C.; Prosposito, P. Fluorescent Silver Nanoclusters Embedded in Hydrogel Matrix and Its Potential Use in Environmental Monitoring. *Appl. Sci.* **2021**, *11*, 3470. [CrossRef]

6. Yuan, X.; Tay, Y.; Dou, X.; Luo, Z.; Leong, D.T.; Xie, J. Glutathione-protected silver nanoclusters as cysteine-selective fluorometric and colorimetric probe. *Anal. Chem.* **2013**, *85*, 1913–1919. [CrossRef]
7. Liu, Y.; Ai, K.; Cheng, X.; Huo, L.; Lu, L. Gold-Nanocluster-Based Fluorescent Sensors for Highly Sensitive and Selective Detection of Cyanide in Water. *Adv. Funct. Mater.* **2010**, *20*, 951–956. [CrossRef]
8. Zhu, X.; Li, Y.; Li, R.; Tu, K.; Li, J.; Xie, Z.; Lei, J.; Liu, D.; Qu, D. Self-assembled N-doped carbon with a tube-in-tube nanostructure for lithium-sulfur batteries. *J. Colloid. Interface Sci.* **2020**, *559*, 244–253. [CrossRef]
9. Shang, L.; Dong, S.; Nienhaus, G.U. Ultra-small fluorescent metal nanoclusters: Synthesis and biological applications. *Nano Today* **2011**, *6*, 401–418. [CrossRef]
10. Diez, I.; Pusa, M.; Kulmala, S.; Jiang, H.; Walther, A.; Goldmann, A.S.; Muller, A.H.; Ikkala, O.; Ras, R.H. Color tunability and electrochemiluminescence of silver nanoclusters. *Angew. Chem. Int. Ed.* **2009**, *48*, 2122–2125. [CrossRef]
11. Diez, I.; Kanyuk, M.I.; Demchenko, A.P.; Walther, A.; Jiang, H.; Ikkala, O.; Ras, R.H. Blue, green and red emissive silver nanoclusters formed in organic solvents. *Nanoscale* **2012**, *4*, 4434–4437. [CrossRef] [PubMed]
12. Niesen, B.; Rand, B.P. Thin film metal nanocluster light-emitting devices. *Adv. Mater.* **2014**, *26*, 1446–1449. [CrossRef]
13. Yuan, Q.; Wang, Y.; Zhao, L.; Liu, R.; Gao, F.; Gao, L.; Gao, X. Peptide protected gold clusters: Chemical synthesis and biomedical applications. *Nanoscale* **2016**, *8*, 12095–12104. [CrossRef]
14. Chen, L.Y.; Wang, C.W.; Yuan, Z.; Chang, H.T. Fluorescent gold nanoclusters: Recent advances in sensing and imaging. *Anal. Chem.* **2015**, *87*, 216–229. [CrossRef]
15. Abbas, M.A.; Kamat, P.V.; Bang, J.H. Thiolated Gold Nanoclusters for Light Energy Conversion. *ACS Energy Lett.* **2018**, *3*, 840–854. [CrossRef]
16. Shiang, Y.-C.; Huang, C.-C.; Chen, W.-Y.; Chen, P.-C.; Chang, H.-T. Fluorescent gold and silver nanoclusters for the analysis of biopolymers and cell imaging. *J. Mater. Chem.* **2012**, *22*, 12972–12982. [CrossRef]
17. Yang, L.; Shang, L.; Nienhaus, G.U. Mechanistic aspects of fluorescent gold nanocluster internalization by live HeLa cells. *Nanoscale* **2013**, *5*, 1537–1543. [CrossRef]
18. Kunwar, P.; Hassinen, J.; Bautista, G.; Ras, R.H.A.; Toivonen, J. Direct Laser Writing of Photostable Fluorescent Silver Nanoclusters in Polymer Films. *ACS Nano* **2014**, *8*, 11165–11171. [CrossRef]
19. Pourreza, N.; Ghomi, M. In Situ synthesized and embedded silver nanoclusters into poly vinyl alcohol-borax hydrogel as a novel dual mode "on and off" fluorescence sensor for Fe (III) and thiosulfate. *Talanta* **2018**, *179*, 92–99. [CrossRef]
20. Chen, B.; Li, C.; Zhang, J.; Kan, J.; Jiang, T.; Zhou, J.; Ma, H. Sensing and imaging of mitochondrial viscosity in living cells using a red fluorescent probe with a long lifetime. *Chem. Commun.* **2019**, *55*, 7410–7413. [CrossRef]
21. Hu, R.; Leung, N.L.; Tang, B.Z. AIE macromolecules: Syntheses, structures and functionalities. *Chem. Soc. Rev.* **2014**, *43*, 4494–4562. [CrossRef] [PubMed]
22. Jia, X.; Li, J.; Wang, E. Cu nanoclusters with aggregation induced emission enhancement. *Small* **2013**, *9*, 3873–3879. [CrossRef] [PubMed]
23. Zhao, T.; Zhang, S.; Bi, Y.; Sun, D.; Kong, F.; Yuan, Z.; Xin, X. Development and characterisation of multi-form composite materials based on silver nanoclusters and cellulose nanocrystals. *Colloid Surf. A* **2020**, *603*, 125257. [CrossRef]
24. Zhou, Y.; Zeng, H.C. Simultaneous synthesis and assembly of noble metal nanoclusters with variable micellar templates. *J. Am. Chem. Soc.* **2014**, *136*, 13805–13817. [CrossRef]
25. Luo, Z.; Yuan, X.; Yu, Y.; Zhang, Q.; Leong, D.T.; Lee, J.Y.; Xie, J. From aggregation-induced emission of Au(I)-thiolate complexes to ultrabright Au(0)@Au(I)-thiolate core-shell nanoclusters. *J. Am. Chem. Soc.* **2012**, *134*, 16662–16670. [CrossRef]
26. Tan, C.; Qi, X.; Liu, Z.; Zhao, F.; Li, H.; Huang, X.; Shi, L.; Zheng, B.; Zhang, X.; Xie, L.; et al. Self-assembled chiral nanofibers from ultrathin low-dimensional nanomaterials. *J. Am. Chem. Soc.* **2015**, *137*, 1565–1571. [CrossRef]
27. Hong, X.; Tan, C.; Liu, J.; Yang, J.; Wu, X.J.; Fan, Z.; Luo, Z.; Chen, J.; Zhang, X.; Chen, B.; et al. AuAg nanosheets assembled from ultrathin AuAg nanowires. *J. Am. Chem. Soc.* **2015**, *137*, 1444–1447. [CrossRef]
28. Xia, Y.; Nguyen, T.D.; Yang, M.; Lee, B.; Santos, A.; Podsiadlo, P.; Tang, Z.; Glotzer, S.C.; Kotov, N.A. Self-assembly of self-limiting monodisperse supraparticles from polydisperse nanoparticles. *Nat. Nanotechnol.* **2011**, *6*, 580–587. [CrossRef]
29. Wu, Z.; Yao, Q.; Zang, S.; Xie, J. Directed Self-Assembly of Ultrasmall Metal Nanoclusters. *ACS Mater. Lett.* **2019**, *1*, 237–248. [CrossRef]
30. Shen, J.; Wang, Z.; Sun, D.; Xia, C.; Yuan, S.; Sun, P.; Xin, X. pH-Responsive Nanovesicles with Enhanced Emission Co-Assembled by Ag(I) Nanoclusters and Polyethyleneimine as a Superior Sensor for Al(3). *ACS. Appl. Mater. Interfaces* **2018**, *10*, 3955–3963. [CrossRef]
31. Wu, Z.; Liu, J.; Gao, Y.; Liu, H.; Li, T.; Zou, H.; Wang, Z.; Zhang, K.; Wang, Y.; Zhang, H.; et al. Assembly-Induced Enhancement of Cu Nanoclusters Luminescence with Mechanochromic Property. *J. Am. Chem. Soc.* **2015**, *137*, 12906–12913. [CrossRef]
32. Chen, W.; Fan, R.; Fan, J.; Liu, H.; Sun, T.; Wang, P.; Yang, Y. Lanthanide Coordination Polymer-Based Composite Films for Selective and Highly Sensitive Detection of $Cr_2O_7^{2-}$ in Aqueous Media. *Inorg. Chem.* **2019**, *58*, 15118–15125. [CrossRef]
33. Liu, L.; Li, Z.; Liu, Y.; Zhang, S.X.-A. Recent advances in film-based fluorescence sensing. *Sci. Sin. Chim.* **2019**, *50*, 39–69. [CrossRef]
34. Bhasin, A.K.K.; Chauhan, P.; Chaudhary, S. A novel sulfur-incorporated naphthoquinone as a selective "turn-on" fluorescence chemical sensor for rapid detection of Ba^{2+} ion in semi-aqueous medium. *Sens. Actuators B* **2019**, *294*, 116–122. [CrossRef]
35. Song, C.; Liu, Y.; Jiang, X.; Zhang, J.; Dong, C.; Li, J.; Wang, L. Ultrasensitive SERS determination of avian influenza A H7N9 virus via exonuclease III-assisted cycling amplification. *Talanta* **2019**, *205*, 120137. [CrossRef]
36. Aysha, T.S.; El-Sedik, M.S.; Mohamed, M.B.I.; Gaballah, S.T.; Kamel, M.M. Dual functional colorimetric and turn-off fluorescence probe based on pyrrolinone ester hydrazone dye derivative for Cu^{2+} monitoring and pH change. *Dyes Pigment.* **2019**, *170*, 107549. [CrossRef]

37. Li, J.J.; Qiao, D.; Zhao, J.; Weng, G.J.; Zhu, J.; Zhao, J.W. Ratiometric fluorescence detection of Hg^{2+} and Fe^{3+} based on BSA-protected Au/Ag nanoclusters and His-stabilized Au nanoclusters. *Methods Appl. Fluoresc.* **2019**, *7*, 045001. [CrossRef]
38. Li, Y.; Li, Y.; Duan, J.; Hou, J.; Hou, Q.; Yang, Y.; Li, H.; Ai, S. Rapid and ultrasensitive detection of mercury ion (II) by colorimetric and SERS method based on silver nanocrystals. *Microchem. J.* **2021**, *161*, 105790. [CrossRef]
39. Katowah, D.F.; Alqarni, S.; Mohammed, G.I.; Al Sheheri, S.Z.; Alam, M.M.; Ismail, S.H.; Asiri, A.M.; Hussein, M.A.; Rahman, M.M. Selective Hg^{2+} sensor performance based various carbon-nanofillers into CuO-PMMA nanocomposites. *Polym. Adv. Technol.* **2020**, *31*, 1946–1962. [CrossRef]
40. Sun, P.; Wang, Z.; Bi, Y.; Sun, D.; Zhao, T.; Zhao, F.; Wang, W.; Xin, X. Self-Assembly-Driven Aggregation-Induced Emission of Silver Nanoclusters for Light Conversion and Temperature Sensing. *ACS Appl. Nano Mater.* **2020**, *3*, 2038–2046. [CrossRef]
41. Sun, P.; Wang, Z.; Sun, D.; Bai, H.; Zhu, Z.; Bi, Y.; Zhao, T.; Xin, X. pH-guided self-assembly of silver nanoclusters with aggregation-induced emission for rewritable fluorescent platform and white light emitting diode application. *J. Colloid Interface Sci.* **2020**, *567*, 235–242. [CrossRef]
42. Cheng, X.; Sun, P.; Zhang, N.; Zhou, S.; Xin, X. Self-assembly of silver nanoclusters and phthalic acid into hollow tubes as a superior sensor for Fe^{3+}. *J. Mol. Liq.* **2021**, *323*, 115032. [CrossRef]
43. Barreiro, E.; Casas, J.S.; Couce, M.D.; Laguna, A.; López-de-Luzuriaga, J.M.; Monge, M.; Sánchez, A.; Sordo, J.; Vázquez López, E.M. A novel hexanuclear silver(i) cluster containing a regular Ag6 ring with short Ag–Ag distances and an argentophilic interaction. *Dalton Trans.* **2013**, *42*, 5916. [CrossRef]
44. Xie, Z.; Sun, P.; Wang, Z.; Li, H.; Yu, L.; Sun, D.; Chen, M.; Bi, Y.; Xin, X.; Hao, J. Metal-Organic Gels from Silver Nanoclusters with Aggregation-Induced Emission and Fluorescence-to-Phosphorescence Switching. *Angew. Chem. Int. Ed.* **2020**, *59*, 9922–9927. [CrossRef]

gels

MDPI

Communication

Stoichiometric Ratio Controlled Dimension Transition and Supramolecular Chirality Enhancement in a Two-Component Assembly System

Penghui Zhang [1,†], Yiran Liu [1,†], Xinkuo Fang [1,†], Li Ma [1,*,†], Yuanyuan Wang [2,*] and Lukang Ji [1,*]

1 Hebei Key Laboratory of Organic Functional Molecules, College of Chemistry and Materials Science, Hebei Normal University, Shijiazhuang 050024, China; 17599090428@stu.hebtu.edu.cn (P.Z.); liuyiran@stu.hebtu.edu.cn (Y.L.); xinkuo86@163.com (X.F.)

2 Department of Pharmacology, College of Basic Medicine, Hebei University of Chinese Medicine, Shijiazhuang 050200, China

* Correspondence: mali0801@hebtu.edu.cn (L.M.); wangyy0830@iccas.ac.cn (Y.W.); jilukang@hebtu.edu.cn (L.J.)

† These authors contributed equally to this work.

Abstract: To control the dimension of the supramolecular system was of great significance. We construct a two component self-assembly system, in which the gelator LHC18 and achiral azobenzene carboxylic acid could co-assembly and form gels. By modulating the stoichiometric ratio of the two components, not only the morphology could be transformed from 1D nanaotube to 0D nanospheres but also the supramolecualr chirality could be tuned. This work could provide some insights to the control of dimension and the supramolecular chirality in the two-component systems by simply modulating the stoichiometric ratio.

Keywords: stoichiometric ratio; dimension transition; supramolecular chirality; two-component gel

Citation: Zhang, P.; Liu, Y.; Fang, X.; Ma, L.; Wang, Y.; Ji, L. Stoichiometric Ratio Controlled Dimension Transition and Supramolecular Chirality Enhancement in a Two-Component Assembly System. *Gels* **2022**, *8*, 269. https://doi.org/10.3390/gels8050269

Academic Editor: Hiroyuki Takeno

Received: 2 April 2022
Accepted: 22 April 2022
Published: 26 April 2022

Publisher's Note: MDPI stays neutral with regard to jurisdictional claims in published maps and institutional affiliations.

1. Introduction

The control of dimension is crucial in nanomaterials and bioscience [1–3]. As the dimension are varied, the properties and functions of the nanomaterials will dramatically change [4–6]. For instance, in carbon nanomaterial aspects, 0D fullerenes, 1D nanotubes, 2D graphene, and 3D graphite had different dimensions and thus lead to distinct electronic and photonic properties (in which 0D, 1D, 2D, 3D represents zero, one, two, and three dimension, respectively). Similarly, in biosciences, proteins form nanostructures including 1D microtubules, 2D bacterial surface layers, and 3D virus capsids with distinct structure and functions, i.e., the dimension information not only had tight relationships with the corresponding functional expression but also had great influence on the pathological changes or diseases in bioscience [7–9]. Supramolecular systems are formed by the weak non-covalent interactions including hydrogen bond, pi-pi stacking and hydrophobic effect [10–15]. By the synergy of one or multi-interactions, nanostructures such as 0D nanospheres, 1D nanotubes and 2D nanoplates could form. Moreover, the dimension of the supramolecular assemblies played a crucial role in the applications, such as the 1D gold nanorod had photothermal effect and had been used in the tumor therapy fields, however the effect cannot be realized by 0D gold nanoparticle [16]. Thus, it is significant to explore the relationships between the dimension transition and functions in supramolecular assemblies.

Benefit by the dynamic properties of the supramolecular systems, many approaches could be utilized to modulate the dimension of the assemblies. For examples, Zhao and co-workers reported that by light irradiation, the morphology of the azobenzene derivative assemblies transform from 2D nanosheet to 1D nanotube and 0D nanoparticles by the *Z-E* isomerization of the azobenzene moiety [17]. Kawai and co-workers reported that the dimension of the assemblies could be tuned by solvents, and the *g* factor of circularly polarized luminescence can be promoted significantly [18]. Therefore it could be concluded that not only the supramolecular structures but also the dimension of the assembly could affect the intensity of the chirality and even the direction of the helicity [19–26].

Two-component self-assembly systems referring to two kinds of materials were involved in the assembly process, such as two types of organic molecule or organic molecule with inorganic nanoparticles et al. [27–32]. In two-component self-assembly systems, new properties or functions could appear, thus inducing many superiorities compared with the traditional one-component systems. For examples, in cyclodextrin, calixarenes, and chiral host-based host-guest systems, by the co-assembly process the chirality could be transferred from the chiral host to the achiral guests [33–43]. In the light harvesting systems, by doping the energy acceptor to the chiral donor assemblies, not only the fluorescent emission wavelength could be changed but also the utilization effect of the light could be promoted [44–46]. Moreover, by co-assembly strategy, the supramolecular chirality of the systems could be modulated, such as enhancement and inversion of the chirality or the helicity of the nanostructures [47–53].

In the two-component self-assembly systems, stoichiometric ratio of the components could be used to tune the supramolecular chirality and the nanostructure of the assemblies. For instance, Liu and co-workers control both the length and the diameter of the nanotubes only by the ratio of the gelator and melamine [54]. Yin and co-workers realized the inversion of the CPL signals by modulating the stoichiometric ratio of the chiral gelator and the achiral AIEgen lumiphores [55]. However, by regulating the stoichiometric ratio of the two-component in the co-assembly systems to realize both the dimension transition and chirality intensity control was rarely reported.

Herein, we construct a two-component supramolecualr system including a chaperone gelator LHC18 and a bola-type azobenzene carboxylic acid. Gelator LHC18 is an amphiphilic L-histidine derivative contains a long alkyl chain and a urea bond, LHC18 can self-assemble to supramolecular nanotwist in the mixed solution of DMF and H_2O by hydrogen bond and hydrophobic effect [56]. Moreover, LHC18 could form two-component gel with azobenzene derivatives by the interaction between the imidazole and carboxylic acid group and exhibit distinct morphology and chiroptical properties. By modulating the stoichiometric ratio of the two components, we will explore that weather the two-component assembly could show dimension transformation and chirality enhancement.

2. Results and Discussion

Carboxylic acid-terminated achiral azobenzene derivatives AZO as shown in Figure 1, could form hetero-hydrogen bond with imidazole moiety of LHC18, while the ratio of AZO was exceeded, AZO itself could form homo-hydrogen bond. Therefore, we suppose that by modulating the stoichiometric ratio of the two components will change the stacking mode thus affect the chirality transfer from LHC18 to AZO and induce the morphology and even dimension transition. LHC18 itself can dissolve in DMF solution, to obtain the supramolecular gel anti-solvent method was applied. 5 mg LHC18 was dissolved and heat in 500 DMF μL, the the boiled Milli-Q water was injected, after about ten minutes an opaque white supramolecular gel was formed proved by inverted tests (Figure S1) and rheological data (Figure S2), and nanotwist structure can be measured as shown in Figure 2a, the enlarged images is listed in Figure 2.

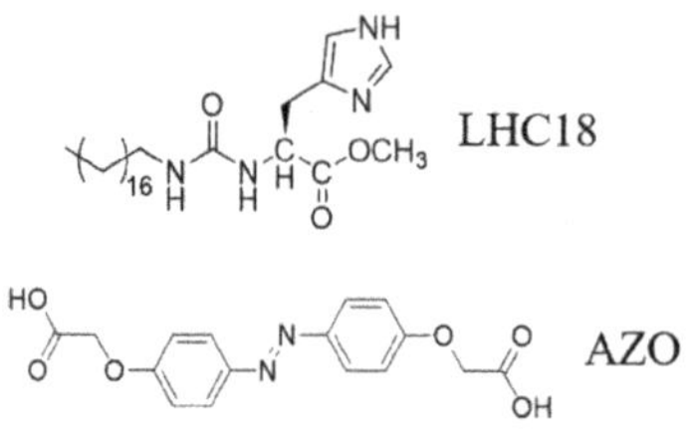

Figure 1. Molecular structure of chiral gelator LHC18 and achiral AZO.

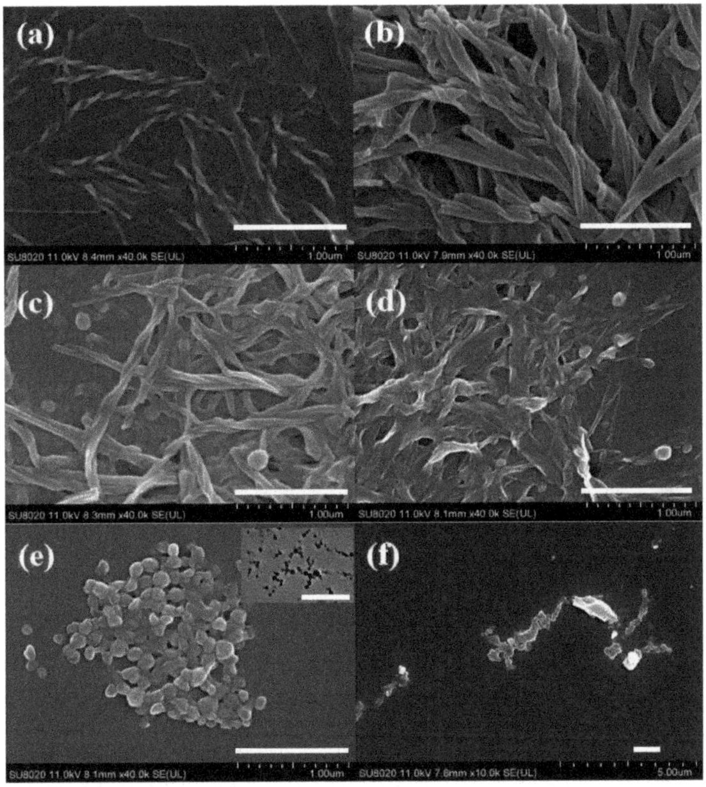

Figure 2. SEM images of (**a**) LHC18 assembly (DMF/H_2O = 500 μL/900 μL), and LHC18/AZO assemblies (DMF/H_2O = 300 μL/600 μL) at different molar ratios: (**b**) 1/0.5, (**c**) 1/1, (**d**) 1/2, (**e**) 1/4. (**f**) AZO precipitate (DMF/H_2O = 300 μL/600 μL). The inserted image was the TEM image of the assembly of 1/4. Scale bar 1 μm.

For the formation process of the co-assemblies, 5 mg LHC18 and various ratios of AZO were added into 300 μL DMF. The mixture was heated to form a transparent solution and then 600 μL boiled Milli-Q water was injected, after about 10 min supramolecular could be obtained. AZO (3.45 mg) itself cannot form ordered nanostructures in the mixed solution, it only forms some precipitates as showed in the SEM image (Figure 2f).

Next, we modulated the stoichiometric ratio of LHC18 and AZO to explore the morphology change. When LHC18 co-assembled with AZO at a molar ratio of 1/0.5, i.e., the molar ratio of the imidazole and the carboxylic acid group was 1/1, an opaque orange gel can be obtained, which was confirmed by inverted test tube experiments (Figure S1) and rheological data (Figure S3), nanotube can be observed by the SEM measurement and further proved by TEM image (Figure S4).

When the proportion of AZO was increased, some new phenomena could be found, at the molar ratio of LHC18/AZO was changed to 1/1 and 1/2, nanofibers and spheres can be observed from the SEM images (Figure 2c,d). However, as the increasing of AZO to the radio 1/4, only spheres can be found in the SEM images and the spheres are solid as proved by the TEM image, Figure 2e top right corner. It should be noted that the nanotwist of pure LHC18 is left-handed (*M*), after adding AZO to the ratio is 1/0.5, the supramolecular chirality of nanostructure was inverted to right-handed (*P*). Moreover, by observing the reagent bottle after co-assembly process, only the assembly of the ratio was 1/0.5 could form gel, the assemblies of other ratios only had the sol phase, Figure S1. The above results indicated that by modulating the stoichiometric ratio of LHC18 and AZO, not only the morphology but also the dimension could be altered.

For LHC18 was a chiral molecule, we also test the chirality transfer from LHC18 to AZO at different molar ratio. As shown in Figure 3, for LHC18 had no obvious Uv-vis absorption at 300–400 nm, the CD spectra had no signals. However, for the co-assembly of LHC18/AZO at molar ratio was 1/0.5, evident CD signals could be observed. The CD spectra showed a positive cotton effect with bisignated splitting. The positive and negative cotton effect centered at 424 nm and 354 nm with a crossover at 401 nm. The *trans*-isomer of AZO had absorption at 364 nm which could be ascribed to the π–π^* absorption band, so the excited couplings of the AZO chromophore were responsible for the CD signals. Moreover, the CD spectra showed a positive cotton effect thus indicating that the AZO moiety adopted a *P*-helicity stacking mode, i.e., the azobenzene chromophore stacked in a clockwise direction [57,58].

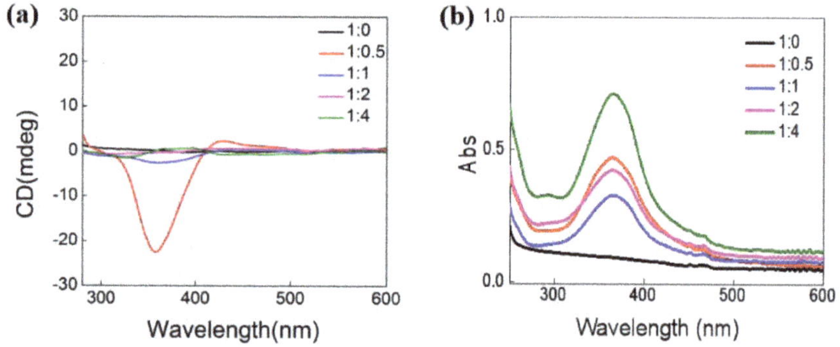

Figure 3. (**a**) CD and (**b**) UV-vis spectrum of LHC18/AZO at different molar radio, LHC18 was kept constant at 5 mg in the co-assemblies.

However, as the molar ratio of the AZO increased, the CD signals was distinctly decreased. As showed in Figure 3, only the co-assembly with molar ratio 1/1 had a weak signal, the co-assemblies of 1/2 and 1/4 nearly kept silence in the CD spectra. The CD tests suggested that only the stoichiometric ratio kept at a suitable value the chirality transfer can be effectively realized. For azobenzene moiety was light responsive which can undergo *trans-cis* isomerization, we also tested the chiroptical properties. As shown in Figure 4, the Uv-vis spectra showed that after UV 365 nm irradiation for 1 h, all the co-assembly spectra had obvious changes, after UV irradiation for one hours the assembly with molar ratio 1/0.5 had only intensity change indicating that the *trans-cis* isomerization was hampered, one AZO molecule was fixed by two LHC18 molecule thus limiting the isomerization. As the amount of AZO increased, AZO could have more flexibility thus the isomerizaiton could proceed. For the assemblies with the molar ratio was 1/1, 1/2 and 1/4, the 364 nm absorption band attributed to the π–π^* of the *trans*-isomer decreased with the emergence of the peak at 310 nm and 440 nm corresponding to the π–π^* and n–π^* bands of the *cis*-isomer, respectively. After visible light irradiation for one hour, the 364 nm absorption band of the *trans*-isomer appeared again accompanied with the vanishment of the 310 nm and 440 nm signals of the *cis*-isomer. The UV-Vis result demonstrated the azobenzene *trans-cis* transition could be triggered by UV or visible light at the molar ratio was 1/1, 1/2, and 1/4.

In order to deeply understand the intermolecular interactions, we further investigated the FT-IR spectra and XRD patterns. All the samples are prepared from either the xerogel or the air-dried suspensions. The FT-IR spectra showed some common characteristics. Firstly, there are wide vibration bands around 1950 cm^{-1} for all the LHC18/AZO samples, which can be attributed to the formation of the hydrogen bond between imidazole and carboxylic acid group. Secondly, all the FT-IR spectra showed strong vibrations at around 1620 cm^{-1}, which demonstrated a strong and ordered hydrogen bonding between the urea groups.

Finally, all the samples of the asymmetric and symmetric stretching vibrations of CH_2 appear at about 2920 cm^{-1} and 2850 cm^{-1}, respectively, indicating that the alkyl chains are in the existence of considerable gauche conformations [59–63]. The XRD patterns of all the samples are also shown in Figure 5. Based on the Bragg's equation, the *d*-spacing value of the samples can be estimated to be 3.51 nm, 5.02 nm and 5.20 nm. Since the distance 3.51 nm is larger than one but much less than twice the molecular length of the LHC18, we indicate that LHC18 forms an interdigitated bilayer structure. When the molar ratio of LHC18/AZO is 1/0.5, 1/1, and 1/2, the distance 5.02 nm and 5.20 nm is larger than one but slightly less than twice the molecular length of the LHC18, we suppose that it was the *d*-value of bilayer LHC18 and one AZO molecule. For the XRD patterns of LHC18/AZO is 1/1, 1/2, 1/4 and 0/1 systems, scattering peak intensity at $2\theta = 16.6$ (0.53 nm) gradually increased, indicating that the proportion of AZO which did not participate the co-assembly was improved. Moreover, the intensity peaks at $2\theta = 26.62$ and 27.46 (0.33 nm and 0.32 nm) also enhanced as the molar ratio of AZO increased, suggesting that the π–π stackings among the azobenzene groups appeared and increased.

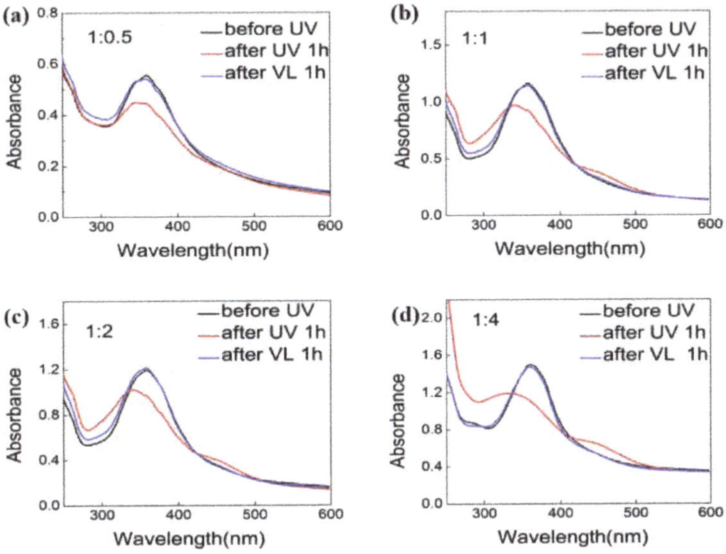

Figure 4. Uv-vis spectra of LHC18/AZO assemblies after UV 365 nm irradiation for one hour at different molar radio, (**a**) 1/0.5, (**b**) 1/1, (**c**) 1/2, and (**d**) 1/4.

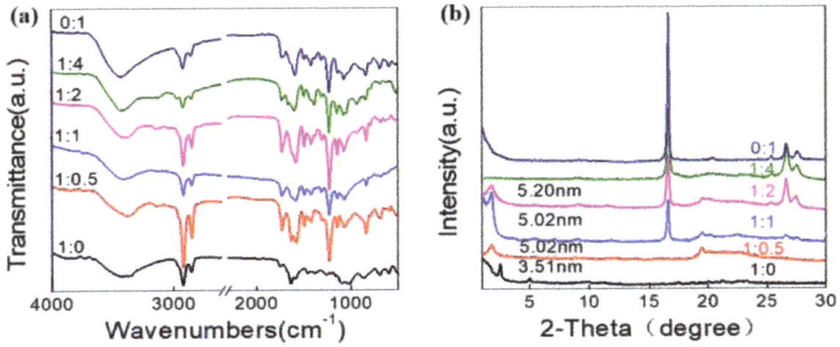

Figure 5. FT-IR spectra (**a**) and XRD patterns (**b**) of LHC18/AZO assemblies with different molar ratios.

Based on the above mentioned results, a possible mechanism to explain the dimension transition and chirality transfer is shown in Figure 6. Firstly, LHC18 itself formed bilayer structure and self-assembled into nanotwists, the hydrophobic effect and the hydrogen bonding between the urea groups is the main driving force to form bilayer structure. When LHC18 co-assembly with AZO with the ratio is 1/0.5, nanotubes could be formed. Hydrogen bond could be formed between imidazole groups of LHC18 and carboxylic acid groups of the AZO, which is verified by the FT-IR spectra. When the molar ratio is higher than 1/0.5, AZO has tendency to form π–π stacking and hydrogen bond among the carboxylic acid groups, thus influence the stacking mode and leading to the nanostructure change. When the ratio is 1/4, the π–π stacking at 0.33 nm and 0.32 nm dramatically enhanced and form nanopheres. Therefore, the π–π stacking and hydrogen bond between AZO synergistically induce the morphology and chirality change.

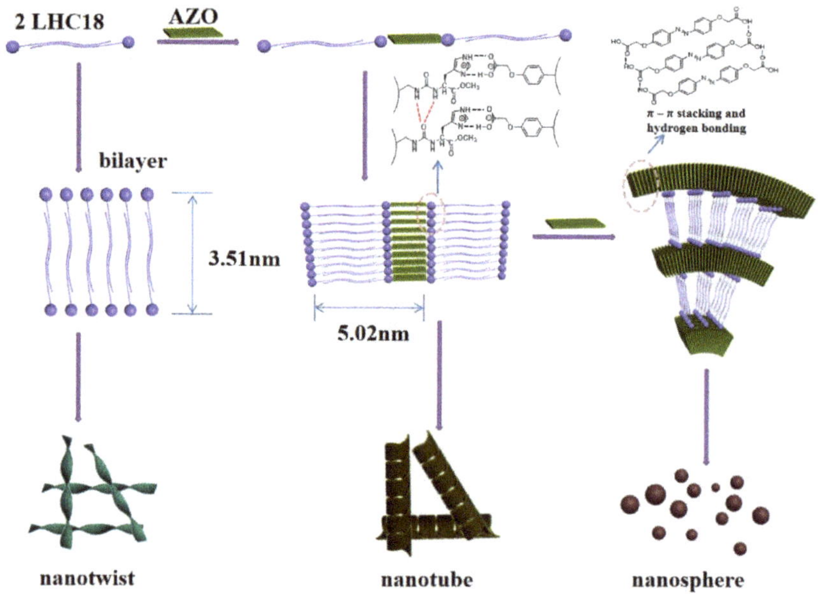

Figure 6. Illustration of the LHC18 and AZO co-assembly process. As the molar ratio of AZO increased, the chirality transfer was weakened, and the dimension transited from 1D nanotube to 0D nanosphere.

3. Conclusions

In conclusion, we construct the two-component self-assembly system. By modulating the stoichiometric ratio of the components of chaperone gelator LHC18 and AZO derivatives, both the dimension transition and chirality transfer could be modulated. It was proved that the π–π stacking and hydrogen bond among AZO was the main reason for the assembly dimension transition and chirality transfer altering. We believe our work will provide some insights and approach into the dimension and chirality control in the self-assembly systems.

4. Materials and Methods

All the starting materials and solvents were obtained from commercial suppliers and used as received. The synthesis route and the characterization were in reference [60] A typical procedure for the assembly formation in mixed solution was as follows: 5.0 mg of LHC18 and 1.8 mg AZO were mixed in 300 µL DMF solution, heat to form a transparent solution, and then injected 600 µL boiled Milli-Q water, the mix solution was cooled to room temperature naturally. UV−Vis spectra were recorded in quartz curvettes (light path 0.1 mm) on a JASCO UV-550 spectrometer.

Circular Dichroism spectra were recorded in quartz cuvettes (light path 0.1 mm) on a JASCO J-810 spectrophotometer. The samples were prepared and cast on quartz plates. FT-IR spectra were recorded on a Bruker Tensor 27 FTIR spectrometer at room temperature. The KBr pellets made from the vacuum-dried samples were used for FT-IR spectra measurements. Scanning Electron Microscopy were cast onto single-crystal silica plates, the solvent was evaporated under the ambient conditions, and then vacuum-dried. The sample surface was coated with a thin layer of Pt to increase the contrast. SEM images were recorded on a Hitachi SU-8020 SEM instrument with an accelerating voltage of 11 kV. Transmission electron microscopy images were obtained on a JEM-1011 electron microscope at an accelerating voltage of 100 kV. The TEM samples were prepared by casting a small amount of sample on carbon-coated copper grids (300 mesh) and dried under strong vacuum. X-ray Diffraction (XRD) analysis was performed on a Rigaku D/Max-2500 X-ray diffract meter (Japan) with Cu Kα radiation (λ = 1.5406 Å), which was operated at a voltage of 40 kV and a current of 200 mA. Samples were cast on glass substrates and vacuum-dried for XRD measurements. Rheology Study were measured on a strain-controlled rheometer (MC-RhR 302, Anton Paar) using cone-plate geometry (25 mm diameter). The experiments were performed at 25 $\pm$ 0.05 °C, and the temperature was controlled with an integrated electrical heater.

Supplementary Materials: The following supporting information can be downloaded at: https://www.mdpi.com/article/10.3390/gels8050269/s1, Figure S1: Phase image of the LHC18/AZO assemblies with different molar ratios. Figure S2: Rheological data for the LHC18 gel [64]. Figure S3: Rheological data for the LHC18/AZO gel. Figure S4: TEM image of the assembly of LHC18/AZO. Figure S5: CD spectra of the gel LHC18/AZO (molar ratio 1/0.5) with rotating the cuvette 90°. Figure S6: The SEM image of the LHC18/AZO assemblies under UV irradiation.

Author Contributions: Conceptualization, L.J. and Y.W.; investigation, P.Z. and Y.L.; data curation, P.Z., Y.L. and X.F.; writing—original draft preparation, P.Z., Y.L. and X.F.; writing—review and editing, L.J., L.M. and Y.W.; visualization, L.J., L.M. and Y.W.; supervision, L.J., L.M. and Y.W.; project administration, L.J., L.M. and Y.W.; funding acquisition, L.J., L.M. and Y.W. All authors have read and agreed to the published version of the manuscript.

Funding: This research was supported by the program of the National Natural Science Foundation of China Grant No. 22002034, the Natural Science Foundation of Hebei Province Grant No. B2021205014, B2021205024 Science and Technology Project of Hebei Education Department Grant No. BJ2021027, QN2022006, Science Foundation of Hebei Normal University Grant No. L2020B06, L2020B11, and Excellent Young Teacher Fundamental Research Grant No. YQ2020009.

Institutional Review Board Statement: Not applicable.

Informed Consent Statement: Not applicable.

Data Availability Statement: Not applicable.

Conflicts of Interest: The authors declare no conflict of interest.

References

1. Zhou, J.H.; Yang, Y.S.; Yang, Y.; Kim, D.S.; Yuan, A.; Tian, X.; Ophus, C.; Sun, F.; Schmid, A.K.; Nathanson, M.; et al. Observing crystal nucleation in four dimensions using atomic electron tomography. *Nature* **2019**, *570*, 500–503. [CrossRef] [PubMed]
2. Tai, J.S.B.; Smalyukh, I.I. Three-dimensional crystals of adaptive knots. *Science* **2019**, *365*, 1449–1453. [CrossRef] [PubMed]
3. Katan, C.; Mercier, N.; Even, J. Quantum and dielectric confinement effects in lower-dimensional hybrid perovskite semiconductors. *Chem. Rev.* **2019**, *119*, 3140–3192. [CrossRef]
4. Huo, D.; Kim, M.J.; Lyu, Z.H.; Shi, Y.F.; Wiley, B.J.; Xia, Y.N. One dimensional metal nanostructures: From colloidal syntheses to applications. *Chem. Rev.* **2019**, *119*, 8972–9073. [CrossRef] [PubMed]
5. Chen, L.F.; Su, B.; Jiang, L. Recent advances in one-dimensional assembly of nanoparticles. *Chem. Soc. Rev.* **2019**, *48*, 8–21. [CrossRef] [PubMed]

6. Zhao, Y.F.; Waterhouse, G.I.N.; Chen, G.B.; Xiong, X.Y.; Wu, L.Z.; Tung, C.H.; Zhang, T.R. Two-dimensional-related catalytic materials for solar-driven conversion of cox into valuable chemical feedstocks. *Chem. Soc. Rev.* **2019**, *48*, 1972–2010. [CrossRef] [PubMed]
7. Selkoe, D.J. Alzheimer's disease: Genes, proteins, and therapy. *Physiol. Rev.* **2001**, *81*, 741–766. [CrossRef] [PubMed]
8. Zong, L.; Li, M.J.; Li, C.X. Intensifying solar-thermal harvest of low-dimension biologic nanostructures for electric power and solar desalination. *Nano Energy* **2018**, *50*, 308–315. [CrossRef]
9. Brouhard, G.J.; Rice, L.M. Microtubule dynamics: An interplay of biochemistry and mechanics. *Nat. Rev. Mol. Cell Biol.* **2018**, *19*, 451–463. [CrossRef]
10. Amabilino, D.B.; Smith, D.K.; Steed, J.W. Supramolecular materials. *Chem. Soc. Rev.* **2017**, *46*, 2404–2420. [CrossRef]
11. Yashima, E.; Ousaka, N.; Taura, D.; Shimomura, K.; Ikai, T.; Maeda, K. Supramolecular helical systems: Helical assemblies of small molecules, foldamers, and polymers with chiral amplification and their functions. *Chem. Rev.* **2016**, *116*, 13752–13990. [CrossRef] [PubMed]
12. Yoshii, T.; Onogi, S.; Shigemitsu, H.; Hamachi, I. Chemically reactive supramolecular hydrogel coupled with a signal amplification system for enhanced analyte sensitivity. *J. Am. Chem. Soc.* **2015**, *137*, 3360–3365. [CrossRef] [PubMed]
13. Zhang, L.; Wang, X.F.; Wang, T.Y.; Liu, M.H. Tuning soft nanostructures in self-assembled supramolecular gels: From morphology control to morphology-dependent functions. *Small* **2015**, *11*, 1025–1038. [CrossRef] [PubMed]
14. Liu, M.H.; Zhang, L.; Wang, T.Y. Supramolecular chirality in self-assembled systems. *Chem. Rev.* **2015**, *115*, 7304–7397. [CrossRef]
15. Xu, Y.Y.; Hao, A.Y.; Xing, P.Y. X···X halogen bond-induced supramolecular helices. *Angew. Chem. Int. Ed.* **2021**, *61*, e202113786.
16. Tsai, M.F.; Chang, S.H.G.; Cheng, F.Y.; Shanmugam, V.; Cheng, Y.S.; Su, C.H.; Yeh, C.S. Au nanorod design as light-absorber in the first and second biological near-infrared windows for in vivo photothermal therapy. *ACS Nano* **2013**, *7*, 5330–5342. [CrossRef]
17. Liu, G.F.; Sheng, J.H.; Teo, W.L.; Yang, G.B.; Wu, H.W.; Li, Y.X.; Zhao, Y.L. Control on dimensions and supramolecular chirality of self-assemblies through light and metal Ions. *J. Am. Chem. Soc.* **2018**, *140*, 16275–16283. [CrossRef]
18. Kumar, J.; Nakashima, T.; Tsumatori, H.; Kawai, T. Circularly polarized luminescence in chiral aggregates: Dependence of morphology on luminescence dissymmetry. *J. Phys. Chem. Lett.* **2014**, *5*, 316–321. [CrossRef]
19. Zhou, P.; Shi, R.F.; Yao, J.F.; Sheng, C.F.; Li, H. Supramolecular self-assembly of nucleotide–metal coordination complexes: From simple molecules to nanomaterials. *Coord. Chem. Rev.* **2015**, *292*, 107–143. [CrossRef]
20. Xing, P.Y.; Zhao, Y.L. Controlling supramolecular chirality in multicomponent self-assembled systems. *Acc. Chem. Res.* **2018**, *51*, 2324–2334. [CrossRef]
21. Liu, Y.Q.; Chen, C.F.; Wang, T.Y.; Liu, M.H. Supramolecular chirality of the two-component supramolecular copolymer gels: Who determines the handedness? *Langmuir* **2016**, *32*, 322–328. [CrossRef] [PubMed]
22. Yang, Y.; Zhang, Y.J.; Wei, Z.X. Supramolecular helices: Chirality transfer from conjugated molecules to structures. *Adv. Mater.* **2013**, *25*, 6039–6049. [CrossRef] [PubMed]
23. Huang, S.; Yu, H.F.; Li, Q. Supramolecular chirality transfer toward chiral aggregation: Asymmetric hierarchical self-assembly. *Adv. Sci.* **2021**, *8*, 2002132. [CrossRef] [PubMed]
24. Wang, Z.E.; Li, Y.Z.; Hao, A.Y.; Xing, P.Y. Multi-modal chiral superstructures in self-assembled anthracene-terminal amino acids with predictable and adjustable chiroptical activities and color evolution. *Angew. Chem. Int. Ed.* **2021**, *60*, 3138–3147. [CrossRef]
25. Wu, A.L.; Guo, Y.X.; Li, X.B.; Xue, H.M.; Fei, J.B.; Li, J.B. Co-assembled supramolecular gel of dipeptide and pyridine derivatives with controlled chirality. *Angew. Chem. Int. Ed.* **2021**, *60*, 2099–2103. [CrossRef] [PubMed]
26. Adawy, A. Functional chirality: From small molecules to supramolecular assemblies. *Symmetry* **2022**, *14*, 292. [CrossRef]
27. Colquhoun, C.; Draper, E.R.; Eden, E.G.B.; Cattoz, B.N.; Morris, K.L.; Chen, L.; McDonald, T.O.; Terry, A.E.; Griffiths, P.C.; Serpell, L.C.; et al. The effect of self-sorting and Co-assembly on the mechanical properties of low molecular weight hydrogels. *Nanoscale* **2014**, *6*, 13719–13725. [CrossRef]
28. Tena-Solsona, M.; Alonso-de Castro, S.; Miravet, J.F.; Escuder, B. Co-assembly of tetrapeptides into complex pH-responsive molecular hydrogel networks. *J. Mater. Chem. B* **2014**, *2*, 6192–6197. [CrossRef]
29. Das, P.; Yuran, S.; Yan, J.; Lee, P.S.; Reches, M. Sticky tubes and magnetic hydrogels co-assembled by a short peptide and melanin-like nanoparticles. *Chem. Commun.* **2015**, *51*, 5432–5435. [CrossRef]
30. Matsumoto, K.; Shundo, A.; Ohno, M.; Saruhashi, K.; Miyachi, N.; Tsuruzoe, N.; Tanaka, K. Sol-gel transition accelerated by the co-assembly of two components in supramolecular hydrogels. *Phys. Chem. Chem. Phys.* **2015**, *17*, 26724–26730. [CrossRef]
31. Khatua, D.; Maiti, R.; Dey, J. A supramolecular hydrogel that responds to biologically relevant stimuli. *Chem. Commun.* **2006**, *47*, 4903–4905. [CrossRef] [PubMed]
32. Liao, X.J.; Chen, G.S.; Liu, X.X.; Chen, W.X.; Chen, F.; Jiang, M. Photoresponsive pseudopolyrotaxane hydrogels based on competition of host−guest interactions. *Angew. Chem. Int. Ed.* **2010**, *49*, 4409–4413. [CrossRef] [PubMed]
33. Yang, C.C.; Chen, W.J.; Zhu, X.F.; Song, X.; Liu, M.H. Self-assembly and circularly polarized luminescence from achiral pyrene-adamantane conjugates by selective inclusion with cyclodextrins. *J. Phys. Chem. Lett.* **2021**, *12*, 7491–7496. [CrossRef] [PubMed]
34. Palmans, A.R.A.; Meijer, E.W. Amplification of chirality in dynamic supramolecular aggregates. *Angew. Chem. Int. Ed.* **2007**, *46*, 8948–8968. [CrossRef]
35. Dijken, D.J.; Beierle, J.M.; Stuart, M.C.A.; Szymański, W.; Browne, W.R.; Feringa, B.L. Autoamplification of molecular chirality through the induction of supramolecular chirality. *Angew. Chem. Int. Ed.* **2014**, *53*, 5073–5077. [CrossRef] [PubMed]

36. Liu, G.F.; Zhu, L.Y.; Ji, W.; Feng, C.L.; Wei, Z.X. Inversion of the supramolecular chirality of nanofibrous structures through Co-assembly with achiral molecules. *Angew. Chem. Int. Ed.* **2016**, *55*, 2411–2415. [CrossRef] [PubMed]
37. Gaeta, M.; Raciti, D.; Randazzo, R.; Gangemi, C.M.A.; Raudino, A.; D'Urso, A.; Fragalà, M.E.; Purrello, R. Chirality enhancement of porphyrin supramolecular assembly driven by a template preorganization effect. *Angew. Chem. Int. Ed.* **2018**, *57*, 10656–10660. [CrossRef] [PubMed]
38. Guo, D.S.; Chen, K.; Zhang, H.Q.; Liu, Y. Nano-supramolecular assemblies constructed from water-soluble bis(calix[5]arenes) with porphyrins and their photoinduced electron transfer properties. *Chem. Asian J.* **2009**, *4*, 436–445. [CrossRef] [PubMed]
39. Gaeta, M.; Sortino, G.; Randazzo, R.; Pisagatti, I.; Notti, A.; Fragalà, M.E.; Parisi, M.F.; D'Urso, A.; Purrello, R. Long-range chiral induction by a fully noncovalent approach in supramolecular porphyrin–calixarene assemblies. *Chem. Eur. J.* **2020**, *26*, 3515–3518. [CrossRef] [PubMed]
40. Mateos-Timoneda, M.A.; Crego-Calama, M.; Reinhoudt, D.N. Supramolecular chirality of self-assembled systems in solution. *Chem. Soc. Rev.* **2004**, *33*, 363–372. [CrossRef] [PubMed]
41. Gattuso, G.; Notti, A.; Pappalardo, S.; Parisi, M.; Pisagatti, I. Recognition in water of bioactive substrates by a sulphonato p-tert-butylcalix[5]arene. *Supramol. Chem.* **2014**, *26*, 597–600. [CrossRef]
42. Borovkov, V.V.; Lintuluoto, J.M.; Sugiura, M.; Inoue, Y.; Kuroda, R. Remarkable stability and enhanced optical activity of a chiral supramolecular bis-porphyrin tweezer in both solution and solid state. *J. Am. Chem. Soc.* **2002**, *124*, 11282–11283. [CrossRef] [PubMed]
43. D'Urso, A.; Marino, N.; Gaeta, M.; Rizzo, M.S.; Cristaldi, D.A.; Fragalà, M.E.; Pappalardo, S.; Gattuso, G.; Notti, A.; Parisi, M.F.; et al. Porphyrin stacks as an efficient molecular glue to induce chirality in hetero-component calixarene–porphyrin assemblies. *New J. Chem.* **2017**, *41*, 8078–8083. [CrossRef]
44. Ji, L.K.; Sang, Y.T.; Ouyang, G.H.; Yang, D.; Duan, P.F.; Jiang, Y.Q.; Liu, M.H. Cooperative chirality and sequential energy transfer in a supramolecular light-harvesting nanotube. *Angew. Chem. Int. Ed.* **2019**, *58*, 844–848. [CrossRef] [PubMed]
45. Kameta, N.; Ishikawa, K.; Masuda, M.; Asakawa, M.; Shimizu, T. Soft nanotubes acting as a light-harvesting antenna system. *Chem. Mater.* **2012**, *24*, 209–214. [CrossRef]
46. Cao, Z.Z.; Hao, A.Y.; Xing, P.Y. Photoresponsive chiral vesicles as a light harvesting matrix with tunable chiroptical properties. *Nanoscale* **2021**, *13*, 700–707. [CrossRef] [PubMed]
47. Yang, D.; Duan, P.F.; Zhang, L.; Liu, M.H. Chirality and energy transfer amplified circularly polarized luminescence in composite nanohelix. *Nat. Commun.* **2017**, *8*, 15727. [CrossRef] [PubMed]
48. Zhao, T.H.; Han, J.L.; Duan, P.F.; Liu, M.H. New perspectives to trigger and modulate circularly polarized luminescence of complex and aggregated systems: Energy transfer, photon upconversion, charge transfer, and organic radical. *Acc. Chem. Res.* **2020**, *53*, 1279–1292. [CrossRef]
49. Cao, Z.Z.; Wang, B.; Zhu, F.; Hao, A.Y.; Xing, P.Y. Solvent-processed circularly polarized luminescence in light-harvesting coassemblies. *ACS Appl. Mater. Interfaces* **2020**, *12*, 34470–34478. [CrossRef]
50. Li, M.; Zhang, C.; Fang, L.; Shi, L.; Tang, Z.Y.; Lu, H.Y.; Chen, C.F. Chiral nanoparticles with full-color and white CPL properties based on optically stable helical aromatic imide enantiomers. *ACS Appl. Mater. Interfaces* **2018**, *10*, 8225–8230. [CrossRef]
51. Nishikawa, T.; Nagata, Y.; Suginome, M. Poly(quinoxaline-2,3-diyl) as a multifunctional chiral scaffold for circularly polarized luminescent materials: Color tuning, energy transfer, and switching of the CPL handedness. *ACS Macro Lett.* **2017**, *6*, 431–435. [CrossRef]
52. Niu, D.; Ji, L.K.; Ouyang, G.H.; Liu, M.H. Histidine proton shuttle-initiated switchable inversion of circularly polarized luminescence. *ACS Appl. Mater. Interfaces* **2020**, *12*, 18148–18156. [CrossRef] [PubMed]
53. Amako, T.; Nakabayashi, K.; Mori, T.; Inoue, Y.; Fujiki, M.; Imai, Y. Sign inversion of circularly polarized luminescence by geometry manipulation of four naphthalene units introduced into a tartaric acid scaffold. *Chem. Commun.* **2014**, *50*, 12836–128399. [CrossRef] [PubMed]
54. Shen, Z.C.; Wang, T.Y.; Liu, M.H. H-bond and π–π stacking directed self-assembly of two-component supramolecular nanotubes: Tuning length, diameter and wall thickness. *Chem. Commun.* **2014**, *50*, 2096–2099. [CrossRef] [PubMed]
55. Qin, P.H.; Wu, Z.; Li, P.Y.; Niu, D.; Liu, M.H.; Yin, M.Z. Triple-modulated chiral inversion of Co-assembly system based on alanine amphiphile and cyanostilbene derivative. *ACS Appl. Mater. Interfaces* **2021**, *13*, 18047–18055. [CrossRef]
56. Chen, J.; Wang, T.Y.; Liu, M.H. Chaperone gelator for the chiral self-assembly of all proteinogenic amino acids and their enantiomers. *Chem. Commun.* **2016**, *52*, 6123–6126. [CrossRef]
57. Berova, N.; Di Bari, L.; Pescitelli, G. Application of electronic circular dichroism in configurational and conformational analysis of organic compounds. *Chem. Soc. Rev.* **2007**, *36*, 914–931. [CrossRef]
58. Cao, H.; Jiang, J.; Zhu, X.F.; Duan, P.F.; Liu, M.H. Hierarchical Co-assembly of chiral lipid nanotubes with an azobenzene derivative: Optical and chiroptical switching. *Soft Matter* **2011**, *7*, 4654. [CrossRef]
59. Duan, P.F.; Li, Y.G.; Li, L.C.; Deng, J.G.; Liu, M.H. Multiresponsive chiroptical switch of an azobenzene-containing lipid: Solvent, temperature, and photoregulated supramolecular chirality. *J. Phys. Chem. B* **2011**, *115*, 3322–3329. [CrossRef] [PubMed]
60. Ji, L.K.; Ouyang, G.H.; Liu, M.H. Binary supramolecular gel of achiral azobenzene with a chaperone gelator: Chirality transfer, tuned morphology, and chiroptical property. *Langmuir* **2017**, *33*, 12419–12426. [CrossRef] [PubMed]

61. Ang, T.P.; Wee, T.S.A.; Chin, W.S. Three-dimensional self-assembled monolayer (3D SAM) of n-alkanethiols on copper nanoclusters. *J. Phys. Chem. B* **2004**, *108*, 11001–11010. [CrossRef]
62. Shi, H.F.; Ying, Z.; Zhang, X.Q.; Yong, Z.; Xu, Y.Z.; Zhou, S.R.; Wang, D.J.; Han, C.C.; Xu, D.F. Packing mode and conformational transition of alkyl side chains in N-alkylated poly(p-benzamide) comb-like polymer. *Polymer* **2014**, *45*, 6299–6307. [CrossRef]
63. Zhang, Z.J.; Verma, A.L.; Yoneyama, M.; Nakashima, K.; Iriyama, K.; Ozaki, Y. Molecular orientation and aggregation in langmuir-blodgett films of 5-(4-*N*-Octadecylpyridyl)-10,15,20-tri-p-tolylporphyrin studied by ultraviolet-visible and infrared spectroscopies. *Langmuir* **1997**, *13*, 4422–4427. [CrossRef]
64. Li, H.; Duan, Z.; Yang, Y.; Xu, F.; Chen, M.; Liang, T.; Bai, Y.; Li, R. Regulable Aggregation-Induced Emission Supramolecular Polymer and Gel Based on Self-sorting Assembly. *Macromolecules* **2020**, *53*, 4255–4263. [CrossRef]

gels

Article

2D Materials (WS_2, MoS_2, $MoSe_2$) Enhanced Polyacrylamide Gels for Multifunctional Applications

Bengü Özuğur Uysal [1,*], Şeyma Nayır [1,2], Melike Açba [1], Betül Çıtır [1], Sümeyye Durmaz [1], Şevval Koçoğlu [1], Ekrem Yıldız [1] and Önder Pekcan [1]

[1] Faculty of Engineering and Natural Sciences, Kadir Has University, Cibali, Fatih, Istanbul 34083, Turkey; nayirs@itu.edu.tr (Ş.N.); 20161704046@stu.khas.edu.tr (M.A.); 20161704081@stu.khas.edu.tr (B.Ç.); 20161709007@stu.khas.edu.tr (S.D.); 20161704072@stu.khas.edu.tr (Ş.K.); 20161701051@stu.khas.edu.tr (E.Y.); pekcan@khas.edu.tr (Ö.P.)

[2] Faculty of Science and Letters, Istanbul Technical University, Maslak, Istanbul 34469, Turkey

* Correspondence: bozugur@khas.edu.tr; Tel.: +90-2125336532 (ext. 1345)

Abstract: Multifunctional polymer composite gels have attracted attention because of their high thermal stability, conductivity, mechanical properties, and fast optical response. To enable the simultaneous incorporation of all these different functions into composite gels, the best doping material alternatives are two-dimensional (2D) materials, especially transition metal dichalcogenides (TMD), which have been used in so many applications recently, such as energy storage units, opto-electronic devices and catalysis. They have the capacity to regulate optical, electronic and mechanical properties of basic molecular hydrogels when incorporated into them. In this study, 2D materials (WS_2, MoS_2 and $MoSe_2$)-doped polyacrylamide (PAAm) gels were prepared via the free radical crosslinking copolymerization technique at room temperature. The gelation process and amount of the gels were investigated depending on the optical properties and band gap energies. Band gap energies of composite gels containing different amounts of TMD were calculated and found to be in the range of 2.48–2.84 eV, which is the characteristic band gap energy range of promising semiconductors. Our results revealed that the microgel growth mechanism and gel point of PAAm composite incorporated with 2D materials can be significantly tailored by the amount of 2D materials. Furthermore, tunable band gap energies of these composite gels are crucial for many applications such as biosensors, cartilage repair, drug delivery, tissue regeneration, wound dressing. Therefore, our study will contribute to the understanding of the correlation between the optical and electronic properties of such composite gels and will help to increase the usage areas so as to obtain multifunctional composite gels.

Keywords: TMDs; gelation; optical properties; polyacrylamide; multifunctional composite gels

Citation: Özuğur Uysal, B.; Nayır, Ş.; Açba, M.; Çıtır, B.; Durmaz, S.; Koçoğlu, Ş.; Yıldız, E.; Pekcan, Ö. 2D Materials (WS_2, MoS_2, $MoSe_2$) Enhanced Polyacrylamide Gels for Multifunctional Applications. *Gels* **2022**, *8*, 465. https://doi.org/10.3390/gels8080465

Academic Editor: Hiroyuki Takeno

Received: 8 June 2022
Accepted: 7 July 2022
Published: 25 July 2022

1. Introduction

Multifunctional polymer composite gels have been developed to address a wide variety of applications, from absorption-dominated electromagnetic-interference shielding [1] to tissue engineering [2], and even catalysts and water purification [3] due to the synergistic effect between polymer and new generation additives. To achieve this multifunctional effect, 2D materials are one of the best materials that can be integrated into polymer gels.

Among the family of 2D materials, graphene, and transition metal dichalcogenides (disulphides and selenides of molybdenum and tungsten, etc.) have started to be used in the production of opto-electronic devices and catalysis, and also, they have attracted attention in the field of energy storage units as nano-filler materials [4,5]. Besides the other transition metal dichalcogenide (TMD) materials, molybdenum disulphide (MoS_2) is the most important one because of its superior electronic behavior and mechanical properties [6,7]. Another representative of this two-dimensional material group is Tungsten disulphide (WS_2), which can be used for advanced applications such as photovoltaic and photocatalysis due to its size dependent tunable band gap energy. Both MoS_2 and WS_2

are classified as semiconductor materials because they both have band gap energies of about 1.2 eV. The indirect bandgap of $MoSe_2$ in its bulk form has a value of 1.1 eV. A direct band gap of 1.5 eV can be obtained by exfoliating $MoSe_2$ into a few layers [8]. The band gap energy of as-grown MoS_2 flakes from chemical vapor deposition can be modulated from 1.86 eV to 1.57 eV [9]. Furthermore, it is reported that the band gap energies of TMDs are in the range of 1.5–1.8 eV [10–12]. However, the magnitude of this band gap energy is insufficient for many electronic applications [13,14]. In fact, the minimum band gap energy required for technological applications must be greater than 1.8 eV. The band gap energy of 1.8 eV is the limit value that allows one to turn between on and off states through the conduction of MoS_2 [15]. Two-dimensional MoS_2 possesses this intrinsic band gap energy that makes it ideal for applications in electronics, optoelectronics and biosensors [16]. Therefore, it is necessary to prepare composites of different materials to provide the desired band gap values of these 2D structures [17].

In addition, it has been thought that the molybdenum dichalcogenides (MoS_2 and $MoSe_2$) have the capacity to increase and regulate mechanical properties when incorporated into basic molecular hydrogels [18,19]. On the other hand, the most common hydrogel is polyacrylamide (PAAm), which is affordable; it can be formed into desired shapes and has the flexibility to match biological materials [20]. It has many realized and potential applications including drug delivery systems and chemo-mechanical devices. PAAm applications have recently moved their focus to using them to create innovative polymer systems with unique structure and functionalities, because PAAm can be chemically infused with other elements or compounds to use in applications of wound dressing, biosensors, drug delivery, tissue regeneration and cartilage repair [21]. The copolymerization process of hydrogel composites made of PAAm and various reinforcing additives is also important for its clinical uses in cell biology and drug delivery applications [22]. Encapsulation of nanoparticles such as silicon, carbon nanotubes, gelatine, cellulose, and other materials improves the strength, bonding, and self-healing properties of the polyacrylamide hydrogel [23–28]. PAAm as promising hosting organic matrix for composite materials not only provides mechanical stability, but also new functionalities after the incorporation of conducting materials with different structures.

Similarly, previous studies [29–32] have shown that different nano-filler materials (carbon nanotube and graphene oxide)-doped polymers such as polyacrylamides (PAAm), latex, polystyrene have also manifested better mechanical and electrical properties. In this work, the amount of dependent electronic and optical properties of MoS_2, $MoSe_2$, WS_2-doped PAAm was investigated. It is normally accepted that the composite hydrogels can effectively connect the unusual properties of the inorganic (TMD) and organic (PAAm) components to get the desired properties for multifunctional materials.

2. Results and Discussion

2.1. Gelation Process and Drying of Composite Gels

The gelation process of PAAm takes place very quickly. For this reason, a part of the solution (approximately 3 mL) in the beaker was poured into the quartz cuvette immediately after N,N,N′,N′-Tetramethylethylenediamine (TEMED), which is used as an essential catalyst, was added to the solution and quickly placed in the spectrophotometer cabinet. As a result, composite gels with different TMD amount have the form of a square prism and disc shape as presented in Figure 1.

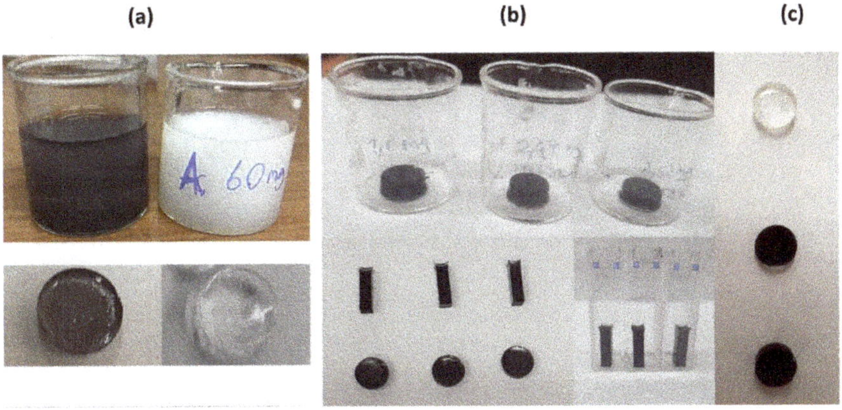

Figure 1. (**a**) **Above:** $MoSe_2$-doped PAAm and PAAm just after the gelation. **Below:** The same gels after drying in air. (**b**) **Above:** Disc-shaped samples after drying in air. From left to right, MoS_2, WS_2, $MoSe_2$-doped PAAm. **Below:** Dry gels in quartz cuvette, and disk-shaped by drying inside the beaker. (**c**) Disk-shaped gels. PAAm at the top, PAAm + 7.5 mg WS_2 in the middle, PAAm + 10 mg WS_2 at the bottom.

The AAm solution, which is white at first, loses its opacity and becomes transparent while it is drying. When the effect of drying on the optical clarity of the gels was examined, it was observed that all the gels subjected to the drying process became more transparent. This characteristic of the gel indicates that there are two different regions as polyacrylamide-rich and aqueous inside the gel. Thus, the gels scatter light vigorously as a result of the fluctuations in refractive index of these two regions. This leads to the decrease in transmission of the gel. The gel is opaque. After drying, water content and in other words, the number of aqueous regions in the gel are decreasing. Therefore, the refractive indices of all regions are well matched. There are no significant fluctuations. As a result, gels become transparent. Figure 1a depicts the white AAm solution in the beaker (above right) and its dried gel (below right). One can see that the gel was opaque and white; after drying in air, it became hard, transparent, and shrunk in size. TMD-doped PAAm gels have a grey color and after drying, they have a dark grey color due to the decrease in the water content. Figure 1b shows MoS_2, WS_2, $MoSe_2$-enhanced PAAm dried gels with different geometrical shapes, from left to right, respectively. For the added TMDs to be compared with each other, they were included in the solution with masses corresponding to the equimolar ratio (the masses are seen on the beakers in the upper figure). After the acrylamide was completely dissolved, TMD was added. A homogeneous solution could not be obtained, since TMDs were gathered and agglomerated in the middle of the gel. Before the insertion of TMD, Polyvinylpyrrolidone (PVP), which is an organic adhesive, was added to the solution so that homogenous and evenly distributed TMD inside the gel was observed.

Many experiments have been carried out in different studies on the interaction of different doping materials with PVP [33–36] and it has been observed that the polymer composites formed reach their most homogeneous state by adding PVP. This would be due to experimental evidence showing that the use of PVP can significantly speed up the exfoliation of TMDs in composite solutions. This results in a homogenous solution and limits the amount of surface imperfections [37,38]. Figure 1c shows PAAm composite gels containing different amounts of TMD additives.

2.2. Optical and Electronic Properties

To describe and understand the gelation process, one can use the following kinetic equation showing the relationship between the rate of monomer consumption, which is the derivative of the concentration of monomer, $[M]$ with respect to time, t, and the rate of the polymerization [39].

$$d[M]/dt = -k_r[M] \tag{1}$$

Here, k_r is the constant rate of gelation. If one needs to solve this equation, it yields

$$[M] = [M_0]\exp(-k_r t) \tag{2}$$

where $[M_0]$ is the concentration of monomer at $t = 0$. During the gelation process, describing the phenomena, Equation (2) is needed. According to Equation (2), the concentration of monomer is decreasing exponentially with time, since the consumption of monomers is creating the microgels that lead to the turbidity. On the other hand, spectroscopic measurements of transmitted light through the gels also allow one to monitor the gelation process. The turbidity in the medium is decreasing and causes a decrease exponentially in the intensity of photon transmission, I_{tr}.

$$I_{tr} = I_0 \exp(-k_r t) \tag{3}$$

where I_0 is the intensity of the incident photon at $t = 0$ [40]. The ratio of I_{tr} and I_0 yields the transmittance of gels.

A slight difference between the light transmission abilities of the PAAm gels doped with different TMDs is observed in Figure 1b. Spectrophotometric measurements are required to distinguish them in terms of light transmittance. After pouring the solution into the quartz cuvette, the transmittance percentages of the solutions before, during, and after gelation were measured with a uv-vis spectrophotometer.

The transmittance of the TMD-doped PAAm composites are presented in Figure 2. It is seen that the intensity decreased dramatically above a certain time indicating that opalescence occurs during gelation for each composite gels.

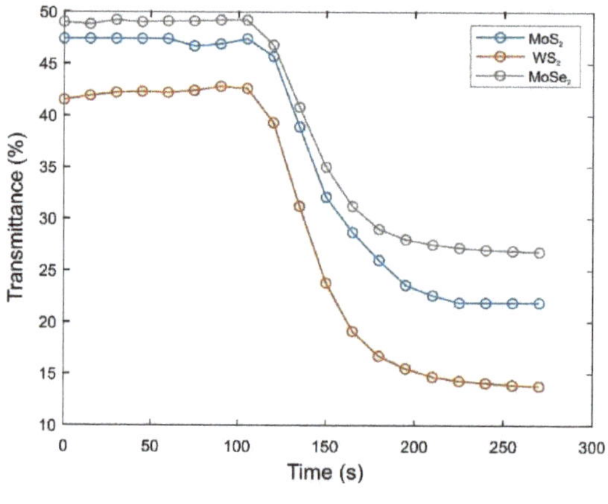

Figure 2. Transmittance versus gelation time graph of 2D materials-doped PAAm gels.

The statistical models have been shown to approximate the results obtained by Tanaka and new effects derived from the Flory–Stockmayer theory pointing to the solution of the site–link correlated percolation problem [41–44]. According to the assumptions, before and after the sol–gel phase transition points, the trends can be different. The curve resembles a sigmoidal and it fits the Boltzmann Sigmoidal Function [45–48]. These curves were used to determine gel points. According to the nature of them, the second derivative of the transmittance with respect to time must be zero around the gel point. The second derivative of the transmittance with time is shown in Figure 3. After 120 s, the curves change the trend from negative to positive. Hence, the gel point of the composite gels around 120 s.

This result is important for hydrogel studies accompanying all biological target materials, since the delivery of hydrogels to the body must occur while they are in liquid form (just at the right time). Afterwards, the hydrogel should turn into a gel inside the body.

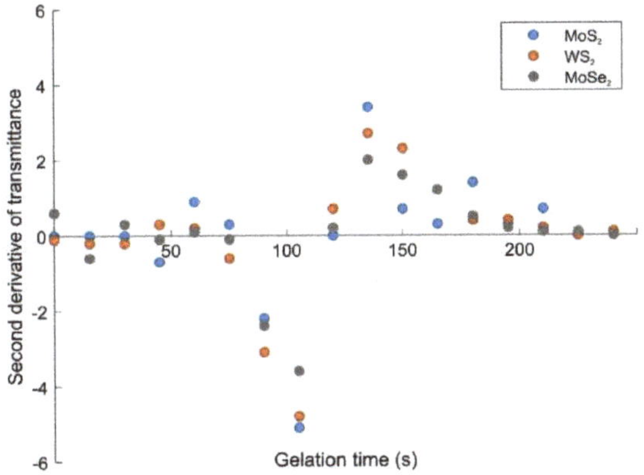

Figure 3. Second derivative of transmittance of the TMD-doped PAAm gels with respect to time.

All composite gels were allowed to dry at room temperature. The relationship between the amount of TMDs and the transmittance response of the composite gels to the incident photon is shown in Figure 4. A decrease in transmittance due to WS_2 content occurs. The change in the color of the gel and its gradual darkening from clear to black are shown in Figure 1c. This observation is in agreement with the change in transmittance in Figure 4. Both MoS_2 and $MoSe_2$-enhanced PAAm composite gels have also the same trendline. Transmittance of $MoSe_2$–PAAm composites with various $MoSe_2$ amounts measured using a uv-vis spectrophotometer is inversely proportional to the $MoSe_2$ amount as expected. For MoS_2-enhanced PAAm composite gels, when the light transmission behavior is examined, a decrease in the form of exponential decay of transmittance is observed as the amount of MoS_2 increases. If these gels are to be used for a multifunctional application, it will not be sufficient to have only transparent, only conductive, or only elastic property. In order to obtain a composite gel that combines as many properties as possible, those with an average transmittance value can be taken. Therefore, if the transmittance values of these gels are 20% and below, they should be eliminated. The absorbance curves of the composite gels prepared with the remaining TMD amounts were evaluated to calculate the band gap energy.

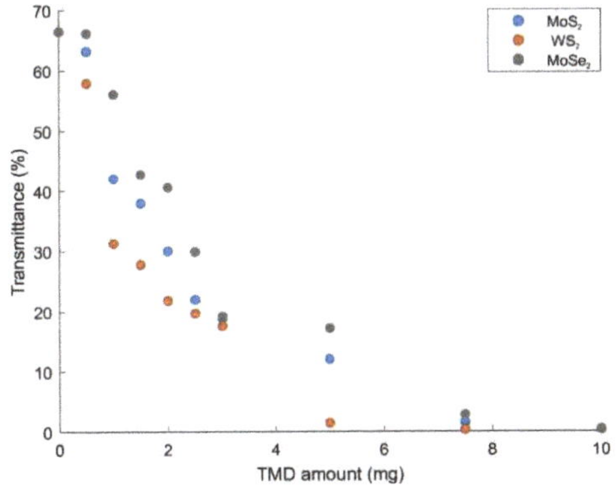

Figure 4. Transmittance plot of PAAm gels reinforced with MoS_2, WS_2 and $MoSe_2$ depending on the amount of TMD.

After drying in air, the absorption response of composite gels was measured by a uv-vis spectrophotometer. Figure 5 represents absorbance versus wavelength curves for all TMDs (WS_2, MoS_2, $MoSe_2$) with various TMD amounts. It was found that the absorption edge shifts to a shorter wavelength with the decreasing TMD amount.

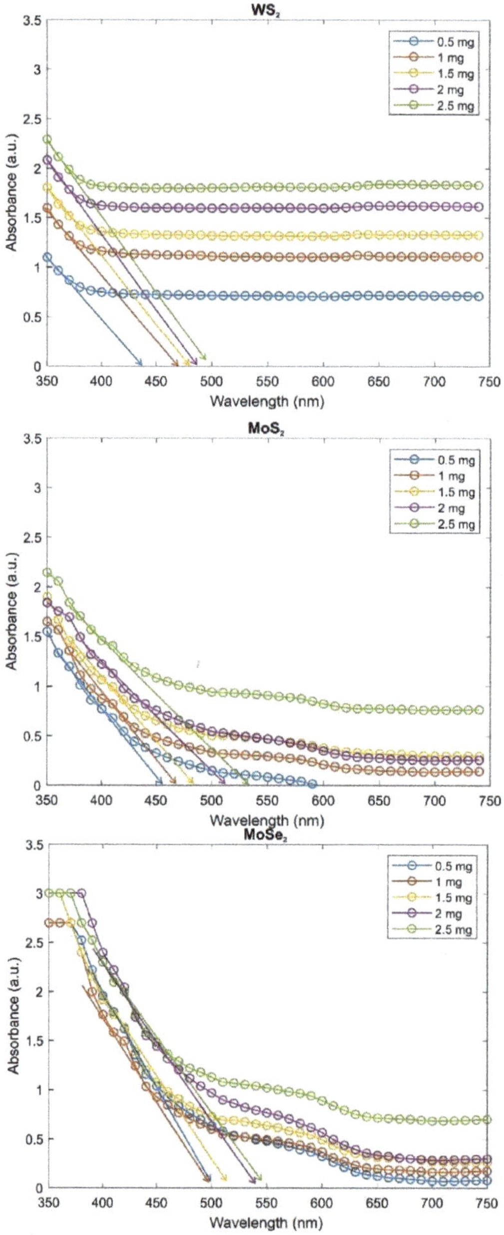

Figure 5. Absorbance versus wavelength graph of WS_2, MoS_2, $MoSe_2$ reinforced PAAm composites from top to bottom, respectively.

Just like in this study, it is not known what kind of interband transition (direct, indirect, allowed or forbidden) of electrons from the occupied state in the conductive state to the empty state in the valence band of composite materials. In this case, it will be difficult to use the Tauc model [49] when calculating the band gap energy values of composite materials.

Therefore, researchers have tried different methods. The most striking of these in recent years is the dielectric loss method [50]. The band transition of electrons is related to the electron–photon interaction. The imaginary part of the dielectric function which corresponds to the dielectric loss describes the interband transition type. This new method can be used to estimate the band gap energy of composites with high accuracy. However, this requires a long and difficult calculation including the extinction coefficient and refractive index [51,52]. On the other hand, the Tauc model is a reliable method to specify the type of electronic transition.

In addition to this information, the band gap energy values at the absorbance edge can be calculated in order to have at least an idea about the band gap energy values of the new TMD-enhanced PAAm composite gels to determine the ones prepared with TMD in a sufficient amount for multifunctional applications. Since the grain size of at least WS_2 is roughly 1 micrometer, it would be accurate to discuss grains considerably larger than nanometers when the polymer composite is obtained, and therefore, examining the shift of the absorbance edge has to be taken into account. In this case, as the amount of TMD increases, the extrapolated wavelength value of the composite, expressed as the absorbance edge, shifts to larger wavelengths.

The extrapolation of the absorbance curve with respect to the wavelength (arrows in Figure 5) gives the band gap energy at the absorption edge. These energy values are given in Table 1, after calculating from the following formula using the wavelength values at which it cuts the curve. The Planck–Einstein equation between photon energy, E, and wavelength, λ, is:

$$E = hc/\lambda \tag{4}$$

Table 1. The calculated band gap energies of all composites at the absorbance edge.

TMD Amount in $WS_2/MoS_2/MoSe_2$ + PAAm Composite Gel (mg)	Bandgap Energy of WS_2 + PAAm (eV, from Absorbance Edge)	Bandgap Energy of MoS_2 + PAAm (eV, from Absorbance Edge)	Bandgap Energy of $MoSe_2$ + PAAm (eV, from Absorbance Edge)
0.5	2.91	2.73	2.56
1	2.73	2.66	2.53
1.5	2.69	2.55	2.49
2	2.59	2.40	2.46
2.5	2.50	2.35	2.43

Here, h is the Planck constant, and c is the speed of light. The increase in TMD amount occurs with a red shift in the absorbance spectrum of composite gels. The absorbance intensities also become strong with an increment in the TMD amount inside the composite gels. The band gap energy at the absorption edge decreases with an increase of the values of TMD amount as expected.

The band gap energy detected from the absorption edge is usually found to be smaller than the energy calculated from the Tauc plot [49]. Considering this, it would be appropriate to prefer composites with a band gap of around 2.7 eV, which is close to the energy value range required for biomedical applications [53]. On the other hand, for electronic applications [54], those with a higher TMD content of around 2.4 eV should be chosen, or if the composite is considered to be used as a transparent conductor material, the composite with the highest transmittance and around 2.8 eV of band gap energy can be selected. When one compares the values in Table 1, one can consider the 2 mg TMD added PAAm composite gel, which has average optical and electronic properties, as the ideal gel after fine tuning. It can be subjected to further investigations regarding the calculation of band gap energy. The absorption coefficients, α, can be calculated by the equation:

$$\alpha(h\nu) = (\log_e 10/t)\ A(h\nu) \quad (5)$$

where t is the thickness of the discs, and A is the absorbance for each photon energy. Then, the band gap energy was calculated from the Tauc's plot by varying the absorbance values with respect to the wavelength for each composite gel.

The band gap energy (E_g) between the maximum of the conduction band and the minimum of the valence band of an amorphous semiconductor is determined by Tauc [49] in 1966. The following can be written for direct transition:

$$(\alpha h\nu)^2 = C\ (h\nu - E_g) \quad (6)$$

The value of the crossing photon energy is taken as the band gap energy. A characteristic Tauc plot of the $WS_2/MoS_2/MoSe_2$ + PAAm composite gel with 2 mg of TMD doping is shown in Figure 6. The band gap energies were obtained by extrapolating the $(\alpha h\nu)^2$ linear fit. The linear fitting equations and their R^2 values are given in Table 2. The reason why composite gels were taken as direct band transition is that TMDs in the form of 2D sheets are exfoliated as separate layers [8,9] in the PAAm gel matrix.

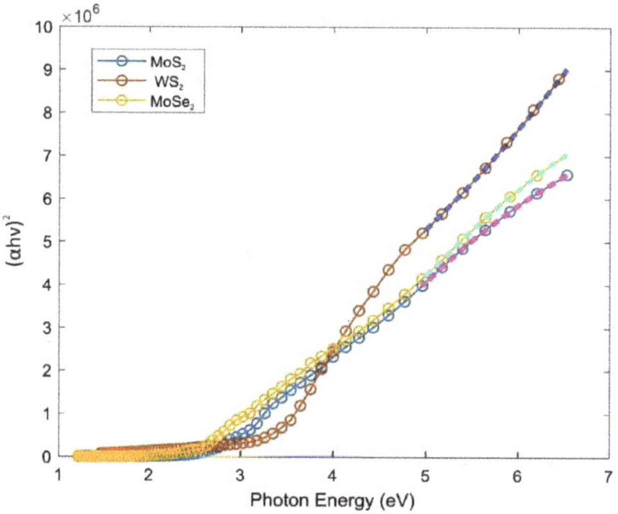

Figure 6. Tauc plots of MoS_2, WS_2, $MoSe_2$ enhanced PAAm composite gels. The linear fitting equations given in Table 2 were used to calculate band gap energies of these composite gels.

Table 2. The calculated band gap energies of all composites from the Tauc plots.

Type of the Composite Gel	Linear Fitting Equations with R^2 Values	Bandgap Energy (eV, from Tauc Plot)
MoS_2 + PAAm	$y = (1.650 \times 10^6)x - 4.087 \times 10^6$ $R^2 = 0.994$	2.48
WS_2 + PAAm	$y = (2.434 \times 10^6)x - 6.923 \times 10^6$ $R^2 = 0.998$	2.84
$MoSe_2$ + PAAm	$y = (1.867 \times 10^6)x - 5.028 \times 10^6$ $R^2 = 0.996$	2.69

Narrow or lower band gap energy is attributed to the high electrical conductivity of the material [55]. Therefore, the increase in the TMD amount leads to the decrease in band gap energy that corresponds to the increase in conductivity. Band gap energies were calculated and found in the range 2.48–2.84 eV. Table 2 summarizes these results for different composite gels.

The values of band gap energy have been found to be in the range of 4.70–5.10 eV for PVP-reinforced PAAm composites [56]. On the other hand, the band gap values calculated for MoS_2/PVP/PAAm composites with different amounts of MoS_2 were found in the range of 3.10–3.81 eV [57]. The estimated band gap energies of WS_2/PVP/PAAm composites with various WS_2 contents were between 3.35 and 4.57 eV [58]. It is observed that the band gap energies of the composites are in good agreement with the previous results. Since the amounts of Bis and PVP used to prepare composites in previous studies were different from the current study, the band gap energy value ranges found were different. By changing the amount of Bis and PVP, band gap energy can be regulated in accordance with the desired optical and electronic properties for multifunctional applications. Using this information, another group of researchers can easily obtain the composite gel by determining a special TMD amount according to the desired band gap energy and transmittance combination for the specific application.

3. Conclusions

Characteristics of polymer-based nanocomposites continue to be an attractive area of interest for materials scientists. This study aims to reveal the relationship between tunable band gap energy and optical properties of 2D materials-doped polyacrylamide composite gels with the effect of the TMD amount. MoS_2, $MoSe_2$, WS_2-doped PAAm composites were prepared using free-radical polymerization. The optical properties of MoS_2, $MoSe_2$, WS_2-doped PAAm were investigated. An increment of the TMD amount caused a decrease in the compactness of the composite's macrogel structure.

The properties of gels can be controlled by changing TMD content. The results indicate that an increase in the amount of TMD leads to a decrease in the gel points. The calculated band gap energies of the composites are in harmony with the literature. The optical results indicated that the absorption edge shifts to a shorter wavelength with the decreasing TMD amount. The properties of the final gels are highly dependent on the amount of TMD, Bis, and PVP present during the formation of gels. Composite gels with high TMD content exhibit stronger absorption in the longer wavelength region compared to gels with low TMD content, which consequently decreases the band gap values of the films. In this case, it can be concluded that composite gels are better semiconductors regarding the conductivity if the correlation between conductivity and band gap energy is considered. We have chosen the 2 mg TMD-added PAAm composite gel, which has average optical and electronic properties, as the ideal gel after fine tuning. The band gap energy of the films was found to be in the range between 2.48 and 2.84 eV. This range fits the characteristic band gap energy range of promising semiconductors. It is observed that the band gap energies of the composites are in good agreement with the results of the previous studies on TMD-doped PAAm. Consequently, the optical and electronic properties of these gels can be altered by the change in the amount of TMD of the composite gels. Due to these characteristics, these composites can be used in electronic and optical applications in addition to biomedical applications. It has been determined by previous studies in the literature that according to the mechanical properties of TMDs, enhanced PAAm gels have already been registered as flexible materials. Optical and electronic properties of these new composite gels have been studied so that they can be used in multifunctional applications. The approach used in this study is promising in modulating the optical and electronic properties of other transition metal dichalcogenide monolayers. This study reveals that the wide optical band gap of composite gels can be significantly tuned by incorporating 2D structures. As future work, the surface resistivity and the conductivity of these composite gels can be investigated using AC current-voltage measurements that are crucial for the electronic applications.

4. Materials and Methods

4.1. Preparation of WS_2, MoS_2, and MoSe-Doped PAAm Composite Gels

Solutions were prepared via free radical crosslinking copolymerization technique at controlled room temperature (22.8–23.1 C) and ambient relative humidity (32–38%) by solving Acrylamide (AAm, Merck, Rahway, NJ, USA), Polyvinylpyrrolidone (PVP, Sigma-Aldrich, Saint Louis, MO, USA), N,N′-Methylenebisacrylamide (BIS, Sigma-Aldrich), Ammonium persulfate (APS, Sigma-Aldrich), and N,N,N′,N′-Tetramethylethylenediamine (TEMED, Sigma-Aldrich), respectively, in distilled water by stirring magnetically and mechanically. This experimental technique is a similar procedure mentioned before in our previous works with different BIS and PVP amounts [48]. Distilled water prepared by the Chemistry Laboratory of KHU was used as the reaction solvent. In order to provide various composite solutions and compare the characteristics of them, TMD content with the same molar value for WS_2, MoS_2, and $MoSe_2$ was added to the solution after Acrylamide and PVP were completely dissolved in distilled water. The whole preparation process was repeated by changing only the amount of TMD so that the amounts of the other chemical contents were kept the same, and composite gels containing different amounts of TMD were prepared.

4.2. Characterization

The gelation of Acrylamide with various amounts of TMD was measured using a Labomed Spectro 22 uv-vis Spectrophotometer. The change in the photon transmission during the gelation of Acrylamide with various amounts of WS_2, MoS_2, and $MoSe_2$ was measured at λ = 650 nm wavelength. The reason why all transmission measurements were performed at a wavelength of 650 nm, is one of the maxima wavelengths of the halogen lamp inside the spectrophotometer. Gelation process and transmittance of photon were monitored in real-time by the camera system. Gel points of the polymer composites, t_{gel}, were obtained. Absorbance and transmittance measurements were taken for single wavelength one by one. Absorbance and transmittance were noted after the calibration of the spectrophotometer for each wavelength.

Author Contributions: B.Ö.U. conceptualized, supervised, edited, and majorly contributed to the writing of the manuscript; Ş.N. majorly contributed to the formal analysis, validation, visualization, writing and reviewing of the original draft; M.A., B.Ç., S.D., Ş.K., E.Y. majorly contributed to the investigation, data curation, and validation of the article and partially contributed to the writing. Ö.P. majorly contributed to the reviewing and supervision of the manuscript. All authors have read and agreed to the published version of the manuscript.

Funding: This research was funded by Personal Research Funds of Kadir Has University (B.Ö.U., and Ö.P.).

Institutional Review Board Statement: Not applicable.

Informed Consent Statement: Not applicable.

Data Availability Statement: The data that support findings of this study are available from the corresponding author upon reasonable request.

Acknowledgments: The authors would like to thank Nihat Berker for providing this collaborative working environment and Sondan Durukanoğlu Feyiz for her support in establishing the Materials Design and Innovation Lab where all the experiments and characterizations of this study were performed.

Conflicts of Interest: The authors declare no conflict of interest.

References

1. Zhu, Y.; Liu, J.; Guo, T.; Wang, J.J.; Tang, X.; Nicolosi, V. Multifunctional $Ti_3C_2T_x$ MXene Composite Hydrogels with Strain Sensitivity toward Absorption-Dominated Electromagnetic-Interference Shielding. *ACS Nano* **2021**, *15*, 1465–1474. [CrossRef]
2. Lin, W.; Kluzek, M.; Iuster, N.; Shimoni, E.; Kampf, N.; Goldberg, R.; Klein, J. Cartilage-inspired, lipid-based boundary-lubricated hydrogels. *Science* **2020**, *370*, 335–338. [CrossRef] [PubMed]

3. Zhu, J.; Zhang, X.; Qin, Z.; Zhang, L.; Ye, Y.; Cao, M.; Gao, L.; Jiao, T. Preparation of PdNPs doped chitosan-based composite hydrogels as highly efficient catalysts for reduction of 4-nitrophenol. *Colloids Surf. A Physicochem. Eng. Asp.* **2021**, *611*, 125889. [CrossRef]
4. Geim, A.K. Graphene: Status and prospects. *Science* **2009**, *324*, 1530–1534. [CrossRef] [PubMed]
5. Chhowalla, M.; Shin, H.S.; Eda, G.; Li, L.-J.; Loh, K.P.; Zhang, H. The chemistry of two-dimensional layered transition metal dichalcogenide nanosheets. *Nat. Chem.* **2013**, *5*, 263–275. [CrossRef]
6. Radisavljevic, B.; Radenovic, A.; Brivio, J.; Giacometti, V.; Kis, A. Single-layer MoS2 transistors. *Nat. Nanotechnol.* **2011**, *6*, 147–150. [CrossRef]
7. Wu, W.; Wang, L.; Li, Y.; Zhang, F.; Lin, L.; Niu, S.; Chenet, D.; Zhang, X.; Hao, Y.; Heinz, T.F.; et al. Piezoelectricity of single-atomic-layer MoS2 for energy conversion and piezotronics. *Nature* **2014**, *514*, 470–474. [CrossRef]
8. Beal, A.R.; Hughes, H.P. Kramers-Kronig analysis of the reflectivity spectra of 2H-MoS2, 2H-MoSe2 and 2H-MoTe2. *J. Phys. C Solid State Phys.* **1979**, *12*, 881–890. [CrossRef]
9. Su, S.-H.; Hsu, Y.-T.; Chang, Y.-H.; Chiu, M.-H.; Hsu, C.-L.; Hsu, W.-T.; Chang, W.-H.; He, J.-H.; Li, L.-J. Band gap-tunable molybdenum sulfide selenide monolayer alloy. *Small* **2014**, *10*, 2589–2594. [CrossRef]
10. Yao, W.; Kang, Z.; Deng, J.; Chen, Y.; Song, Q.; Ding, X.L.; Lu, F.; Wang, W. Synthesis of 2D $MoS_{2(1-x)}Se_{2x}$ semiconductor alloy by chemical vapor deposition. *RSC Adv.* **2020**, *10*, 42172–42177. [CrossRef]
11. Dong, J.; Zhao, Y.; Ouyang, G. The effect of alloying on the band engineering of two-dimensional transition metal dichalcogenides. *Phys. E Low-Dimens. Syst. Nanostruct.* **2019**, *105*, 90–96. [CrossRef]
12. Singh, A.K.; Kumbhakar, P.; Krishnamoorthy, A.; Nakano, A.; Sadasivuni, K.K.; Vashishta, P.; Roy, A.K.; Kochat, V.; Tiwary, C.S. Review of strategies toward the development of alloy two-dimensional (2D) transition metal dichalcogenides. *iScience* **2021**, *24*, 103532. [CrossRef] [PubMed]
13. Yerga, R.M.N.; Alvarez-Galvan, M.C.; Del Valle, F.; De La Mano, J.A.V.; Fierro, J.L.G. Water Splitting on Semiconductor Catalysts under Visible-Light Irradiation. *ChemSusChem* **2009**, *2*, 471–485. [CrossRef] [PubMed]
14. Maeda, K. Photocatalytic water splitting using semiconductor particles: History and recent developments. *J. Photochem. Photobiol. C Photochem. Rev.* **2011**, *12*, 237–268. [CrossRef]
15. Lopez-Sanchez, O.; Lembke, D.; Kayci, M.; Radenovic, A.; Kis, A. Ultrasensitive photodetectors based on monolayer MoS2. *Nat. Nanotechnol.* **2013**, *8*, 497–501. [CrossRef] [PubMed]
16. Shi, Y.; Huang, J.-K.; Jin, L.; Hsu, Y.-T.; Yu, S.F.; Li, L.-J.; Yang, H.Y. Selective decoration of Au nanoparticles on monolayer MoS2 single crystals. *Sci. Rep.* **2013**, *3*, 1839. [CrossRef] [PubMed]
17. Notley, S.M. High yield production of photoluminescent tungsten disulphide nanoparticles. *J. Colloid Interface Sci.* **2013**, *396*, 160–164. [CrossRef]
18. Dey, B.; Mondal, R.K.; Mukherjee, S.; Satpati, B.; Mandal, A.; Senapati, D.; Babu, S.P.S. A supramolecular hydrogel for generation of a benign DNA-hydrogel. *RSC Adv.* **2015**, *5*, 105961–105968. [CrossRef]
19. Dhibar, S.; Dey, A.; Ghosh, D.; Mandal, A.; Dey, B. Mechanically tuned Molybdenum dichalogenides (MoS2 and MoSe2) dispersed supramolecular hydrogel scaffolds. *J. Mol. Liq.* **2019**, *276*, 184–193. [CrossRef]
20. Byron, M.L.; Variano, E.A. Refractive-index-matched hydrogel materials for measuring flow-structure interactions. *Exp. Fluids* **2013**, *54*, 1456. [CrossRef]
21. Chen, X.; Yan, H.; Bao, C.; Zhu, Q.; Liu, Z.; Wen, Y.; Li, Z.; Zhang, T.; Lin, Q. Fabrication and evaluation of homogeneous alginate/polyacrylamide–chitosan–gelatin composite hydrogel scaffolds based on the interpenetrating networks for tissue engineering. *Polym. Eng. Sci.* **2022**, *62*, 116–128. [CrossRef]
22. Kalshetti, P.P.; Rajendra, V.B.; Dixit, D.N.; Parekh, P.P. Hydrogels as a drug delivery system and applications: A review. *Int. J. Pharm. Pharm. Sci.* **2012**, *4*, 1–7.
23. Voronova, M.I.; Surov, O.V.; Afineevskii, A.V.; Zakharov, A.G. Properties of polyacrylamide composites reinforced by cellulose nanocrystals. *Heliyon* **2020**, *6*, e05529. [CrossRef] [PubMed]
24. Rukmanikrishnan, B.; Ramalingam, S.; Lee, J. Quaternary ammonium silane-reinforced agar/polyacrylamide composites for packaging applications. *Int. J. Biol. Macromol.* **2021**, *182*, 1301–1309. [CrossRef] [PubMed]
25. Yang, R.M.; Liu, Y.H.; Dong, G.X.; Zhao, D.J. Fabrication and Properties of Composites of Linear Polyacrylamide and Functionalized Carbon Nanotubes. *Adv. Mater. Res.* **2014**, *936*, 12–16. [CrossRef]
26. Kausar, A. Properties of Polyacrylamide and Functional Multi-walled Carbon Nanotube Composite. *Am. J. Nanosci. Nanotechnol. Res.* **2016**, *4*, 1–9.
27. Narimani, A.; Kordnejad, F.; Kaur, P.; Bazgir, S.; Hemmati, M.; Duong, A. Rheological and thermal stability of interpenetrating polymer network hydrogel based on polyacrylamide/hydroxypropyl guar reinforced with graphene oxide for application in oil recovery. *J. Polym. Eng.* **2021**, *41*, 788–798. [CrossRef]
28. Awasthi, S.; Gaur, J.K.; Bobji, M.S.; Srivastava, C. Nanoparticle-reinforced polyacrylamide hydrogel composites for clinical applications: A review. *J. Mater. Sci.* **2022**, *57*, 8041–8063. [CrossRef]
29. Aktaş, D.K.; Evingür, G.A.; Pekcan, Ö. Critical exponents of gelation and conductivity in Polyacrylamide gels doped by multiwalled carbon nanotubes. *Compos. Interfaces* **2010**, *17*, 301–318. [CrossRef]
30. Kara, S.; Arda, E.; Dolastir, F.; Pekcan, O. Electrical and optical percolations of polystyrene latex–multiwalled carbon nanotube composites. *J. Colloid Interface Sci.* **2010**, *344*, 395–401. [CrossRef]

31. Arda, E.; Mergen, Ö.B.; Pekcan, Ö. Electrical and optical percolations in PMMA/GNP composite films. *Phase Transit.* **2018**, *91*, 546–557. [CrossRef]
32. Uğur, Ş.; Yargı, Ö.; Pekcan, Ö. Percolation and Film Formation Behaviors of MWNT/PS Nanocomposites. *Procedia Eng.* **2011**, 1709–1717. [CrossRef]
33. Zhai, L.; Shi, T.; Wang, H. Preparation of polyvinylpyrrodione microspheres by dispersion polymerization. *Front. Chem. China* **2009**, *4*, 83–88. [CrossRef]
34. Murray, P.G.; Ramesh, M. Chapter 4—Water Soluble Polymers Produced by Homogeneous Dispersion Polymerization. In *Specialty Monomers and Polymers*; ACS Symposium Series; American Chemical Society: Washington, DC, USA, 2000. [CrossRef]
35. Wang, G.-X.; Lu, M.; Hou, Z.-H.; Liu, L.-C.; Liang, E.-X.; Wu, H. Living radical polymerization of polyacrylamide with submicrometer size by dispersion polymerization. *e-Polymers* **2015**, *15*, 75–79. [CrossRef]
36. Umaña, E.; Ougizawa, T.; Inoue, T. Preparation of new membranes by complex formation of itaconic acid-acrylamide copolymer with polyvinylpyrrolidone: Studies on gelation mechanism by light scattering. *J. Membr. Sci.* **1999**, *157*, 85–96. [CrossRef]
37. Cho, K.; Pak, J.; Chung, S.; Lee, T. Recent Advances in Interface Engineering of Transition-Metal Dichalcogenides with Organic Molecules and Polymers. *ACS Nano* **2019**, *13*, 9713–9734. [CrossRef]
38. Liu, J.; Zeng, Z.; Cao, X.; Lu, G.; Wang, L.-H.; Fan, Q.-L.; Huang, W.; Zhang, H. Preparation of MoS_2 -Polyvinylpyrrolidone Nanocomposites for Flexible Nonvolatile Rewritable Memory Devices with Reduced Graphene Oxide Electrodes. *Small* **2012**, *8*, 3517–3522. [CrossRef]
39. Voyutsky, S. *Colloid Chemistry*; Mir Publishers: Moscow, Russia, 1978; pp. 38–43.
40. Young, R.J. *Introduction to Polymers*; Chapman and Hall: New York, NY, USA, 1983.
41. Tanaka, T. Collapse of Gels and the Critical Endpoint. *Phys. Rev. Lett.* **1978**, *40*, 820–823. [CrossRef]
42. Flory, P.J. *Principles of Polymer Chemistry*; Cornell University Press: Ithaca, NY, USA, 1953.
43. Stockmayer, W.H. Theory of Molecular Size Distribution and Gel Formation in Branched—Chain Polymers. *J. Chem. Phys.* **1943**, *11*, 45–55. [CrossRef]
44. Navarro-Verdugo, A.L.; Goycoolea, F.M.; Romero-Meléndez, G.; Higuera-Ciapara, I.; Argüelles-Monal, W. A modified Boltzmann sigmoidal model for the phase transition of smart gels. *Soft Matter* **2011**, *7*, 5847. [CrossRef]
45. Pekcan, Ö.; Kara, S. Gelation Mechanisms. *Mod. Phys. Lett. B* **2012**, *26*, 1230019. [CrossRef]
46. Ziff, R.M.; Stell, G. Kinetics of polymer gelation. *J. Chem. Phys.* **1980**, *73*, 3492–3499. [CrossRef]
47. Thakur, V.K.; Thakur, M.K. (Eds.) *Polymer Gels*; Gels Horizons: From Science to Smart Materials; Springer: Berlin/Heidelberg, Germany, 2018. [CrossRef]
48. Nayır, Ş.; Kıvrak, S.; Kara, İ.; Uysal, B.Ö.; Pekcan, Ö. Tungsten disulfide (WS_2) Doped Polyacrylamide (PAAm) Composites: Gelation and Optical Studies. *Optik Int. J. Light Electron Optik* **2021**, *245*, 167673. [CrossRef]
49. Tauc, J. Optical properties and electronic structure of amorphous Ge and Si. *Mater. Res. Bull.* **1968**, *3*, 37–46. [CrossRef]
50. Aziz, S.B.; Mamand, S.M.; Saed, S.R.; Abdullah, R.M.; Hussein, S.A. New Method for the Development of Plasmonic Metal-Semiconductor Interface Layer: Polymer Composites with Reduced Energy Band Gap. *J. Nanomater.* **2017**, *2017*, 8140693. [CrossRef]
51. Yu, L.; Li, D.; Zhao, S.; Li, G.; Yang, K. First Principles Study on Electronic Structure and Optical Properties of Ternary GaAs:Bi Alloy. *Materials* **2012**, *5*, 2486–2497. [CrossRef]
52. Biskri, Z.E.; Rached, H.; Bouchear, M.; Rached, D.; Aida, M.S. A Comparative Study of Structural Stability and Mechanical and Optical Properties of Fluorapatite (Ca5(PO4)3F) and Lithium Disilicate (Li2Si2O5) Components Forming Dental Glass–Ceramics: First Principles Study. *J. Electron. Mater.* **2016**, *45*, 5082–5083. [CrossRef]
53. Aziz, S.B.; Abdullah, O.G.; Hussein, A.M.; Ahmed, H.M. From Insulating PMMA Polymer to Conjugated Double Bond Behavior: Green Chemistry as a Novel Approach to Fabricate Small Band Gap Polymers. *Polymers* **2017**, *9*, 626. [CrossRef]
54. Aziz, S.B.; Rasheed, M.A.; Ahmed, H.M. Synthesis of Polymer Nanocomposites Based on [Methyl Cellulose] $_{(1-x)}$:$(CuS)_x$ (0.02 M $\geq x \geq$ 0.08 M) with Desired Optical Band Gaps. *Polymers* **2017**, *9*, 194. [CrossRef] [PubMed]
55. Kittel, C. *Introduction to Solid State Physics*, 8th ed.; Wiley Publishing: Hoboken, NJ, USA, 2004.
56. Rawat, A.; Mahavar, H.K.; Chauhan, S.; Tanwar, A.; Singh, P.J. Optical band gap of Polyvinylpyrrolidone/Polyacrylamide blend thin films. *Indian J. Pure Appl. Phys.* **2012**, *50*, 100–104.
57. Uysal, B.Ö.; Evingür, G.A.; Pekcan, Ö. Polyacrylamide mediated polyvinyl pyrrolidone composites incorporated with aligned molybdenum disulfide. *J. Appl. Polym. Sci.* **2022**, *139*, 52061. [CrossRef]
58. Evingür, G.A.; Sağlam, N.A.; Çimen, B.; Uysal, B.Ö.; Pekcan, Ö. The WS_2 dependence on the elasticity and optical band gap energies of swollen PAAm composites. *J. Compos. Mater.* **2021**, *55*, 71–76. [CrossRef]

gels

MDPI

Article

Incorporation of Natural and Recombinant Collagen Proteins within Fmoc-Based Self-Assembling Peptide Hydrogels

Mattia Vitale, Cosimo Ligorio, Ian P. Smith, Stephen M. Richardson, Judith A. Hoyland and Jordi Bella *

Division of Cell Matrix Biology & Regenerative Medicine, School of Biological Sciences, Faculty of Biology, Medicine and Health, Manchester Academic Health Sciences Centre, The University of Manchester, Manchester M13 9PT, UK; mattia.vitale@manchester.ac.uk (M.V.); cosimo.ligorio@manchester.ac.uk (C.L.); ian.smith-9@postgrad.manchester.ac.uk (I.P.S.); s.richardson@manchester.ac.uk (S.M.R.); judith.a.hoyland@manchester.ac.uk (J.A.H.)

* Correspondence: jordi.bella@manchester.ac.uk; Tel.: +44-(0)-161-275-5467

Abstract: Hydrogel biomaterials mimic the natural extracellular matrix through their nanofibrous ultrastructure and composition and provide an appropriate environment for cell–matrix and cell–cell interactions within their polymeric network. Hydrogels can be modified with different proteins, cytokines, or cell-adhesion motifs to control cell behavior and cell differentiation. Collagens are desirable and versatile proteins for hydrogel modification due to their abundance in the vertebrate extracellular matrix and their interactions with cell-surface receptors. Here, we report a quick, inexpensive and effective protocol for incorporation of natural, synthetic and recombinant collagens into Fmoc-based self-assembling peptide hydrogels. The hydrogels are modified through a diffusion protocol in which collagen molecules of different molecular sizes are successfully incorporated and retained over time. Characterization studies show that these collagens interact with the hydrogel fibers without affecting the overall mechanical properties of the composite hydrogels. Furthermore, the collagen molecules incorporated into the hydrogels are still biologically active and provide sites for adhesion and spreading of human fibrosarcoma cells through interaction with the α2β1 integrin. Our protocol can be used to incorporate different types of collagen molecules into peptide-based hydrogels without any prior chemical modification. These modified hydrogels could be used in studies where collagen-based substrates are required to differentiate and control the cell behavior. Our protocol can be easily adapted to the incorporation of other bioactive proteins and peptides into peptide-based hydrogels to modulate their characteristics and their interaction with different cell types.

Keywords: peptide hydrogel; recombinant collagen; diffusion protocol; fibrosarcoma cell; integrin

Citation: Vitale, M.; Ligorio, C.; Smith, I.P.; Richardson, S.M.; Hoyland, J.A.; Bella, J. Incorporation of Natural and Recombinant Collagen Proteins within Fmoc-Based Self-Assembling Peptide Hydrogels. *Gels* **2022**, *8*, 254. https://doi.org/10.3390/gels8050254

Academic Editor: Hiroyuki Takeno

Received: 5 April 2022
Accepted: 16 April 2022
Published: 21 April 2022

Publisher's Note: MDPI stays neutral with regard to jurisdictional claims in published maps and institutional affiliations.

1. Introduction

Hydrogels are three-dimensional (3D) networks made of hydrophilic polymers that can entrap and hold water while forming self-supporting systems through chemical or physical cross-linking of individual polymer chains [1]. Importantly, they can serve as scaffolds for tissue engineering applications due to their structural and functional similarity with the extracellular matrix (ECM). In fact, natural ECM is a complex meshwork consisting of macromolecules and in some tissues minerals where resident cells interact with their surroundings in a 3D network [2]. One of the challenges when developing new hydrogel-based biomaterials is to design an ECM-mimicking nanofibrous network while keeping the system simple, biocompatible and able to modulate cellular behaviour with minimal chemical modification. Peptide-based hydrogels that self-assemble into 3D networks represent a valid approach to this problem [3]. However, the hydrogelator peptides themselves may be non-interactive to cells or provide low cell adhesion. Therefore, biologically active components can be attached by chemical modification of these peptides without interfering with the self-assembly mechanism. The biologically active groups are then exposed on the surface of the hydrogels and hence are available for interaction with cells [4,5]. It has been shown that incorporation of small functional groups such as amines and phosphates can induce stem cell differentiation towards bone and adipose cells [6], whereas addition of integrin-binding motifs, such as RGD, can improve cell adhesion and viability [7,8]. Similarly, hydrogels can be modified with inorganic components, used as "nanofillers", to form peptide-nanocomposite hydrogels with enhanced mechanical properties and improved cell differentiation activity [9–15].

Hydrogel modification through incorporation of collagen molecules has gained widespread popularity in tissue engineering applications. Collagen is the most abundant protein in the body, providing strength and structural stability to tissues as well as cell-binding motifs [16,17]. Moreover, its abundance in the ECM makes it an ideal candidate for hydrogel modifications. For instance, it has been shown that hydrogels modified with collagen peptides showed enhanced cell-adhesion properties and control of stem cell differentiation [18,19]. Indeed, collagen modification of hydrogels can be challenging due to heterogeneity of fabrication protocols used by different research groups and, most importantly, potential undesirable properties of the collagen source [20]. Animal-derived collagen, for example, is most commonly harvested from bovine and porcine sources, which present significant risks of zoonotic disease transmission and immunogenicity [21]. Extraction of collagen from animal connective tissues is also hampered by the large proportion of material that is crosslinked and insoluble [22]. Thus, animal-derived collagens often require use of enzymes such as pepsin, which may digest the collagen proteins to some extent and decrease the mechanical properties of the final biomaterial. These inherent limitations have fuelled the exploration of alternative sources to animal-derived collagen such as the chemical synthesis of collagen-like peptides or the production of collagens or collagen-like proteins using recombinant techniques [23,24]. In fact, the production of collagen via recombinant technology represents a valuable, cost-effective and most importantly, safe alternative for collagen production [25].

In this work we used a minimalistic approach to incorporate natural and recombinant collagen proteins within 9-fluorenylmethoxycarbonyl (Fmoc)-based self-assembling peptide hydrogels (SAPHs). It has been shown that short Fmoc-protected peptides are effective low molecular weight gelators that form rigid nanotube structures that lead to hydrogel formation without further need of crosslinking agents. Furthermore, by tailoring the amount of peptide powder, the hydrogel's resulting stiffness can be modulated accordingly [26–28]. This makes Fmoc-based hydrogels potentially attractive for different tissue engineering applications. Here, we used Fmoc-diphenylalanine/Fmoc-serine (Fmoc-FF/S) peptide hydrogels to develop new customised peptide/collagen composite hydrogels. We used our prior expertise on the successful incorporation of inorganic crystalline materials into Fmoc-based hydrogels [13] to develop a quick and effective way of modifying these hydrogels with natural, synthetic or recombinant peptides/proteins of a wide range of molecular weights and sizes. We demonstrate here effective incorporation of collagen proteins via diffusion within peptide hydrogels and report the mechanical characterization of collagen-modified hydrogels via rheology and their biological characterisation through cytocompatibility studies. Our data demonstrate that our protocol for protein incorporation produces hybrid peptide/collagen hydrogels where the added proteins are well integrated and retained within the host peptide-based scaffolds without the need of chemical modification. Furthermore, incorporated collagens do not interfere with the self-assembly mechanism of the hydrogels or with the mechanical properties of the resulting biomaterial but serve to provide extra integrin-binding sites that improve cell adhesion and spreading. Furthermore, we believe that this protocol can be adapted to the incorporation of both small and large proteins within Fmoc-based peptide hydrogels providing a simple, yet effective way of creating peptide/protein composite hydrogels to be used in tissue engineering and regeneration applications.

2. Results and Discussion

2.1. Characterization of the Proteins Used in Hydrogel Modification

Three different collagen molecules were chosen for incorporation within the self-assembling peptide hydrogels (Table 1): Rat Tail Collagen (RTC); a 42-amino acid collagen peptide containing a GFOGER integrin-binding motif (GFOGER peptide); and an engineered recombinant collagen-like mini protein produced in house (DCol1). In addition, recombinant enhanced green fluorescent protein (eGFP) was used as a fluorescent reporter to follow the protein incorporation and retention within the hydrogel during optimization of the modification protocol [29]. RTC has been extensively used as a positive control to study cell–matrix interaction within 3D hydrogels [30,31], and it has been chosen as the largest protein (approximately 300 kDa molecular weight for the full-length trimer) to be incorporated within our system. The GFOGER peptide is the shortest version of collagen used here (11.24 kDa molecular weight for the trimer), containing a GFOGER motif (O is 4-hydroxyproline) that binds integrins $\alpha1\beta1$ and $\alpha2\beta1$ [32]. DCol1 is a designed recombinant collagen-like protein containing a GFPGER motif which is also able to bind $\alpha1\beta1$ and $\alpha2\beta1$ integrins. Proteins were characterized by SDS-PAGE and circular dichroism (CD) (Figure 1). Purification of DCol1 via nickel-affinity chromatography showed high production and purification yields (Figure 1A). The triple-helical conformation of the three collagen molecules was studied by CD spectroscopy. The CD spectra of RTC, GFOGER and DCol1 at 4 °C in CD buffer (Figure 1B) showed the characteristic features of triple helical collagen: a band of positive ellipticity (around +3000 to +5000 deg cm^2 $dmol^{-1}$) with a maximum at around 220 nm and a deep band of negative ellipticity (around −35,000 deg cm^2 $dmol^{-1}$) with a minimum around 198 nm [33–35]. Both these features are associated with the polyproline II conformation [36], and they are characteristic of the collagen triple helix.

Table 1. Proteins and peptides used for hydrogel formation and modification. Sequences are shown with standard single amino acid symbols, plus O for 4-hydroxyproline. Capping groups: Fmoc, fluorenylmethoxycarbonyl protecting group; Ac, N-terminal acetylation; NH_2, C-terminal amidation.

Molecule	Sequence/Access IDs (Integrin Binding Sites in Bold Type)	Amino Acids (aa)	M_w (kDa)	Isoelectric Point (pI)
Fmoc-FF/S hydrogel Fmoc-FF peptide Fmoc-S peptide	Fmoc-FF Fmoc-S	2 1	0.53 0.33	7.81 7.81
Rat tail collagen [1] α_1 (I) chain α_2 (I) chain	P02454, NP_445756 P02466, NP_445808	1056 [3] 1040 [3]	300 [2]	9.52
GFOGER peptide	Ac-GPCGPPGPPGPPGPPGFOGERGPPGPPGPPGPPGPPGPC-NH_2	42	11.2 [2]	6.96
DCol1 recombinant protein	MGSHHHHHHSGLVPRGSGPPGPPGPQGPAGPRGEPGPAGPKGEPGPAG PPGFPGERGPPGPQGPAGPIGPKGEPGPIGPQGPKGDPGETQIRFRLGPASII ETNSHGWFPGTDGALITGLTFLAPKDATRVQVFFQHLQVRFGDGPWQDV KGLDEVGSDTGRTGE	165	50.0 [2]	6.97
eGFP recombinant protein	MGSSHHHHHHSSGLVPRGSHMVSKGEELFTGVVPILVELDGDVNGHKFS VSGEGEGDATYGKLTLKFICTTGKLPVPWPTLVTTLTYGVQCFSRYPDHMK QHDFFKSAMPEGYVQERTIFFKDDGNYKTRAEVKFEGDTLVNRIELKGIDF KEDGNILGHKLEYNYNSHNVYIMADKQKNGIKVNFKIRHNIEDGSVQLA DHYQQNTPIGDGPVLLPDNHYLSTQSALSKDPNEKRDHMVLLEFVTAAG ITLGMDELYK	259	29.1	6.61

[1] Rat tail collagen is predominantly type I collagen, a heterotrimer made of two α_1 (I) chains and one α_2 (I) chain. [2] Molecular weights of the trimeric collagen molecules. [3] The amino acid number counts correspond to the processed, mature chains, of type I collagen.

2.2. Protein Incorporation into Self-Assembly Peptide Hydrogels

Proteins with a wide range of molecular weights and sizes were successfully incorporated into Fmoc-FF/S hydrogels by simple diffusion, without need for any chemical modifications (Figure 2A), as demonstrated by SDS-PAGE analysis (Figure 2B) (full SDS-PAGE blot is available in Figure S1). Protein incorporation was comparable to the control bands (Figure 2B). We believe that, although samples obtained from hydrogel sections (Figure 2B, lane 2) showed a darker band than the controls (Figure 2B, lane 1), the stronger intensity of the band is caused by the hydrogels acting as a "concentrator" of protein in a constrained space. Indeed, the protein in the stock solution passed from a liquid form to being concentrated within a spheroid due to a volume shrinkage. A similar phenomenon was also shown by Kim and co-workers as they demonstrated that the fluorescence intensity of graphene oxide was much greater in a limited space as the cell pellet [37]. Importantly, hydrogels incorporating the collagen protein showed a significantly stronger band than the remaining protein solutions in the well (Figure 2B, lane 3), suggesting a clear protein uptake. The resulting spheroids were stable and formed a self-supporting hydrogel structure, as shown in Figure 2C. Interestingly, hydrogels incorporating eGFP and fluorescently labelled GFOGER showed a bright, green color when exposed to UV light (Figure 2C(II)) indicating that the proteins were diffused throughout the whole scaffold without affecting their chromophore fluorescence. Finally, we analyzed the protein retention over time. As shown in Figure 2D, SDS-PAGE showed consistent dark bands for all of the tested constructs, suggesting that all the proteins were successfully retained for at least the first 72 h. Different co-assembled peptide hydrogels systems of using alternative mechanisms have been explored [5,38]. In particular Stupp and co-workers have described a similar phenomenon when self-assembling peptide amphiphiles are mixed with high molecular weight polysaccharide hyaluronic acid. Liquid–liquid interaction results in the formation of hierarchically orientated sacs and membranes due a mixture of both osmotic pressure of ions and strong electrostatic interactions [38]. However, in this case, electrostatic interactions are not expected to play a significant role in the incorporation of the proteins into the Fmoc-FF/S peptide hydrogels, at least for the three collagen molecules (see Table 1 for their predicted values of pI). Thus, we believe that the incorporation of collagen proteins within our system is mostly driven by diffusion of the proteins into the hydrogel mesh, with the two components (i.e., Fmoc-FF/S and collagens) co-assembling upon contact.

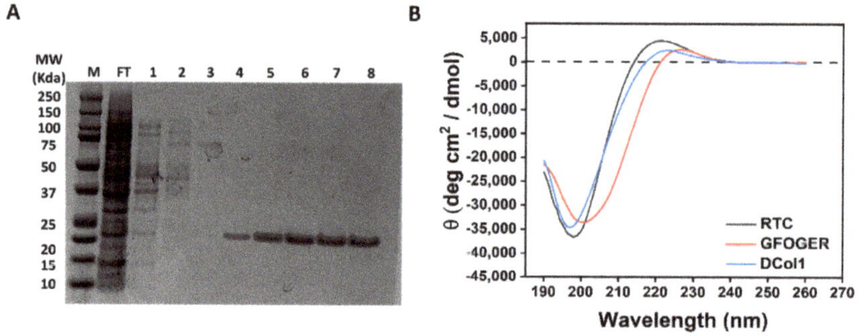

Figure 1. Protein characterization. (**A**) SDS-PAGE analysis of the purification of DCol1 by nickel-affinity chromatography. Lanes: M, molecular weight markers; FT, flow through; 1–2, wash fractions with 60 mM imidazole; 3–4, elution fractions with 250 mM imidazole; 5–8 elution fractions with 1 M imidazole. (**B**) CD spectra at 4 °C of RTC (grey), DCol1 (blue) and GFOGER peptide (red). The vertical axis measures mean residue ellipticity θ in degrees cm^2 $dmol^{-1}$. CD data were collected between 190 and 260 nm.

Moreover, interactions of a hydrophobic nature may also be involved when forming the hydrogels. As shown by the fluorescence spectroscopy analysis, each protein shows a fluorescence emission peak in the 320–380 nm region, where the fluorescence of tyrosine and tryptophan side groups is observed (Figure 3) [39]. The intensity of this peak is significantly lower when the proteins are mixed 1:1 v/v with the Fmoc-FF/S pre-gel solution. The dramatic reduction of fluorescence intensity reveals the existence of a strong interaction between the fluorescent groups in these collagen proteins and the Fmoc-FF/S pre-gel solution. We believe that as the self-assembly mechanism occurs, the hydrogels act as a "sink", where the proteins passively diffuse through the hydrogels' mesh until an equilibrium is reached after 24 h (Figure 2B) as previously demonstrated by Sassi et al. [40].

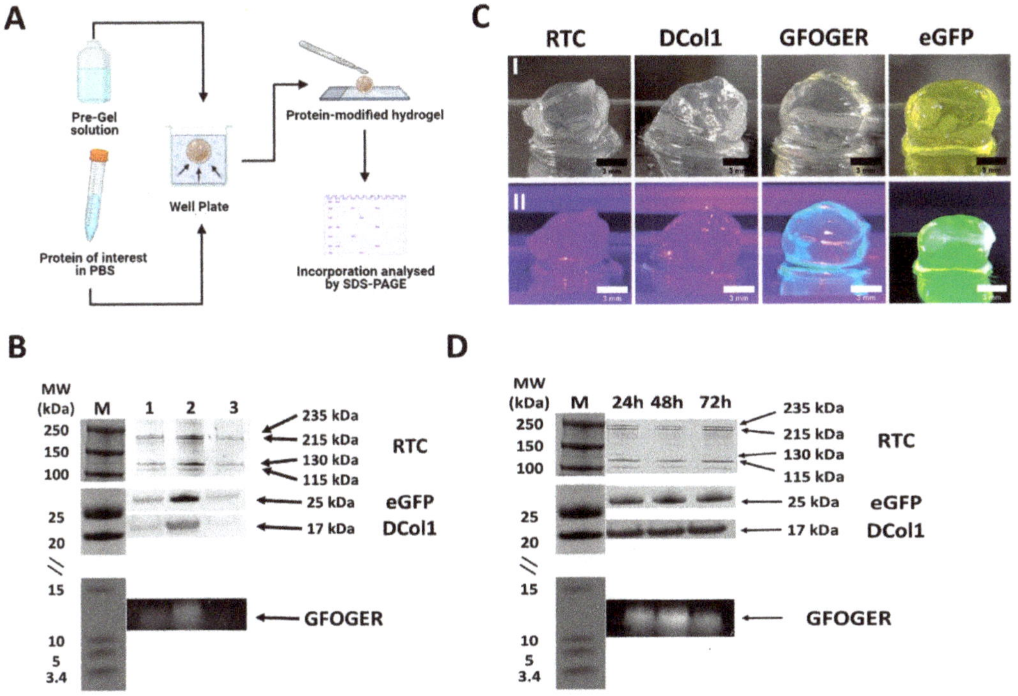

Figure 2. Hydrogel modification. (**A**) Schematic representation of the optimised protocol used to incorporate proteins within hydrogels. (**B**) SDS-PAGE analysis of the proteins incorporated into the Fmoc-FF/S hydrogels. Lanes: M, molecular weight markers; 1, protein stock solution; 2, protein incorporated into the hydrogel; 3, remaining protein into the well. Arrows (right) indicates protein bands at their corresponding molecular weights. (**C**) Photograph of spheroid-like hydrogels incorporating proteins under visible (**I**) and UV (**II**) light. (**D**) SDS-PAGE analysis of protein retention inside the hydrogels over time. Lanes: M, molecular weight markers; 1, 2, 3 show protein retention after 24, 48, 72 h, respectively. Arrows (right) indicates protein bands at their corresponding molecular weights.

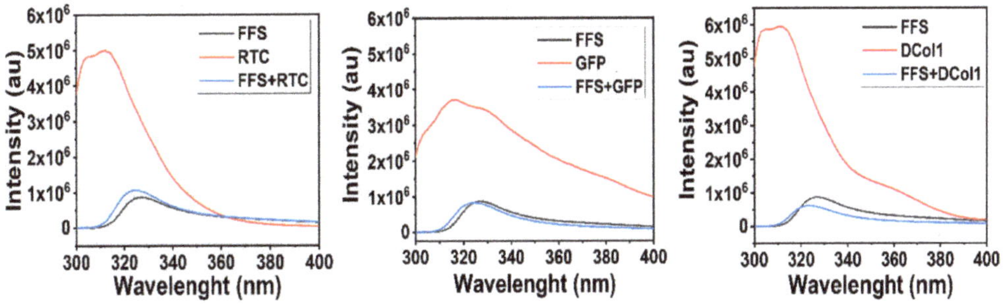

Figure 3. Fluorescence spectroscopy spectra of Fmoc-FF/S peptide (grey), collagen (red) and Fmoc-FF/S-collagen solutions (blue) prepared in PBS (pH = 7.4). Excitation wavelength 280 nm; emission wavelength recorded within the 300–450 nm range.

2.3. Hydrogels Microstrusture

Hydrogel microstructures were also assessed using scanning electron microscopy (SEM). In addition to imaging the outer hydrogel surface, the gels were sliced to expose and allow imaging of the inner surface (Figure 4). Unmodified Fmoc-FF/S hydrogel shows the typical nanofibrous structure consisting of bundles of supramolecular stacks [27]. Additionally, by exposing its inner surface, we observed concentric lamellae as typically shown for the Fmoc-FF/S system where diffusion is involved [41]. The incorporation of different collagen proteins changed the hydrogels' surface features and morphologies according to protein size. The incorporation of the biggest protein, RTC, showed differences with Fmoc-FF/S only on the inner surface of the scaffold where this appeared to be much more disorganised than the rest of the hydrogel. In contrast, the outer surface of the Fmoc-FF/S/RTC showed morphological similarity with the microstructures observed on the outer surface of Fmoc-FF/S. We speculate that smaller collagen peptides such as GFOGER and DCol1 may diffuse more efficiently into the co-assembled systems, as the final structures of the hybrid hydrogels appeared more compact and less disorganised. For this reason, we envisage that the size and shape of the diffusing proteins will dictate a more or less efficient incorporation.

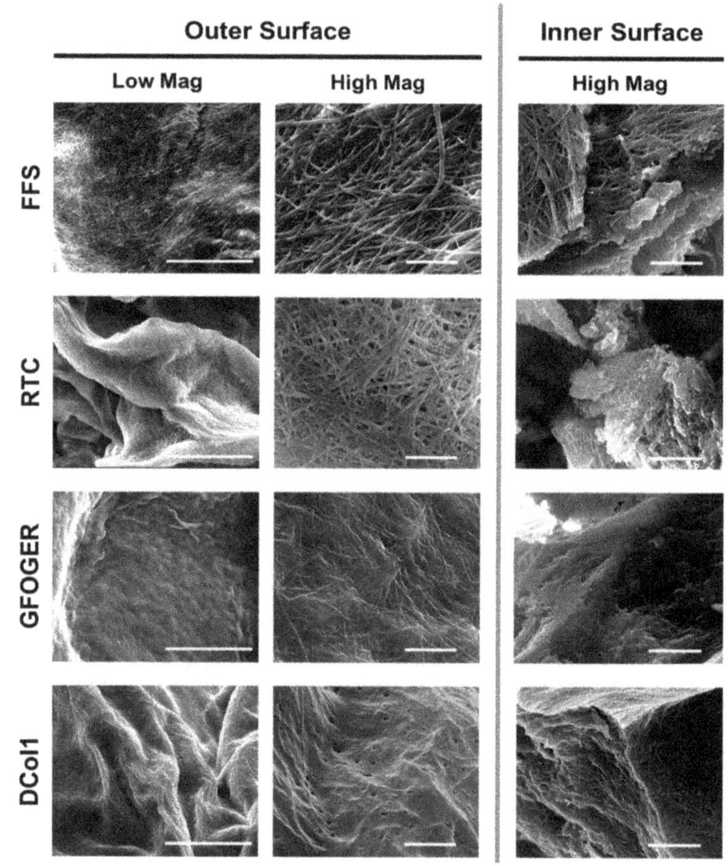

Figure 4. Scanning electron microscopy images showing the outer (**left**) surface morphology of Fmoc-FF/S and the collagen-modified hydrogels and a cross section showing the inner (**right**) surface morphology at lower and higher magnification, respectively. Low magnification scale bar is 50 μm; high magnification scale bar is 5 μm.

2.4. Mechanical Characterization of the Hydrogels

In order to assess the effect of protein incorporation on the final hydrogels' mechanical properties, we analyzed the viscoelastic properties of the protein-modified hydrogels via rheology using unmodified Fmoc-FF/S hydrogel as control. Fmoc-FF/S with and without RTC, GFOGER, eGFP and DCol1 at different concentrations (1, 50 and 100 µg/mL) were subjected to an amplitude sweep experiment (strain range 0 to 100%, frequency 1 Hz, gap size 500 µm) at 37 °C. As can be seen from Figure 5A, no significant difference in the storage modulus G′ (used here as a measure of stiffness) was reported when hybrid hydrogels were formed. Moreover, the G′ was consistently higher than the loss modulus (G″) for all the formulation tested (Figure S2). This indicates that Fmoc-FF/S was able to incorporate all the proteins of interest while still self-assembling and forming self-supporting spheroid shaped hydrogels (Figure 5B). The incorporation of the different proteins did not affect the "bulk" properties of the final peptide/protein hybrid hydrogels, as the stiffness values showed neither a positive nor a detrimental effect to the final mechanical properties, compared to the unmodified Fmoc-FF/S peptide hydrogel.

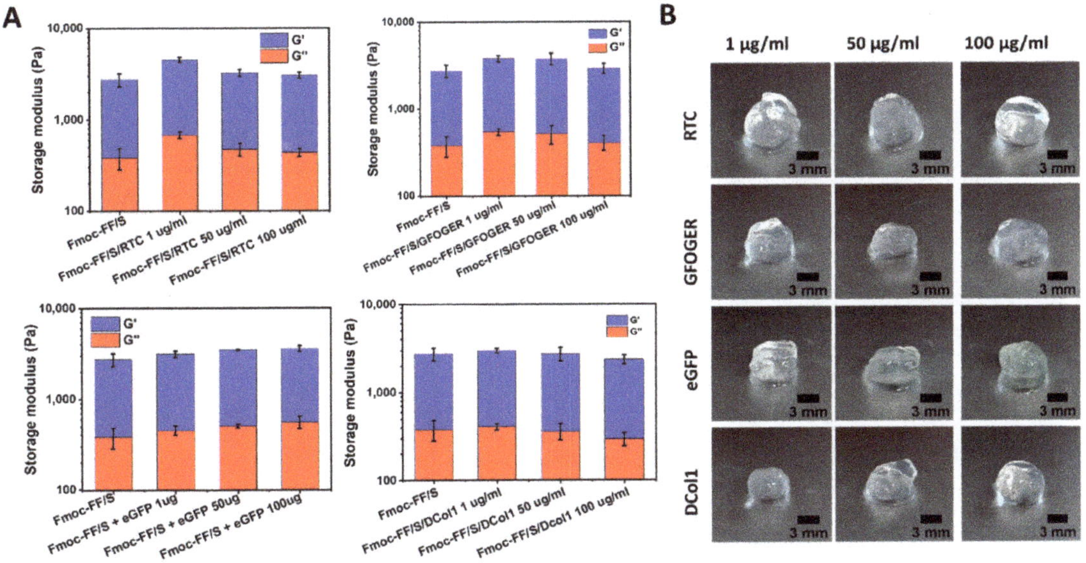

Figure 5. Analysis of the mechanical properties of the hydrogels studied. (**A**) Storage and loss moduli (0.02% strain, 1 Hz) of Fmoc-FF/S hydrogels with and without collagen at different concentrations (1, 50, 100 µg/mL). (**B**) Photographs of spheroids of Fmoc-FF/S with different collagen construct at different concentrations.

2.5. Collagen-Modified Hydrogels as Platform for Cell Culture

To assess the biological activity of the collagen-modified hydrogels, HT1080 cells were used. They are a human fibrosarcoma cell line that is often used for collagen adhesion experiments, and they are known to express high levels of $\alpha 2\beta 1$ integrin on their surface, which is a major cellular receptor for collagen [42,43]. Hydrogels modified with RTC, GFOGER peptide and DCol1 were tested. eGFP was not used for this experiment as it does not provide any cell binding site [44]. Two major parameters were evaluated to assess the properties of the hybrid hydrogels: cell adhesion and cell spreading. One representative protein concentration (100 µg/mL) was used throughout this experiment. Compared to the unmodified Fmoc-FF/S hydrogels, hybrid hydrogels incorporating collagen provided a significantly higher cell adhesion, as can be seen from Figure 6A. As expected, RTC-modified hydrogel provided the maximum cell adhesion (51 ± 1% vs. 14 ± 6.7%, $p < 0.05$). Similarly, Fmoc-FF/S + GFOGER provided greater cell attachment than unmodified peptide systems (47 ± 3.7% vs. 14 ± 6.7%, $p < 0.05$). Among all the collagen mimics incorporated, DCol1, provided the lowest cell adhesion properties (36 ± 5.9%), although still significantly higher than the unmodified Fmoc-FF/S hydrogel (14 ± 6.7%, $p < 0.05$) (Figure 6B, blue bars). We believe that two main factors may be affecting this: firstly, the effect is measured after 24 h, and it could be mitigated due the physiological adaptation of the cells into a new microenvironment [45]; secondly, we hypothesize that, due to small size of DCol1 (17 kDa), easy access to integrin binding sites could be hindered within the hydrogel mesh.

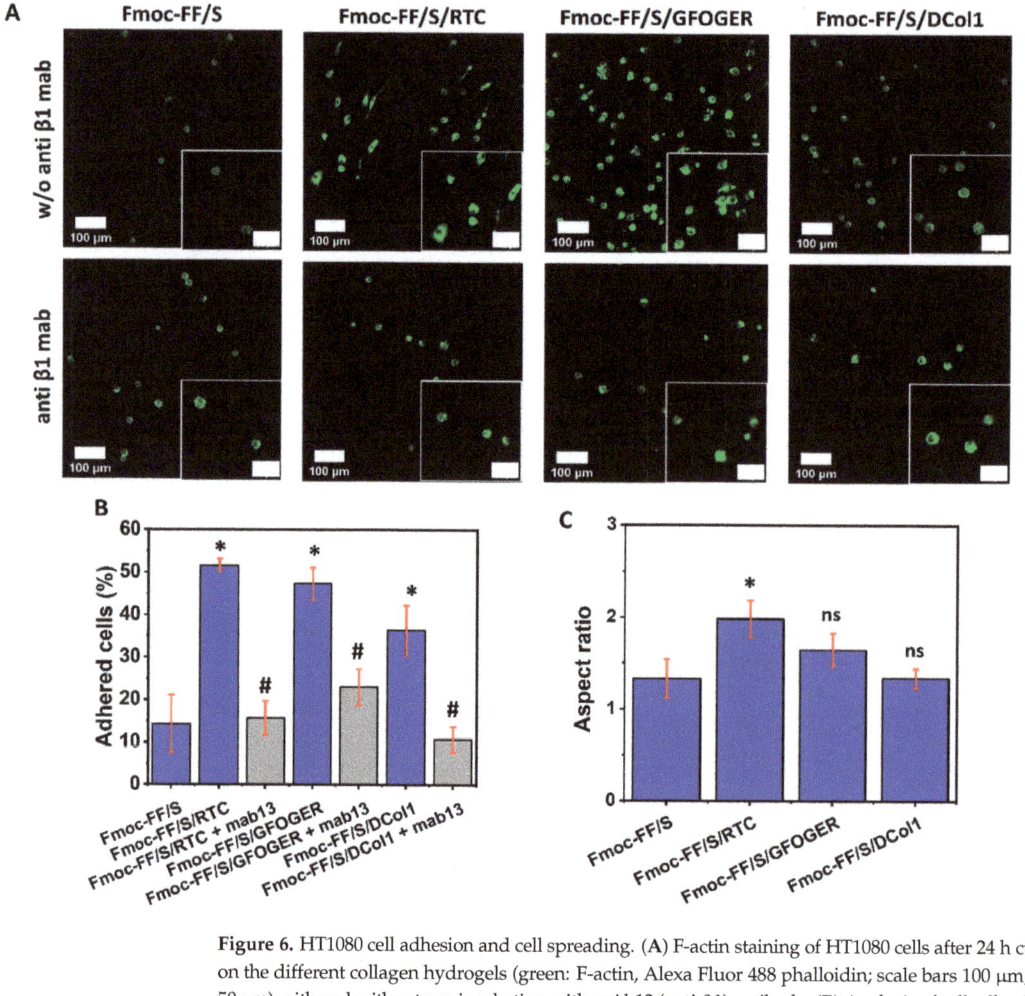

Figure 6. HT1080 cell adhesion and cell spreading. (**A**) F-actin staining of HT1080 cells after 24 h culture on the different collagen hydrogels (green: F-actin, Alexa Fluor 488 phalloidin; scale bars 100 µm, inset 50 µm) with and without pre-incubation with mAb13 (anti β1) antibody. (**B**) Analysis of cell adhesion on the different collagen-modified hydrogels without (blue bars) and with (grey bars) pre-incubation with mAb13 antibody. (**C**) Spread HT1080 cells show a mean aspect ratio above 1 while rounded shaped cells are reflected by an aspect ratio 1. Data shown as mean ± SD, $n = 43$, */# $p < 0.05$, ns (not significant). Significance for each group is relative to Fmoc-FF/S (*), with the exception of groups shown in the grey columns which are in comparison to their respective blue columns (#).

Secondly, we analyzed the capability of HT1080 cells to spread on the modified hydrogels. Analysis of the aspect-ratio (Figure 6C) showed that RTC and GFOGER-modified hydrogels were able to support cell-spreading, as also shown in the upper panel of Figure 6A. However, HT1080 cells showed no spreading and a rounded morphology when they were cultured on the DCol1-modified hydrogels. This effect was already seen in previous experiments on TCP (data not shown) and the reasons for this are still unknown and will require further investigation. We hypothesize that it could be due to an intrinsic toxic component within the DCol1 structure, whose effect becomes more pronounced when the protein is more concentrated [46]. This aspect is currently under investigation and would be the subject of a future work. Finally, to determine whether the cells were directly binding to the collagen decorated on the hydrogel via their β1-integrin, a blocking antibody was used. Optimal concentration of anti-β1 antibody was previously optimized as illustrated in Figure S3. As can be seen from the lower panels in Figure 6A, when cells were pre-incubated with mAb13 antibody, they showed less or no cell spreading, maintaining a well-rounded morphology compared to the Fmoc-FF/S hydrogel. Additionally, cell adhesion was significantly lower than that of the untreated cells. In fact, shown in Figure 6B (grey bars), a significant reduction of ~30% of cell adhesion was observed for RTC-modified hydrogel (51 ± 1% vs. 15 ± 4% $p < 0.05$) as well as a ~50% decrease for Fmoc-FF/S/GFOGER hydrogels (47 ± 3.7% vs. 23 ± 4.2%, $p < 0.05$) compared to the untreated cells. Similarly, DCol1-modified hydrogels showed a ~27% decrease in cell adhesion (36 ± 5.9% vs. 10 ± 3%, $p < 0.05$). Nevertheless, they did not show any significant difference with the unmodified Fmoc-FF/S peptide hydrogel. It has been demonstrated that the integrin α2β1 is one of the major receptors for collagens protein [47,48]. Our findings are consistent with these studies suggesting that the β1-integrin subunit acted as a mediator, linking cells to the collagen modified hydrogels for cell-biomaterial interactions.

3. Conclusions

In this study, we have demonstrated a new and efficient diffusion-based approach to incorporate natural, synthetic and recombinant collagen proteins within Fmoc-based self-assembling peptide hydrogels. Our protocol of incorporation is effective in creating new collagen-containing hydrogels without the need for a prior chemical modification. We believe that a passive diffusion occurs when the collagen protein of interest makes contact with the pre-gel peptide solution creating a co-assembled system, with the protein becoming entrapped in the hydrogel nanofibrillar mesh as the self-assembly mechanism occurs. Moreover, hydrophobic interactions may be involved between the two counterparts, stabilizing the overall networks of peptide/collagen hydrogels. Furthermore, the size of the incorporated collagen impacts the nanostructure of the composite hydrogel. As shown by our EM images, the larger molecules of type I collagen generate more disorganized co-assembled systems than the smaller recombinant collagen or synthetic peptide. Nevertheless, our material characterization indicates that the incorporation of collagen does not affect the mechanical properties of the resulting hydrogels at any chosen concentration (the difference in G' is not significant). We consider this a positive result as we did not want to alter the original stiffness of the hydrogels, as such alteration could have had consequences for cellular mechanotransduction signaling and cell behavior. [49,50].

Furthermore, the addition of collagen to the Fmoc-FF/S introduces additional biological features that are missing in the unmodified hydrogels. Firstly, our collagen-modified hydrogels show enhanced cell adhesion with a significantly higher number of adhered cells compared to the unmodified hydrogels. This confirms that the protocol developed here is able to successfully incorporate collagen protein molecules that remain biologically active. Secondly, HT1080 cells appear to spread on the collagen-modified hydrogels, apart from Fmoc-FF/S/DCol1. This is something already expected and in line with our preliminary data on 2D plastic. Finally, we have demonstrated that the cell adhesion is mediated via the β1-integrin subunit as mab13 pre-treated cells lose their ability to attach to the hydrogels and do not spread. Numerous methods can be used to modify hydrogels in order to incorporate bioactive adhesion molecules [51]. We believe that our protocol of incorporation can be exploited to incorporate collagen-like proteins, as well as other functional peptides containing the RGD cell-binding motif, and bioactive components such as anabolic growth factors, without modifying the chemical structure of the hydrogels and without using toxic cross linkers. As such, we have demonstrated a cost-effective way of creating peptide/protein co-assembled, composite hydrogels for use in tissue engineering and cell culture applications.

4. Materials and Methods

4.1. Materials

Table 1 summarises the different peptides and proteins used in this work for hydrogel preparation and modification. Fmoc-FF/S (a mixture of Fmoc-diphenylalanine and Fmoc-serine peptides, 1:1 ratio) was obtained from Biogelx Ltd., Motherwell, UK (batch number FFS052RM). Peptide quality (97% purity) was assessed at Biogelx via High Performance Liquid Chromatography (HPLC). Rat Tail Collagen (RTC, C3867, Sigma-Aldrich, Welwyn Garden City, UK) was obtained as an aqueous solution in 20 mM acetic acid, at a stock concentration of 3 mg/mL and 95% purity. The 42-amino acid integrin-binding GFOGER peptide [52] (Table 1) was obtained from Cambcol Laboratories Ltd., Cambridge, UK as lyophilized powder and dissolved to a final concentration of 1 mg/mL in Dulbecco's Phosphate Buffered Saline (PBS, Sigma-Aldrich). For simplicity we will include the GFOGER peptide in the general category of "proteins" used here for hydrogel modification. Recombinant eGFP (Table 1) was already available in purified form, previously produced from an in house pET15b-eGFP expression vector [53].

4.2. Recombinant Collagen Design and Purification

A 165-amino acid recombinant collagen was designed in house by fusing a short collagen sequence (72 amino acids) with a trimerization domain from a collagen-like protein from *E. coli* [46]. The amino acid sequence of designed collagen DCol1 is shown in Table 1. Gene synthesis, subcloning and expression tests with different *E. coli* strains was carried out by ProteoGenix, Schiltigheim, France [54]. Best expression conditions were obtained with the protein expression vector pET28b using T7 Express cells. Bacterial pellets containing expressed DCol1 were re-suspended in 20 mL lysis buffer (PBS, lysozyme, 5 mM Imidazole, pH 7.5) with a protease inhibitor tablet (cOmplete Mini EDTA-free protease inhibitor, Roche, Basel, Switzerland). Cells were homogenized using a French cell press (Thermo IEC, FA-078A, Waltham, MA, USA) with a miniature pressure cell (FA-003) working at 20,000 psi. Disrupted cells were collected on ice before centrifugation for 2 h at 12,500 rpm and 4 °C. The pellets of cell debris were discarded, and the supernatant containing soluble protein was mixed with 2 mL of Nickel-nitrilotriacetic resin (HisPur™ Ni-NTA Thermo Scientific, Waltham, MA, USA) previously equilibrated with 10 mL binding buffer (PBS, 5 mM Imidazole, pH 7.5). The resin-protein suspension was incubated on a roller unit overnight at 4 °C, with continuous mixing to maximise binding. The following day gravity columns were prepared with the resin-protein suspension. The unbound fraction was collected. The column was then washed twice with 20 mL of washing buffer (PBS, 60 mM Imidazole, pH 7.5) to remove contaminants. The column was further washed with 250 mM Imidazole in PBS and the protein finally was eluted using 1 M Imidazole in PBS. All fractions were analysed by SDS-PAGE to determine which of them contained the purified protein. Dialysis tubing (Biodesign™ D100, 8000 MWCO, Thermo Scientific, Waltham, MA, USA) was used to remove the unwanted imidazole from the desired fractions by dialysing them against fresh PBS.

4.3. SDS-PAGE

Pre-cast NuPAGE™ 4–12% Bis-Tris mini protein gels (ThermoFisher Scientific), 1.0 mm gel thickness and 10 wells, were used with Invitrogen™ mini gel tanks. Samples were prepared by diluting 15 µL of analyte in 10 µL NuPAGE™ 4X LDS loading buffer (ThermoFisher Scientific) before heating at 95 °C for 5 min on a heating block (HB120-S, Scilogex, Rocky Hill, CT, USA). Hydrogels samples were mixed with loading buffer up to a final volume of 100 µL to help dissolving the gel. After heating, 10 µL of each sample was loaded alongside 5 µL of prestained protein ladder, 10 to 250 kDa (PageRuler™ Plus, ThermoFisher Scientific). All gels were run for 1 h at a constant voltage (120 V) using NuPAGE™ MES SDS running buffer (ThermoFisher Scientific). Gels were stained overnight using Coomassie blue (InstantBlue™, Expedeon, Heidelberg, Germany) and imaged using a compact scanner (CanoScan LiDE 220, Tokyo, Japan).

4.4. Circular Dichroism Spectroscopy

Secondary structures of the RTC and DCol1 proteins and the GFOGER peptide were analysed by circular dichroism (CD) spectroscopy using a Jasco® J-810 spectropolarimeter equipped with a Peltier temperature controller. Samples were diluted to a concentration of ~0.5 mg/mL in CD phosphate buffer (10 mM K_2HPO_4, 10 mM KH_2PO_4, 150 mM KF, pH 7.4) [55]. CD spectra were measured between 190 nm and 260 nm at 4 °C using a 1 mm pathlength CD-matched quartz cuvette (Starna Scientific, Ilford, UK). Data were collected every 0.2 nm with a 1 nm bandwidth. Spectral baselines were corrected by subtracting the spectrum of CD phosphate buffer (blank) collected under the same conditions.

4.5. Hydrogel Modification Protocol

Fmoc-FF/S peptide powder (13 mg) was dissolved in 1 mL of sterile deionized H_2O to a concentration of 15 mM, hereafter referred as pre-gel solution. Collagen proteins were incorporated within the hydrogels by diffusion. In order to follow the incorporation of GFOGER, the peptide was prior labelled with 10 µM NHS-Fluorescein (Sigma-Aldrich) for 30 min at room temperature. Briefly, the proteins of interest were diluted up to the desired concentration in PBS; 1.5 mL of this solution was then placed into a 24-well plate (Corning, New York, NY, USA). A 1.5 mL solution of PBS was used as a protein-free control well. Approximately 300 µL of pre-gel solution were pipetted into the same well forming a spheroid-shape hydrogel which encapsulated the protein as the crosslinking process took place. After 24 h, the resulting spheroids were scooped out of the well, and protein incorporation/retention was analysed by cutting a slice of each hydrogel and by loading them into an SDS-PAGE as described in Section 4.3.

4.6. Fluorescence Spectroscopy

Fluorescence spectroscopy measurements were performed on RTC, eGFP and DCol1 on PBS (100 µg/mL, pH 7.4), on Fmoc-FF/S pre-gel solutions (15 mM, 13.2 mg/mL) and on 1:1 *v/v* mixtures of each protein in PBS and Fmoc-FF/S pre-gel solutions. Measures were carried out at room temperature using a FluoroMax-4 spectrofluorometer (HORIBA, Northampton, UK). Samples were loaded into 0.2 cm path length quartz cuvettes. Fluorescence spectra were acquired using a 280 nm excitation wavelength and emission recorded in the 300–450 nm range.

4.7. Scanning Electron Microscopy (SEM)

The morphologies of Fmoc-FF/S hydrogels with and without incorporated collagens were analysed by Scanning Electron Microscopy (SEM, Thermo Fisher Scientific, Loughborough, UK). Briefly, hydrogels were prepared by pipetting ~300 µL of the pre-gel solutions into Thin-Cert well inserts (0.4 µm pore size Greiner Bio-One Ltd., Stonehouse, UK). The inserts were then placed into 24-well plates and incubated at 37 °C with a total volume of 1.3 mL PBS containing the protein of interest to fully crosslink the hydrogels. After 24 h, hydrogels were fixed in 2.5% (*w/v*) glutaraldehyde (Sigma-Aldrich, Welwyn Garden City, UK) and 4% (*w/v*) paraformaldehyde (Sigma-Aldrich, Welwyn Garden City, UK) in 0.1 M HEPES buffer (Sigma-Aldrich, Welwyn Garden City, UK). After rinsing the samples in PBS, all samples were dehydrated in a graded ethanol (EtOH) series (25, 50, 75, 95, and 100% *v/v* EtOH/water). Samples were maintained at 100% EtOH and dried in a K850 Critical Point Drier (CPD, Quorum Technologies, Lewes, UK). After the CPD step, samples were transferred into metallic pins and coated with gold palladium alloy using an SC7620 Mini Sputter Coater (Quorum, Lewes, UK). Samples were then imaged on a Quanta 250 FEG SEM (Thermo Fisher Scientific, Loughborough, UK) at 20 kV.

4.8. Mechanical Properties of the Hydrogels

The rheological properties of Fmoc-FF/S hydrogels with and without incorporated proteins (RTC, GFOGER, DCol1 or eGFP) were analysed via a rheological amplitude sweep test on a Discovery HR-2 rheometer (TA instruments, New Castle, DE, USA). Each hydrogel sample was tested in the 0–100% shear strain range with a frequency of 1 Hz, gap size of 500 µm, temperature of 37 °C, and rheometer's plate diameter of 20 mm. The rheometer upper head was lowered to the desired gap size and a soak time of 180 sec was used for equilibration. A solvent trap was employed to minimise sample evaporation. Once the rheological spectra were collected, representative storage and loss moduli at 0.02% and frequency of 1 Hz were selected for the summary rheology plots.

4.9. Hydrogel Cell Adhesion and Spreading

Human Fibrosarcoma HT1080 cells (ATCC CCL-121) were maintained in tissue culture polystyrene (TCPS) using Dulbecco's Modified Eagle Medium (DMEM, Gibco, Loughborough, UK) containing 10% (*v/v*) foetal bovine serum (FBS) and 5% (*v/v*) Penicillin-Streptomycin-Amphotericin antibiotic mixture (PSA, 100 units/mL penicillin, 100 µg/mL streptomycin, 0.25 µg/mL amphotericin) (Sigma-Aldrich, Welwyn Garden City, UK). When reaching confluency, cells were gently detached from the tissue culture flask by adding 4 mL of Trypsin-EDTA (Sigma-Aldrich Welwyn Garden City, UK) and pelleted by centrifugation (400× *g* for 5 min). After cell counting, fresh culture medium was added to obtain the desired cell density. Collagen-modified hydrogels were prepared 24 h in advance as described in Section 4.5. However, when used for cell culture, ~250 µL of pre-gel solution was pipetted into the inner well of a 35 mm glass bottom dish for confocal microscopy (VWR, Leicestershire, UK, 734–2905), followed by the addition of 2 mL of the protein of interest in PBS to allow crosslinking and collagen incorporation. Cell adhesion and cell spreading analysis were evaluated as described by Humphries et al. [56]. Briefly, non-specific bindings were blocked by adding 1 mL of heat-denatured high grade Bovine Serum Albumin (BSA, Sigma-Aldrich, Welwyn Garden City, UK), at 10 mg/mL concentration for 1 h. Then, 2 mL of cell suspension (4×10^5 cell/mL) was pipetted onto each hydrogel and incubated for 24 h at 37 °C and 5% CO_2. The following day, the cells were fixed in 4% (*w/v*) paraformaldehyde (Sigma-Aldrich, Welwyn Garden City, UK) for 30 min and permeabilised in 0.5% (*v/v*) Triton X-100 solution (Sigma-Aldrich, Welwyn Garden City, UK) in PBS for 15 min. Cell morphology and cytoskeleton arrangement was assessed using Alexa Fluor™ 488 Phalloidin (Invitrogen™, A12379, Loughborough UK) as previously described [13]. RGB images were split in their channels and green channel images were used for morphological analysis by using ImageJ, version 1.51 [57]. In particular, images were threshold using the Huang's algorithm, and touching cells were separated through a watershed algorithm. Cell adhesion and spreading were evaluated in terms of number of spread cells (%) and aspect ratio (major cell axis/minor cell axis).

4.10. Integrin-Dependent Cell Adhesion Assay

Cell adhesion on different modified hydrogels was also assessed in the presence of 10 µg/mL of the function-blocking monoclonal antibody mAb13 (Sigma-Aldrich, Welwyn Garden City, UK, MABT821) which inhibits the interaction between collagen and the β1-integrin subunit. Following the procedure described by Tuckwell [47], the antibody was diluted 10-fold in warm serum-free DMEM, and the cells were incubated for 30 min in the presence of antibody, before seeding. The effect of the inhibition on the cell spreading assay was evaluated as described in Section 4.9.

4.11. Statistical Analysis

All quantitative values are presented as mean ± standard deviation. All experiments were performed using at least three replicates. Data were plotted using Origin® 2019b [58] and compared using an unpaired t test, unless stated otherwise. One level of significance was used: $p < 0.05$ (* or #, where appropriate).

Supplementary Materials: The following supporting information can be downloaded at: https://www.mdpi.com/article/10.3390/gels8050254/s1, Figure S1: SDS-PAGE analysis of the proteins incorporated into the Fmoc-FF/S hydrogels, Figure S2: Rheological amplitude sweep test of Fmoc-FF/S (shear strain range of 0–100% 1 Hz) combined with different proteins, Figure S3: Optimisation of the concentration for mab13 anti-β1 subunit antibody.

Author Contributions: Conceptualization, M.V. and J.B.; data curation, M.V., C.L., I.P.S. and S.M.R.; formal analysis, M.V.; funding acquisition, S.M.R., J.A.H. and J.B.; methodology, M.V.; supervision, S.M.R., J.A.H. and J.B.; writing—original draft, M.V.; writing—review and editing, M.V., C.L., I.P.S., S.M.R., J.A.H. and J.B. All authors have read and agreed to the published version of the manuscript.

Funding: This research has been supported by the Medical Research Council (MRC) through a CASE PhD studentship to MV (reference MR/R105767/1) and a Proximity to Discovery (P2D) award (MC_PC_19046). It has also been supported by the Biotechnology and Biological Sciences Research Council (BBSRC) through an Impact Acceleration Account (BB/S506692/1) and a Flexible Talent Mobility Account (BB/S507969/1). CL acknowledges the financial support from the Engineering and Physical Sciences Research Council (EPSRC) Doctoral Prize Fellowship (EP/T517823/1). IPS acknowledges the financial support from the BBSRC through a Doctoral Training Partnership studentship (BB/T008725/1).

Institutional Review Board Statement: Not applicable.

Informed Consent Statement: Not applicable.

Acknowledgments: The authors thank David Smith and the staff in the EM Core Facility in the Faculty of Biology, Medicine and Health for their assistance, and the Wellcome Trust for equipment grant support to the EM Core Facility. The authors thank Tom Jowitt and the staff in the Biomolecular Analysis Core Facility in the Faculty of Biology, Medicine and Health for their assistance.

Conflicts of Interest: The authors declare no conflict of interest.

References

1. Ahmed, E.M. Hydrogel: Preparation, characterization, and applications: A review. *J. Adv. Res.* **2015**, *6*, 105–121. [CrossRef] [PubMed]
2. Mann, S.; Webb, J.; Williams, R.J.P.; Falini, G.; Albeck, S.; Weiner, S.; Addadi, L.; Aizenberg, J.; Hanson, J.; Koetzle, T.F.; et al. Porphyrin Amphiphiles as Templates for the Nucleation of Calcium Carbonate. 1987. Available online: https://pubs.acs.org/sharingguidelines (accessed on 7 February 2022).
3. Zhang, S. Fabrication of novel biomaterials through molecular self-assembly. *Nat. Biotechnol.* **2003**, *21*, 1171–1178. [CrossRef] [PubMed]
4. Guler, M.O.; Hsu, L.; Soukasene, S.; Harrington, D.A.; Hulvat, J.F.; Stupp, S.I. Presentation of RGDS Epitopes on Self-Assembled Nanofibers of Branched Peptide Amphiphiles. *Biomacromolecules* **2006**, *7*, 1855–1863. [CrossRef] [PubMed]
5. Storrie, H.; Guler, M.O.; Abu-Amara, S.N.; Volberg, T.; Rao, M.; Geiger, B.; Stupp, S.I. Supramolecular crafting of cell adhesion. *Biomaterials* **2007**, *28*, 4608–4618. [CrossRef] [PubMed]
6. Benoit, D.; Schwartz, M.; Durney, A.R.; Anseth, K.S. Small functional groups for controlled differentiation of hydrogel-encapsulated human mesenchymal stem cells. *Nat. Mater.* **2008**, *7*, 816–823. [CrossRef]
7. Lutolf, M.P.; Lauer-Fields, J.L.; Schmoekel, H.G.; Metters, A.T.; Weber, F.E.; Fields, G.B.; Hubbell, J.A. Synthetic matrix metalloproteinase-sensitive hydrogels for the conduction of tissue regeneration: Engineering cell-invasion characteristics. *Proc. Natl. Acad. Sci. USA* **2003**, *100*, 5413–5418. [CrossRef]
8. Salinas, C.N.; Anseth, K.S. The influence of the RGD peptide motif and its contextual presentation in PEG gels on human mesenchymal stem cell viability. *J. Tissue Eng. Regen. Med.* **2008**, *2*, 296–304. [CrossRef]
9. Ligorio, C.; Zhou, M.; Wychowaniec, J.K.; Zhu, X.; Bartlam, C.; Miller, A.F.; Vijayaraghavan, A.; Hoyland, J.A.; Saiani, A. Graphene oxide containing self-assembling peptide hybrid hydrogels as a potential 3D injectable cell delivery platform for intervertebral disc repair applications. *Acta Biomater.* **2019**, *92*, 92–103. [CrossRef]
10. Bai, X.; Gao, M.; Syed, S.; Zhuang, J.; Xu, X.; Zhang, X.-Q. Bioactive hydrogels for bone regeneration. *Bioact. Mater.* **2018**, *3*, 401–417. [CrossRef]

11. Xavier, J.R.; Thakur, T.; Desai, P.; Jaiswal, M.K.; Sears, N.; Cosgriff-Hernandez, E.; Kaunas, R.; Gaharwar, A.K. Bioactive nanoengineered hydrogels for bone tissue engineering: A growth-factor-free approach. *ACS Nano* **2015**, *9*, 3109–3118. [CrossRef]
12. Hu, Y.; Chen, J.; Fan, T.; Zhang, Y.; Zhao, Y.; Shi, X.; Zhang, Q. Biomimetic mineralized hierarchical hybrid scaffolds based on in situ synthesis of nano-hydroxyapatite/chitosan/chondroitin sulfate/hyaluronic acid for bone tissue engineering. *Colloids Surf. B Biointerfaces* **2017**, *157*, 93–100. [CrossRef] [PubMed]
13. Vitale, M.; Ligorio, C.; McAvan, B.; Hodson, N.W.; Allan, C.; Richardson, S.M.; Hoyland, J.A.; Bella, J. Hydroxyapatite-decorated Fmoc-hydrogel as a bone-mimicking substrate for osteoclast differentiation and culture. *Acta Biomater.* **2021**, *138*, 144–154. [CrossRef] [PubMed]
14. Ligorio, C.; O'Brien, M.; Hodson, N.W.; Mironov, A.; Iliut, M.; Miller, A.F.; Vijayaraghavan, A.; Hoyland, J.A.; Saiani, A. TGF-β3-loaded graphene oxide—Self-assembling peptide hybrid hydrogels as functional 3D scaffolds for the regeneration of the nucleus pulposus. *Acta Biomater.* **2021**, *127*, 116–130. [CrossRef] [PubMed]
15. Ligorio, C.; Vijayaraghavan, A.; Hoyland, J.A.; Saiani, A. Acidic and basic self-assembling peptide and peptide-graphene oxide hydrogels: Characterisation and effect on encapsulated nucleus pulposus cells. *Acta Biomater.* **2022**, *92*, 92–103. [CrossRef]
16. Gordon, M.K.; Hahn, R.A. Collagens. *Cell Tissue Res.* **2010**, *339*, 247–257. [CrossRef]
17. Kadler, K.E.; Baldock, C.; Bella, J.; Boot-Handford, R.P. Collagens at a glance. *Cell Sci. Glance* **2005**, *247*, 1–6. [CrossRef]
18. Wojtowicz, A.M.; Shekaran, A.; Oest, M.; Dupont, K.M.; Templeman, K.L.; Hutmacher, D.W.; Guldberg, R.E.; García, A.J. Coating of biomaterial scaffolds with the collagen-mimetic peptide GFOGER for bone defect repair. *Biomaterials* **2010**, *31*, 2574–2582. [CrossRef]
19. Chiu, L.-H.; Chen, S.-C.; Wu, K.-C.; Yang, C.-B.; Fang, C.-L.; Lai, W.-F.T.; Tsai, Y.-H. Differential effect of ECM molcules on re-expression of cartilaginous markers in near quiescent human chondrocytes. *J. Cell. Physiol.* **2010**, *226*, 1981–1988. [CrossRef]
20. Drury, J.L.; Mooney, D.J. Hydrogels for tissue engineering: Scaffold design variables and applications. *Biomaterials* **2003**, *24*, 4337–4351. [CrossRef]
21. Peng, Y.Y.; Glattauer, V.; Ramshaw, J.A.M.; Werkmeister, J.A. Evaluation of the immunogenicity and cell compatibility of avian collagen for biomedical applications. *J. Biomed. Mater. Res. Part A* **2009**, *93*, 1235–1244. [CrossRef]
22. Terzi, A.; Gallo, N.; Bettini, S.; Sibillano, T.; Altamura, D.; Madaghiele, M.; De Caro, L.; Valli, L.; Salvatore, L.; Sannino, A.; et al. Sub and supramolecular X-ray characterization of engineered tissues from equine tendons, bovine dermis and fish skin type-I collagen. *Macromol. Biosci.* **2020**, *20*, 2000017. [CrossRef] [PubMed]
23. Toman, P.D.; Chisholm, G.; McMullin, H.; Giere, L.M.; Olsen, D.R.; Kovach, R.J.; Leigh, S.D.; Fong, B.E.; Chang, R.; Daniels, G.A.; et al. Production of Recombinant Human Type I Procollagen Trimers Using a Four-gene Expression System in the Yeast *Saccharomyces cerevisiae*. *J. Biol. Chem.* **2000**, *275*, 23303–23309. [CrossRef] [PubMed]
24. Brodsky, B.; Kaplan, D.L. Shining Light on Collagen: Expressing Collagen in Plants. *Tissue Eng. Part A* **2013**, *19*, 1499–1501. [CrossRef] [PubMed]
25. Ruggiero, F.; Koch, M. Making recombinant extracellular matrix proteins. *Methods* **2008**, *45*, 75–85. [CrossRef] [PubMed]
26. Reches, M.; Gazit, E. Casting Metal Nanowires Within Discrete Self-Assembled Peptide Nanotubes. *Science* **2003**, *300*, 625–627. [CrossRef] [PubMed]
27. Jayawarna, V.; Ali, M.; Jowitt, T.; Miller, A.F.; Saiani, A.; Gough, J.E.; Ulijn, R.V. Nanostructured Hydrogels for Three-Dimensional Cell Culture Through Self-Assembly of Fluorenylmethoxycarbonyl-Dipeptides. *Adv. Mater.* **2006**, *18*, 611–614. [CrossRef]
28. Diaferia, C.; Morelli, G.; Accardo, A. Fmoc-diphenylalanine as a suitable building block for the preparation of hybrid materials and their potential applications. *J. Mater. Chem. B* **2019**, *7*, 5142–5155. [CrossRef]
29. Remington, S.J. Green fluorescent protein: A perspective. *Protein Sci.* **2011**, *20*, 1509–1519. [CrossRef]
30. Elsdale, T.; Bard, J. Collagen substrata for studies on cell behavior. *J. Cell Biol.* **1972**, *54*, 626–637. [CrossRef]
31. Rajan, N.; Habermehl, J.; Coté, M.-F.; Doillon, C.J.; Mantovani, D. Preparation of ready-to-use, storable and reconstituted type I collagen from rat tail tendon for tissue engineering applications. *Nat. Protoc.* **2006**, *1*, 2753–2758. [CrossRef]
32. Knight, C.G.; Morton, L.F.; Onley, D.J.; Peachey, A.R.; Messent, A.J.; Smethurst, P.A.; Tuckwell, D.S.; Farndale, R.W.; Barnes, M.J. Identification in Collagen Type I of an Integrin α2β1-binding Site Containing an Essential GER Sequence. *J. Biol. Chem.* **1998**, *273*, 33287–33294. [CrossRef] [PubMed]
33. Greenfield, N.J. Analysis of the kinetics of folding of proteins and peptides using circular dichroism. *Nat. Protoc.* **2006**, *1*, 2891–2899. [CrossRef]
34. Drzewiecki, K.E.; Grisham, D.R.; Parmar, A.S.; Nanda, V.; Shreiber, D.I. Circular Dichroism Spectroscopy of Collagen Fibrillogenesis: A New Use for an Old Technique. *Biophys. J.* **2016**, *111*, 2377–2386. [CrossRef] [PubMed]
35. Kelly, S.; Price, N. The Use of Circular Dichroism in the Investigation of Protein Structure and Function. *Curr. Protein Pept. Sci.* **2000**, *1*, 349–384. [CrossRef] [PubMed]
36. Sreerama, N.; Woody, R.W. Poly (Pro) II helixes in globular proteins: Identification and circular dichroic analysis. *Biochemistry* **1994**, *33*, 10022–10025. [CrossRef]
37. Yoon, H.H.; Bhang, S.H.; Kim, T.; Yu, T.; Hyeon, T.; Kim, B.-S. Dual Roles of Graphene Oxide in Chondrogenic Differentiation of Adult Stem Cells: Cell-Adhesion Substrate and Growth Factor-Delivery Carrier. *Adv. Funct. Mater.* **2014**, *24*, 6455–6464. [CrossRef]
38. Capito, R.M.; Azevedo, H.S.; Velichko, Y.S.; Mata, A.; Stupp, S.I. Self-Assembly of Large and Small Molecules into Hierarchically Ordered Sacs and Membranes. *Science* **2008**, *319*, 1812–1816. [CrossRef]

39. Teale, F.W.J.; Weber, G. Ultraviolet fluorescence of the aromatic amino acids. *Biochem. J.* **1957**, *65*, 476–482. [CrossRef]
40. Sassi, A.P.; Blanch, H.W.; Prausnitz, J.M. Phase equilibria for aqueous protein/polyelectrolyte gel systems. *AIChE J.* **1996**, *42*, 2335–2353. [CrossRef]
41. Inostroza-Brito, K.E.; Collin, E.; Siton-Mendelson, O.; Smith, K.; Monge-Marcet, A.; Ferreira, D.S.; Rodriguez, R.P.; Alonso, M.; Rodriguez-Cabello, J.C.; Reis, R.L.; et al. Co-assembly, spatiotemporal control and morphogenesis of a hybrid protein–peptide system. *Nat. Chem.* **2015**, *7*, 897–904. [CrossRef]
42. Tuckwell, D.S.; Reid, K.B.M.; Barnes, M.J.; Humphries, M.J. The A-Domain of Integrin alpha2 Binds Specifically to a Range of Collagens but is not a General Receptor for the Collagenous Motif. *J. Biol. Inorg. Chem.* **1996**, *241*, 732–739. [CrossRef] [PubMed]
43. Tuckwell, D.; Calderwood, D.A.; Green, L.J.; Humphries, M. Integrin alpha 2 I-domain is a binding site for collagens. *J. Cell Sci.* **1995**, *108*, 1629–1637. [CrossRef] [PubMed]
44. Shimomura, O.; Johnson, F.H.; Saiga, Y. Extraction, Purification and Properties of Aequorin, a Bioluminescent Protein from the Luminous Hydromedusan, Aequorea. *J. Cell. Comp. Physiol.* **1962**, *59*, 223–239. [CrossRef] [PubMed]
45. Ghosh, K.; Pan, Z.; Guan, E.; Ge, S.; Liu, Y.; Nakamura, T.; Ren, X.-D.; Rafailovich, M.; Clark, R.A. Cell adaptation to a physiologically relevant ECM mimic with different viscoelastic properties. *Biomaterials* **2007**, *28*, 671–679. [CrossRef]
46. Ghosh, N.; McKillop, T.J.; Jowitt, T.; Howard, M.; Davies, H.; Holmes, D.F.; Roberts, I.S.; Bella, J. Collagen-like Proteins in Pathogenic, *E. coli* Strains. *PLoS ONE* **2012**, *7*, e37872. [CrossRef] [PubMed]
47. Tuckwell, D.S.; Smith, L.; Korda, M.; Askari, J.A.; Santoso, S.; Barnes, M.J.; Farndale, R.W.; Humphries, M.J. Monoclonal antibodies identify residues 199–216 of the integrin alpha2 vWFA domain as a functionally important region within alpha2beta1. *Biochem. J.* **2000**, *350*, 485–493. [CrossRef]
48. Gullberg, D. I Domain Integrins. In *Advances in Experimental Medicine and Biology*; Springer: Dordrecht, The Netherland, 2014; Volume 819, pp. 41–60. [CrossRef]
49. Ingber, D.E. Cellular mechanotransduction: Putting all the pieces together again. *FASEB J.* **2006**, *20*, 811–827. [CrossRef]
50. Caliari, S.R.; Burdick, J.A. A practical guide to hydrogels for cell culture. *Nat. Methods* **2016**, *13*, 405–414. [CrossRef]
51. Slaughter, B.V.; Fisher, O.Z. Hydrogels in regenerative medicine. *Adv Mater.* **2009**, *21*, 3307–3329. [CrossRef]
52. Knight, C.G.; Morton, L.F.; Peachey, A.R.; Tuckwell, D.S.; Farndale, R.W.; Barnes, M.J. The collagen-binding a-domains of integrins $\alpha 1/\beta 1$ and $\alpha 2/\beta 1$ recognize the same specific amino acid sequence, GFOGER, in native (triple-helical) collagens. *J. Biol. Chem.* **2000**, *275*, 35–40. [CrossRef]
53. John, T. Eyes. Ph.D. Thesis, University of Manchester, Manchester, UK, 2014.
54. ProteoGenix. Available online: https://www.proteogenix.science/ (accessed on 18 March 2022).
55. Greenfield, N.J. Using circular dichroism spectra to estimate protein secondary structure. *Nat. Protoc.* **2006**, *1*, 2876–2890. [CrossRef] [PubMed]
56. Humphries, M.J. Cell-substrate adhesion assays. *Curr. Protoc. Cell Biol.* **1998**, *9*, 9.1.1–9.1.11. [CrossRef] [PubMed]
57. Schindelin, J.; Arganda-Carreras, I.; Frise, E.; Kaynig, V.; Longair, M.; Pietzsch, T.; Preibisch, S.; Rueden, C.; Saalfeld, S.; Schmid, B.; et al. Fiji: An open-source platform for biological-image analysis. *Nat. Methods* **2012**, *9*, 676–682. [CrossRef] [PubMed]
58. OriginLab—Origin and OriginPro—Data Analysis and Graphing Software. Available online: https://www.originlab.com/ (accessed on 9 September 2021).

gels

MDPI

Review

Review of Microgels for Enhanced Oil Recovery: Properties and Cases of Application

Yulia A. Rozhkova [1], Denis A. Burin [1], Sergey V. Galkin [2,*] and Hongbin Yang [3]

[1] Faculty of Chemical Technologies, Industrial Ecology and Biotechnology, Perm National Research Polytechnic University, 614990 Perm, Russia; ketova.pstu@gmail.com (Y.A.R.); burinwork@gmail.com (D.A.B.)
[2] Mining and Oil Faculty, Perm National Research Polytechnic University, 614990 Perm, Russia
[3] School of Petroleum Engineering, China University of Petroleum (East China), Qingdao 266580, China; yhb0810@126.com
* Correspondence: gnfd@pstu.ru; Tel.: +7-(342)-2-198-118

Abstract: In todays' world, there is an increasing number of mature oil fields every year, a phenomenon that is leading to the development of more elegant enhanced oil recovery (EOR) technologies that are potentially effective for reservoir profile modification. The technology of conformance control using crosslinked microgels is one the newest trends that is gaining momentum every year. This is due to the simplicity of the treatment process and its management, as well as the guaranteed effect in the case of the correct well candidate selection. We identified the following varieties of microgels: microspheres, thermo- and pH-responsible microgels, thin fracture of preformed particle gels, colloidal dispersed gels. In this publication, we try to combine the available chemical aspects of microgel production with the practical features of their application at oil production facilities. The purpose of this publication is to gather available information about microgels (synthesis method, monomers) and to explore world experience in microgel application for enhanced oil recovery. This article will be of great benefit to specialists engaged in polymer technologies at the initial stage of microgel development.

Keywords: microgels; preformed particle gels; enhanced oil recovery; reservoir conformance control; pH-sensitive microgels; temperature sensitive microgels

Citation: Rozhkova, Y.A.; Burin, D.A.; Galkin, S.V.; Yang, H. Review of Microgels for Enhanced Oil Recovery: Properties and Cases of Application. *Gels* **2022**, *8*, 112. https://doi.org/10.3390/gels8020112

Academic Editor: Hiroyuki Takeno

Received: 13 December 2021
Accepted: 7 February 2022
Published: 11 February 2022

1. Introduction

It has been estimated that an average of 210 million barrels of water and 75 million barrels of oil are produced worldwide every day. An excessive amount of water leads to undesirable consequences, including corrosion, scale formation, and a decrease in well efficiency [1]. The promising method for reducing the water cut of production oil well and decreasing residual oil saturation is the application of polyacrylamide gels.

Polymer gels are widely used in mature reservoirs due to their excellent profile control ability, easy preparation, and disproportionate permeability reduction property [2]. A new trend in gel preparation for EOR is the synthesis of preformed particle gels (PPG) having significant potential for conformance control [3]. PPGs do not have the disadvantages of other in-situ gels such as gelation time, insufficient strength in the presence of formation water [4], gel structure deformation due to shear degradation and changing of gelant composition induced by contact with reservoir minerals and fluids [5].

PPG is a macrogel with a particle size of more than 200 microns in dry form. These gels swell several times in water, and are injected directly into the well. Due to its viscoelastic properties, the gel's particles are able to penetrate the high permeable layers [6]. However, the effectiveness of PPG macrogel use is limited by the high permeability of the reservoir, they are mainly effective at operational facilities with a permeability of 500 mD or more. To increase the applicability of PPG-based technologies in the oil industry, it is relevant to search for ways to use particles of a smaller granulometric composition. In this regard, a promising scientific field involves the development of reagents based on pre-cross-linked microgels having a particle size of up to 100 microns. From international experience, microgel application is effective in oil fields with a permeability starting from 10 mD. In this review article, we gather the available information on microgels developed and applied for EOR.

2. Methodology

In today's world, due to machine learning [7] and other modern data processing technologies, multiple methods for search and systematization of relevant information for review research [8–10] now exist. One prospective approach is the PICO model. This method is used most commonly in health sciences, nursing, and medicine, but can also be adapted for application in other areas. The PICO recognition model consists of the following elements: P—problem, I—intervention, C—comparison, O—outcome [7,11]. It allows researchers to organize their work, break the topic into its key components, and make the article more precise. In our case, we will use the PICO model for research.

This article was organized based on the PICO technique. The first part of the article describes the problem of reservoirs that have been in operation for a long time. We consider this issue in connection with the problems experienced in Perm Krai, the territory of our research interest. The second part of the article is dedicated to microgels used for oil reservoir conformance control. We gather information on developed and in-development microgels. In the third part, we summarize the chemical approaches to microgel synthesis The article may be interesting for researchers at the initial stage of development of microgels for reservoir conformance control.

A search using the keywords "microgels for enhanced oil recovery" gives 942 articles published on the Science Direct website between 1998 and 2021. In the past year, the number of published articles has increased by more than 1.5 times, with 91 articles for 2020 and 154 for 2021. The graph in Figure 1 shows a constant increase in the number of articles, especially for the past year, which speaks to increasing interest in the topic of microgel application for EOR. The quality of work is guaranteed by the research data taken from the databases of Science Direct, ACS Publications, One Petro, etc. The main keywords used are "microgels", "microgels for enhanced oil recovery", and "preformed particle gels".

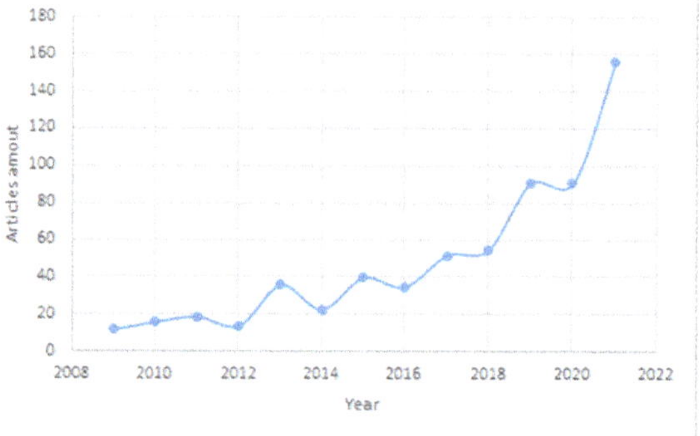

Figure 1. Increase in the number of publications since 2008 on the topic of microgels for enhanced oil recovery.

2.1. Problem Description

The number of mature oil fields is constantly increasing [12,13]. One effective instrument for maintaining the production levels of these oil fields is by applying oil recovery enhancement methods, particularly reservoir conformance control methods [14,15]. One of the cheapest and environmentally friendly [16] reagents that is used for this is polyacrylamide (PAM), which costs about USD 2.00–4.00 per kg [17]. There are several approaches for PAM application in mature reservoirs: polymer flooding [18–20], bulk gels [21,22], preformed particle gels [17,23], and their combinations. The following comprehensive reviews are dedicated to polymer gels systems, including preformed particle gels and its application for enhanced oil recovery [17,24–26].

Preformed particle gels (PPG) were developed in 1996 by PetroChina [15,24]. PPGs are hydrogels synthesized by free-radical polymerization of acrylamide, cross-linker (usually N,N′-methylenebisacrylamide (BIS)), and other additives [27]. A PPG suspension prepared using any available water is injected into the reservoir. PPGs are able to swell up to 200 times in water [17]. The swelling particles of the hydrogel penetrate to the highly permeable fractured zones and block them. This helps redistribute injected water flows to the lowly-permeable oil-saturated interlayers of the reservoir. PPG technology allows reducing the water cut of the produced wells and increase the oil-well exploitation period.

PPG technology has been intensively developed in the last two decades. There are now more than 10,000 successful cases of its application [28]. It has allowed to overcome many difficulties and drawbacks of other PAM-application-based methods [25]. Experts developing this technology admit the following advantages:

- high selectivity: particles preferentially enter fractures and fracture-like channels and are unable to penetrate to low permeable oil saturated zones;
- simplicity of treatment: the suspension is usually prepared using only water (any available water with a wide salt concentration range is acceptable) and PPG; particles are easily dispersed in water;
- PPG properties: particles can be assigned their strength and size during synthesis on the surface; PPG particles have predictable properties in reservoir conditions due to their three-dimensional structure, while hydrogel particles are stable up to 120 °C [15,18,28,29].

Depending on size, PPGs can be divided into macrogels (more than 100 μm to mm) and microgels (less than 100 μm) [17]. The different approaches to microgel and macrogel synthesis are presented in Figure 2. Macrogels and microgels have different application conditions due to their difference in size. Macrogels are used for reservoir conformance control in the formation near the wellbore, while microgels are designed for the highly-permeable zones deep in the reservoir (see Figure 3).

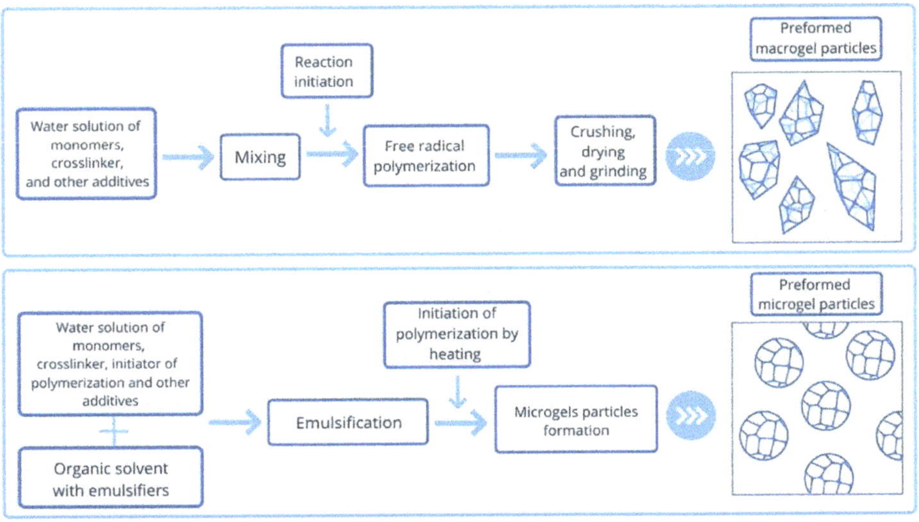

Figure 2. Types of preformed particle gels (macrogels and microgels) and principal schemes of their synthesis.

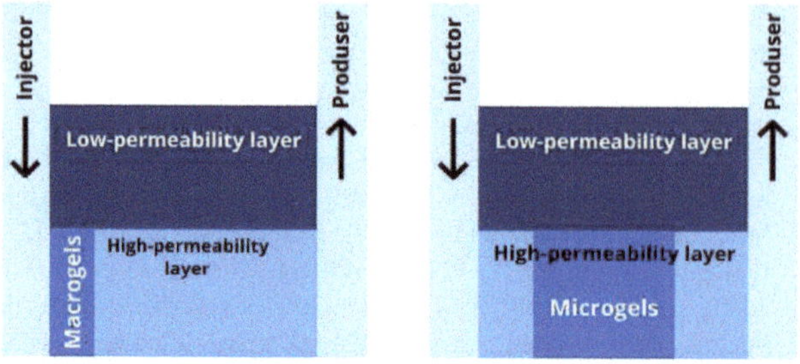

Figure 3. Zones of preferential reservoir conformance control by macrogels (**left**) and microgels (**right**).

It is known that the effectiveness of PPG macrogels is limited by the high permeability of the reservoir. As laboratory studies and pilot projects show, macrogels are predominantly effective at production facilities with a permeability of 500 mD or more [17,30]. To increase the applicability of PPG technologies in local oil fields, it is important to search for ways to use particles with a smaller particle size distribution. In this regard, a promising scientific area involves the development of microgels with a particle size of up to 100 microns. International experience shows that microgel application is effective in oil fields with a permeability starting from 10 mD [31].

The authors of this article are engaged in extensive research on PPG technology, particularly for application in the Volgo-Ural province. Oilfields have been under continuous development in the area for more than 50 years now. Extended oilfield exploitation leads to an increase in the water content in well production. We obtained a PPG adapted to the reservoir conditions of the Volga-Ural oil-and-gas province characterized by a low oil reservoir temperature ($T < 30$ °C) and a high mineralization of formation water (200–230 g/cm^3). The PPG was synthesized in a concentrated solution of polyacrylamide with the addition of acrylic acid. During polymerization, the polymer chains cross-linked by imidization reactions between –COOH and $-NH_2$ groups. PPG obtained using this method has a salt-water absorption capacity of 35–45 g depending on the salt concentration [32].

Analysis of the characteristics of 600 fields in the Perm Region, which can be considered a sample area for the Volga-Ural region, shows that only about 10% of production facilities have a permeability of more than 500 mD. With a decrease in permeability to 50 mD, the number of facilities sharply increases to 70% of the total fund. Therefore, the development of microgel compositions multiplies the number of potential facilities for the application. In this regard, a promising scientific field is the development of reagents based on pre-crosslinked microgels with a swollen particle size of less than 100 μm.

In this publication, we tried to combine the available chemical aspects of microgel production with the practical features of their application at oil production facilities. The purpose of this publication is to study the trends in microgel development (synthesis method, monomers) and explore world experience in microgel application for enhanced oil recovery. This article will be of great benefit to experts engaged in polymer technologies at the initial stage of PPG development.

2.2. *Microgels for Reservoir Conformance Control*

Microgels are particles of cross-linked polymers (gels) 0.1–100 μm in size (according to the IUPAC Gold book) having a three-dimensional structure and capable of swelling in a solvent [33]. The swelling process is caused by conformational changes of the cross-linked polymer network [34]. The interest in microgels is due to their unique properties, as they combine the properties of three groups of compounds: colloidal substances, polymers, and surfactants (see Figure 4) by compiling the unique properties of each class, including structural integrity, functionalization, softness, deformability, permeability, and others [35].

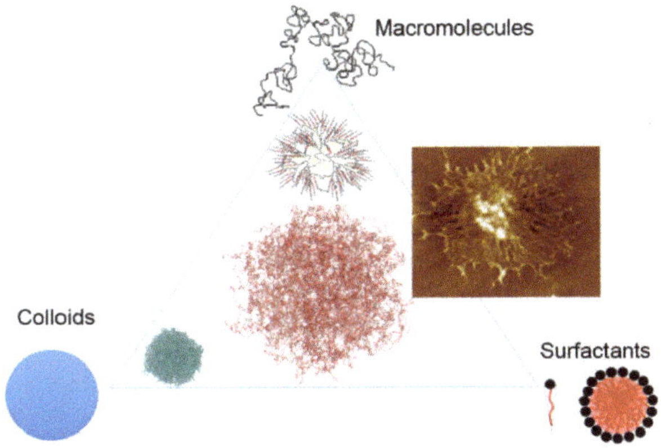

Figure 4. Schematic representation of microgels as systems with combined properties of three fundamental classes of colloids: flexible macromolecules, surfactants, and rigid colloids [35].

The most important characteristic of microgels is their average crosslink density that determines parameters such as the swelling ability (absorption capacity). Microgel swelling in an aqueous medium is determined by the balance between the solvent's entropy, the energy of its interaction with the polymer chain, and the rigidity of the polymer chain [36]. The swelling capacity can also be improved by incorporating different functional groups into the microgel structure [37]. However, there are many factors affecting microgel absorption capacity, particularly temperature, ionic strength of the continuous phase, and pH. These factors can be used to control microgel size and depth of penetration into the reservoir. Several microgels with different properties are applied for EOR. In this part of the article, we will consider each of them.

2.2.1. Colloidal Dispersed Gels (CDG)

CDG are microgels formed in-situ. Gel aggregates form in a lowly-concentrated solution of partially hydrolyzed PAM with a high molecular weight (more than 22 million Da) and a cross-linker (usually aluminum citrate of chrome citrate) [17,27]. The PAM concentration must be below the critical overlap concentration of the polymer, usually 100–1200 ppm. At this concentration, the polymer chains undergo intramolecular cross-linking, forming polymer coils. The ratio of the polymer to crosslinker concentrations vary from 20:1 to 80:1 [35]. The CDG globules formed may reach 1–150 nm in size [38]. The end of the CDG globules formation process is identified by a decrease in the solution's viscosity [39]. CDG gels have been tested successfully in the fields of Argentina, China, and the United States [40–42]. An analysis of 31 cases of pilot tests is presented in the paper [43], the authors of which summarize the main parameters for implementing the technology on the well. The temperature of reservoirs where CDG was applied was 25–100 °C, reservoir permeability varied from 10 to 4200 μm^2, and the oil viscosity of the treated deposits was 5–30 cPs.

During CDG injection, it is important to avoid any sudden pressure surges that could lead to gel transfer into the production well. The injection pressure can be controlled using the following parameters: injection rate, gel concentration, and polymer-to-crosslinker concentration ratio. In cases where the reservoir has the pronounced heterogeneity, CDG treatment is carried out after preliminary in-situ gel injection [44]. Depending on the injection pressure, CDG treatment can be changed in stages. In cases where the initial permeability of the formation is high, a high-concentration gel slurry is injected; after increasing the injection pressure, the PAM concentration is reduced [44].

CDGs are in-situ gels, i.e., microgel globule formation takes place inside the reservoir. The CDG technology requires low concentrations of polymer and crosslinker having inherent drawbacks of poor water production control. Moreover, factors such as shear degradation during injection, dilution by reservoir water, and interaction with minerals leads to a decrease in effective polymer concentration [45]. All these factors in a case of CDG application could lead to weak reservoir conformance control. The microgels considered further in this article have a significant difference in that they are synthesized before being injected into the dedicated equipment.

2.2.2. Dispersed Particle Gel (DPG)

DPGs are uniform spherical particles with the size adjusted from nm to mm (see Figure 5). The DPG receiving process occurs in two stages. The first is bulk gel formation. Here, gel strength and thermostability can be adjusted using the PAM with the suitable degree of hydrolyzation and a suitable cross-linker. The second stage is gel cutting by imposing high-speed shearing forces for several minutes [46]. A peristaltic pump [47] or colloid mill [46] can be used for the shearing. In the paper [46], research on DPG made of PAM and phenolic resin is described. The preparation procedure conditions are as follows: the first stage involves the formation of bulk gel at a temperature 75 °C. Thereafter, the gel is mixed with water (in similar proportions) and grinded using a colloid mill (3000 rpm, 3 min) to produce uniform particles 2.5 μm in size. In the article [38], the author used chromium acetate as a cross-linker. The results of experiments showed that Cr-DPG demonstrates good salinity resistance at 30 °C. The experiment for determining the thermostability showed that DPG size distribution dramatically changes after 15 days at 90 °C. The size of the lowest particles was halved (from 186.6 nm to 400 nm), and that of the highest particles increased by more than 5 times (from 796.2 nm to 4450 nm). Lab sand pack core flooding experiments on samples with a permeability ranging from 0.47 μm^2 to 8.89 μm^2 showed that DPG has good injectivity, which makes it an effective in-depth plugging agent.

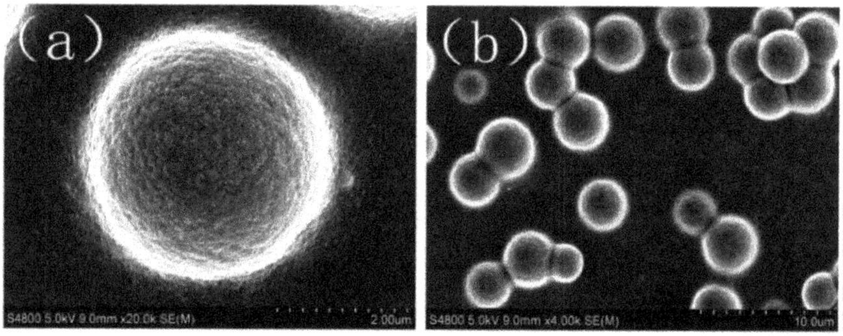

Figure 5. SEM images of dispersed particle gels (DPG) [48]. (**a**) spherical DPG particle, the size of particles may vary from 1.5 to 4 μm; (**b**) SEM photo demonstrate that DPG particles are able to form aggre-gates. The authors attributed that with the high surface energy of the particles.

In the article [49], a novel strengthened dispersed particle gel (SDPG) is presented. Silica nanoparticles (SiO_2) were used as reinforced material to improve resistance to the high temperatures and high-salt content in the reservoir water. A non-ionic PAM of molecular weight 9,650,000 g/mol and a phenolic resin crosslinker were used. The SDPG obtained demonstrated their stability at 110 °C and a total salinity of about 213 g/L [50]. Several research describe the effective combination of DPG with surfactants [45,49,50]. The synergetic effect of the combination was confirmed during core flooding tests. The current research of the team of scientists who developed DPG focuses on self-growing hydrogel particles capable of growing after migration to deep fractures [51].

2.2.3. Preformed Micro-Size Particle Gel

As mentioned above, PPG is a particle gel obtained from drying and grinding of the bulk gel [52]. These gels have a three-dimensional structure that forms during synthesis by cross-linking polymer chains with covalent polar bonds [30]. Preformed particle gels are polyacrylamide-based gels that absorb water and become soft and elastic. Their properties allow particles to penetrate into highly-permeable intervals of the formation. PPGs can be applied both on the fracture and on sandpack reservoirs [52–54]. Depending on their structures, PPGs have different mechanical properties that determine particle penetration into the permeable interlayers of rock. Weak PPGs have a better penetration capability and form a permeable crust on the surface of the lowly-permeable interlayer [55]. PPGs block the fractures partially because they are able to form channels for water passage. It is known that weak PPGs create internal channels more easily than strong gels [56]. Under harsh conditions (high salinity and temperature), ordinary PPGs shrink as a result of amide group hydrolysis and crosslinking by polyvalent metal from the water [57,58]. Nanocomposite PPGs with a superior stability and improved mechanical properties are presented in the paper [57]. The modified PPGs contained an equimolar ratio of the acrylamide, vinylpyrrolidone, 2-acrylamido-2-methylpropane sulfonic sodium salt; BIS as a crosslinker. The mechanical properties of the PPGs were enhanced by adding a dispersion of modified bentonite (MB) to the formulation. The nanocomposite PPS obtained demonstrated stability over 3 months at 130 °C. The swelling capacity in 25% total dissolved solids (TDS) solution was 9.53 g/g and that in fresh water 53.43 g/g [57].

The authors of the article [59] studied the matching factor of PPG in coreflooding experiments. Matching factor is the ratio of the PPG's average diameter to the average pore-throat diameter. PPGs with a swelling particle average size of 9.1 μm were used in the experiments. Several cemented quartz cores were used with different permeability characteristics. The plugging behaviors of the PPG particles were summarized as three basic patterns: complete plugging (core permeability 26.26 mD); plugging-passing through in a deformation or broken state–deep migration (strong plugging in test on core with permeability 46.63 mD; general plugging in cores with 180.34 mD and 240.77 mD, weak plugging in core with 327.74 mD); inefficient plugging—smoothly passing through—stable flow (core permeability of 430.93, 633.29, and 857.86 mD).

Field trials of micro-sized PPG particles are represented in the paper [60]. Microgel size is described as, for example, 28 pm, which means the particles size is less than 28 μm. Particles were obtained from the grinding of bulk gel pieces. Microgel absorption capability at 125 °C in fresh water is about 23, and decreases to 6.5–7.0 when the brine TDS is more that 5%. Microgels are stable during at least one year at a temperature of 125 °C. For the trial treatment, a mature oil field in Northwestern China was chosen. The basic reservoir characteristics include a severe vertical and lateral heterogeneity and an average permeability of 230 md (max permeability was about 1500 md, water-cut of production wells was 95%). Treatment lasted 10 months, 169 tonnes of microgel were injected. The post-treatment effect lasted 18 months, and the quantity of additional oil was about 29.6 thousand tonnes, i.e., 175 tonnes of oil per 1 tonne microgel particles.

2.2.4. SMG Microgels (Small Microgels)

SMG Microgels (Small Microgels) are a great example of acrylamide-based covalent cross-linked polymeric gels [31,61–63]. SMG are nontoxic [62]. Particle size varies between 0.3 and 2 μm. Their rigidity depends on the chemical composition, particularly of the cross-linker concentration. The three-dimensional structure gives the particles mechanical, thermal, and chemical stability. High shear-rate treatment (15,000 s^{-1}) experiments have demonstrated stable viscosity for 16 min. SMGs are almost two times stable in brine containing H_2S than in an ordinary PAM solution of a similar concentration. Thermal stability tests at 120 °C have demonstrated stable viscosity over three months. First, SMGs were considered for water shut off [64]; however, lab experiments on the SiC granular pack and natural sandstones demonstrated that SMGs have a great in-depth propagation. It has also been established that due to capillary forces, SMGs may be absorbed on pore walls forming thick layers, leading to water permeability reduction. The authors note that in these conditions, oil permeability remains unaffected. As conventional polymers, SMGs behave in the same way as relative permeability modifiers (RPM) [61]. The thickness of the absorbed layer can be regulated by varying the injected flow rate, microgel size, and concentration [31]. The paper [62] presents studies on how the high salinity of water affect SMG properties. These experiments showed satisfactory dissolution of microgels in brine with a salinity 215 g/L of TDS (sodium and calcium chloride). Layers adsorbing microgels tend to swell in low-salinity solutions and shrink when the salinity increases due to a charge screen on the gel particles surface. During coreflooding tests, the permeability reduction was fixed for a wide range of SMG suspension brine concentrations (20–108 g/L of TDS). When testing a suspension with a high salinity (200 g/L of sodium chloride and 15 g/L of calcium chloride), the post-treatment core permeability was found to be less. However, post-flush experiments allowed observing the hysteresis effect of microgel swelling behavior [62]. The first industrial case of successful SMG application was in 2005 for the treatment of an underground gas-reservoir storage [61,62]. In the paper [62], the authors present the first conformance control treatment using SMG. The chosen test well was an injection well surrounded by 7 producers. The basic reservoir had the following characteristic: permeability varying from 10–1000 mD (average around 200 mD), reservoir temperature 48 °C and reservoir water salinity 8000 ppm of TDS. A microgel suspension having a concentration of 500 ppm (1500 ppm of commercial solution) was injected over three months for a total volume of 9000 m^3 (0.1 pore volume). Although the treatment pressure increased, it remained below the maximum authorized pressure. One year after treatment, the amount of additional oil was 1570 m^3, and water production had been reduced by 23,830 m^3 [31]. Two years later, the volume of additional oil was still increasing, reaching 5440 m^3 after 26 months. 2.5 kg of microgel was injected for each tonne of additional oil [63].

2.2.5. Microspheres

The authors of the paper [65] synthesized microgels by free-radical polymerization in inverse emulsion using diesel oil in a continuous phase. The chemical structure of the microspheres is formed by polyacrylamide cross-linked by BIS; a mixture of Tween-60 (polyethylene glycol sorbitan monostearate) and Span-80 (sorbitan monooleate) was used as an emulsifier. The absorption capability was determined by changing the particle diameter: the average diameter of the original macrogels was 50 nm, and after swelling, it reached several μm. The authors discovered that the emulsifiers used during the synthesis in combination with an additional surfactant or NaOH sharply reduce the oil/water interfacial tension during microsphere injection, leading to increasing residual and remaining oil saturation. Core flooding test on sandpacks models showed that microgel injection gives about 20% of additional oil. It was proved that microspheres have a great potential in EOR, particularly in reservoir profile control.

Other examples of microspheres based on polyacrylamide and synthesized using the invers emulsion method are represented by different groups of researchers in the papers [66–69]. The BIS crosslinked elastic microspheres described in [66] swell in 3 days. The average microsphere diameter is 12.05 μm, the size of the swelling particle may increase to 25 μm depending on the temperature. In the salt solution (15–20 g/L), the size of particles is about 16–17 μm. Core flooding experiments on a sand pack demonstrated that the ideal matching factor (microsphere diameter/pore size ratio) is 1.35–1.55. At this ratio of microsphere diameter to pore size, gel particles are able to move while embedded deep in the sand pack. Authors of the article [68] presented a visualization of the process of pore filling by microspheres. A micro-visual model with a pore-throat size of 200–1000 μm was created. After microsphere dispersion pumping, some microspheres were accumulated and squeezed in the pore throat of the model, being used for plugging (see Figure 6). This observation proves that microgels are capable of changing the direction of the injected water to a reservoir's oil-bearing interval [67].

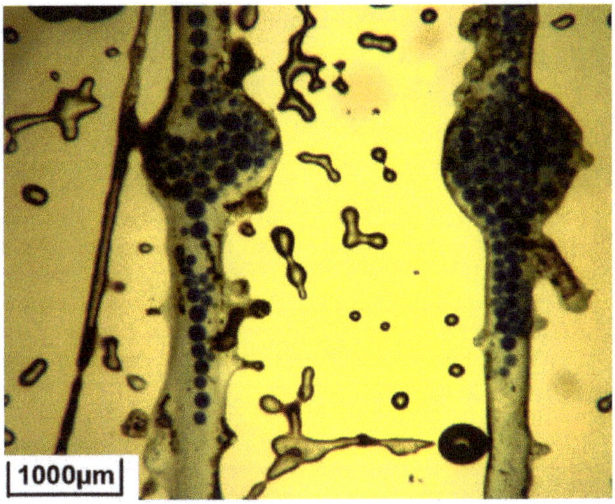

Figure 6. Distribution of microspheres in the micro-visual model [68].

Authors of the article [67] describe microsphere synthesis using the invers emulsion method; however, the difference is in the absence of any cross-linker in the co-monomers mixture (only acrylamide and acrylic acid). The average diameter of the microspheres obtained was 5 μm (Figure 5). After swelling, the particle size increased 5 times, which is more than with BIS cross-linked microgels described in the article [66]. Core flooding tests on the sand pack also proved the ability to redistribute injected water flows in the low permeable zones of the reservoir.

To adapt microsphere properties to the reservoir conditions, different types of microspheres with different viscoelastic properties were developed. The teams of authors of the papers [70–72] presented low elastic polymer microspheres given names such as L-EPM (i.e.) also synthesized using the invers emulsion method. The co-monomers of microspheres are acrylamide, acrylic acid, 2-acrylamido-2-metilpropansulfonic acid (AMPS), and BIS. Aviation kerosene was used as the continuous medium. The emulsifier was made from a mixture of Span-80 and Tween-60. The authors consider one important characteristic of microgels responsible for particle deformability and injection ability as storage modulus. The storage modulus G′ of L-EPM is 23.6 Pa. Experiments for determining microsphere behavior in the core pore space are presented in the paper [73]. For the test, sandstone cores with different porosities and pore sizes were taken. Depending on the microsphere-diameter-to-pore-size ratio, the following mechanisms of L-EPM penetration in the core were identified: (1) direct passing through the pore throat; (2) microsphere adhesion in the pores; (3) dehydration, stretching, extrusion, and retention to original form; (4) squeezing and breakage into pieces under pressure and its forward migration; (5) microspheres stack at the injection end of the sand pack, forming an external filter cake [48]. The coreflooding experiments showed that L-EPMs have a high selective profile control performance in remote heterogeneous reservoir zones [71].

Microsphere modifications are micron-size silica-reinforced polymer microspheres synthesized using the above-mentioned inverse suspension polymerization with the addition of 3-(methacrylyloxy)-propyl-trimethoxysilane (MPS) and silicon dioxide (nano sized) (authors call these microspheres such as PNSCMs) [74–76]. By adjusting the content of MPS-modified SiO_2, the microsphere swelling ratio can be regulated well and the sensitivity of the swelling behavior to the environment is weakened. With increasing SiO_2 loading, the microsphere mechanical stability, thermal stability, viscoelasticity, and dispersion stability were correspondingly improved [74]. PNSCM particle sizes vary between 10 and 100 μm. The swelling capability of SO_2-modified microspheres was 35.5 g/g (for comparing the swelling capability of conventional microspheres was 47 g/g). Temperature has less effect on the swelling capability of silica-reinforced microspheres than that of conventional ones. This can be attributed to the introduction of modified SiO_2. The maximum degradation temperature of silica-reinforced polymer microspheres was 430 °C (11 °C better than conventional microgels). In the article [76], the results of sand pack core flooding tests are represented. The main parameters of the experiment are as follows: the initial permeability of the core was 2.17 μm^2, porosity about 30%, the resulting permeability decreases to 0.35 μm^2. SO_2-modified microspheres demonstrated excellent plugging properties in the micron-size pore throat, and the authors recommend it for deep conformance control application.

To detect microspheres in the reservoir-produced fluid, a new type of microspheres that fluoresce under ultraviolet irradiation was synthesized using an inverse suspension polymerization method [77–80]. The following fluorescent co-monomers were used for microgel synthesis: acryloyloxy coumarin [77,80], allyl-rhodamine B (RhB) [77,78], oxyfluorescein [77] (see Figure 7, Table 1). Since the concentration of fluorescent monomers was quite small, there were no significant changes in the initial particle size and the swelling property of the polymer microspheres.

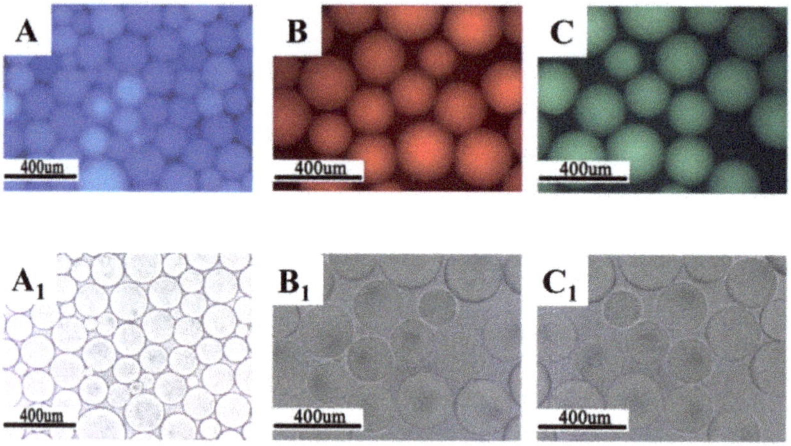

Figure 7. Morphology of fluorescent polymer microspheres (**A**) (co-monomer—cryloyloxy coumarin), (**B**) (co-monomer—allyl-rhodamine B), (**C**) (co-monomer—oxyfluorescein): under the uv light; (**A_1–C_1**): under the ordinary light) [77].

Table 1. Comparison of PPG and DPG synthesis.

Comparison Point	**PPG**	**DPG**
Synthesis feature	(1) synthesis from the monomers and crosslinker mixture; (2) initiation of free radical reaction of polymerization; (3) drying and cutting;	(1) gel formation from solution of partially hydrolyzed PAM and crosslinker; (2) heating (temperature depends on type of crosslinker); (3) mechanical cutting;
Commercial product	Dry powder	Suspension in water
Particle size	30 μm and higher	0.4–2.5 μm
Swelling process	Particles swell during suspension preparation of the oil field	Particles swell in the product

2.2.6. Thermal-Activated Microgels. Brightwater™ System (or "Popping" Microgel)

These microgels were developed by a consortium of BP, Chevron, Texaco, Mobil and Nalco Exxon Energy Services representatives. The idea of thermal-activated microgels is based on the fact that water injected into an oil reservoir is often cooler than the reservoir rock, which leads to formation of a temperature front somewhere between injection and the production wells [81]. Brightwater microgels are synthesized using the invers emulsion method in light mineral oil. The commercial form is 30% wt concentrate in light mineral oil. The chemical structure is based on highly cross-linked sulfonate-containing polyacrylamide. The three-dimensional structure is formed by two types of cross-linkers: stable and unstable. If the temperature rises the unstable cross-linker dissevers, particles absorb more surrounding water and expand. The author calls it the "popcorn effect". Correct selection of cross-linkers provides particles with a sensitivity to the required temperature. Microgel particle sizes may vary from 0.1 to 3 μm. After popping, the particle size increases to 15–20 μm. Core fluid experiments on kernel particles have demonstrated that swelled microgels are able to penetrate the core with a permeability higher than 124 mD. The resistance factor and residual resistance factor values depend on the microgel concentration and core porosity [81]. Brightwater has been applied since 2001 starting from the Minas Field (Indonesia) [82]. There have also been trials in Brazil (Salema Field, Campos Basin) [83], and Alaska [84]. The main criteria for choosing the test oil-well for treatment include available movable oil reserves, early water breakthrough to high water-cut, porosity of highest permeable zones more that 17%, permeability in thief zones more than 100 mD, minimal reservoir fracturing, reservoir temperature 50–150 °C, and salinity of injection water not higher than 70,000 ppm [82]. The paper [84] presents tests of the Brightwater product in Alaska, Mylne Point field. The reservoir was generally homogeneous, with several macro-fractures mapped over the area. The tests were carried out on the area with three wells: one injection and two production wells. The production wells had a water cut of about 90%, the recovery factor was 20%. For processing, a microgel with a dry particle size of 0.1 to 1 μm was used, in swollen form—5 μm. Laboratory tests of the microgel demonstrated its stability for 2 years at elevated temperatures and a water salinity of 120 g/L. The reagent injection was divided into three sequential stages: (1) rapid injection of particles to pass the particles across the near wellbore formation zone; (2) filling the most permeable intervals with particles, heating to the formation temperature, destruction of temporary crosslinking; (3) popping and swelling of particles at a temperature of 50–75 °C. The temperature of the suspension when it acquires the perforation zone in the injection well was about 45 °C. The formation temperature at the production well in the same interval reached 80 °C. The treatment was carried out within 21 days. 60.8 tonnes of microgel suspension with a concentration of 3300 ppm were injected and 30.4 tonnes of surfactant were additionally pumped together. No change in injectivity was observed during the pumping. The first decrease in injectivity occurred 9 months after treatment. During the same period, recovery of additional oil was started for one oil production well. For another well, the production response was 11 months after treatment. The effect lasted 2 years. The total incremental oil volume was about 8000 m^3 [84]. Experiments conducted on the test oil field demonstrated that Brigthwater is a thermally reactive particulate system that functions as an in-depth reservoir conformance control agent.

2.2.7. pH-Activated Microgels

This type of microgel is highly sensitive to pH changes: the microgel suspension has low viscosity at low pH, with the viscosity increasing with an increase in the pH value. This gel is used for conformance control in remote formation zones: a microgel suspension is injected into a low-pH environment, small particles penetrate and move across the near-wellbore zone. During this process, acid from the microgel suspension reacts with rock minerals, leading to an increase in pH, microgels swell and block the highly-permeable zones deep in the reservoir. The pH-responsivity is provided by incorporation of co-monomers possessing carboxyl functional groups (e.g., acrylic acid) [85].

With increasing pH, the carboxylic groups of polyacrylate networks are hydrolyzed, and the charged groups repel each other. As a result, polymers chains are stretched and microgels swell [86]. To guarantee microgel penetration into the deep reservoir zone on the first stage of the reservoir treatment, pre-flush of formation by acids is recommended [86]. The experiments showed that acetic acid is better as a pre-flush agent [86,87]. The second stage is microgel injection, and the last stage of treatment is post water flooding. The higher water mineralization, the lower the pH value of weak acids, and the less swelling capability of microgels [87]. The most widely used pH-sensitive microgel is Carbopol® (Lubrizol, Calvert, KY, USA). There have been several studies on the efficiency of microgels made using sandstone and carbonate cores [86–91]. The average size of Carbopol dry microgel particles is 2–7 μm. The permeability of cores used in the experiment was about 2.3 μm^2 [87]. All experiments showed that at a pH of 2 (pH of microgel suspension), microgels block the injection end of the core. pH correction using NaOH helps overcome this problem [88]. During the flooding experiments, it was shown that these microgels efficiently plug the fractures and high permeable zones and redistribute filtration flows of water into the rock matrix. [86,87].

2.3. Microgel Synthesis Approaches

Several review articles have been published outlining different microgel production methods [89–91]. For microgels applied for EOR, the following basic approaches are used: balk gels synthesis and its mechanical grinding, precipitation polymerization, inverse emulsion polymerization.

The simplest microgel preparation method is mechanical grinding of bulk gel. Table 1 below describes two approaches. The resulting particle size differs at least 10 times. Therefore, the obtained product is aimed at different properties of the reservoir.

Another microgel synthesis method is precipitation polymerization. Three components are involved in the synthesis: monomer, crosslinker, and initiator [92]. The process is followed by a radical mechanism. At the polymerization temperature (50–70 °C), the water-soluble initiator (a compound based on peroxide or persulfates) decomposes on free radicals. In the case of persulfate, decomposition of the initiator leads to the formation of sulfate radicals that attack water-soluble monomers with subsequent propagation of radicals and chain growth [92]. When the microgel particles reach a critical size, they are stabilized by electrostatic stabilization mechanism. When microgel particles are formed, electrostatic repulsion prevents particles coagulation [93].

The precipitation method of monomer synthesis does not appear in publications on microgels for reservoir conformance control application. However, intra- and intermolecular crosslinking of partially-hydrolyzed PAM polymer chains may be considered a special case of this method. An example of chemical crosslinking of polymer chains is CDG formation. Chemical cross linkers of partially-hydrolyzed PAM may be inorganic and organic in nature [94]. For CDG, aluminum citrate is the most common crosslinker. Chromium triacetate is used in fields with a high salinity of reservoir water. The polymer-to-crosslinker ratio may vary from 20:1 to 80:1 [95].

Crosslinking of PAM polymer chains may be achieved using not only chemical methods, but also a physical approach, particularly irradiation treatment. An example of crosslinking by radiation is a PPG-resembling technology used on Russian oil fields (Western Siberia and Tatarstan), and named similar to the polymer-gel system (PGS) Temposcreen [96]. This product is actually a macrogel, however, the approach could be considered for microgels synthesis in future. The three-dimensional structure of the polymer gel particles is formed by ionizing radiation at a dose of 10 kGy of PAM with a molecular weight of 20×10^6 Da and a degree of hydrolysis of 30% [97]. The powder particle size distribution is 0.5–2 mm; in water, the particle can expand up to 1–10 mm in diameter. Depending on the size of the residual reserves and the geological structure of the reservoir, use of the Temposcreen PGM can yield 2–8 additional tonnes of oil. The technology was applied in high-temperature fields (85–95 °C) [97].

The widespread microgel synthesis method for EOR application is inverse emulsion polymerization that has a different modification and allows obtaining different-sized particles (see Figure 8). Depending on the conditions (emulsifier concentration, stirring speed, initiator and dispersant concentrations), inverse emulsion-based polymerization may occur as suspension polymerization, microemulsion and nanoemulsion, giving microgels of different particle sizes. Although all types of emulsion are prepared using the same reagents (hydrocarbon solvent, water, and surfactant), the difference between these methods is the thermodynamic stability of both types of emulsions, which may influence the sizes of the microspheres obtained [90]. The inverse suspension method has many advantages: reaction heating control, granular product can be obtained without the grinding process; the product is easy to dry, and the resulting microgels have an excellent water absorption capability [80].

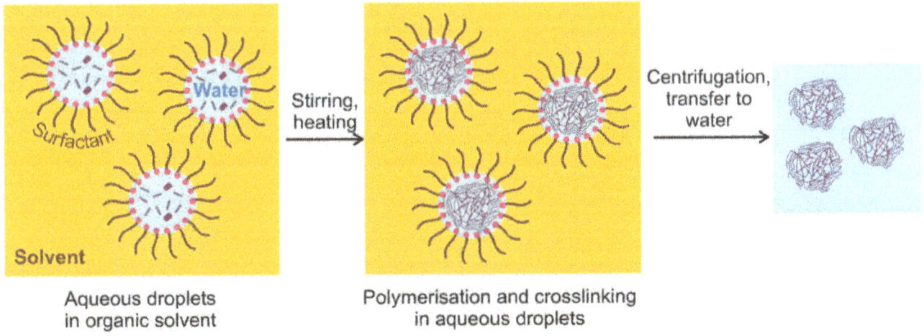

Figure 8. Microgel synthesis in water-in-oil emulsion [33].

The main steps of the synthesis are: choosing the continuous phase (organic solvent, usually mineral oil or refined kerosene); selecting emulsifier systems; selecting a mixture of co-monomer and cross-linkers that will form physical and chemical properties of microgels; selecting an initiator system (for example, chemical initiation, then free radical are produced after chemical interaction, e.g., ammonium persulfate and sodium sulfite) [65,70,98,99]. The most important stage of inverse emulsion preparation is selection of the emulsifier mixture. The hydrophilic-lipophilic balance (HLB) of the emulsifier mixture must match the organic solvent HLB. Combinations of the following emulsifiers are usually used for inverse emulsion synthesis: Tween 60, Tween 80, Span 80, and other, [66,71,99]. As already mentioned above, such microgels such as microspheres and temperature- and pH-sensitive microgels have been synthesized using this method. The commercial products of microgels obtained using this approach is an organic solvent suspension (usually 30% mass). Figure 9 presents a diagram of the laboratory synthesis unit [66]. Table 1 lists compounds that are usually used for microgel synthesis using the invers emulsion approach. Examples of the synthesis of microspheres for reservoir conformance control are represented by the authors of the publications [66–68].

The microgel structure consists of the following fragments: polymer chains network, cross-linkers, and functionalized fragments embedded in the polymer backbone. Table 2 lists some compounds that are used for microgel synthesis and functionalization.

Table 2. Compounds used for microgels synthesis using the inverse emulsion method.

Compound	Formula	Function
Acrylamide	$CH_2{=}CH{-}C(=O)NH_2$	Scaffolding monomer
Acrylic acid	$CH_2{=}CH{-}C(=O)OH$	Scaffolding monomer, improving of hydrophilic properties

Table 2. *Cont.*

Compound	Formula	Function
2-acrylamido-2-metilpropansulfonic acid	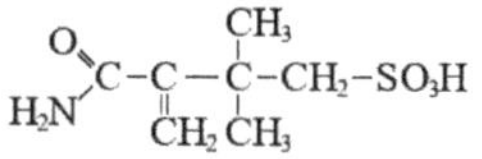	Scaffolding monomer, resistance to the high temperatures
N,N′-methylene-bis(acrylamide) BIS	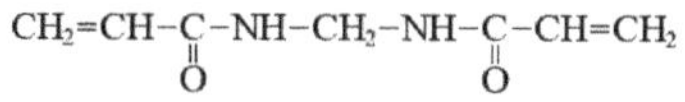	Cross-linking
3-(methacrylyloxy) propyl trimethoxysilane	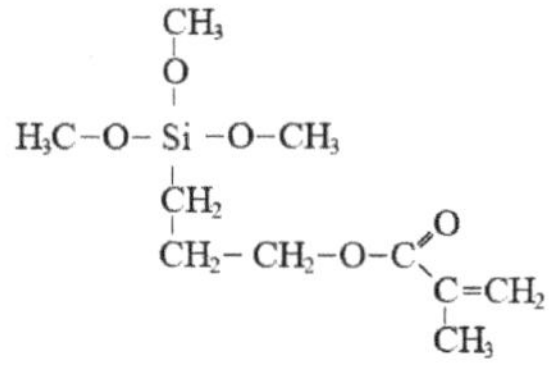	Reinforcing co-monomer for SiO_2 encapsulation
Acryloyloxy coumarin	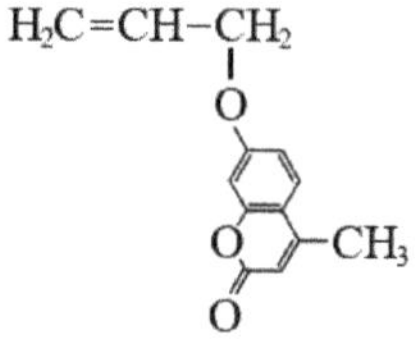	Fluorescent violet color
Allyl-Rhodamine B	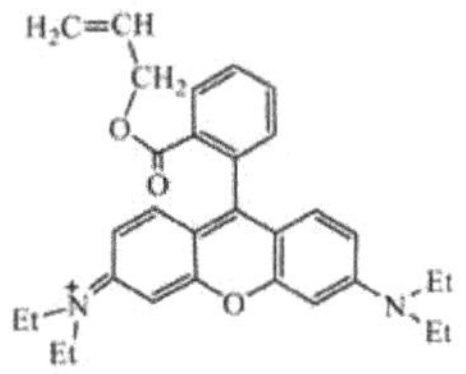	Fluorescent red color
Oxyfluorescein	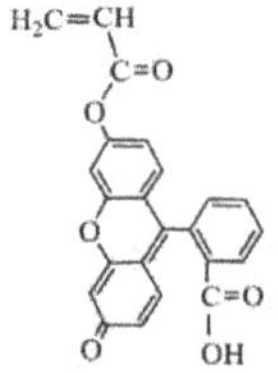	Fluorescent green color

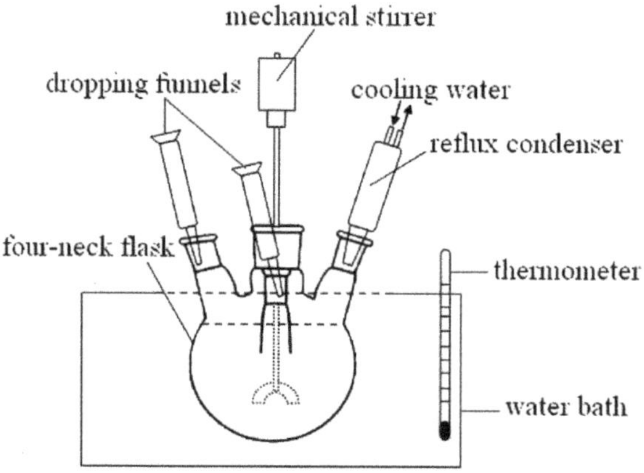

Figure 9. Diagram of laboratory unit for microgel synthesis using the invers emulsion method [78].

3. Conclusions

Several types of microgels have currently been developed. Table 3 summarizes the data about reservoir conditions where each type of microgel could be applied. Many laboratory and trial tests have demonstrated and proved the high efficiency of microgels for in-deep reservoir conformance control that is important today as the number of mature oil fields grows every year.

Table 3. Microgels for EOR.

Type of Microgel	Synthesis Method	Particle Size	Target Reservoir Characteristics		
			Permeability	Water Salinity Limitation	Reservoir Temperature
Colloidal dispersed gels (CDG)	Precipitation polymerization	1–150 nm	10 to 4200 μm^2	Depend on crosslinker type	25–100 °C
Dispersed particle gel (DPG)	Balk gels formation and its mechanical grinding	0.4–2.5 μm	0.47 μm^2 to 8.89 μm^2	213 g/L	stable under 110 °C
Preformed particle gel of micro size	Balk gels synthesis and its mechanical grinding	30 μm and higher	230–1500 md (average 230 μm^2	wide range	stable under 125 °C
SMG Microgels (Small Microgels)	Inverse emulsion polymerization	0.3–2 μm	10–1000 μm^2 (average around 200 μm^2)	215 g/L	stable under 120 °C
Microspheres	Inverse suspension polymerization, inverse emulsion polymerization	About 12 μm and higher by inverse suspension polymerization and 0.3–2 μm by inverse emulsion polymerization	Wide range	wide range	stable under 120 °C
BrightwaterTM thermal-activated microgels	Inverse emulsion polymerization	0.1 to 3 μm	higher than 124 μm^2	120 g/L	50–150 °C
pH-activated microgels	Inverse emulsion polymerization	2–7 μm	more than 10 μm^2	Low salinity is preferable	-

Incorporation of additional monomers into the microgels' structure make it possible to obtain new unique characteristics: high strength [80–82], fluorescent [93,94,100], etc. Additional characteristics expand the microgels' application ways. For example, fluorescent microgels can be used as markers to determine the formation lateral permeability.

Microgels have several advantages over other technologies for profile control with PAM application. These include high treatment selectivity, possibility of technology adaptation to reservoir conditions, easy treatment control, and guaranteed effect if the processing conditions are correctly followed. The current progress in microgel development demonstrates many possibilities for improving the technology relating to changing the mechanical properties (i.e., low elastic microspheres and SiO_2-reinforced microgels) and incorporation of fluorescent monomers, which could improve lateral reservoir conformance control. Microgels can be considered the only component of injected suspension and in synergetic combination with surfactants and polymers solution that has a complex effect on the formation in terms of reducing residual and remaining oil saturation.

Based on the conducted review, the authors believe that technology of conformance control using microgels is a promising one having great prospects, especially in the development of mature oil fields.

Author Contributions: Conceptualization, methodology, research, draft preparation Y.A.R.; research, draft preparation D.A.B.; formal analysis, supervision, data curation S.V.G.; review and editing, H.Y. All authors have read and agreed to the published version of the manuscript.

Funding: This research was carried out with the financial support of the Ministry of Science and Higher Education of the Russian Federation under the program of activities of the Perm Scientific and Educational Center—Rational Subsoil Use.

Institutional Review Board Statement: Not applicable.

Informed Consent Statement: Not applicable.

Conflicts of Interest: The authors declare no conflict of interest.

References

1. Yang, Y.; Peng, W.; Zhang, H.; Wang, H.; He, X. The oil/water interfacial behavior of microgels used for enhancing oil recovery: A comparative study on microgel powder and microgel emulsion. *Colloids Surfaces A Physicochem. Eng. Asp.* **2020**, *632*, 127731. [CrossRef]
2. Salehi, M.B.; Soleimanian, M.; Moghadam, A.M. Examination of disproportionate permeability reduction mechanism on rupture of hydrogels performance. *Colloids Surfaces A Physicochem. Eng. Asp.* **2019**, *560*, 1–8. [CrossRef]
3. Salehi, M.B.; Moghadam, A.M.; Jarrahian, K. Effect of Network Parameters of Preformed Particle Gel on Structural Strength for Water Management. *SPE Prod. Oper.* **2020**, *35*, 362–372. [CrossRef]
4. Aqcheli, F.; Salehi, M.B.; Pahlevani, H.; Taghikhani, V. Rheological properties and the micromodel investigation of nanosilica gel-reinforced preformed particle gels developed for improved oil recovery. *J. Pet. Sci. Eng.* **2020**, *192*, 107258. [CrossRef]
5. Lashari, Z.A.; Yang, H.; Zhu, Z.; Tang, X.; Cao, C.; Iqbal, M.W.; Kang, W. Experimental research of high strength thermally stable organic composite polymer gel. *J. Mol. Liq.* **2018**, *263*, 118–124. [CrossRef]
6. Aqcheli, F.; Salehi, M.B.; Taghikhani, V.; Pahlevani, H. Synthesis of a custom-made suspension of preformed particle gel with improved strength properties and its application in the enhancement of oil recovery in a micromodel scale. *J. Pet. Sci. Eng.* **2021**, *207*, 109108. [CrossRef]
7. Brockmeier, A.J.; Ju, M.; Przybyła, P.; Ananiadou, S. Improving reference prioritisation with PICO recognition. *BMC Med. Inform. Decis. Mak.* **2019**, *19*, 256. [CrossRef]
8. Gupta, S.; Rajiah, P.; Middlebrooks, E.H.; Baruah, D.; Carter, B.W.; Burton, K.R.; Chatterjee, A.R.; Miller, M.M. Systematic Review of the Literature: Best Practices. *Acad. Radiol.* **2018**, *25*, 1481–1490. [CrossRef]
9. Aromataris, E.; Fernandez, R.; Godfrey, C.M.; Holly, C.; Khalil, H.; Tungpunkom, P. Summarizing systematic reviews: Methodological development, conduct and reporting of an umbrella review approach. *Int. J. Evid.-Based Healthc.* **2015**, *13*, 132–140. [CrossRef]
10. O'Connor, A.M.; Anderson, K.M.; Goodell, C.K.; Sargeant, J.M. Conducting systematic reviews of intervention questions I: Writing the review protocol, formulating the question and searching the literature. *Zoonoses Public Health* **2014**, *61*, 28–38. [CrossRef]
11. Roever, L. PICO: Model for Clinical Questions. *Evid.-Based Med.* **2018**, *3*, 2. [CrossRef]
12. Bakker, S.J.; Kleiven, A.; Fleten, S.-E.; Tomasgard, A. Mature offshore oil field development: Solving a real options problem using stochastic dual dynamic integer programming. *Comput. Oper. Res.* **2021**, *136*, 105480. [CrossRef]
13. Liu, Y.Z.; Bai, B.; Li, Y.X.; Coste, J.-P.; Guo, X.H. Optimization Design for Conformance Control Based on Profile Modification Treatments of Multiple Injectors in Reservoir. In Proceedings of the International Oil and Gas Conference and Exhibition in China, Beijing, China, 7–10 November 2000; pp. 1–10. [CrossRef]
14. Liu, Y.; Bai, B.; Wang, Y. Applied Technologies and Prospects of Conformance Control Treatment in China. *Oil Gas Sci. Technol.–Rev. IFP Energies Nouv.* **2010**, *65*, 859–878. [CrossRef]
15. Lamas, L.F.; Botechia, V.E.; Schiozer, D.J.; Rocha, M.L.; Delshad, M. Application of polymer flooding in the revitalization of a mature heavy oil field. *J. Pet. Sci. Eng.* **2021**, *204*, 108695. [CrossRef]
16. Xiong, B.; Loss, R.D.; Shields, D.; Pawlik, T.; Hochreiter, R.; Zydney, A.; Kumar, M. Polyacrylamide degradation and its implications in the environment systems. *NPJ Clean Water* **2018**, *1*, 17. [CrossRef]
17. Zhu, D.; Bai, B.; Hou, J. Polymer Gel Systems for Water Management in High-Temperature Petroleum Reservoirs: A Chemical Review. *Energy Fuels* **2017**, *31*, 13063–13087. [CrossRef]
18. Kamala, M.S.; Sultan, A.S.; Al-Mubaiyedha, U.A.; Hussein, I.A. Review on Polymer Flooding: Rheology, Adsorption, Stability, and Field Applications of Various Polymer Systems. *Polym. Rev.* **2015**, *55*, 491–530. [CrossRef]
19. Silva, I.P.G.; Aguiar, A.A.; Rezende, V.P.; Monsores, A.L.M.; Lucas, E.F. A polymer flooding mechanism for mature oil fields: Laboratory measurements and field results interpretation. *J. Pet. Sci. Eng.* **2018**, *161*, 468–475. [CrossRef]
20. Aldhaheri, M.; Wei, M.; Zhang, N.; Bai, B. Field design guidelines for gel strengths of profile-control gel treatments based on reservoir type. *J. Pet. Sci. Eng.* **2020**, *194*, 107482. [CrossRef]
21. Gussenov, I.; Nuraje, N.; Kudaibergenov, S. Bulk gels for permeability reduction in fractured and matrix reservoirs. *Energy Rep.* **2019**, *5*, 733–746. [CrossRef]

22. Farasat, A.; Younesian-Farid, H.; Sadeghnejad, S. Conformance control study of preformed particle gels (PPGs) in mature waterflooded reservoirs: Numerical and experimental investigations. *J. Pet. Sci. Eng.* **2021**, *203*, 108575. [CrossRef]
23. Pu, J.; Bai, B.; Alhuraishawy, A.; Schuman, T.; Chen, Y.; Sun, X. A Recrosslinkable Preformed Particle Gel for Conformance Control in Heterogeneous Reservoirs Containing Linear-Flow Features. *SPE J.* **2019**, *24*, 1714–1725. [CrossRef]
24. Bai, B.; Zhou, J.; Yin, M. A comprehensive review of polyacrylamide polymer gels for conformance control. *Pet. Explor. Dev.* **2015**, *42*, 525–532. [CrossRef]
25. Esfahlan, M.S.; Khodapanah, E.; Tabatabaei-Nezhad, S.A. Comprehensive review on the research and field application of preformed particle gel conformance control technology. *J. Pet. Sci. Eng.* **2021**, *202*, 108440. [CrossRef]
26. O'Brein, J.; Sayavedra, L.; Mogollon, J.L.; Lokhandwala, T.; Lakani, R. Maximizing Mature Field Production—A Novel Approach to Screening Mature Fields Revitalization Options. In Proceedings of the SPE Europec Featured at 78th EAGE Conference and Exhibition, Vienna, Austria, 30 May 2016. [CrossRef]
27. Abdulbaki, M.; Huh, C.; Sepehrnoori, K.; Delshad, M.; Varavei, A. A critical review on use of polymer microgels for conformance control purposes. *J. Pet. Sci. Eng.* **2014**, *122*, 741–753. [CrossRef]
28. Bai, B.; Li, L.; Liu, Y.; Liu, H.; Wang, Z.; You, C. Preformed Particle Gel for Conformance Control: Factors Affecting Its Properties and Applications. *SPE Reserv. Eval. Eng.* **2007**, *10*, 415–422. [CrossRef]
29. Bai, B.; Wei, M.; Liu, Y. Field and Lab Experience with a Successful Preformed Particle Gel Conformance Control Technology. In Proceedings of the SPE Production and Operations Symposium, Oklahoma City, OK, USA, 23–26 March 2013. [CrossRef]
30. Dai, Z.; Ngai, T. Microgel Particles: The Structure-Property Relationships and Their Biomedical Applications. *J. Polym. Sci. Part A Polym. Chem.* **2013**, *51*, 2995–3003. [CrossRef]
31. Rousseau, D.; Chauveteau, G.; Renard, M.; Tebary, R.; Zaitoun, A.; Mallo, P.; Braun, O.; Omari, A.; Bordeaux, U. Rheology and Transport in Porous Media of New Water Shutoff/Conformance Control Microgels. In Proceedings of the SPE International Symposium on Oilfield Chemistry, The Woodlands, TX, USA, 2–4 February 2005. [CrossRef]
32. Ketova, Y.; Galkin, S.; Kolychev, I. Evaluation and X-Ray tomography analysis of super-absorbent polymer for water management in high salinity mature reservoirs. *J. Pet. Sci. Eng.* **2021**, *196*, 107998. [CrossRef]
33. Zhilin, D.M.; Pich, A. Nano- and microgels: A review for educators. *Chem. Teach. Int.* **2021**, *3*, 155–167. [CrossRef]
34. Kang, W.; Kang, X.; Lashari, Z.A.; Li, Z.; Zhou, B.; Yang, H.; Sarsenbekuly, B.; Aidarova, S. Progress of polymer gels for conformance control in oilfield. *Adv. Colloid Interface Sci.* **2021**, *289*, 102363. [CrossRef]
35. Plamper, F.A.; Richtering, W. Functional Microgels and Microgel Systems. *Acc. Chem. Res.* **2017**, *50*, 131–140. [CrossRef] [PubMed]
36. Daoud, M.; Bouchaud, E.; Jannink, G. Swelling of polymer gels. *Macromolecules* **1986**, *19*, 1955–1960. [CrossRef]
37. Lang, P.R.; Vlassopoulos, D.; Richtering, W. Polymer/Colloid Interactions and Soft Polymer Colloids. *Polym. Sci. A Compr. Ref.* **2012**, *1*, 315–338. [CrossRef]
38. You, Q.; Tang, Y.; Dai, C.; Shuler, P.; Lu, Z.; Zhao, F. Research on a New Profile Control Agent: Dispersed Particle Gel. In Proceedings of the SPE Enhanced Oil Recovery Conference, Kuala Lumpur, Malaysia, 19–21 July 2011. [CrossRef]
39. Khamees, T.; Flori, R. Investigating the Propagation of the Colloidal Dispersion Gel (CDG) in Thick Heterogeneous Reservoirs Using Numerical Simulation. *Am. J. Sci. Eng. Technol.* **2019**, *4*, 1–17. [CrossRef]
40. Chang, H.L.; Sui, X.; Guo, Z.; Yao, Y.; Yiao, Y.; Chen, G.; Song, K.; Mack, J.C. Successful Filed Pilot of In-Depth Colloidal Dispersion Gel (CDG) Technology in Daqing Oilfield. *SPE Reserv. Eval. Eng.* **2006**, *9*, 664–673. [CrossRef]
41. Diaz, D.; Saez, N.; Cabrera, M.; Manrique, E.; Romero, J.; Kazempour, M.; Aye, N. CDG in a Heterogeneous Fluvial Reservoir in Argentina: Pilot and Field Expansion Evaluation. In Proceedings of the SPE/EOR, Kuala Lumpur, Malaysia, 11–13 August 2015. [CrossRef]
42. Smith, J.E.; Liu, H.; Guo, Z.D. Laboratory Studies of In-Depth Colloidal Dispersion Gel Technology for Daqing Oil Field. In Proceedings of the SPE/AAPG Western Regional Meeting, Long Beach, CA, USA, 19–22 June 2000. [CrossRef]
43. Manrique, E.; Reyes, S.; Romero, J.; Aye, N.; Kiani, M.; North, W.; Tomas, C.; Kazempour, M.; Izadi, M.; Roostapour, A.; et al. Colloidal Dispersion Gels (CDG): Field Projects Review. In Proceedings of the SPE EOR Conference at Oil and Gas West Asia, Muscat, Oman, 31 March–2 April 2014. [CrossRef]
44. Ricks, G.V.; Portwood, J.T. Injection-side Application of MARCIT Polymer Improves Waterflood Sweep Efficiency, Decreases Water-Oil Ratio, and Enhances Oil Recovery in the McElroy Field, Upton Country, Texas. In Proceedings of the SPE Permian Basin Oil and Gas Recovery Conference, Midland, TX, USA, 21–23 March 2000. [CrossRef]
45. Zhao, G.; Li, J.; Gu, C.; Li, L.; Sun, Y.; Dai, C. Dispersed Particle Gel Strengthened Polymer/Surfactant as a Novel Combination Flooding System for Enhanced Oil Recovery. *Energy Fuels* **2018**, *32*, 11317–11327. [CrossRef]
46. Dai, C.; Zhao, G.; Zhao, M.; You, Q. Preparation of Dispersed Particle Gel (DPG) through a Simple High Speed Shearing Method. *Molecules* **2012**, *17*, 14484–14489. [CrossRef]
47. You, Q.; Tang, Y.; Dai, C.; Zhao, M.; Zhao, F. A Study on the Morphology of a Dispersed Particle Gel Used as a Profile Control Agent for Improved Oil Recovery. *J. Chem.* **2014**, *2014*, 150256. [CrossRef]
48. Yang, H.; Kang, W.; Yin, X.; Tang, X.; Song, S.; Lashari, Z.A.; Bai, B.; Sarsenbekuly, B. Research on matching mechanism between polymer microspheres with different storage modulus and pore throats in the reservoir. *Powder Technol.* **2017**, *313*, 191–200. [CrossRef]
49. Dai, C.; Chena, W.; You, Q.; Wang, H.; Zhe, Y.; He, L.; Jiao, B.; Wu, Y. A novel strengthened dispersed particle gel for enhanced oil recovery application. *J. Ind. Eng. Chem.* **2016**, *41*, 175–182. [CrossRef]

50. Liu, Y.; Zou, C.; Zhou, D.; Li, H.; Gao, M.; Zhao, G.; Dai, C. Novel Chemical Flooding System Based on Dispersed Particle Gel Coupling In-Depth Profile Control and High Efficient Oil Displacement. *Energy Fuels* **2019**, *33*, 3123–3132. [CrossRef]
51. Liu, J.; Li, L.; Xu, Z.; Chen, J.; Dai, C. Self-growing Hydrogel Particles with Applications for Reservoir Control: Growth Behaviors and Influencing Factors. *J. Phys. Chem. B* **2021**, *125*, 9870–9878. [CrossRef]
52. Bai, B.; Li, L.; Liu, Y.; Wan, Z.; Liu, H. Preformed Particle Gel for Conformance Control: Factors Affecting its Properties and Applications. *SPE/DOE Symp. Improv. Oil Recover. Tulsa OK* **2004**, *17*, 21. [CrossRef]
53. Bai, B.; Liu, Y.; Coste, J.-P.; Li, L. Preformed Particle Gel for Conformance Control: Transport Mechanism through Porous Media. *SPE Reserv. Eval. Eng.* **2007**, *10*, 176–184. [CrossRef]
54. Goudarzi, A.; Zhang, H.; Varavei, A.; Hu, Y.; Delshad, M.; Bai, B.; Seperhroori, K. Water Management in Mature Oil Fields using Preformed Particle Gels. In Proceedings of the SPE Western Regional & AAPG Pacific Section Meeting 2013 Joint Technical Conference, Monterey, CA, USA, 19–25 April 2013. [CrossRef]
55. Elsharafi, M.; Bai, B. Minimizing Formation Damage for Preformed Particle Gels in Mature Reservoirs. In Proceedings of the SPE Asia Pacific Enhanced Oil Recovery Conference, Kuala Lumpur, Malaysia, 11–13 August 2015. [CrossRef]
56. Imqam, A.; Bai, B. Optimizing the strength and size of preformed particle gels for better conformance control treatment. *Feul* **2015**, *148*, 178–185. [CrossRef]
57. Duran-Valencia, C.; Bai, B.; Reyes, H.; Fajardo-Lopez, R.; Barragan-Aroche, F.; Lopez-Ramirez, S. Development of enhanced nanocomposite preformed particle gels for control of High-temperature and high-salinity oil reservoirs. *Polym. J.* **2014**, *46*, 277–284. [CrossRef]
58. Yang, H.; Kang, W.; Liu, S.; Bai, B.; Zhao, J.; Zhang, B. Mechanism and Influencing Factors on the Intitial Particle Size and Swelling Capability of Viscoelastic Microspheres. *J. Dispers. Sci. Technol.* **2015**, *36*, 1673–1684. [CrossRef]
59. Yuan, C.; Pu, W.; Varfolomeev, M.A.; Wei, J.; Zhao, S.; Cao, L.-N. Deformable Micro-Gel for EOR in High-Temperature and Ultra-High-Salinity Reservoirs: How to Design the Particle Size of Micro-Gel to Achieve its Optimal Match with Pore Throat of Porous Media. *SPE J.* **2019**, *26*, 2053–2067. [CrossRef]
60. Qiu, Y.; Wei, M.; Geng, J.; Wu, F. Successful Field Application of Microgel Treatment in High Temperature High Salinity Reservoir in China. In Proceedings of the SPE Improved Oil Recovery Conference, Tulsa, OK, USA, 11–13 April 2016. [CrossRef]
61. Dupuis, G.; Lesuffleur, T.; Desbois, M.; Bouillot, J.; Zaitoun, A. Water Conformance Treatment using SMG Microgels: A Successful Field Case. In Proceedings of the SPE EOR Conference at Oil and Gas West Asia, Muscat, Oman, 21–23 March 2016. [CrossRef]
62. Cozic, C.; Rousseau, D.; Tabary, R. Novel Insights into Microgel Systems for Water Control. *SPE Prod. Oper.* **2009**, *24*, 590–601. [CrossRef]
63. Zaitoun, A.; Dupuis, G. Conformance Control Using SMG Microgels: Laboratory Evaluation and First Field Results. In Proceedings of the SPE Europec Featured at 79th EAGE Conference and Exhibition, Paris, France, 12–15 June 2017. [CrossRef]
64. Zaitoun, A.; Tabary, R.; Rousseau, D.; Pichery, T.R.; Nouyoux, S.; Mallo, P.; Braun, O. Using Microgels to Shut Off Water in Gas Storage Well. In Proceedings of the International Symposium on Oilfield Chemistry, Houston, TX, USA, 28 February–2 March 2007. [CrossRef]
65. Wang, L.; Zhang, G.C.; Ge, J.J.; Li, G.H.; Zhang, J.Q.; Ding, B.D. Preparation of Microgel Nanospheres and Their Application in EOR. In Proceedings of the International Oil and Gas Conference and Exhibition in China, Beijing, China, 8–10 June 2010. [CrossRef]
66. Yao, C.; Lei, G.; Gao, X.; Li, L. Controllable Preparation, Rheology, and Plugging Property of Micron-Grade Polyacrylamide Microspheres as a Novel Profile Control and Flooding Agent. *J. Appl. Polym. Sci.* **2013**, *130*, 1124–1130. [CrossRef]
67. Wang, B.; Lin, M.; Guo, J.; Wang, D.; Xu, F.; Li, M. Plugging properties and profile control effects of crosslinked polyacrylamide microspheres. *J. Appl. Polym. Sci.* **2016**, *133*, 43666. [CrossRef]
68. Lin, M.; Zhang, G.; Hua, Z.; Zhao, Q.; Sun, F. Conformation and plugging properties of crosslinked polymer microspheres for profile control. *Colloids Surfaces A Physicochem. Eng. Asp.* **2015**, *477*, 49–54. [CrossRef]
69. Yang, H.; Kang, W.; Yu, Y.; Yin, X.; Wang, P.; Zhang, X. A new approach to evaluate the particle growth and sedimentation of dispersed polymer microsphere profile control system based on multiple light scattering. *Powder Technol.* **2017**, *315*, 477–485. [CrossRef]
70. Yang, H.; Kang, W.; Tang, X.; Gao, Y.; Zhu, Z.; Wang, P.; Zhang, X. Gel kinetic characteristics and creep behavior of polymer microspheres based on bulk gel. *J. Dispers. Sci. Technol.* **2018**, *39*, 1808–1819. [CrossRef]
71. Yang, H.; Zhou, B.; Zhu, T.; Wang, P.; Zhang, X.; Wang, T.; Wu, F.; Zhang, L.; Kang, W.; Ketova, Y.A.; et al. Conformance control mechanism of low elastic polymer microspheres in porous medium. *J. Pet. Sci. Eng.* **2021**, *196*, 107708. [CrossRef]
72. Yang, H.; Shao, S.; Zhu, T.; Chen, C.; Liu, S.; Zhou, B.; Hou, X.; Zhang, Y.; Kang, W. Shear resistance performance of low elastic polymer microspheres used for conformance control treatment. *J. Ind. Eng. Chem.* **2019**, *79*, 295–306. [CrossRef]
73. Yang, H.; Kang, W.; Wu, H.; Yu, Y.; Zhu, Z.; Wang, P.; Zhang, X.; Sarsenbekulyb, B. Stability, rheological property and oil-displacement mechanism of a dispersed low-elastic microsphere system for enhanced oil recovery. *RSC Adv.* **2017**, *7*, 8118–8130. [CrossRef]
74. Tang, X.; Kang, W.; Zhou, B.; Gao, Y.; Cao, C.; Guo, S.; Iqbal, M.W.; Yang, H. Characteristics of composite microspheres for in-depth profile control in oilfields and the effects of polymerizable silica nanoparticles. *Powder Technol.* **2020**, *359*, 205–215. [CrossRef]

75. Tang, X.; Zhou, B.; Chen, C.; Sarsenbekuly, B.; Yang, H.; Kang, W. Regulation of polymerizable modification degree of nano-SiO_2 and the effects on performance of composite microsphere for conformance control. *Colloids Surfaces A Physicochem. Eng. Asp.* **2020**, *585*, 124100. [CrossRef]
76. Tang, X.; Yang, H.; Gao, Y.; Lashari, Z.A.; Cao, C.; Kang, W. Preparation of a micron-size silica-reinforced polymer microsphere and evaluation of its properties as a plugging agent. *Colloids Surfaces A Physicochem. Eng. Asp.* **2018**, *547*, 8–18. [CrossRef]
77. Yang, H.; Hu, L.; Chen, C.; Zhao, H.; Wang, P.; Zhu, T.; Wang, T.; Zhang, L.; Fan, H.; Kang, W. Influence mechanism of fluorescent monomer on the performance of polymer microspheres. *J. Mol. Liq.* **2020**, *308*, 113081. [CrossRef]
78. Yang, H.; Hu, L.; Tang, X.; Gao, Y.; Shao, S.; Zhang, X.; Wang, P.; Zhu, Z.; Kang, W. Preparation of a fluorescent polymer microsphere and stability evaluation of its profile control system by a fluorescence stability index. *Colloids Surfaces A Physicochem. Eng. Asp.* **2018**, *558*, 512–519. [CrossRef]
79. Yang, H.; Hu, L.; Chen, C.; Gao, Y.; Tang, X.; Yin, X.; Kang, W. Synthesis and plugging behavior of fluorescent polymer microspheres as a kind of conformance control agent in reservoirs. *RSC Adv.* **2018**, *8*, 10478–10488. [CrossRef]
80. Kang, W.; Hu, L.; Zhang, X.; Yang, R.-M.; Fan, H.-M.; Geng, J. Preparation and performance of fluorescent polyacrylamide microspheres as a profile control and tracer agent. *Pet. Sci.* **2015**, *12*, 483–491. [CrossRef]
81. Frampton, H.; Morgan, J.C.; Cheung, S.K.; Munson, L.; Chang, K.T.; Williams, D. Development of a Novel Waterflood Conformance Control System. In Proceedings of the SPE/DOE Symposium on Improved Oil Recovery, Tulsa, OK, USA, 17–21 April 2004. [CrossRef]
82. Pritchett, J.; Frampton, H.; Brinkman, J.; Cheung, S.; Morgan, J.; Chang, K.T.; Williams, D.; Goodgame, J. Field Application of a New In-Depth Waterflood Conformance Improvement Tool. In Proceedings of the SPE International Improved Oil Recovery Conference in Asia Pacific, Kuala Lumpur, Malaysia, 20–21 October 2003. [CrossRef]
83. Roussennac, B.; Toschi, C. Brightwater® Trial in Salema Field (Campos Basin, Brazil). In Proceedings of the SPE EUROPEC/EAGE Annual Conference and Exhibition, Barcelona, Spain, 14–17 June 2010. [CrossRef]
84. Ohms, D.S.; McLeod, J.D.; Graff, C.J.; Frampton, H.; Morgan, J.; Cheung, S.K.; Yancey, K.E.; Chang, K.-T. Incremental Oil Success From Waterflood Sweep Improvement in Alaska. In Proceedings of the SPE International Symposium on Oilfield Chemistry, The Woodlands, TX, USA, 20–22 April 2009. [CrossRef]
85. Ashrafizadeh, M.; Tam, K.C.; Javadi, A.; Abdollahi, M.; Sadeghnejad, S.; Bahramian, A. Synthesis and physicochemical properties of dual-responsive acrylic acid/butyl acrylate crosslinked nanogel systems. *J. Colloid Interface Sci.* **2019**, *556*, 313–323. [CrossRef]
86. Teimouri, A.; Sadeghnejad, S.; Dehaghani, A.H.S. Investigation of acid pre-flushing and pH-sensitive microgel injection in fractured carbonate rocks for conformance control purposes. *Oil Gas Sci. Technol.–Rev. IFP Energies Nouv.* **2020**, *75*, 52. [CrossRef]
87. Koochakzadeh, A.; Younesian-Farid, H.; Sadeghnejad, S. Acid pre-flushing evaluation before pH-sensitive microgel treatment in carbonate reservoirs: Experimental and numerical approach. *Fuel* **2021**, *297*, 120670. [CrossRef]
88. Al-Wahaibi, Y.; Al-Wahaibi, T.; Abdelgoad, M. Characterization of pH-sensitive Polymer microgel transport in porous media for improving oil recovery. *Energy Sources Part A Recovery Util. Environ. Eff.* **2011**, *33*, 1048–1057. [CrossRef]
89. Yen, W.S.; Coscia, A.T.; Kohen, S.I. Polyacrylamides Chapter 8. *Dev. Pet. Sci.* **1989**, *17*, 189–218. [CrossRef]
90. Hamzah, Y.B.; Hashim, S.; Rahman, W.A.W.A. Synthesis of polymeric nano/microgels: A review. *J. Polym. Res.* **2017**, *24*, 134. [CrossRef]
91. Chern, C.S. Emulsion polymerization mechanisms and kinetics. *Prog. Polym. Sci.* **2006**, *31*, 443–486. [CrossRef]
92. Pich, A.; Richtering, W. Microgels by Precipitation Polymerization: Synthesis, Characterization, and Functionalization. *Adv. Polym. Sci.* **2010**, *234*, 1–37. [CrossRef]
93. Tauer, K.; Hernandez, H.; Kozempel, S.; Lazareva, O.; Nazaran, P. Towards a consistent mechanism of emulsion polymerization—New experimental details. *Colloid Polym. Sci.* **2008**, *286*, 499–515. [CrossRef] [PubMed]
94. Bjørsvik, M.; Høiland, H.; Skauge, A. Formation of colloidal dispersion gels from aqueous polyacrylamide solutions. *Colloids Surfaces A Physicochem. Eng. Asp.* **2008**, *317*, 504–511. [CrossRef]
95. Castro-Garcia, R.-H.; Maya-Toro, G.A.; Sandoval-Munoz, J.E.; Cohen-Paternina, L.-M. Colloigal Despersion Gels (CDG) to Improve Volumetric Sweep Efficiency in Waterflooding Processes. *Lat. Am. J. Oil Gas Altern. Energies* **2013**, *5*, 61–78. [CrossRef]
96. Kaushanski, D.A.; Batyrbaev, M.D.; Duzbaev, S.K.; Demyanovski, V.B. The results of the use of Temposcreen technology in the fields of the Republic of Kazakhstan (using the example of PF Embamunaigas). *Geol. Geophys. Dev. Oil Gas Fields* **2006**, *9*, 51–58.
97. Kaushanski, D.A. Multifunctional innovative technology of enhanced oil recovery at a late stage of development "Temposcreen-lux". *Georesour. Geoenergy Geopolit.* **2014**, *1*, 9. Available online: http://oilgasjournal.ru/vol_9/kaush-adv.pdf (accessed on 12 December 2021).
98. Chen, L.-W.; Yang, B.-Z.; Wu, M.-L. Synthesis and kinetics of microgel in inverse emulsion polymerization of acrylamide. *Prog. Org. Coat.* **1997**, *31*, 393–399. [CrossRef]
99. Hajighasem, A.; Kabiri, K. Novel crosslinking method for preparation of acrylic thickener microgels through inverse emulsion polymerization. *Iran. Polym. J.* **2015**, *24*, 1049–1056. [CrossRef]
100. Barabanov, V.L.; Demyanoski, V.B.; Kaushanski, D.A. Study of geological heterogeneity of liquid systems on the example of polyacrylamide dispersion gels swollen in water. *Actual Probl. Oil Gas* **2016**, *1*, 13. [CrossRef]

gels

MDPI

Article

Fabrication and Characterization of Xanthan Gum-cl-poly(acrylamide-co-alginic acid) Hydrogel for Adsorption of Cadmium Ions from Aqueous Medium

Gaurav Sharma [1,2,3,*], Amit Kumar [1,2], Ayman A. Ghfar [4], Alberto García-Peñas [5], Mu. Naushad [4,*] and Florian J. Stadler [1]

1 College of Materials Science and Engineering, Shenzhen Key Laboratory of Polymer Science and Technology, Guangdong Research Center for Interfacial Engineering of Functional Materials, Nanshan District Key Laboratory for Biopolymers and Safety Evaluation, Shenzhen University, Shenzhen 518060, China; mittuchem83@gmail.com (A.K.); fjstadlr@szu.edu.cn (F.J.S.)

2 International Research Centre of Nanotechnology for Himalayan Sustainability (IRCNHS), Shoolini University, Solan 173212, Himachal Pradesh, India

3 School of Science and Technology, Glocal University, Saharanpur 247001, Uttar Pradesh, India

4 Department of Chemistry, College of Science, King Saud University, P.O. Box 2455, Riyadh 11451, Saudi Arabia; aghafr@ksu.edu.sa

5 Departamento de Ciencia e Ingeniería de Materiales e Ingeniería Química (IAAB), Universidad Carlos III de Madrid, Leganés, 28911 Madrid, Spain; alberto.garcia.penas@uc3m.es

* Correspondence: gaurav8777@gmail.com (G.S.); mnaushad@ksu.edu.sa (M.N.)

Citation: Sharma, G.; Kumar, A.; Ghfar, A.A.; García-Peñas, A.; Naushad, M.; Stadler, F.J. Fabrication and Characterization of Xanthan Gum-cl-poly(acrylamide-co-alginic acid) Hydrogel for Adsorption of Cadmium Ions from Aqueous Medium. *Gels* **2022**, *8*, 23. https://doi.org/10.3390/gels8010023

Academic Editor: Hiroyuki Takeno

Received: 2 December 2021
Accepted: 23 December 2021
Published: 27 December 2021

Abstract: The present research demonstrates the facile fabrication of xanthan gum-cl-poly(acrylamide-co-alginic acid) (XG-cl-poly(AAm-co-AA)) hydrogel by employing microwave-assisted copolymerization. Simultaneous copolymerization of acrylamide (AAm) and alginic acid (AA) onto xanthan gum (XG) was carried out. Different samples were fabricated by changing the concentrations of AAm and AA. A sample with maximum swelling percentage was chosen for adsorption experiments. The structural and functional characteristics of synthesized hydrogel were elucidated using diverse characterization tools. Adsorption performance of XG-cl-poly(AAm-co-AA) hydrogel was investigated for the removal of noxious cadmium (Cd(II)) ions using batch adsorption from the aqueous system, various reaction parameters optimized include pH, contact time, temperature, and concentration of Cd(II) ions and temperature. The maximum adsorption was achieved at optimal pH 7, contact time 180 min, temperature 35 °C and cadmium ion centration of 10 mg·L^{-1}. The XG-cl-poly(AAm-co-AA) hydrogel unveiled a very high adsorption potential, and its adsorption capacities considered based on the Langmuir isotherm for Cd(II) ions was 125 mg·g^{-1} at 35 °C. The Cd(II) ions adsorption data fitted nicely to the Freundlich isotherm and pseudo-first-order model. The reusability investigation demonstrated that hydrogel retained its adsorption capacity even after several uses without significant loss.

Keywords: xanthum gum; polyacrylamide; cadmium ions; adsorption; Langmuir; Freundlich

1. Introduction

Wastewater remediation is attaining importance in the present era due to the severe scarcity of water in various countries. Various types of contaminants are added to the water bodies without any pretreatments such as dyes, heavy metal ions, pigments, pesticides, pharmaceutical effluents, personal care products, and radioactive wastes, etc. Among these, heavy metals are very common pollutants added from diverse sources such as electroplating, batteries manufacturing, metal treating in processing plants, sweltering of coal in power plants, incineration of petroleum, nuclear power stations, plastics, textiles, microelectronics, paper processing plants, and wood preservation, etc. [1–3].

Heavy metals are also known as trace elements as they are present in very less quantity few heavy metals such as iron, copper, chromium, zinc, magnesium, selenium, manganese and molybdenum, etc. are important and essential for several physiological and biochemical functions [4]. Whereas other heavy metals as arsenic, aluminum, lead, antinomy, mercury, barium, indium, beryllium, nickel, bismuth, cadmium, gold, gallium, germanium, lithium, platinum, tin, silver, titanium, strontium, tellurium, vanadium, thallium, and uranium do not have any such recognized physiological or biological functions and thus are non-essential [5–8]. Out of these heavy metals, cadmium is recorded as a very common contaminant added to groundwater and soil. Cadmium is extremely lethal and may threaten the aquatic ecosystem and humans. Acquaintance with cadmium for long periods of time can severely affect health and may result in diseases such as osteoporosis, cardiac failure, cancer and itai-itai disease, etc. [9,10]. The presence of cadmium in the environment has become a topic of concern, as it cannot be decayed by microorganisms, thus it unceasingly accumulates, transforms, and migrates in food chains. The cadmium bioavailability and migration in the ecosystem are influenced by its magnitude of adsorption with solid segments. The adsorption extent of cadmium ions depends upon its interaction between aqueous and solid phases and is entirely controlled by properties such as temperature, pH, surface area, ionic strength, and surface charge, etc. [11–13].

Numerous methods were implemented for remediation of noxious cadmium ions from wastewater these include phytoremediation, ion exchange, chemical precipitation, solvent extraction, coagulation, filtration, adsorption and membrane techniques, etc. [14,15]. As adsorption is a most effective and easy to handle technique it has been extensively used for heavy metals remediation. Numerous researchers are working on designing and fabricating highly efficient adsorbents which include modified activated carbon, zeolites, metal-organic frameworks, carbon nanotubes, biochar, MXenes, carbon nitride, and hydrogels, etc. [16–19]. Hydrogels are very efficient adsorbents as they possess greater functionalities, better swelling, good surface area and biocompatibility, etc. Diverse synthetic and natural polymer or gums have been used for the fabrication of superabsorbent hydrogel, these include chitosan, pectin, carrageenan, starch, gelatin, cellulose, chitin, sodium alginate, guar gum, gum arabica, tragacanth gum, xanthan gum, polyacrylamide (PAM), polyalginic acid, polyacrylic acid(PAA), poly(hydroxyethyl methacrylate) (PHEMA), Poly(glyceryl methacrylate) (PGMA), 1,1,1-trimethylolpropane trimethacrylate (TMPTMA), polyvinyl pyrrolidone (PVP), poly(ethylene glycol) dimethacrylate (PEGDMA), polymethacrylamide (PMAM), triethylene glycol dimethacrylate (TEGDMA), polyethylene glycol (PEG), polyvinyl alcohol (PVA), Poly(hydroxypropyl methacrylate) (PHPMA), and poly(ε-caprolactone) (PCL), etc. [20–24]. The high swelling ability of hydrogels helps in enriching the structure with high absorption and adsorption characteristics. When present in the aqueous solution, the polymer chains loosen up so as to swell more and more solvent inside it. Being eco-friendly in nature, hydrogels are now utilized for diverse applications ranging from water purification to biomedical. The crosslinking of monomers and polymer chains helps in enriching the structure with high swelling capacity, enhanced adsorption or absorption ability, and mechanical strength.

So, herein, an efficient adsorbent was synthesized using xantham gum, acrylamide, and acrylic acid by a green method in which microwave radiations were used for the synthesis. The synthesized adsorbent hydrogel was used positively for the adsorption of Cd(II) ions from an aqueous solution. It is worth mentioning that the hydrogel swelling shoot up the adsorption process and lessened the needed time to reach 90% of optimum adsorption from 240 to 180 min. The adsorption process followed linear forms of the PSO and Freundlich model. The free energy of Cd(II) ions after adsorption disclosed the physical nature of adsorption, while PSO anticipated chemical interactions. Consequently, it can be established that the interactions are physiochemical in nature.

2. Results and Discussion

The three different samples of XG-cl-poly(AAm-co-AA) hydrogel were prepared by varying the concentration of acrylamide and alginic acid, i.e., 1XG:1AAm:1 AA, 1XG:1AAm:0.5 AA, 1XG:1AAm:0.5 AA. The swelling experiments were performed for 36 hrs and swelling percentages were found to be 84%, 79%, and 96%, respectively. Thus sample with reactant ratio1XG:1AAm:0.5 AA was chosen for adsorption studies as it possesses the maximum swelling. The swelling test is the characteristic of any hydrogel, thus, the higher the swelling ability, the better the sorption activity will be.

The results of FTIR spectra of XG-cl-poly(AAm-co-AA) hydrogel before and after Cd(II) ion adsorption are presented in Figure 1. The FTIR spectrum in Figure 1a represents the peaks obtained for XG-cl-poly(AAm-co-AA) hydrogel. It exhibits a prominent broad band at 3420 cm^{-1} signifying the stretching vibrations for the -OH groups of acrylic acid and xanthan gum [25]. The peak for the C-H stretching vibrations for the aliphatic units present in the hydrogel is detected at 2923 cm^{-1}. The peaks at 1796 cm^{-1} and 1676 cm^{-1} are owing to the stretching of carbonyl groups and reveal symmetrical stretching for the carboxylate group [26]. Thus, these peaks are indicative of the C=O groups present in poly(acrylic acid), NMBA, and XG [27]. Another band at 1440 cm^{-1} can be ascertained to the −CH symmetric bending vibrations due to the existence of −CHOH− groups in the hydrogel [28]. The peak at 1385 cm^{-1} is due to the bending vibration of C-H of the isopropyl group, 1188 cm^{-1} is due to C-N stretching, 1124 cm^{-1} is owing to the -CO stretching, and 876 cm^{-1} and 714 cm^{-1} peaks can be consigned to N-H bending vibrations [29]. The band at 1036 cm^{-1} may possibly be owing to the stretching of the C-O bond of glycosidic bonds. Alike bands are present in the XG-cl-poly(AAm-co-AA) hydrogel after adsorption of Cd(II) ions with variation in peaks intensities and some additional shifts in the peaks. For example, the broad band observed at 3420 cm^{-1} in hydrogel was shifted to higher wavenumber 3492 cm^{-1}, which could probably be due to the possible intermolecular interactions between the hydrogel and Cd(II) ions [30]. Furthermore, the characteristic peak of C=O at 1676 cm^{-1} was shifted to 1735 cm^{-1} suggesting the probable complexation of carbonyl units and Cd(II) ions. Thus, it can be concluded that the changes in the FTIR spectrum of hydrogel after Cd(II) ion adsorption is due to the physicochemical interactions. Considering the FTIR results, it can be concluded that the major groups that participated in the complex formation (Cd(II) adsorbed XG-cl-poly(AAm-co-AA) hydrogel), were –COOH, –C=O, and –OH groups. Thus, a probable scheme is presented (Scheme 1) that shows the expected structure of the complex formed.

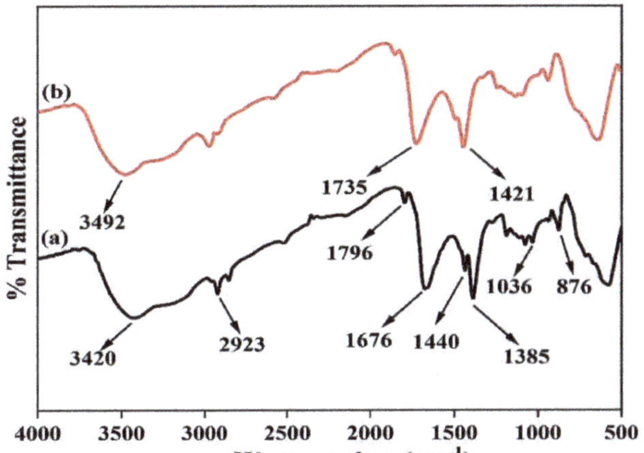

Figure 1. FTIR of spectra of xanthan gum-cl-poly(acrylamide-co-alginic acid) hydrogel (**a**) before adsorption and (**b**) after adsorption.

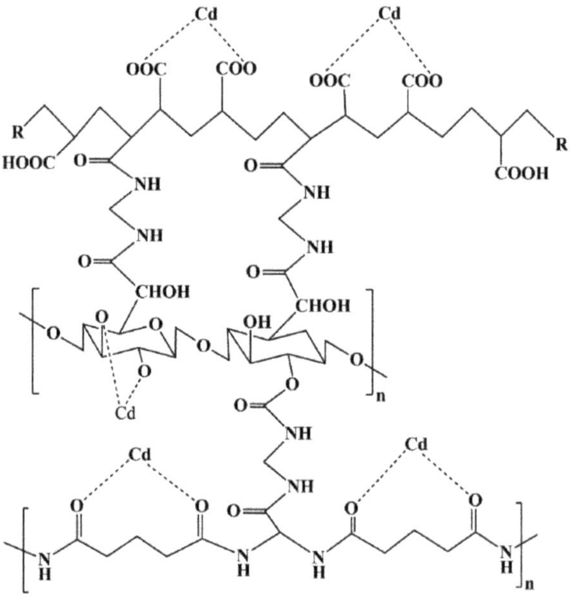

Scheme 1. Probable scheme for the complex formation by the adsorption of Cd(II) ions onto XG-cl-poly(AAm-co-AA) hydrogel.

The X-ray diffractogram of XG-cl-poly(AAm-co-AA) hydrogel is shown in Figure 2. X-ray diffractometer PAN analytical X′Pert PRO was used during the study. The X-ray diffractometric examination was performed to disclose the phase (crystallinity or amorphous) of the XG-cl-poly(AAm-co-AA) hydrogel. Figure 2 depicts that the diffractogram of hydrogel was found to be semi-crystalline in nature. The previous studies of crude xanthan gum diffractogram show XRD peaks at a 2θ value of 20° [31]. The XG-cl-poly(AAm-co-AA) hydrogel diffractogram shows a peak at a 2θ value of 20.72° for xanthan gum. It was observed that in hydrogel formation crystallinity of the native xanthan gum increases. It was observed that XG-cl-poly(AAm-co-AA) hydrogel displayed eighteen observable and discrete diffraction peaks at 2θ values of 19.23°, 20.72°, 23.13°, 25.18°, 27.22°, 28.02°, 29.27°, 30.24°, 31.58°, 33.71°, 40.49°, 44.22°, 45.29°, 56.42°, 57.84°, 60.15°, 66.20°, 75.10°. These peaks displayed the grafted acrylamide and alginic acid onto crude xanthan gum. This outcome confirms that the optimized grafting happened when acrylamide and alginic acid reacted with xanthan gum and the obtained end product was established to be a highly crosslinked hydrogel. Thus intensification in crystallinity of xanthan gum could be ascribed to the impact made by acrylamide and alginic acid. As the synthesis was carried out with microwave irradiation which further significantly improved the crystallinity of xanthan gum-based hydrogel. A similar upsurge in crystallinity with the subsequent microwave treatment of xanthan gum was also observed by Sharma et.al., (2011) Singh et al. (2009) and Anjum et al. (2015) [32–34].

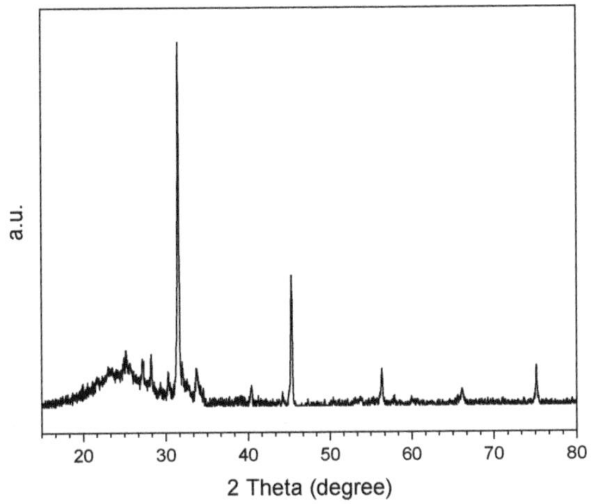

Figure 2. XRD of xanthan gum-cl-poly(acrylamide-co-alginic acid) hydrogel.

The SEM images of that XG-cl-poly(AAm-co-AA) hydrogel were displayed in Figure 3. Figure 3a,b depicts the smooth surface of hydrogel whereas Figure 3c displays the highly folded surface with pores at higher magnification. Figure 3d represents the SEM image of XG-cl-poly(AAm-co-AA) hydrogel after adsorption of Cd(II) ions depicting rough surface which may be due to the addition of Cd(II) ions to the surface.

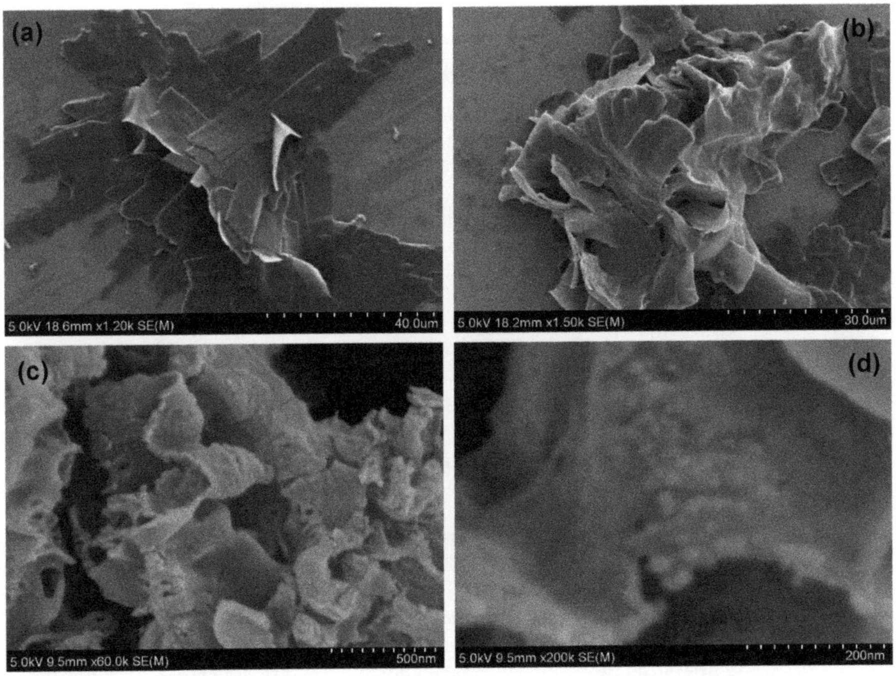

Figure 3. SEM micrographs of xanthan gum-cl-poly(acrylamide-co-alginic acid) hydrogel (**a**–**c**) before adsorption and (**d**) after adsorption.

2.1. Adsorption of Cadmium Ions by XG-cl-Poly(AAm-co-AA) Hydrogel

2.1.1. Effect of Various Factors

Figure 4 shows the effect of various parameters such as contact interval, solution pH, temperature, and concentration of adsorbate on the adsorption of Cd(II) ions onto XG-cl-poly(AAm-co-AA) hydrogel. Figure 4a displays the outcome of contact time on the adsorption rate as an escalation in time duration enhances the probability of interactions between the adsorbate and adsorbent surface. The result shows the initial increase in adsorption rate up to 180 min which is probably due to the high accessibility of free or active adsorbent sites. However, afterward, a nearly constant rate was obtained due to the partial saturation of accessible active sites.

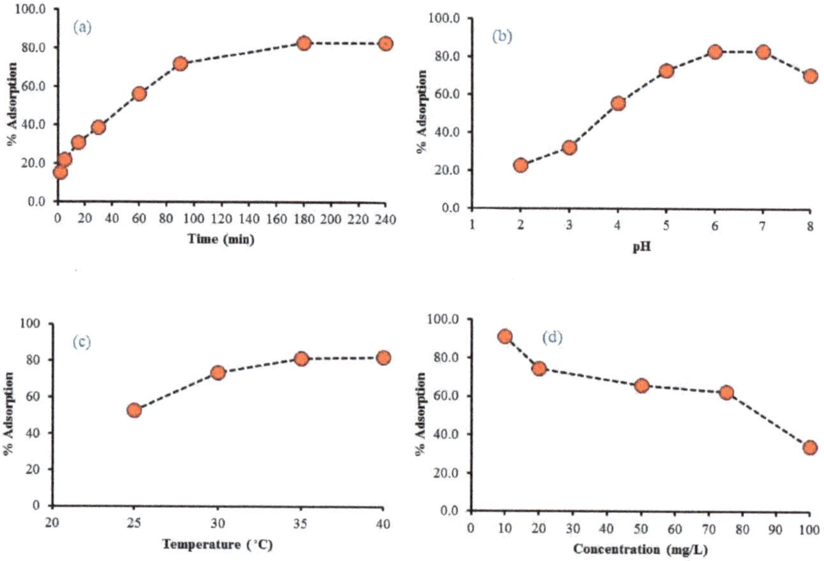

Figure 4. Effect of influential parameters on the adsorption of cadmium ions (**a**) Contact time, (**b**) solution pH, (**c**) temperature, and (**d**) concentration of adsorbate.

The effect of solution pH on the adsorption rate is presented in Figure 4b and was studied at different pH values such as 2, 4, 5, 6, 7, and 8. Results show that the rate first increased up to 7 and then decreased to 8, thus, working pH was found to be 7. The % adsorption of the XG-cl-poly(AAm-co-AA) hydrogel first improved with the initial upsurge in pH value because of the ionization of the hydrophilic polymer network. Major functional units existing in the structure of hydrogel are, -NH_2, COOH, and –OH. The adsorption rate enhanced with the rise in pH because of the ionization of –NH_2 and –COOH groups at pH greater than the pka value. This favored the intermolecular interactions among the cadmium (II) ions and hydrophilic XG-cl-poly(AAm-co-AA) hydrogel networks.

The influence of temperature on the adsorption rate was also analyzed and considered at four altered temperatures; 25 °C, 30 °C, 35 °C, and 40 °C. The temperature influence generalizes the chemical or physical nature of the adsorption process. Results presented in Figure 4c show the maximum adsorption rate of 84% was obtained at 35 °C suggesting the inclination of Cd(II) ion adsorption towards chemical behavior. Further increase in temperature resulted in a constant adsorption rate indicating the saturation at 35 °C. This study indicated that the temperature played a critical part in the undertaken Cd(II) ions adsorption.

The significance of Cd(II) ion concentration on the adsorption rate was analyzed in the range of 10–100 mg·L^{-1}. The results presented in Figure 4d show the fall in % adsorption with the upsurge in Cd(II) concentration and the maximum rate was obtained at 10 mg·L^{-1}. The increase in concentration overloaded the solution with Cd(II) ions which made it difficult for the hydrogel to adsorb the ions, as a result of which the adsorption decreased. Maximum adsorption of 92% was obtained at 10 mg·L^{-1} Cd(II) concentration.

2.1.2. Adsorption Kinetics

Adsorption kinetics includes the variation in the adsorption characteristics of the system w.r.t. time, where the amount of surface covered delivers significant insight on the rate of the process. The rating mechanism of Cd(II) ion adsorption onto XG-cl-poly(AAm-co-AA) hydrogel was considered using two kinetic models, pseudo-first-order (PFO) and pseudo-second-order (PSO). These models were applied to the kinetic data obtained for a time at an interval of 5–180 min. The linear form of the models was used and the equations representing them were [35]:

$$\text{Log}\ (q_e - q_t) = \text{Log}\ q_e - K_1 t\ \text{Pseudo} - \text{first} - \text{order} \quad (1)$$

$$\frac{t}{q_t} = \frac{1}{k_2 {q_e}^2} + \frac{t}{q_e}\ \text{Pseudo} - \text{second} - \text{order} \quad (2)$$

where q_e and q_t (mg·g^{-1}) denote the Cd (II) ion amount at equilibrium (e) and at time t. k_1 (min^{-1}) and k_2 (g·mg^{-1}·min^{-1}) are the pseudo-first-order and pseudo-second-order rate constant. The calculated parameter values of the two models are presented in Table 1. Assessment of the R^2 values disclosed that the pseudo-first-order model better fitted the kinetic data than that of the pseudo-second-order model. Additionally, the experimental qe value (91.26 mg·g^{-1}) was found to best correlate with the calculated qe value from pseudo-first-order (95.9 mg·g^{-1}). In addition, the best fit was attained at a higher initial concentration of 60 mg·L^{-1} signifying the reason for its better fit [36]. Undertaken adsorption process signifies that the Cd(II) ions adsorption was quite a tedious process of synchronized action of several reactions, and might encompass various interactions among adsorbent and adsorbate.

Table 1. Adsorption kinetic parameters.

Kinetic Models	Parameters	10 mg·L^{-1}	20 mg·L^{-1}	60 mg·L^{-1}
Pseudo-first-order	q_e (mg·g^{-1})	34.8	68.5	95.9
	k_1 (min^{-1})	1.70×10^{-2}	1.68×10^{-2}	1.65×10^{-2}
	R^2	0.990	0.986	0.992
Pseudo-second-order	q_e (mg·g^{-1})	39.8	76.9	111.1
	k_2 (g·mg^{-1}·min^{-1})	13.4×10^{-3}	5.6×10^{-4}	4.2×10^{-4}
	R^2	0.929	0.923	0.904

2.1.3. Adsorption Isotherms

Figure 5c,d shows the linear Langmuir, and Freundlich isotherm models for the adsorption of Cd(II) ions onto hydrogel at two temperatures, viz., 25 °C and 35 °C. The equations used for the linear isotherm models are [37]:

$$\frac{1}{q_e} = \frac{1}{(qmK_L)C_e} + \frac{1}{qm} \quad \text{Langmuir} \quad (3)$$

$$\text{Log}\ q_e = \text{Log}\ K_F + \frac{1}{n}\text{Log}\ C_e \quad \text{Freundlich} \quad (4)$$

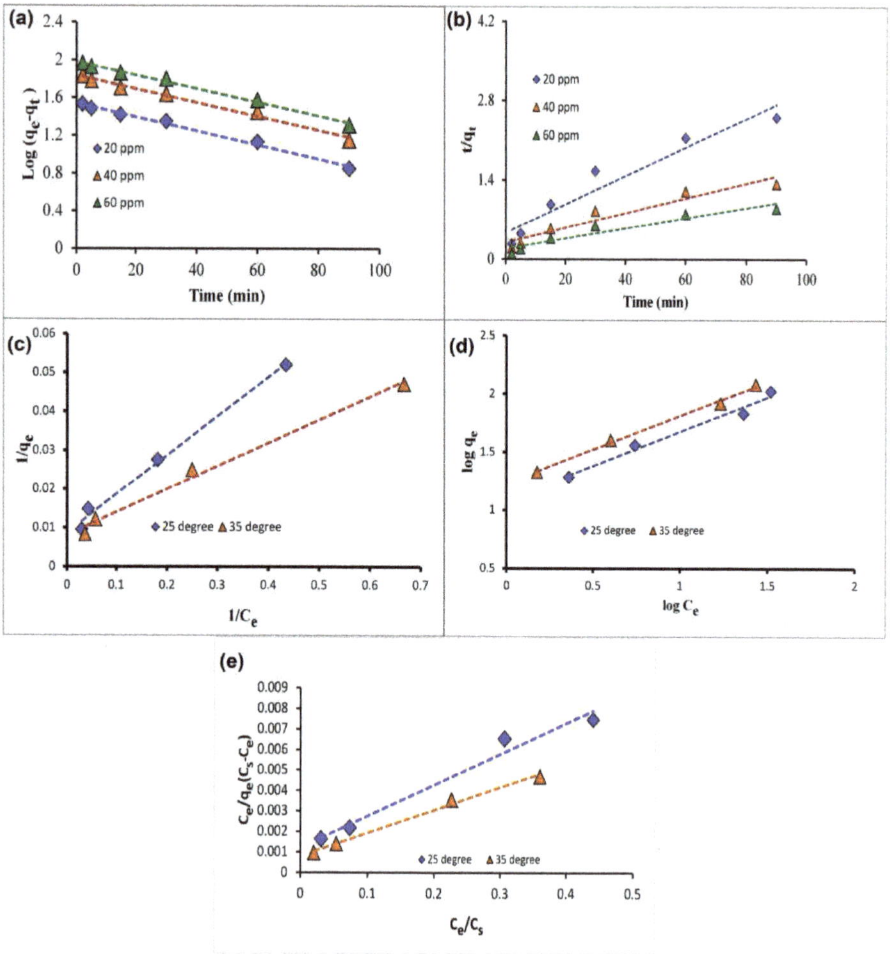

Figure 5. Adsorption kinetic models (**a**) Pseudo-first-order, (**b**) Pseudo-second-order, and Adsorption isotherm models (**c**) Langmuir isotherm, (**d**) Freundlich isotherm and (**e**) BET isotherm.

In addition, the multilayer adsorption isotherm model, BET, was also applied to the isotherm data. BET isotherm was first established in 1938 by Brunauer and his coworkers and it is considered as one of the most proficient models for expressing the adsorption phenomena. It helps in determining various parameters of the undertaken adsorption such as heat of adsorption, adsorption capacity, and multilayer adsorption behavior. The linear equation used for the analysis is [38]:

$$\frac{C_e}{q_e(C_s - C_e)} = \frac{1}{q_s C_{BET}} + \frac{(C_{BET} - 1)}{q_s C_{BET}}\left(\frac{C_e}{C_s}\right) \quad \text{BET} \tag{5}$$

Here C_e and C_s denote the equilibrium concentration ($mg \cdot L^{-1}$) and monolayer saturation concentration ($mg \cdot L^{-1}$). C_{BET} and qs represent the BET adsorption isotherm constant ($L \cdot mg^{-1}$) and isotherm saturation capacity ($mg \cdot g^{-1}$), respectively.

The values obtained for different parameters of the three isotherm models are displayed in Table 2. The comparison of three fitted isotherm models by correlation coefficient (R^2) showed that the Freundlich isotherm fitted the best to the undertaken adsorption experiment at 35 °C. This suggested the Cd(II) ions followed multilayer adsorption onto the XG-cl-poly(AAm-co-AA) hydrogel. Isotherm analysis also suggested that the Cd(II) ions were physically adsorbed onto the surface. Maximum adsorption capacity achieved was 114.94 and 125.00 mg·g^{-1} at 25 °C, and 35 °C, respectively, which was found to be quite high as compared to other adsorbents used for Cd(II) ion adsorption testified in the literature such as 27.3 mg·g^{-1} by commercial activated carbon [39], 2.9 mg·g^{-1} for Fe_3O_4/AC [40], 7.4 mg·g^{-1} by magnetic oak bark biochar [41], 17.54 mg·g^{-1} by $CuFe_2O_4$ nano-particles [42], and 63 mg·g^{-1} by EDTA@Fe_3O_4/SC nanocomposite [43]. The factor representing the binding affinity (K_L) was found to be quite favorable suggesting that the adsorption process relied on high binding affinity among the hydrogel and Cd(II) ions. The value of separation factor (R_L) obtained is less than 1 indicate the favorable shape of the isotherm. The BET model showed temperature-dependent activity in which better fit was obtained at higher temperatures. Furthermore, it can be generalized that the adsorption of Cd(II) ions onto the XG-cl-poly(AAm-co-AA) hydrogel did not follow the monolayer adsorption.

Table 2. Adsorption isotherm parameters.

Equilibrium Model	Parameters	25 °C	35 °C
Langmuir isotherm	q_m (mg·g^{-1})	114.9	125
	b (L·mg^{-1})	8.3×10^{-2}	9.1×10^{-2}
	R_L	0.37	0.35
	R^2	0.981	0.990
Freundlich isotherm	K_F (L·mg^{-1})	12.21	17.14
	n	1.69	1.72
	R^2	0.990	0.995
BET	Qs (mg·g^{-1})	75.75	84.74
	C_{BET} (L·mg^{-1})	13.225	14.755
	R^2	0.963	0.988

Comparison data are also given in Table 3 that highlighted the superiority of the synthesized XG-cl-poly(AAm-co-AA) hydrogel as adsorbent for Cd(II) ions. Although, the comparison of adsorption capacity of any adsorbent is not possible since it is governed by various influential factors such as composition, reaction conditions, temperature, solution pH, and functionality extent, etc. So, in the present case, we gave a generalized comparison of various hydrogel-based adsorbents used for Cd(II) adsorption.

Table 3. Comparison of monolayer adsorption capacity of XG-cl-poly(AAm-co-AA) hydrogel with other adsorbents reported in the literature for Cd(II) adsorption.

Adsorbent	Adsorption Capacity (q_m, mg·g^{-1})	References
S(H)-PAA hydrogel	109.8	[44]
SCHBs	95.6	[45]
NNCA hydrogel	9.54	[46]
Bentonite/alginate composite beads	53.2	[47]
Thiol-functionalized mesoporous silica	78	[48]
sodium alginate-meso-2,3-dimercaptosuccinic acid hybrid aerogel	91.2	[49]
β-cyclodextrin-based hydrogel	98.8	[50]
HMO-P(HMAm/HEA) hydrogel	93.86	[51]
XG-cl-poly(AAm-co-AA) hydrogel	125	**Present work**

2.1.4. Thermodynamic Analysis

Thermodynamic studies were examined at different temperatures, viz., 298 K, 303 K, 308 K and 313 K. Equations employed for determining Gibb's energy, enthalpy, and entropy change of the undertaken adsorption process were:

$$\Delta G^0 = -RT\ln k_C \tag{6}$$

$$\ln k_C = -\frac{\Delta H^0}{RT} + \frac{\Delta S^0}{R} \tag{7}$$

where ΔG^0 was determined from Equation (6) and ΔH^0 and ΔS^0 were determined from Equation (7) by the linear plot of ln kC versus 1/T(K).

The calculated values are depicted in Table 4. Results indicated that Gibb's free energy became more negative with the increase in temperature and maximum was obtained at 318 K. This generalized the spontaneous nature of the Cd(II) ion adsorption onto XG-cl-poly(AAm-co-AA) hydrogel. Additionally, enthalpy change was found to be +24.5 $J\cdot mol^{-1}$ suggesting the endothermic nature of the adsorption process. This is also in accordance with the isotherm results in which the better fit was obtained at higher temperatures. The entropy change showed a positive variation too.

Table 4. Thermodynamic parameters for the adsorption of Cd(II) onto XG-cl-poly(AAm-co-AA) hydrogel.

C_o ($mg\cdot L^{-1}$)	ΔH^0 ($J\cdot mol^{-1}$)	ΔS^0 ($J\cdot mol^{-1}\cdot K^{-1}$)	$-\Delta G^0$ ($J\cdot mol^{-1}$)			
			298 K	303 K	308 K	313 K
20	24.5	0.09	2.32	2.77	3.22	3.67

2.1.5. Reusability of Hydrogel

The commercial-scale applicability of any adsorbent is crucially reliant on its reusable ability. During this study, the adsorbed ions are desorbed from the adsorbent surface to reactivate the adsorption active sites, so that they may possibly be utilized for the next adsorption cycle. The desorption studies of Cd(II) ions were performed in 0.2 M HCl solution. The high concentration of competitive H^+ ions compete with the Cd(II) ions for the active sites on the XG-cl-poly(AAm-co-AA) hydrogel and will thus, help in the desorption of the adsorbed Cd(II) ions. The results presented in Figure 6 show that a high desorption rate of 91% was obtained during the initial cycle. However, consecutively, the desorption rate decreased to 84% after five cycles. This decrease can possibly be due to the blockage of active sites. Furthermore, the reusability of the synthesized XG-cl-poly(AAm-co-AA) hydrogel was also tested for consecutive five cycles and the rate obtained is depicted in Figure 6. Results showed a decline in adsorption rate from 92% (first cycle) to 86% (after 5th cycle). A likely decline in adsorption rate can be linked to the coverage of some of the active sites by the Cd(II) ions which were difficult to remove from the XG-cl-poly(AAm-co-AA) hydrogel surface.

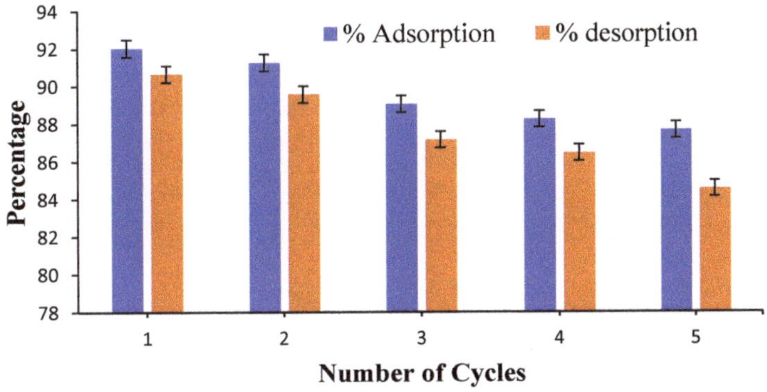

Figure 6. Desorption of Cd(II) ions and reusability of XG-cl-poly(AAm-co-AA) hydrogel for the adsorption.

3. Conclusions

XG-cl-poly(AAm-co-AA) hydrogel is an effective adsorbent for the confiscation of cadmium ions. The FTIR characterization specified the existence of carboxyl, amides, and hydroxyl groups which are recognized as the active sites on the XG-cl-poly(AAm-co-AA) hydrogel for the uptake of cadmium ions from the aqueous system. The kinetic outcomes established that bulk diffusion is the primary mechanism of cadmium ions adsorption. Comparison of the R^2 values disclosed that the pseudo-first-order model better fitted the kinetic data than that of the pseudo-second-order model. Additionally, the experimental q_e value (91.26 mg·g^{-1}) was found to best correlate with the calculated q_e value from the pseudo-first-order (95.9 mg·g^{-1}). The hydrated cadmium ions can merely be adsorbed onto the surface of the XG-cl-poly(AAm-co-AA) hydrogel through physicochemical interactions. Furthermore, the reusability studies demonstrated the usefulness of hydrogel. Fortuitously, all reactants are in the adsorbent are biocompatible, biodegradable, and harmless.

4. Materials and Methods

4.1. Materials

Xanthan gum, acrylamide and alginic acid were acquired from Sigma-Aldrich, India. Crosslinker N, N-methylene-bis-acrylamide, initiator ammonium persulphate and cadmium nitrate were purchased from Loba Chemie India.

4.2. Synthesis of Xanthan Gum-cl-Poly(Acrylamide-co-Alginic Acid)

The 250 mg xanthan gum was dissolved in 100 mL double distilled water. The preparation of XG-cl-poly(AAm-co-AA) hydrogel was carried out with small amendment in procedure stated in our former works. To the above solution of xanthan gum, ammonium persulphate and *N,N*-bismethyleneacrylamide were added with uninterrupted stirring at room temperature, the resultant mixture was stirred for 20 min to achieve a homogeneous phase and the accomplished gel was then positioned in microwave oven at 60 W for 1 min. Then, varying amounts of acrylamide and alginic acid (in ratio of 1:1:1; 1:0.5:1; 1:1:0.5) were engrossed into the above gel and stirred for 10 min. This mixture was yet once more placed in microwave oven operated in cyclic mode (on/off) at 60 W, till hydrogel was achieved. The obtained hydrogel was filtered and washed several times to confiscate impurities using distilled water. Finally, hydrogel was freeze-dried. The investigate swelling behavior XG-cl-poly(AAm-co-AA) hydrogel was submerged into distilled water for 24 h to attain maximum swelling. The XG-cl-poly(AAm-co-AA) hydrogel sample with maximum swelling ability was chosen for further investigation. The % swelling was measured by applying the formula discussed in earlier studies [33].

4.3. Characterization

FTIR spectrum of XG-cl-poly(AAm-co-AA) hydrogel was recorded using Fourier Transform Infrared Spectrophotometer (Shimadzu IR AFFINITY-I, Japan). The spectrum XG-cl-poly(AAm-co-AA) was measured in the wave number region 4000–400 cm^{-1} by preparing the KBr pellet. The hydrogel samples were prepared using 1.0% KBr pellets and all spectra were recorded with 20 runs per minute at a resolution of 4.0 cm^{-1}. The normalization of the peaks was performed by vector normalization method. X-ray pattern of powdered XG-cl-poly(AAm-co-AA) hydrogel was analyzed by X-ray diffractometer with Cu K-α radiation (λ = 1.54 Å) at 45 kV (PAN analytical X′Pert PRO). The surface morphology analysis of dry hydrogel and cadmium ion adsorbed hydrogel were perceived using a model HITACHI S-4800 scanning electron microscope at an accelerating voltage of 25 kV.

4.4. Adsorption Experiments

Cadmium nitrate was acquired from Loba Chemie India. Cadmium nitrate was solubilized in deionized water. The cadmium ions adsorption tests were accomplished on a temperature-controlled shaker at 120 rpm. After every test, the suspension was centrifuged to isolate the hydrogel from the aqueous medium. The cadmium ion concentrations of the adsorbate solution were determined using ICP-OES. All the subsequent experiments were comprehended using dosage of 0.5 $g \cdot L^{-1}$ of the XG-cl-poly(AAm-co-AA) hydrogel. First, the pH influence was examined for diverse initial pH values (2, 3, 4, 5, 6, 7 and 8), adjusted with NaOH or HCL. For this, the XG-cl-poly(AAm-co-AA) hydrogel was added to 50 mL of cadmium ions solution (10 $mg \cdot L^{-1}$) and shaken for 180 min at room temperature. Next, effect of time was investigated for the adsorption assays at the time intervals of 20,40, 80, 120, 180 and 240 min with initial cadmium ions concentrations of 10 mg L^{-1}, all the flasks were agitated at constant rpm of 120. Similarly influence of four different temperatures, i.e., 25 °C, 30 °C, 35 °C and 40 °C for 180 min, to confirm that the adsorbent/adsorbate system attain the equilibrium. Lastly, kinetic studies were executed, using initial cadmium ions concentrations of 10, 20, and 60 $mg \cdot L^{-1}$ at 298 K, aliquots were withdrawn at diverse times intervals ranging from 0–90 min. Langmuir and Freundlich, models were chosen to analyze the adsorption mechanism. A method suggested in previous studies was used to determine the thermodynamic parameters [52–55]. To disclose the effectiveness of XG-cl-poly(AAm-co-AA) hydrogel, the reusability was tested by performing numerous adsorption–desorption cycles.

Author Contributions: G.S.: Conceptualization, Methodology, Investigation, Writing—review and editing, Project administration, A.K.: Conceptualization, Methodology, Software, Validation, Formal analysis, Investigation, Writing—original draft, Visualization. A.A.G.: Writing—review and editing, A.G.-P.: Writing—review and editing, M.N.: Conceptualization, Resources, Writing—review and editing, Funding acquisition, F.J.S.: Writing—review and editing. All authors have read and agreed to the published version of the manuscript.

Funding: The authors acknowledge the Researchers Supporting Project number (RSP-2021/8), King Saud University, Riyadh, Saudi Arabia for the financial support.

Acknowledgments: The authors acknowledge the Researchers Supporting Project number (RSP-2021/8), King Saud University, Riyadh, Saudi Arabia for the financial support.

Conflicts of Interest: The authors declare no conflict of interest.

References

1. Tchounwou, P.B.; Yedjou, C.G.; Patlolla, A.K.; Sutton, D.J. Heavy metal toxicity and the environment. In *Molecular, Clinical and Environmental Toxicology*; Springer: Basel, Switzerland, 2012; pp. 133–164.
2. Rajendran, S.; Priya, T.; Khoo, K.S.; Hoang, T.K.; Ng, H.-S.; Munawaroh, H.S.H.; Karaman, C.; Orooji, Y.; Show, P.L. A critical review on various remediation approaches for heavy metal contaminants removal from contaminated soils. *Chemosphere* **2022**, *287*, 132369. [CrossRef] [PubMed]

3. Zhang, S.; Wang, J.; Zhang, Y.; Ma, J.; Huang, L.; Yu, S.; Chen, L.; Song, G.; Qiu, M.; Wang, X. Applications of water-stable metal-organic frameworks in the removal of water pollutants: A review. *Environ. Pollut.* **2021**, *291*, 118076. [CrossRef] [PubMed]
4. World Health Organization. *Weekly Epidemiological Record: Relevé Épidémiologique Hebdomadaire*; World Health Organization: Geneva, Switzerland, 1948.
5. Bradl, H. Sources and origins of heavy metals. In *Interface Science and Technology*; Elsevier: Amsterdam, The Netherlands, 2005; pp. 1–27.
6. Roy, A.; Bharadvaja, N. Efficient removal of heavy metals from artificial wastewater using biochar. *Environ. Nanotechnol. Monit. Manag.* **2021**, *16*, 100602. [CrossRef]
7. Kong, Q.; Shi, X.; Ma, W.; Zhang, F.; Yu, T.; Zhao, F.; Zhao, D.; Wei, C. Strategies to improve the adsorption properties of graphene-based adsorbent towards heavy metal ions and their compound pollutants: A review. *J. Hazard. Mater.* **2021**, *415*, 125690. [CrossRef] [PubMed]
8. Talaiekhozani, A.; Rezania, S. Application of photosynthetic bacteria for removal of heavy metals, macro-pollutants and dye from wastewater: A review. *J. Water Process Eng.* **2017**, *19*, 312–321. [CrossRef]
9. Shaker, M.A. Dynamics and thermodynamics of toxic metals adsorption onto soil-extracted humic acid. *Chemosphere* **2014**, *111*, 587–595. [CrossRef]
10. Yu, Y.; Liu, J.; Yang, Y.; Ding, J.; Zhang, A. Experimental and theoretical studies of cadmium adsorption over Fe_2O_3 sorbent in incineration flue gas. *Chem. Eng. J.* **2021**, *425*, 131647. [CrossRef]
11. Panuccio, M.R.; Sorgonà, A.; Rizzo, M.; Cacco, G. Cadmium adsorption on vermiculite, zeolite and pumice: Batch experimental studies. *J. Environ. Manag.* **2009**, *90*, 364–374. [CrossRef] [PubMed]
12. Petrović, M.; Kaštelan-Macan, M.; Horvat, A. Interactive sorption of metal ions and humic acids onto mineral particles. *Water Air Soil Pollut.* **1999**, *111*, 41–56. [CrossRef]
13. Sajid, M.; Nazal, M.K.; Baig, N.; Osman, A.M. Removal of heavy metals and organic pollutants from water using dendritic polymers based adsorbents: A critical review. *Sep. Purif. Technol.* **2018**, *191*, 400–423. [CrossRef]
14. Razmi, B.; Ghasemi-Fasaei, R. Investigation of Taguchi optimization, equilibrium isotherms, and kinetic modeling for phosphorus adsorption onto natural zeolite of clinoptilolite type. *Adsorpt. Sci. Technol.* **2018**, *36*, 1470–1483. [CrossRef]
15. Es-said, A.; Nafai, H.; Lamzougui, G.; Bouhaouss, A.; Bchitou, R. Comparative adsorption studies of cadmium ions on phosphogypsum and natural clay. *Sci. Afr.* **2021**, *13*, e00960. [CrossRef]
16. Sharma, G.; AlGarni, T.S.; Kumar, P.S.; Bhogal, S.; Kumar, A.; Sharma, S.; Naushad, M.; Othman, Z.A.A.L.; Stadler, F.J. Utilization of Ag_2O–Al_2O_3–ZrO_2 decorated onto rGO as adsorbent for the removal of Congo red from aqueous solution. *Environ. Res.* **2021**, *197*, 111179. [CrossRef] [PubMed]
17. Qing, Z.; Wang, L.; Liu, X.; Song, Z.; Qian, F.; Song, Y. Simply synthesized sodium alginate/zirconium hydrogel as adsorbent for phosphate adsorption from aqueous solution: Performance and mechanisms. *Chemosphere* **2021**, 133103. [CrossRef]
18. Song, Y.; Gotoh, T.; Nakai, S. Synthesis of Oxidant Functionalised Cationic Polymer Hydrogel for Enhanced Removal of Arsenic (III). *Gels* **2021**, *7*, 197. [CrossRef]
19. Yilmaz, M.S. Graphene oxide/hollow mesoporous silica composite for selective adsorption of methylene blue. *Microporous Mesoporous Mater.* **2022**, *330*, 111570. [CrossRef]
20. Madduma-Bandarage, U.S.; Madihally, S.V. Synthetic hydrogels: Synthesis, novel trends, and applications. *J. Appl. Polym. Sci.* **2021**, *138*, 50376. [CrossRef]
21. Sharma, G.; Thakur, B.; Naushad, M.; Kumar, A.; Stadler, F.J.; Alfadul, S.M.; Mola, G.T. Applications of nanocomposite hydrogels for biomedical engineering and environmental protection. *Environ. Chem. Lett.* **2018**, *16*, 113–146. [CrossRef]
22. Ghauri, Z.H.; Islam, A.; Qadir, M.A.; Ghaffar, A.; Gull, N.; Azam, M.; Mehmood, A.; Ghauri, A.A.; Khan, R.U. Novel pH-responsive chitosan/sodium alginate/PEG based hydrogels for release of sodium ceftriaxone. *Mater. Chem. Phys.* **2022**, *277*, 125456. [CrossRef]
23. Sharma, G.; Kumar, A.; Chauhan, C.; Okram, A.; Sharma, S.; Pathania, D.; Kalia, S. Pectin-crosslinked-guar gum/SPION nanocomposite hydrogel for adsorption of m-cresol and o-chlorophenol. *Sustain. Chem. Pharm.* **2017**, *6*, 96–106. [CrossRef]
24. Sharma, G.; Khosla, A.; Kumar, A.; Kaushal, N.; Sharma, S.; Naushad, M.; Vo, D.-V.N.; Iqbal, J.; Stadler, F.J. A comprehensive review on the removal of noxious pollutants using carrageenan based advanced adsorbents. *Chemosphere* **2021**, *289*, 133100. [CrossRef]
25. Pervaiz, F.; Mushtaq, R.; Noreen, S. Formulation and optimization of terbinafine HCl loaded chitosan/xanthan gum nanoparticles containing gel: Ex-vivo permeation and in-vivo antifungal studies. *J. Drug Deliv. Sci. Technol.* **2021**, *66*, 102935. [CrossRef]
26. Hua, D.; Gao, S.; Zhang, M.; Ma, W.; Huang, C. A novel xanthan gum-based conductive hydrogel with excellent mechanical, biocompatible, and self-healing performances. *Carbohydr. Polym.* **2020**, *247*, 116743. [CrossRef]
27. Amaral, C.N.; Oliveira, P.F.; Pedroni, L.G.; Mansur, C.R. Viscoelastic behavior of hydrogel-based xanthan gum/aluminum lactate with potential applicability for conformance control. *J. Appl. Polym. Sci.* **2021**, *138*, 50640. [CrossRef]
28. Özbaş, F.; Tüzün, E.; Yıldız, A.; Karakuş, S. Sonosynthesis and characterization of konjac gum/xanthan gum supported ironoxide nanoparticles. *Int. J. Biol. Macromol.* **2021**, *183*, 1047–1057. [CrossRef]
29. Sharma, G.; Naushad, M.; Kumar, A.; Rana, S.; Sharma, S.; Bhatnagar, A.; Stadler, F.J.; Ghfar, A.A.; Khan, M.R. Efficient removal of coomassie brilliant blue R-250 dye using starch/poly(alginic acid-cl-acrylamide) nanohydrogel. *Process. Saf. Environ. Prot.* **2017**, *109*, 301–310. [CrossRef]

30. Vilela, P.B.; Matias, C.A.; Dalalibera, A.; Becegato, V.A.; Paulino, A.T. Polyacrylic acid-based and chitosan-based hydrogels for adsorption of cadmium: Equilibrium isotherm, kinetic and thermodynamic studies. *J. Environ. Chem. Eng.* **2019**, *7*, 103327. [CrossRef]
31. Elella, M.H.A.; Goda, E.S.; Gamal, H.; El-Bahy, S.M.; Nour, M.A.; Yoon, K.R. Green antimicrobial adsorbent containing grafted xanthan gum/SiO_2 nanocomposites for malachite green dye. *Int. J. Biol. Macromol.* **2021**, *191*, 385–395. [CrossRef]
32. Anjum, F.; Bukhari, S.A.; Siddique, M.; Shahid, M.; Potgieter, J.H.; Jaafar, H.Z.; Ercisli, S.; Zia-Ul-Haq, M. Microwave irradiated copolymerization of xanthan gum with acrylamide for colonic drug delivery. *BioResources* **2015**, *10*, 1434–1451. [CrossRef]
33. Singh, V.; Singh, S.; Pandey, S.; Sanghi, R. Synthesis and characterization of guar gum templated hybrid nano silica. *Int. J. Biol. Macromol.* **2011**, *49*, 233–240. [CrossRef]
34. Sharma, R.K. Synthesis and characterization of graft copolymers of N-Vinyl-2-Pyrrolidone onto guar gum for sorption of Fe^{2+} and Cr^{6+} ions. *Carbohydr. Polym.* **2011**, *83*, 1929–1936. [CrossRef]
35. Sharma, S.; Sharma, G.; Kumar, A.; AlGarni, T.S.; Naushad, M.; Othman, Z.A.A.L.; Stadler, F.J. Adsorption of cationic dyes onto carrageenan and itaconic acid-based superabsorbent hydrogel: Synthesis, characterization and isotherm analysis. *J. Hazard. Mater.* **2022**, *421*, 126729. [CrossRef]
36. Sharma, G.; Thakur, B.; Kumar, A.; Sharma, S.; Naushad, M.; Stadler, F.J. Atrazine removal using chitin-cl-poly (acrylamide-co-itaconic acid) nanohydrogel: Isotherms and pH responsive nature. *Carbohydr. Polym.* **2020**, *241*, 116258. [CrossRef]
37. Sharma, S.; Sharma, G.; Kumar, A.; Dhiman, P.; AlGarni, T.S.; Naushad, M.; Othman, Z.A.A.L.; Stadler, F.J. Controlled synthesis of porous Zn/Fe based layered double hydroxides: Synthesis mechanism, and ciprofloxacin adsorption. *Sep. Purif. Technol.* **2022**, *278*, 119481. [CrossRef]
38. Ebadi, A.; Mohammadzadeh, J.S.S.; Khudiev, A. What is the correct form of BET isotherm for modeling liquid phase adsorption? *Adsorption* **2009**, *15*, 65–73. [CrossRef]
39. Asuquo, E.; Martin, A.; Nzerem, P.; Siperstein, F.; Fan, X. Adsorption of Cd (II) and Pb (II) ions from aqueous solutions using mesoporous activated carbon adsorbent: Equilibrium, kinetics and characterisation studies. *J. Environ. Chem. Eng.* **2017**, *5*, 679–698. [CrossRef]
40. Jain, M.; Yadav, M.; Kohout, T.; Lahtinen, M.; Garg, V.K.; Sillanpää, M. Development of iron oxide/activated carbon nanoparticle composite for the removal of Cr (VI), Cu (II) and Cd (II) ions from aqueous solution. *Water Resour. Ind.* **2018**, *20*, 54–74. [CrossRef]
41. Mohan, D.; Kumar, H.; Sarswat, A.; Alexandre-Franco, M.; Pittman, C.U., Jr. Cadmium and lead remediation using magnetic oak wood and oak bark fast pyrolysis bio-chars. *Chem. Eng. J.* **2014**, *236*, 513–528. [CrossRef]
42. Tu, Y.-J.; You, C.-F.; Chang, C.-K. Kinetics and thermodynamics of adsorption for Cd on green manufactured nano-particles. *J. Hazard. Mater.* **2012**, *235*, 116–122. [CrossRef]
43. Kataria, N.; Garg, V. Green synthesis of Fe_3O_4 nanoparticles loaded sawdust carbon for cadmium (II) removal from water: Regeneration and mechanism. *Chemosphere* **2018**, *208*, 818–828. [CrossRef] [PubMed]
44. Yang, Z.; Yang, T.; Yang, Y.; Yi, X.; Hao, X.; Xie, T.; Liao, C.J. The behavior and mechanism of the adsorption of Pb(II) and Cd(II) by a porous double network porous hydrogel derived from peanut shells. *Mater. Today Commun.* **2021**, *27*, 102449. [CrossRef]
45. Wang, F.; Li, J.; Su, Y.; Li, Q.; Gao, B.; Yue, Q.; Zhou, W. Adsorption and recycling of Cd(II) from wastewater using straw cellulose hydrogel beads. *J. Ind. Eng. Chem.* **2019**, *80*, 361–369. [CrossRef]
46. Tao, X.; Wang, S.; Li, Z.; Zhou, S. Green synthesis of network nanostructured calcium alginate hydrogel and its removal performance of Cd^{2+} and Cu^{2+} ions. *Mater. Chem. Phys.* **2021**, *258*, 123931. [CrossRef]
47. Ayouch, I.; Barrak, I.; Kassab, Z.; el Achaby, M.; Barhoun, A.; Draoui, K. Improved recovery of cadmium from aqueous medium by alginate composite beads filled by bentonite and phosphate washing sludge. *Colloids Surf. A Physicochem. Eng. Asp.* **2020**, *604*, 125305. [CrossRef]
48. Bagheri, S.; Amini, M.M.; Behbahani, M.; Rabiee, G. Low cost thiol-functionalized mesoporous silica, KIT-6-SH, as a useful adsorbent for cadmium ions removal: A study on the adsorption isotherms and kinetics of KIT-6-SH. *Microchem. J.* **2019**, *145*, 460–469. [CrossRef]
49. Wang, Z.; Wu, S.; Zhang, Y.; Miao, L.; Zhang, Y.; Wu, A. Preparation of modified sodium alginate aerogel and its application in removing lead and cadmium ions in wastewater. *Int. J. Biol. Macromol.* **2020**, *157*, 687–694. [CrossRef]
50. Huang, Z.; Wu, Q.; Liu, S.; Liu, T.; Zhang, B. A novel biodegradable β-cyclodextrin-based hydrogel for the removal of heavy metal ions. *Carbohydr. Polym.* **2013**, *97*, 496–501. [CrossRef]
51. Zhu, Q.; Li, Z. Hydrogel-supported nanosized hydrous manganese dioxide: Synthesis, characterization, and adsorption behavior study for Pb^{2+}, Cu^{2+}, Cd^{2+} and Ni^{2+} removal from water. *Chem. Eng. J.* **2015**, *281*, 69–80. [CrossRef]
52. Sharma, G.; Naushad, M. Adsorptive removal of noxious cadmium ions from aqueous medium using activated carbon/zirconium oxide composite: Isotherm and kinetic modelling. *J. Mol. Liq.* **2020**, *310*, 113025. [CrossRef]
53. Chen, Y.; Tang, J.; Wang, S.; Zhang, L. Facile preparation of a remarkable MOF adsorbent for Au (III) selective separation from wastewater: Adsorption, Regeneration and Mechanism. *J. Mol. Liq.* **2021**, 118137. [CrossRef]
54. Hao, C.; Li, G.; Wang, G.; Chen, W.; Wang, S. Preparation of acrylic acid modified alkalized MXene adsorbent and study on its dye adsorption performance. *Colloids Surf. A Physicochem. Eng. Asp.* **2022**, *632*, 127730. [CrossRef]
55. Sharma, G.; Thakur, B.; Kumar, A.; Sharma, S.; Naushad, M.; Stadler, F.J. Gum acacia-cl-poly (acrylamide)@ carbon nitride nanocomposite hydrogel for adsorption of ciprofloxacin and its sustained release in artificial ocular solution. *Macromol. Mater. Eng.* **2020**, *305*, 2000274. [CrossRef]

gels

Article

Preparation of Various Nanomaterials via Controlled Gelation of a Hydrophilic Polymer Bearing Metal-Coordination Units with Metal Ions

Daisuke Nagai [1,*], Naoki Isobe [1], Tatsushi Inoue [2], Shusuke Okamoto [1], Yasuyuki Maki [3] and Takeshi Yamanobe [2]

1 School of Food and Nutritional Science, University of Shizuoka, 52-1 Yada, Shizuoka 422-8526, Shizuoka, Japan; isobe_45@icloud.com (N.I.); sokamoto@u-shizuoka-ken.ac.jp (S.O.)
2 Division of Molecular Science, Faculty of Science and Technology, Gunma University, 1-5-1 Tenjin-cho, Kiryu 376-8515, Gunma, Japan; t15301013@gunma-u.ac.jp (T.I.); yamanobe@gunma-u.ac.jp (T.Y.)
3 Department of Chemistry, Graduate School of Science, Kyusyu University, 744 Motooka, Fukuoka 819-0395, Fukuoka, Japan; maki@chem.kyushu-univ.jp
* Correspondence: daisukenagai@u-shizuoka-ken.ac.jp; Tel.: +81-54-264-5729

Abstract: We investigated the gelation of a hydrophilic polymer with metal-coordination units (HPMC) and metal ions (Pd^{II} or Au^{III}). Gelation proceeded by addition of an HPMC solution in *N*-methyl-2-pyrrolidone (NMP) to a metal ion aqueous solution. An increase in the composition ratio of the metal-coordination units from 10 mol% to 34 mol% (HPMC-34) increased the cross-linking rate with Au^{III}. Cross-linking immediately occurred after dropwise addition of an HPMC-34 solution to the Au^{III} solution, generating the separation between the phases of HPMC-34 and Au^{III}. The cross-linking of Au^{III} proceeded from the surface to the inside of the HPMC-34 droplets, affording spherical gels. In contrast, a decrease in the ratio of metal-coordination units from 10 mol% to 4 mol% (HPMC-4) decreased the Pd^{II} cross-linking rate. The cross-linking occurred gradually and the gels extended to the bottom of the vessel, forming fibrous gels. On the basis of the mechanism for the formation of gels with different morphologies, the gelation of HPMC-34 and Au^{III} provided nanosheets via gelation at the interface between the Au^{III} solution and the HPMC-34 solution. The gelation of HPMC-4 and Pd^{II} afforded nanofibers by a facile method, i.e., dropwise addition of the HPMC-4 solution to the Pd^{II} solution. These results demonstrated that changing the composition ratio of the metal-coordination units in HPMC can control the gelation behavior, resulting in different types of nanomaterials.

Keywords: gelation; polymer; palladium; gold; coordination; nanosheet; nanofiber

Citation: Nagai, D.; Isobe, N.; Inoue, T.; Okamoto, S.; Maki, Y.; Yamanobe, T. Preparation of Various Nanomaterials via Controlled Gelation of a Hydrophilic Polymer Bearing Metal-Coordination Units with Metal Ions. *Gels* **2022**, *8*, 435. https://doi.org/10.3390/gels8070435

Academic Editor: Jean-Michel Guenet

Received: 26 May 2022
Accepted: 8 July 2022
Published: 11 July 2022

1. Introduction

Organic–inorganic hybrid materials, which consist of organic polymers containing inorganic metals dispersed at the nanometer scale, have generated a great deal of interest due to their unique properties such as flexibility, high transparency, high reactivity, and mechanical and thermal stabilities [1–5]. For example, nanofibers containing metal ions or nanoparticles are promising candidates for various applications including tissue engineering, blood vessels, drug delivery, protective clothing, filtration, catalysis, and sensors [6–10]. The most common method of fabricating such nanofiber is an electrospinning method using a polymer solution containing metal ions or nanoparticles. However, electrospinning requires expensive instruments, cumbersome operations, and high voltage, and thus it runs the risk of electrical shock. Dispersing metal nanoparticles in the polymer solution is also difficult [11–13]. Therefore, the development of a facile method for fabricating nanofibers is greatly desired. Nanosheet materials also have unique physical and chemical properties, which are derived from their two-dimensional nature [14–18]. Nanosheets can be synthesized with a bottom-up method [19–23], which occurs at the interface between an organic polydentate ligand and an aqueous layer containing metal ions. This approach has advantages compared to the top-down approach, which produces nanosheets by exfoliation of bulk layered materials such as graphene. First, the composition, structure, and other properties can be adjusted by selection of the ligand molecules and metal ions. Second, the produced nanosheets are not limited to layers of bulk materials. Therefore, a bottom-up synthesis broadens the diversity and utility of nanosheets.

We previously investigated the gelation behavior of a hydrophilic polymer bearing metal-coordination units (denoted as HPMC) with metal ions (Pd^{II} or Au^{III}) upon addition of a dispersed aqueous solution of HPMC-8 to an aqueous solution of metal ions [24,25]. HPMC-8 consists of thiocarbonyl groups (8 mol%) for metal coordination and hydroxyl groups (92%) for hydrophilicity (Figure 1a). The gelation of HPMC-8 with Pd^{II} or Au^{III} afforded spherical and fibrous gels, respectively (Figure 1b,c). Consequently, gels with different morphologies were found to be formed depending on the metal ions. The formation of different morphologies can be explained by the cross-linking rate. The cross-linking with Pd^{II} occurred immediately after dropwise addition of the dispersed aqueous solution of HPMC-8 to the Pd^{II} solution, generating the separation between aqueous phases of HPMC-8 and Pd^{II} (Figure 1b). The cross-linking of Pd^{II} proceeded from the surface to the inside of the droplets of HPMC, resulting in the formation of spherical gels. In contrast, the cross-linking with Au^{III} occurred gradually and the gels extended to the bottom due to the slower cross-linking rate, forming fibrous gels (Figure 1c). On the basis of this mechanism for the formation of gels with different morphologies, the gelation of HPMC-8 with Au^{III} provided nanofiber containing uniformly dispersed Au nanoparticles by a facile method, i.e., dropwise addition of a dispersed aqueous solution of HPMC-8 to an aqueous solution of Au^{III} ions [24]. In contrast, the faster gelation of HPMC-8 with Pd^{II} provided nanosheets containing uniformly dispersed Pd^{II} ions via gelation at the interface between the aqueous phases of Pd^{II} and HPMC-8 [25].

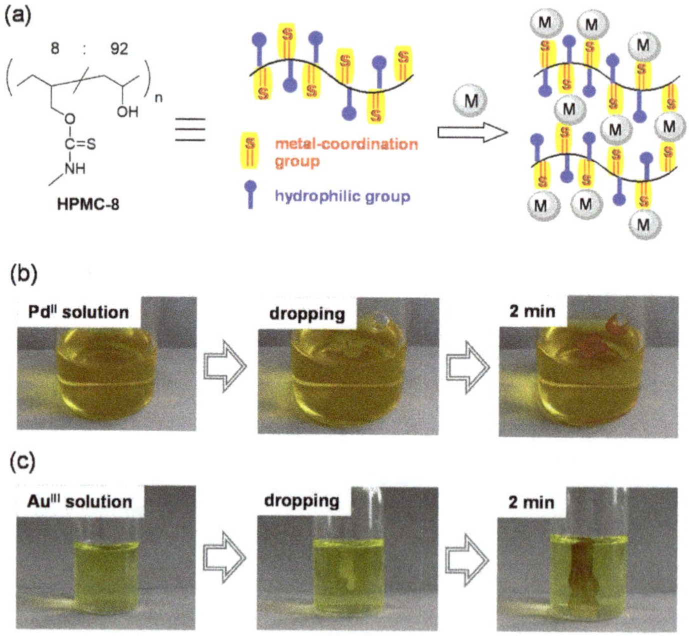

Figure 1. (**a**) Mechanism for gelation of HPMC—8 with metal ions. (**b**) Gelation behavior of HPMC—8 and Pd^{II} ions. (**c**) Gelation behavior of HPMC—8 and Au^{III} ions.

Encouraged by these results, we attempted to control the gelation behavior and synthesize different types of nanomaterials by changing the composition ratio of the metal-coordination units in HPMC. An increase in the composition ratio increased the cross-linking rate with Au^{III}, resulting in the formation of spherical gels and Au nanosheets instead of nanofibers. A decrease in the ratio decreased the cross-linking rate with Pd^{II}, affording fibrous gels and Pd nanofiber instead of nanosheets. Thus, changing the composition ratio of the metal-coordination units can provide contrasting gelation behavior and nanomaterials. We expect that this procedure will become a controlled manufacturing method for various types of nanomaterials containing various metals.

2. Results and Discussion

2.1. *Synthesis of HPMC and Its Gelation Behavior with Metal Ions*

HPMC containing thiocarbonyl and hydroxyl groups was synthesized according to our previous report [24]. HPMC with metal-coordination unit content of 10% (denoted as HPMC-10) was synthesized by reacting poly(vinyl alcohol) and methyl isothiocyanate in dimethyl sulfoxide at 40 °C for 20 h (Scheme 1). HPMC with metal-coordination unit content of 34% (HPMC-34) and HPMC with metal-coordination unit content of 4% (HPMC-4) were synthesized by the above similar method.

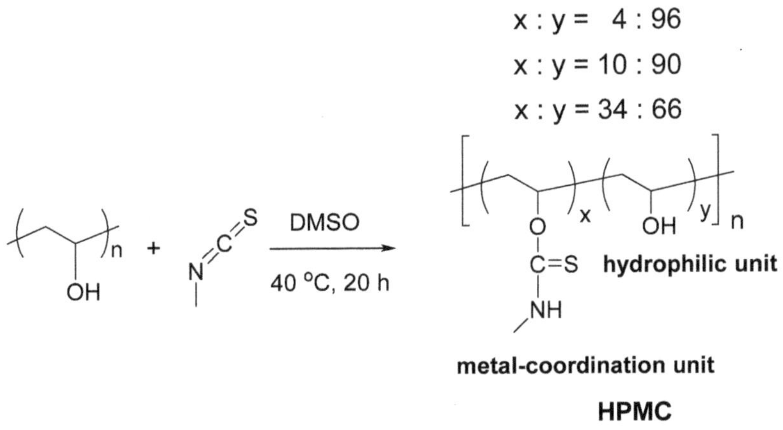

Scheme 1. Synthesis of HPMCs by reactions of poly(vinyl alcohol) with methyl isothiocyanate.

The gelation behavior of HPMC-34 and Au^{III} was compared to that of HPMC-10 and Au^{III} to examine the effect of the increase in the composition ratio of the metal-coordination unit. *N*-Methyl-2-pyrrolidone (NMP) solutions of HPMC-10 or HPMC-34 (13 wt%, 0.2 mL) were added to 4.0 mM $NaAuCl_4$ aqueous solutions (5 mL). In the gelation of HPMC-10, Au^{III} ions gradually cross-linked from the surface to the inside phase of HPMC-10 and the gels extended to the bottom of the container, forming fibrous gels (Figure 2a). In contrast, gelation of HPMC-34 with Au^{III} generated instant separation between the phases of HPMC-34 and Au^{III} (Figure 2b). The separation originated from the immediate cross-linking reaction at the interface and the higher hydrophobicity of the HPMC-34 phase than that of the Au^{III} phase. The cross-linking with Au^{III} ions proceeded from the surface to the inside of the HPMC-34 droplets, resulting in the formation of spherical gels. To observe a microscopic region of the resulting gels, scanning electron microscope (SEM) observations were conducted. Similar to the gel shapes in the photographs (Figure 2a,b), SEM analysis showed fibrous shapes from HPMC-10-Au and a rough surface from HPMC-34-Au (Figure 2c,d). To determine the coordination sites, the IR measurements of HPMC-34 and HPMC-34-Au were carried out (Figure 2e). The absorption peak around 1535 cm^{-1} assigned to the C=S stretching vibration shifted to 1555 cm^{-1}, and the peak intensity became smaller after the gelation. Therefore, it was found that different gelation behaviors and gel shapes were obtained depending on the composition ratio of the metal-coordination units.

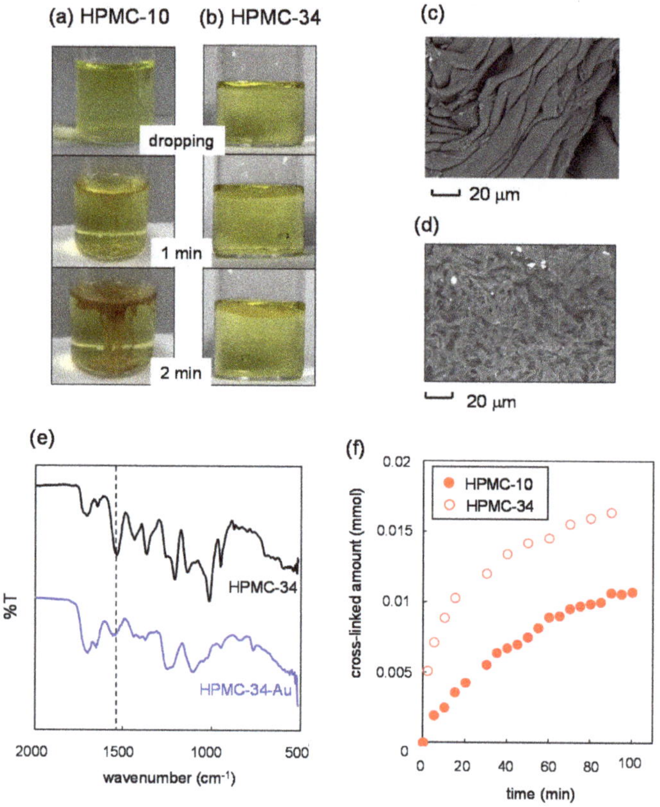

Figure 2. Gelation behavior of Au^{III} with (**a**) HPMC—10 and (**b**) HPMC—34. SEM images of (**c**) HPMC—10—Au and (**d**) HPMC—34—Au. (**e**) IR spectra of HPMC—34 and HPMC—34—Au. (**f**) Cross-linking rates of HPMC—10 and HPMC—34.

A kinetic study and the gel fraction were examined to explain the mechanism for the formation of the different shaped gels. Cross-linking rates of HPMC-34 and HPMC-10 with Au^{III} ions were compared. Figure 2f shows the time course of the cross-linking amount determined by the method in the literature [24]. As shown in Figure 2f, the cross-linking rate of HPMC-34 was faster than that of HPMC-10 due to the increase in the metal-coordination units. The experimental kinetic data were fitted with a pseudo-first-order kinetic equation [26,27]:

$$\log(q_e - q_t) = kt/2.303 \quad (1)$$

where q_e and q_t are the amounts of metal ion cross-linked (g_{metal}/g_{poly}, metal amount adsorbed per gram of polymer) at equilibrium and at t, and k is the pseudo-first-order rate constant (min^{-1}). In the case of cross-linking reaction of HPMC-34, k was estimated to be 10.8×10^{-2} min^{-1} (R^2 = 0.9712), which was faster than that for HPMC-10 (7.02×10^{-2} min^{-1}, R^2 = 0.9821). The gel fraction indicates the cross-linking density of the gels determined by removing soluble parts using Soxhlet extraction. Gel fractions of the gels from HPMC-34 and HPMC-10 were 0.76 and 0.14, respectively, indicating that the cross-linking density of HPMC-34-Au was higher than that of HPMC-10-Au.

On the basis of the results, a mechanism for the formation of the different morphologies was proposed. The cross-linking rate of HPMC-34 and Au^{III} was faster than that of HPMC-10. In the cross-linking of HPMC-34, gelation with Au^{III} occurred immediately at the surface of the droplets after the dropwise addition of the HPMC solution to the Au^{III} solution. Gelation proceeded by immersing Au^{III} inside of the droplets, forming the higher cross-linking density and spherical gels. Contrastingly, due to the slower cross-linking rate of HPMC-10, the cross-linking occurred gradually with the diffusion of Au^{III} from the surface to the inside of the HPMC-10 phase, resulting in the formation of the lower cross-linking density and fibrous gels.

Next, to examine the effect of the decrease in the ratio of metal-coordination units on the gelation, the gelation behavior of HPMC-4 and Pd^{II} was compared to that of HPMC-10 with Pd^{II}. NMP solutions of HPMC-10 or HPMC-4 (13 wt%, 0.2 mL) were added to 4.0 mM Na_2PdCl_4 aqueous solutions (5 mL). In the gelation of HPMC-10, immediate separation occurred between the HPMC-10 and Pd^{II} phases (Figure 3a), whose separation originated from the fast cross-linking at the interface. The cross-linking with Pd^{II} proceeded from the surface to the inside of the HPMC-10 droplets, affording the spherical gels. In contrast, gelation of HPMC-4 and Pd^{II} occurred gradually from the surface to the inside of HPMC-4 phase, and the gels were extended to the bottom of the container, forming the fibrous gels. To observe a microscopic region, SEM analysis of the obtained gels was conducted. Similar to the gel shapes (Figure 3a,b), the SEM analysis revealed a rough surface from HPMC-10-Pd and fibrous shapes from HPMC-4-Pd (Figure 3c,d). To determine the coordination site, IR measurements of HPMC-4 and HPMC-4-Pd were conducted (Figure 3e). The absorption peak at 1550 cm^{-1} attributable to the C=S stretching vibration became smaller after cross-linking, indicating that the sulfur of the thiocarbonyl group was coordinated to Pd^{II}.

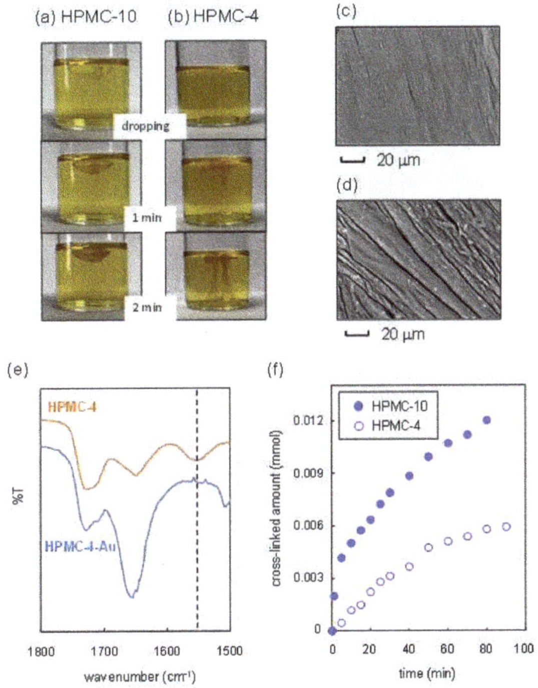

Figure 3. Gelation behavior of Pd^{II} with (**a**) HPMC—10 and (**b**) HPMC—4. SEM images of (**c**) HPMC—10—Pd and (**d**) HPMC—4—Pd. (**e**) IR spectra of HPMC—4 and HPMC—4—Au. (**f**) Cross-linking rates of HPMC—4 and HPMC—10.

As shown in Figure 3f, the cross-linking rate of HPMC-4 was slower than that of HPMC-10 due to the decrease in the metal-coordination units. The pseudo-first-order kinetic rate constants, *k*s of HPMC-4 and HPMC-10 were estimated to be 6.38 $\times$ 10^{-2} min^{-1} (R^2 = 0.9821) and 8.62 $\times$ 10^{-2} min^{-1} (R^2 = 0.9712), respectively. The gel fractions of the gels from HPMC-4 and HPMC-10 were 0.01 and 0.17, respectively, indicating that the cross-linking density of HPMC-4 was lower than that of HPMC-10. Consequently, the cross-linking reaction of HPMC-4 gradually proceeded with the diffusion of Pd^{II} from the surface to the inside of the HPMC phase due to the slower cross-linking rate, providing lower cross-linking density and fibrous gels as opposed to the gelation of HPMC-10 (spherical gels).

2.2. Synthesis of Nanosheets

As described above, the dropwise addition of the NMP solution of HPMC-34 to the Au^{III} aqueous solution allowed instant separation of the HPMC-34 and Au^{III} phases (Figure 2b). The liquid/liquid separation originated from the fast cross-linking at the interface and the higher hydrophobicity of the HPMC-34 phase than the Au^{III} phase. The Au^{III} ions cross-linked from the surface to the inside of the HPMC-34 droplets, affording spherical gels. This feature prompted us to utilize the cross-linking at the liquid–liquid interface between the HPMC-34 and Au^{III} phases for the synthesis of nanosheets.

The synthesis of nanosheets was attempted by the generation of the interface using solutions with different specific gravities and fast cross-linking between thiocarbonyl groups of HPMC-34 and Au^{III} ions (i.e., dropwise addition of an aqueous solution of Au^{III} ions with a lower specific gravity (1.02 g/cm^3) to an NMP solution of HPMC-34 with a higher specific gravity (1.63 g/cm^3)). When the Au^{III} aqueous solutions (16 mmol/L, 0.4 mL) were gently added to the NMP solutions of HPMC-34 with different concentrations (13, 23, and 27 wt%), the upper Au^{III} solution was miscible with the lower HPMC-34 concentration due to the slow cross-linking rate. In contrast, the increase in the concentration of HPMC-34 to 31, 35, and 37 wt% allowed instant cross-linking leading to the liquid/liquid separation, resulting in the formation of film-shaped gels at the interface (Figure 4a). Next, to examine the effect of Au^{III} concentration, aqueous solutions of Au^{III} with different concentrations (12, 16, and 20 mmol) were added to the HPMC-34 solutions (35 wt%). In every case, liquid/liquid separations were observed (Figure 4b). The thin film that formed at the interface between the Au^{III} (20 mmol) and HPMC (35 wt%) phases was transferred onto a Petri dish using tweezers, followed by washing with NMP and drying under reduced pressure. SEM and transmission electron microscope (TEM) images revealed the formation of a sheet structure (Figure 4c,d). Atomic force microscopy (AFM) showed a thickness of approximately 203 nm (Figure 4e). These results demonstrate the successful formation of the nanosheets at the interface between the Au^{III} and HPMC phases. The obtained nanosheet was characterized structurally. As shown in Figure 2c, the coordination site of Au was through the thiocarbonyl groups. The XPS wide-scan spectrum showed a peak of Au 4f around 84.0 eV and no peak of Cl 2p around 200 eV (see Supplementary Materials Figure S1a). The XPS narrow-scan spectrum showed Au $4f_{7/2}$ and Au $4f_{5/2}$ peaks at 83.9 and 87.6 eV, respectively, which are typical of Au^0 species [28–30]. (see Supplementary Materials Figure S1b). These results indicate that the Au^{III} was reduced to Au^0 during gelation, similar to our previously proposed mechanism [24].

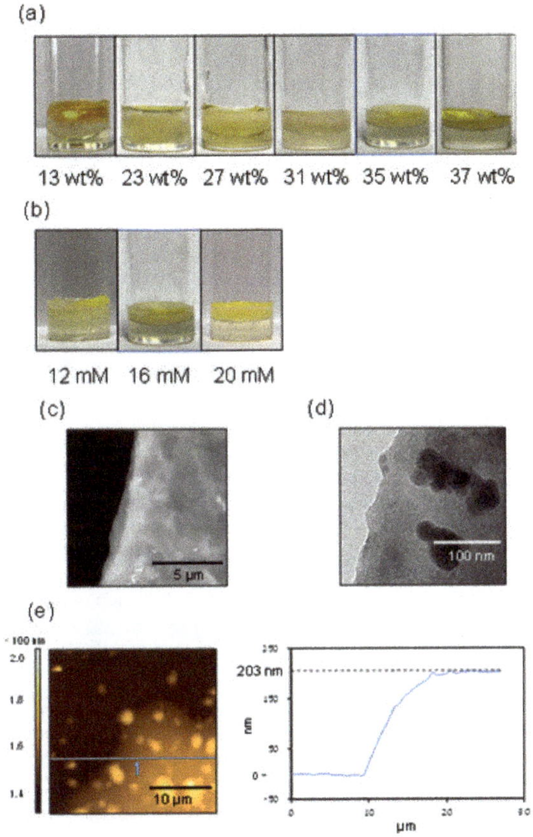

Figure 4. Photographs of (**a**) bottom-up synthesis of nanosheets by the addition of Au^{III} aqueous solutions (16 mM) to NMP solutions of HPMC—34 (13—37 wt%) and (**b**) bottom-up synthesis of nanosheets by the addition of Au^{III} aqueous solutions (12—20 mM) to NMP solutions of HPMC—34 (35 wt%). (**c**) SEM image of HPMC—34—Au nanosheet. (**d**) TEM image of HPMC—34—Au nanosheet. (**e**) AFM image of HPMC—34—Au nanosheet on a Si substrate.

2.3. *Synthesis of Nanofiber*

As mentioned above, the gelation of HPMC-4 with Pd^{II} ions provided fibrous gels with a lower cross-linking rate and the stretching force to the bottom of the container induced by the gel weight. Nanofibers containing metal ions are generally synthesized by the electrospinning method using a polymer solution containing metals. However, electrospinning has various problems, as described in Section 1. Thus, the synthesis of nanofibers was attempted by a facile method, i.e., dropwise addition of an NMP solution of HPMC-4 to an aqueous solution of Pd^{II} ions. HPMC-4 solutions (6, 8, and 11 wt%) were added to a $NaAuCl_4$ aqueous solution (4 mM) in a test tube. In all cases, elongated gels were obtained (Figure 5a–c). Elongation of gels increased with decreasing polymer concentration. The TEM observation of the gels obtained at 6 wt% of polymer concentration revealed the formation of fibrous gels of 100~200 nm diameter. The XPS narrow-scan spectrum of the nanofiber showed Pd $3d_{5/2}$ and Pd $3d_{3/2}$ peaks at 336.8 eV and 342.1 eV, respectively, which are typical of Pd^{II} species (Figure S2a) [28,31]. EDX/SEM measurement showed the presence of Pd and Cl species (Figure S2b). The Cl species was ascribed to Na_2PdCl_4; therefore, $Pd^{II}Cl_2$ was contained in the nanofibers. Thus, nanofibers cross-linked with Pd^{II} ions were successfully synthesized by this dropwise addition method.

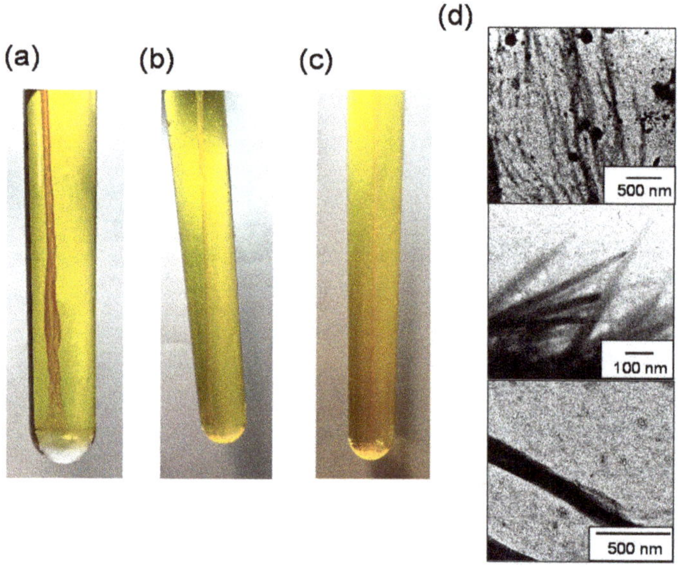

Figure 5. Photographs of gels produced upon addition of NMP solutions of (**a**) 11 wt%, (**b**) 8 wt%, and (**c**) 6 wt% of HPMC—4 (0.2 mL) to 4 mM aqueous solutions of Pd^{II} ions (20 mL). (**d**) TEM images of HPMC-4-Pd nanofiber.

3. Conclusions

In conclusion, we demonstrated the control of gelation behavior for the synthesis of different types of nanomaterials by changing the composition ratio of the metal-coordination unit in HPMC. An increase in the composition ratio of the metal-coordination unit from 10 mol% to 34 mol% increased the cross-linking rate with Au^{III}, resulting in the formation of spherical gels and Au nanosheets. A decrease in the ratio of the metal-coordination unit from 10 mol% to 4 mol% decreased the cross-linking rate with Pd^{II}, affording fibrous gels and Pd nanofibers. Changing the composition ratio of the metal-coordination unit allowed the contrasting gelation behavior to form various types of nanomaterials with metal ions. This procedure will enable the controlled synthesis of various types of nanomaterials containing various metals, which is now under investigation.

4. Materials and Methods

4.1. Materials

Polyvinyl alcohol (PVA, average polymerization degree = 1200) (Wako Pure Chemical,) was used as received. Methyl isothiocyanate (Tokyo Kasei Kogyo, >98.0%) was distilled prior to use. Sodium tetrachloropalladate (II) (Na_2PdCl_4, Tokyo Kasei Kogyo, >98.0%) and sodium tetrachloroaurate (III) dehydrate ($NaAuCl_4$, Wako Pure Chemical, >95.0%) were commercially available and used as received. Dimethylsulfoxide (Wako Pure Chemical, >99.0%) was distilled under CaH_2. N-Methylpyrrolidone (NMP, Wako Pure Chemical, >99.0%) was used as received.

4.2. Instruments

^{1}H NMR spectra were measured with a JEOL JNM ECA-500 using tetramethylsilane (TMS) as an internal standard; δ values are given in parts per million (ppm). IR spectra were measured with a SHIMADZU FTIR IRPrestige-21 spectrometer, and the values are provided in cm^{-1}. Flame atomic absorption spectrometry was conducted with a Hitachi Z-2310 polarized Zeeman atomic absorption spectrometer (AAS). X-ray photoelectron spectroscopy (XPS) was performed with a Kratos AXIS-NOVA instrument. Scanning electron microscopy

(SEM) was performed using a HITACHI S-2500 instrument at an acceleration voltage of 1.5 kV. Energy-dispersive X-ray analysis (EDX/SEM) was conducted using a HITACHI S-3400N/BRUKER Quantax 200 System. Transmission electron microscopy (TEM) was conducted using a HITACHI HT-7700 instrument at an acceleration voltage of 20 kV. Samples for TEM analysis were deposited onto a Cu grid. Atomic force microscopy (AFM) was conducted with a KEYENCE VN-8010 instrument using a silicon substrate in the high amplitude mode (tapping mode) under an ambient condition.

4.3. Gelation of HPMC and Metal Ions

NMP solutions of HPMC (0.200 mL, 13 wt%) were added to 4.00 mM aqueous solutions of Na_2PdCl_4 (10.0 mL) or $NaAuCl_4$ (10.0 mL). Gelation was conducted at room temperature for 2 min, and the obtained gel was dried to constant weight at 60 °C in vacuo.

The gel fraction of the cross-linked HPMC was determined gravimetrically. The dry gels were washed with refluxed distilled water in a Soxhlet extraction to remove the soluble parts. The washed gels were dried to constant weight at 80 °C in vacuo. The gel fraction was calculated from the following equation:

$$\text{Gel fraction} = W_{\text{wash—dry}} / W_{\text{dry}}$$

where W_{dry} is the weight of the dried gel before Soxhlet extraction and $W_{\text{wash—dry}}$ is the weight of the dried gel after Soxhlet extraction.

4.4. Synthesis of Au Nanosheets

An aqueous solution of Au^{III} ions (20.0 mM, 0.800 mL) was carefully added to an NMP solution of HPMC-34 (31.0 wt%, 0.800 mL) in a vial bottle. After 3 s, the obtained nanosheets were transferred from the interface between the Au^{III} ion layer and HPMC-34 layer onto a Petri dish. The nanosheets were washed with NMP, followed by drying under vacuum.

4.5. Synthesis of Pd Nanofibers

An NMP solution of HPMC-4 (0.200 mL, 6.00 wt%) was added to a 25.0 mL test tube containing 4.00 mM Na_2PdCl_4 aqueous solution (20.0 mL), and the mixture was allowed to stand at room temperature for 4 min. The resulting gels were handled using tweezers and placed in a Petri dish. The gels were washed with NMP, followed by drying to obtain the fibrous gels.

Supplementary Materials: The following supporting information can be downloaded at: https://www.mdpi.com/article/10.3390/gels8070435/s1, Figure S1: Stability of HPMC-10-Au; Figure S2: Stability of HPMC-34-Au; Figure S3: TEM images of HPMC-34-Au nanosheet; Figure S4: XPS spectra of Au^0 nanosheet: (a) wide-scan spectrum and (b) narrow-scan spectrum; Figure S5: (a) EDX spectrum of PdII nanofiber. (b) XPS narrow-scan spectrum of PdII nanofiber.

Author Contributions: Conceptualization: D.N.; methodology: S.O. and T.Y.; kinetic study: Y.M.; investigation and experiment: N.I. and T.I. All authors have read and agreed to the published version of the manuscript.

Funding: This research was funded by JSPS KAKENHI 21K19872.

Acknowledgments: The authors thank Hideyuki Sato (University of Shizuoka) for his advice on the TEM analysis (Figures 4d and 5d).

Conflicts of Interest: The authors declare no conflict of interest.

References

1. Masalamani, N.; Bakhsh, E.M.; Khan, S.B.; Danish, E.Y.; Akhtarm, K.; Fagieh, T.M.; Su, X.T.; Asiri, A.M. Chitosan@carboxymethylcellulose/CuO-Co_2O_3 nanoadsorbent as a super catalyst for the removal of water pollutants. *Gels* **2022**, *8*, 91. [CrossRef] [PubMed]
2. Valot, L.; Maumus, M.; Brunel, L.; Martinez, J.; Amblard, M.; Noel, D.; Mehdi, A.; Subra, G. A Collagen-mimetic organic-inorganic hydrogel for cartilage engineering. *Gels* **2021**, *7*, 73. [CrossRef] [PubMed]
3. Paraskevopoulou, P.; Raptopoulos, G.; Leontaridou, F.; Papastergiou, M.; Sakellari, A.; Karavoltsos, S. Evaluation of polyurea-crosslinked alginate aerogels for seawater decontamination. *Gels* **2021**, *7*, 27. [CrossRef] [PubMed]
4. Bellotto, O.; Cringoli, M.C.; Perathoner, S.; Fornasiero, P.; Marchesan, S. Peptide gelators to template inorganic nanoparticle formation. *Gels* **2021**, *7*, 14. [CrossRef]
5. Weinberger, C.; Kuchling, D.; Tiemann, M. Hydrogels as porogens for nanoporous inorganic materials. *Gels* **2018**, *4*, 83. [CrossRef]
6. Riva, L.; Lotito, A.D.; Punta, C.; Sacchetti, A. Zinc- and copper-loaded nanosponges from cellulose nanofibers hydrogels: New heterogeneous catalysts for the synthesis of aromatic acetals. *Gels* **2022**, *8*, 54. [CrossRef]
7. Takeno, H.; Suto, N. Robust and highly stretchable chitosan nanofiber/alumina-coated silica/carboxylated poly(vinyl alcohol)/borax composite hydrogels constructed by multiple crosslinking. *Gels* **2022**, *8*, 6. [CrossRef]
8. Zhao, Q.; Mu, S.; Liu, X.; Qiu, G.; Astruc, D.; Gu, H. Gallol-tethered injectable AuNP hydrogel with desirable self-healing and catalytic properties. *Macromol. Chem. Phys.* **2019**, *220*, 1800427. [CrossRef]
9. Pena, N.; Maldonao, M.; Bonham, A.J.; Aguado, B.A.; Dominguez-Alfaro, A.; Laughter, M.; Rowland, T.J.; Bardill, J.; Farnsworth, N.L.; Ramon, N.A.; et al. Gold nanoparticle-functionalized reverse thermal gel for tissue engineering applications. *ACS Appl. Mater. Interfaces* **2019**, *11*, 18671–18680. [CrossRef]
10. Li, W.; Chu, K.; Liu, L. Zwitterionic gel coating endows gold nanoparticles with ultrastability. *Langmuir* **2019**, *35*, 1369–1378. [CrossRef]
11. Kim, J.; Chan Hong, S.; Bae, G.N.; Jung, J.H. Electrospun magnetic nanoparticle-decorated nanofiber filter and its application to high-efficiency air filtration. *Environ. Sci. Technol.* **2017**, *51*, 11967–11975. [CrossRef] [PubMed]
12. Chee, W.K.; Lim, H.N.; Zainal, Z.; Harrison, I.; Huang, N.M.; Andou, Y.; Chong, K.F.; Pendikumar, A. Electrospun nanofiber membranes as ultratin flexible supercapacitors. *RSC Adv.* **2017**, *7*, 12033–12040. [CrossRef]
13. Wang, Y.; Li, Y.; Sun, G.; Zhang, G.; Liu, H.; Du, J.; Yang, S.; Bai, J.; Yang, Q. Fabrication of Au/PVP nanofiber composites by electrospinning. *J. Appl. Polym. Sci.* **2007**, *105*, 3618–3622. [CrossRef]
14. Ding, H.; Khan, S.T.; Liu, J.J.; Sun, L.Y. Gelation based on host-guest interactions induced by multi-functionalize nanosheets. *Gels* **2021**, *7*, 106. [CrossRef] [PubMed]
15. Yang, C.; Wang, Y.; Wu, Z.; Zhang, Z.; Hu, N.; Peng, C. Three-dimentional MoS_2/reduced graphene oxide nanosheets/graphene quantum dots hybrids for high-performance room-temperature NO_2 gas sensors. *Nanomaterials* **2022**, *12*, 901. [CrossRef]
16. Wang, X.; Wang, Y.; Zhao, X. Nanosheet-assembled MnO-integrated electrode based on the low-temperature and green chemical route. *Crystals* **2022**, *12*, 115. [CrossRef]
17. Zhao, M.; Huang, Y.; Peng, Y.; Huang, Z.; Ma, Q.; Zhang, H. Two-dimentional metal-organic framework nanosheets: Synthesis and applications. *Chem. Soc. Rev.* **2018**, *47*, 6267–6295. [CrossRef]
18. Tan, C.; Cao, X.; Xu, X.-J.; He, Q.; Yang, J.; Zhang, H. Recent advances in ultrathin two-dimentional nanomaterials. *Chem. Rev.* **2017**, *117*, 6225–6331. [CrossRef]
19. Liu, Y.; Xie, Z.; Wong, W.-Y. Synthesis and characterization of a large-sized π-conjugated copper(II) complex nanosheet. *J. Inorg. Organomet. Polym. Mater.* **2020**, *30*, 254–258. [CrossRef]
20. Tsukamoto, T.; Takada, K.; Sakamoto, R.; Matsuoka, R.; Toyoda, R.; Maeda, H.; Yagi, T.; Nishikawa, M.; Shinjo, N.; Amano, S.; et al. Coordination nanosheets based on terpyridine-zinc(II) complexes: As photoactive host materials. *J. Am. Chem. Soc.* **2017**, *139*, 5359–5366. [CrossRef]
21. Sakamoto, R.; Hoshiko, K.; Liu, Q.; Yagi, T.; Nagayama, T.; Kusaka, S.; Tsuchiya, M.; Kitagawa, Y.; Wong, W.-Y.; Nishihara, H. A photofunctional bottom-up bis(dipyrrinato)zinc(II) complex nanosheet. *Nat. Commun.* **2015**, *6*, 6713. [CrossRef] [PubMed]
22. Takada, K.; Sakamoto, R.; Yi, S.-T.; Katagiri, S.; Kambe, T.; Nishihara, H. Electrochromic bis(terpyridine)metal complex nanosheets. *J. Am. Chem. Soc.* **2015**, *137*, 4681–4689. [CrossRef] [PubMed]
23. Kambe, T.; Sakamoto, R.; Hoshiko, K.; Takada, K.; Miyachi, M.; Ryu, J.-H.; Sasaki, S.; Kim, J.; Nakazato, K.; Takata, M.; et al. π-Conjugated nickel bis(dithiolene) complex nanosheet. *J. Am. Chem. Soc.* **2013**, *135*, 2462–2465. [CrossRef] [PubMed]
24. Nagai, D.; Kubo, A.; Morita, M.; Shimazaki, N.; Maki, Y.; Takeno, H.; Mori, M.; Uehara, H.; Yamanobe, T. Pd-and Au-Induced circular fibrous and polymer gelation via thiocarbonyl groups and high Pd catalyst activity. *ACS Appl. Polym. Mater.* **2020**, *2*, 2211–2219. [CrossRef]
25. Nagai, D.; Morita, M.; Yamanobe, T. Synthesis of nanosheets containing uniformly dispersed Pd^{II} ions at an aqueous/aqueous interface: Development of a highly active nanosheet catalyst for Mizoroki-Heck reaction. *ACS Omega* **2020**, *5*, 18484–18489. [CrossRef]
26. Park, J.; Won, S.W.; Mao, I.; Kwak, I.S.; Yun, Y.S. Recovery of Pd(II) from hydrochloric solution using polyallylamine hydrochloride-modified *Escherichia coli* biomass. *J. Hazard. Mater.* **2010**, *181*, 2211–2219. [CrossRef]
27. Huang, M.R.; Peng, Q.Y.; Li, X.G. Rapid and effective adsorption of lead ions on fine poly(phenylenediamine) microparticles. *Chem.-Eur. J.* **2006**, *12*, 4341–4350. [CrossRef]

28. NIST X-Ray Photoelectron Spectroscopy Database (NIST XPS Database, Selected Element Search Menu). Available online: https://srdata.nist.gov/xps/ (accessed on 20 May 2022).
29. Liu, Z.; Yuan, X.; Yu, Y.; Zhang, Q.; Leong, D.T.; Lee, J.Y.; Xie, J. From aggregation-induced emission of Au(I)-thiolate complexes to ultrabright Au(0)@Au(I)-thiolate core-shell nanoclusters. *J. Am. Chem. Soc.* **2012**, *134*, 16662–16670.
30. Shin, H.-S.; Huh, S. Au/Au@polythiophene core/shell nanoparticles for heterogeneous catalysis of nitroarenes. *ACS Appl. Mater. Interfaces* **2012**, *4*, 6324–6331. [CrossRef]
31. Gniewek, A.; Trzecial, A.; Ziolkowski, J.; Kepinski, L.; Wrzyszcz, J.; Tylus, W. Pd-PVP colloid as catalyst for Heck and carbonylation reactions: TEM and XPS studies. *J. Catal.* **2005**, *229*, 332–343. [CrossRef]

gels

MDPI

Article

Development of Flexible Plasticized Ion Conducting Polymer Blend Electrolytes Based on Polyvinyl Alcohol (PVA): Chitosan (CS) with High Ion Transport Parameters Close to Gel Based Electrolytes

Niyaz M. Sadiq [1], Shujahadeen B. Aziz [1,2,*] and Mohd F. Z. Kadir [3]

1 Hameed Majid Advanced Polymeric Materials Research Lab., Physics Department, College of Science, University of Sulaimani, Qlyasan Street, Sulaimani 46001, Iraq; niyaz.sadiq@univsul.edu.iq
2 Department of Civil Engineering, College of Engineering, Komar University of Science and Technology, Sulaimani 46001, Iraq
3 Centre for Foundation Studies in Science, University of Malaya, Kuala Lumpur 50603, Malaysia; mfzkadir@um.edu.my
* Correspondence: shujahadeenaziz@gmail.com

Abstract: In the current study, flexible films of polyvinyl alcohol (PVA): chitosan (CS) solid polymer blend electrolytes (PBEs) with high ion transport property close enough to gel based electrolytes were prepared with the aid of casting methodology. Glycerol (GL) as a plasticizer and sodium bromide (NaBr) as an ionic source provider are added to PBEs. The flexible films have been examined for their structural and electrical properties. The GL content changed the brittle and solid behavior of the films to a soft manner. X-ray diffraction (XRD) and Fourier transform infrared (FTIR) methods were used to examine the structural behavior of the electrolyte films. X-ray diffraction investigation revealed that the crystalline character of PVA:CS:NaBr declined with increasing GL concentration. The FTIR investigation hypothesized the interaction between polymer mix salt systems and added plasticizer. Infrared (FTIR) band shifts and fluctuations in intensity have been found. The ion transport characteristics such as mobility, carrier density, and diffusion were successfully calculated using the experimental impedance data that had been fitted with EEC components and dielectric parameters. CS:PVA at ambient temperature has the highest ionic conductivity of 3.8×10 S/cm for 35 wt.% of NaBr loaded with 55 wt.% of GL. The high ionic conductivity and improved transport properties revealed the suitableness of the films for energy storage device applications. The dielectric constant and dielectric loss were higher at lower frequencies. The relaxation nature of the samples was investigated using loss tangent and electric modulus plots. The peak detected in the spectra of tanδ and M″ plots and the distribution of data points are asymmetric besides the peak positions. The movements of ions are not free from the polymer chain dynamics due to viscoelastic relaxation being dominant. The distorted arcs in the Argand plot have confirmed the viscoelastic relaxation in all the prepared films.

Keywords: polymer blend electrolytes; NaBr salt; glycerol plasticizer; XRD and FTIR methods; circuit design; ion transport parameters; dielectric properties

Citation: Sadiq, N.M.; Aziz, S.B.; Kadir, M.F.Z. Development of Flexible Plasticized Ion Conducting Polymer Blend Electrolytes Based on Polyvinyl Alcohol (PVA): Chitosan (CS) with High Ion Transport Parameters Close to Gel Based Electrolytes. *Gels* **2022**, *8*, 153. https://doi.org/10.3390/gels8030153

Academic Editors: Wei Ji and Hiroyuki Takeno

Received: 25 January 2022
Accepted: 24 February 2022
Published: 2 March 2022

1. Introduction

Renewable energy sources are those derived from naturally replenishing sources, including the sun, wind, storms, seas, seeds, algae, geothermal, and biodegradable polymer materials, and they have attracted great interest due to the growing oil crisis and environmental concerns [1,2]. Polymer electrolyte (PE) science encompasses a wide range of disciplines such as polymer science, organic chemistry, electrochemistry, and inorganic chemistry [3]. Polymers are a prominent issue in material science lately, particularly solid state solutions, which are an example of ion conducting polymers [4]. Because they are important in energy storage devices, including fuel cells, hybrid power sources, supercapacitors, and batteries, solid PEs (SPEs) have been attracting a lot of recent support [5–8]. Polar polymers can coordinate with the cations in the salt, causing it to dissolve. This is due to the presence of functional groups with high electronegativity [9]. To keep pace with rapid technological change, a new generation of highly efficient energy sources must be developed. Because of its use as an electrolyte, polymer-based ion conducting materials have sparked large interest in lithium batteries. Super-ionic conductors are solid substances in which charged atoms, known as ions, carry electric current [10]. These materials' electrical conductivity is vastly different from that of conventional semiconductors. This is because the conductivity of semiconductors is determined by the mobility of light electrons. The conductivity of super-ionic conductors, on the other hand, is the result of ion mobility. Ions have a significant amount of mass and volume [11]. As a result, electric charge transport is linked to mass transfer in super-ionic conductors [10,11].

Chitin is the world's second most abundant biopolymer, derived from fungus and insect cell walls, as well as crustacean exoskeletons [12]. Chitosan (CS), a biopolymer derived from chitin and used in a variety of medicinal and electrochemical devices, is safe, nontoxic, and biodegradable [13]. The backbone structure of CS differs from other biopolymers because it contains amino and hydroxyl functional groups [14]. Poly (vinyl alcohol) (PVA), on the other hand, is a water-soluble polymeric substance with high dielectric strength, strong charge storage capacity, and fascinating optical features [15]. In fact, the PVA molecule has a hydrophobic chain and a hydrophilic end group, the hydrophobic chains occupy space at the solid-liquid interface, while the hydrophilic end groups are extended in the outer aqueous phase. Thus, the steric hindrance and increase in the energy barrier would prevent the aggregation of the crystallinity [16]. The existence of polar groups with a high electron affinity in polar polymers is adequate to form coordination with the cation or surface groups of the fillers, resulting in a uniform nanocomposite [9,17]. PVA has a carbon chain backbone with hydroxyl groups attached, which can help build polymer composites by facilitating hydrogen bonding. Because of its high transparency and ability to form an oxygen barrier, this polymer is a suitable option for use in multilayer coatings for organic solar cells [18].

Ionic conductivity, dimensional stability, and mechanical stability are some of the latest economic and commercialization challenges in membrane technology research. Crystallinity and low ionic conductivity are two of the most significant drawbacks of SPEs [19]. Chemists and engineers are interested in the ionic conductivity of PEs since it is used in commercial electrochemical devices [20]. Both crystalline and amorphous phases exist in PEs. Polymers utilized as host materials in PEs are commonly semi-crystalline, despite the fact that ion transport occurs more frequently in amorphous rather than crystalline phases, as has long been recognized [21]. To overcome the drawbacks of SPEs and enhance conductivity, polymer blending and plasticizer addition are employed to improve ambient ionic conductivity. PVA and CS-based polymer blend complexes are simple to manufacture because of their well-controlled chemical and physical characteristics, such as toughness, homogeneity, and heat stability [22]. Blending CS and PVA is possible since they are both miscible [22]. A result of this interaction is that hydroxyl and amine groups in PVA and CS combine to form an ionic compound. For example, there are many previous works which enhance ionic conductivity by blending CS and PVA [22–25]. An XRD diffractogram revealed that the most amorphous blend host was a 50/50 mix of PVA and CS. Plasticizers may increase SPE's DC electrical conductivity by dissociating PE ion aggregates and boosting the PE's amorphous content [26].

GL, a colorless and odorless liquid, is widely available as an unavoidable by-product of the transesterification of vegetable oils used to make biodiesel, and it can also be obtained from more sustainable sources such as microalgae or cellulose. Because of its inexpensive cost, low toxicity, and unique physicochemical features, such as water solubility and hygroscopicity, it is commonly utilized in pharmaceutical formulations [27]. Increased biopolymer electrical conductivity may be achieved with the use of GL, a plasticizer rich in hydroxyl groups, which reduces the number of internal hydrogen connections between polymer chains. This opens up new applications, such as solid PE films. Electrochemical applications such as humidity sensors, rechargeable batteries, and fuel cells might benefit from this feature's potential [28,29].

It is widespread to use dielectric relaxation spectroscopy to learn relaxation processes in complex systems from a basic perspective. Studying the dielectric characteristics of ion conducting polymers may help researchers understand more about ionic and molecular interactions. The nature of additives and temperature have a huge effect on the dielectric characteristics of ion conducting polymers [30]. The nature of charge transport in polymers has been investigated in order to better understand how these materials conduct electricity [31]. AC impedance spectroscopy is a method to look at the electrical and dielectric properties of materials. The goal of this research is to create a novel type of PE system based on a 50 percent PVA and 50 percent CS (1:1) blend that will act as a good polymer host. This study will look at the structural (XRD and FTIR), conductivity, and relaxation processes that make ions move. Particular attention is given to the exploration of the ion's relaxation and movement in the PVA:CS-based ion conducting mix electrolyte membranes. Polymer films with high content of plasticizer will offer DC conductivities close sufficient to gel-like electrolytes. Because most polymer membranes with ions that are used in devices have a high conductivity, the system made in this study could be used as an electrolyte and separator for electrochemical device applications.

2. Results and Discussion

2.1. FTIR Study

The FTIR spectroscopy is used to investigate the chemical structure of the composite films and probable interactions between the functional groups of PVA and CS in PVA:CS polymer bleed films. Additionally, in this technique, the interaction of NaBr salt with the blended PVA:CS host polymer by modifying the location, intensity, and shape of the IR transmittance bands in the wavenumber range from 500 to 4000 cm^{-1} is explored. The vibrational peaks of OH, C–O, C–H, CH_2, and C=O are used to identify the distinctive bands of PVA and CS polymers [32]. Blend polymer compatibility is shown by changes in the vibrational frequency of the peaks and improved amorphous phase in the PVA:CS system. Figure 1 shows the FTIR spectra of pure CS, PVA, and their blends. In Table 1, the FTIR peaks and their assignments are listed for pure polymers. Figure 1 shows that the OH bands in the PVA:CS system grow in size with decreased intensity, which indicates a reduction in PVA crystallinity. This is evident for the occurrence of complexation between the functional groups of the two polymers, which is clearly related to the OH band of the blend system [33]. The results of the XRD investigation are consistent with this discovery. By shifting the O–H peak position and intensity, the PVA hydrogen atoms and the CS oxygen atom are shown to establish hydrogen bonds with one other. Blending CS with PVA results in an expansion of the crystalline peak that corresponds to the C–O stretching mode and a decrease in its intensity [34]. It has been shown that adding CS to a mix diminishes the absorption band intensity at 834 cm^{-1}, which corresponds to PVA's C–C stretching.

Figure 2 illustrates the FTIR spectra of various PVA : CS : NaBr : Gl in the wavenumber range 500–4000 cm^{-1}. [35] All samples show the key characteristic absorption peaks of CS, for instance the vibration of the amino group (NH_2), O=C–NHR, and amine NH symmetric. The PVA structure has been connected to C–O plane bending, which commonly occurs at 1015–1031 cm^{-1}. Furthermore, the peak at 1600–1700 cm^{-1} is due to C=O stretching of PVA's acetate group, which is pushed to a lower wavenumber in doped samples [35].

There is motionless a noteworthy wide band at 3300–3500 cm^{-1}, despite the possibility of overlap between the N–H and O–H stretching vibrations. Figure 2 shows that as the plasticizer is increased, the bands become more intense and the wavenumber decreases. In the bands of amine (NH2) and (O–H) groups, there is a trend towards lower wavenumbers, as seen in Table 2. This strongly suggests that complexation is happened between the electrolyte's constituents [23,36]. The shift and reduction in relative strength of these bands is due to the electrostatic interaction between the ions and the functional groups of the CS:PVA polymer blend [37]. Furthermore, the N–H stretching vibration's shift to lower wavenumbers shows that CS's intermolecular and intramolecular hydrogen bonds have decreased [38]. Recent research has shown that the carboxyl (–C=O), hydroxyl (–OH), and amine (–NH) groups all have a role in salt interaction [39]. This band's shifting and changing strength specify a superior interaction between the electrolyte's blended host polymer, salt, and plasticizer. The vibrational and stretching modes in the FTIR spectra vary as a consequence of the interaction between the electrolyte components (shown in the table below) [22,40]. This enhanced interaction promotes ion dissociation, which is beneficial to the electrolytes' ionic conductivity [40,41]. The rise in ionic conductivity is corroborated by this increase in intensity (see Figure 2).

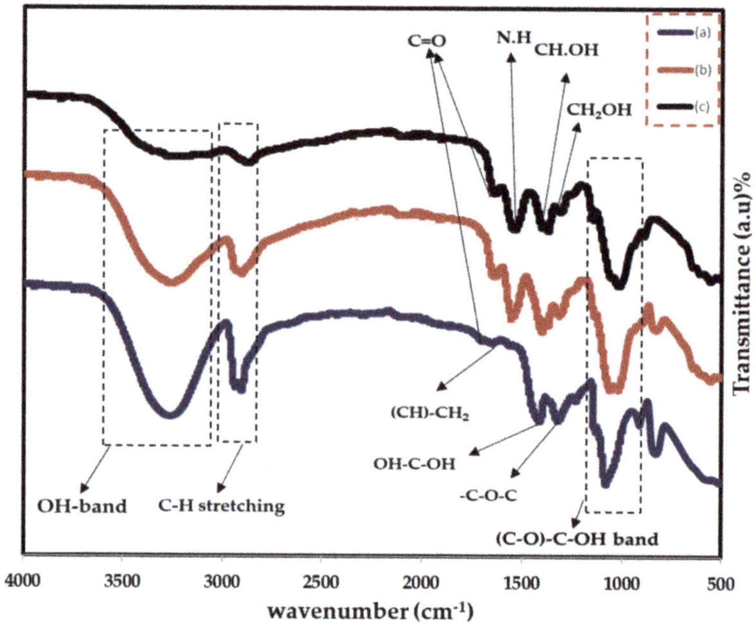

Figure 1. FTIR spectra for (**a**) pure PVA, (**b**) PVA:CS (0.5:0.5) blend, (**c**) pure CS.

Table 1. Purified (PVA), pure (CS) and blended pure (PVA:CS) FTIR data are shown.

Vibrational Modes	Pure PVA	Vibrational Modes	PVA:CS	Vibrational Modes	Pure CS
O–H stretching	3259.34	O–H stretching	3261.67	O–H stretching	3261.42
C–H stretching	2937.98	C–H stretching	2906.38	C–H stretching	2872.47
C = O	1710	C = O	1646.29	C = O	1636.33
(C–H)–CH2	1653.51	N–H	1557.41	N–H	1540.59
OH–C–OH	1417.33	CH·OH	1405.89	CH·OH	1403.24
−C–O–C	1324.90	CH2–OH	1376.61	CH2–OH	1375.46
(C–O)–C–OH	1084.88	(C–O)–C–OH	1027.48	(C–O)–C–OH	1022.98
C–C	834.16	C–C	834.08	C–C	–

2.2. XRD Study

Figure 3 shows XRD results for room-temperature PVA, CS, and CS:PVA (a–c). In contrast to the semi-crystalline character of pure PVA (Figure 3a), the CS exhibits crystallinity maxima at some 2θdegree values (Figure 3b) [42]. Since OH groups are present everywhere throughout PVA and CS's main chain, it is able to form strong inter- and intramolecular hydrogen bonds. Amorphous phases in PVA are responsible for the large peak at 2θ = 40.7° which is due to existing high water content [43]. The XRD pattern of CS: PVA (Figure 3c) revealed two hallows and smaller crystalline peaks in the current study. It is worth noting that, as the vast hallows demonstrate, CS: PVA is less crystalline than pure PVA or CS alone, and its structure is practically amorphous [44,45]. According to previous research, the amorphous structure of PEs is linked to large diffraction peaks [46].

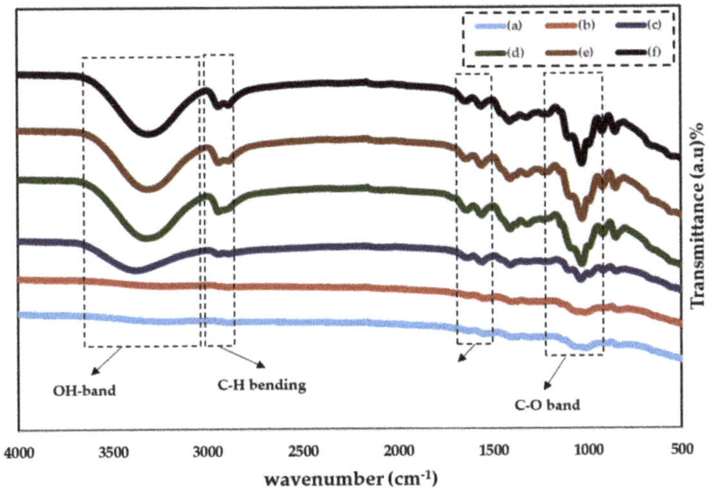

Figure 2. FTIR spectra of (a)CSPVNG0, (b) CSPVNG11, (c) CSPVNG22, (d) CSPVNG33, (e) CSPVNG44 and (f) CSPVNG55.

Table 2. Assignments of FTIR bands for PVA:CS: NaBr:GL solid PEs.

Samples	Wavenumbers (cm^{-1})			
	C–O Band	NH_2	C–H Band	O–H Band
CSPVNG0	1015.61	1545.66	-	-
CSPVNG11	1020.04	1558.79	-	3374.36
CSPVNG22	1039.06	1559.40	2938.77	3374.69
CSPVNG33	1032.10	1559.33	2938.70	3317.19
CSPVNG44	1031.21	1559.41	2937.25	3316.19
CSPVNG55	1031.29	1559.47	2936.4	3312.77

Meanwhile, when GL was added to these samples, the strength of the CS : PVA : NaBr peaks decreased, and the wide nature of the peaks improved, as seen in Figure 4a–f. These findings support the hypothesis that the PE has an amorphous structure that improves conductivity by increasing ionic diffusivity. Furthermore, the NaBr salt dissociates completely in the PE, leaving no peak associated with pure NaBr, as illustrated in Figure 5. The absence of hydrogen bonds between polymer chains is a likely cause for the intensity drop and broadening, indicating the presence of the amorphous phase in the samples [47]. Plasticized PEs are a type of PE made by adding low molecular weight chemicals to the polymer host, such as ethylene carbonate, propylene carbonate, and poly ethylene glycol (PEG) [48]. It is possible for plasticizers to reduce the number of active centers in polymer chains, hence decreasing the interactions between and within molecules [49]. Consequently, the three-dimensional structure generated during drying loses stiffness and the mechanical and thermomechanical characteristics of the films it produces are affected [49,50]. To facilitate charge carrier transfer, low molecular weight plasticizers have been added, resulting in a reduction in crystallinity, an increase in salt dissociation capacity and producing a jelly thin film, as seen in Figure 6a,b.

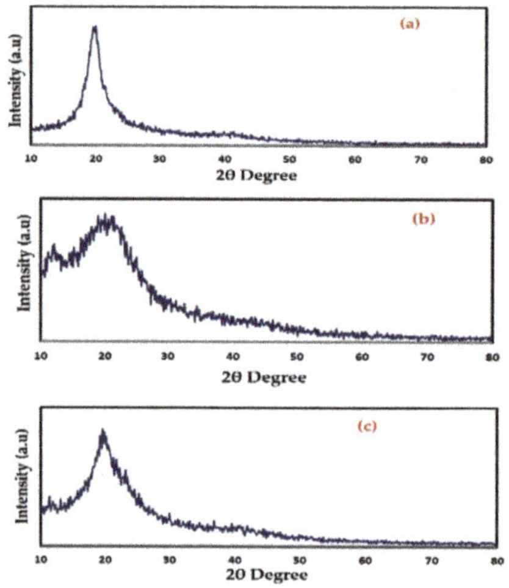

Figure 3. XRD pattern of (**a**) pure PVA film, (**b**) pure CS and (**c**) PVA: CS blend (50:50) films.

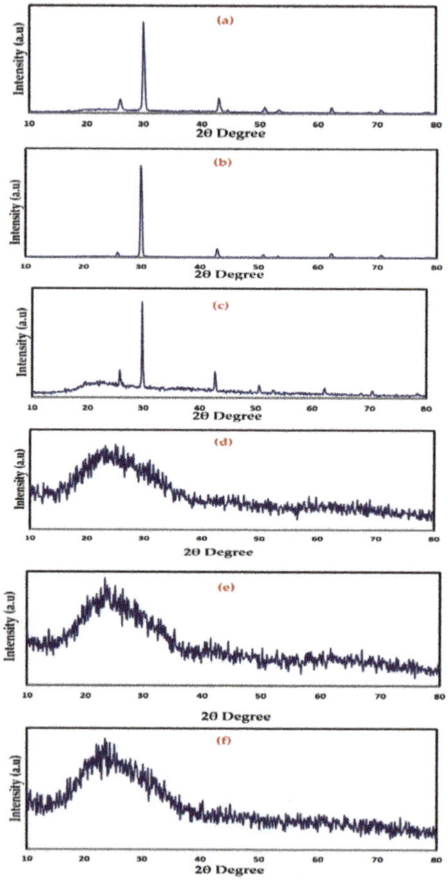

Figure 4. XRD Pattern of (**a**) CSPVNG0, (**b**) CSPVNG11, (**c**) CSPVNG22, (**d**) CSPVNG33, (**e**) CSPVNG44 and (**f**) CSPVNG55.

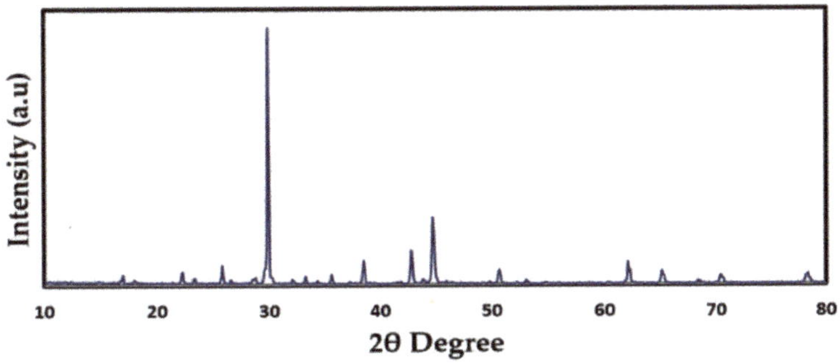

Figure 5. XRD pattern of pure NaBr.

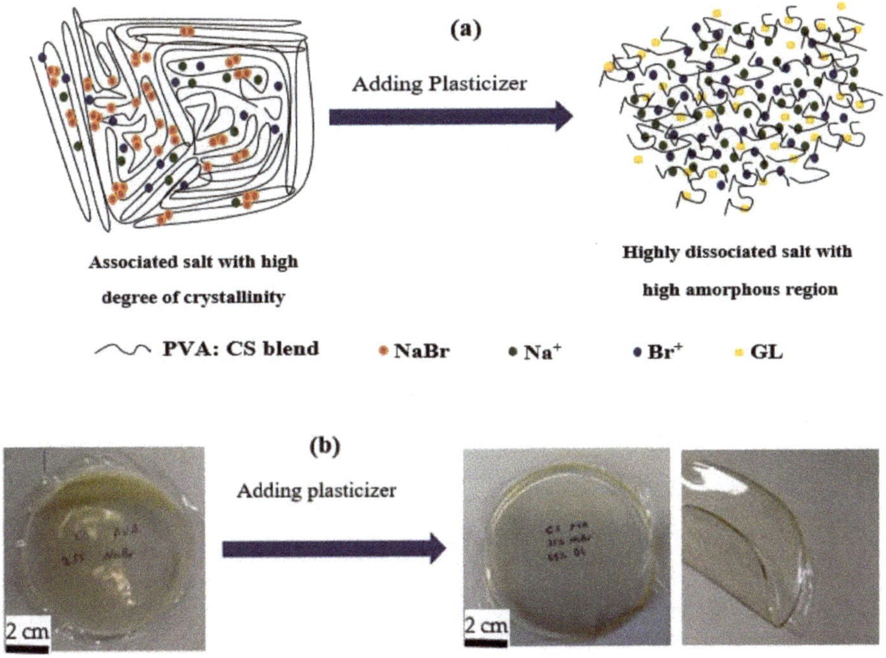

Figure 6. Role of plasticizer on (a) the ion dissociation and increase of flexibility and amorphous phase, and (b) the film with high content of plasticizer can easily be bending without deformation or tear. The last photograph shows the ability of the film to bending.

2.3. Complex Impedance Spectroscopy (CIS) and Ion Transport Study

It is required to conduct non-destructive testing in order to distinguish a wide range of materials. As a tool for studying heterogeneous systems, dielectric impedance spectroscopy, which measures conductivity and permittivity as a function of frequency at different temperatures, may reveal their structure [51]. The CIS response is seen in the Nyquist plot of all the samples, as shown in Figure 7a–e. In the high frequency and low frequency regions, the responses take the shape of a semicircle and a tail (spike), respectively. On the one hand, at high frequencies, the incomplete semicircle is primarily associated with bulk resistance (a bulk property). The development of a spike, on the other hand, denotes the establishment of double layer capacitance at the electrode/sample interface at low frequencies [52].

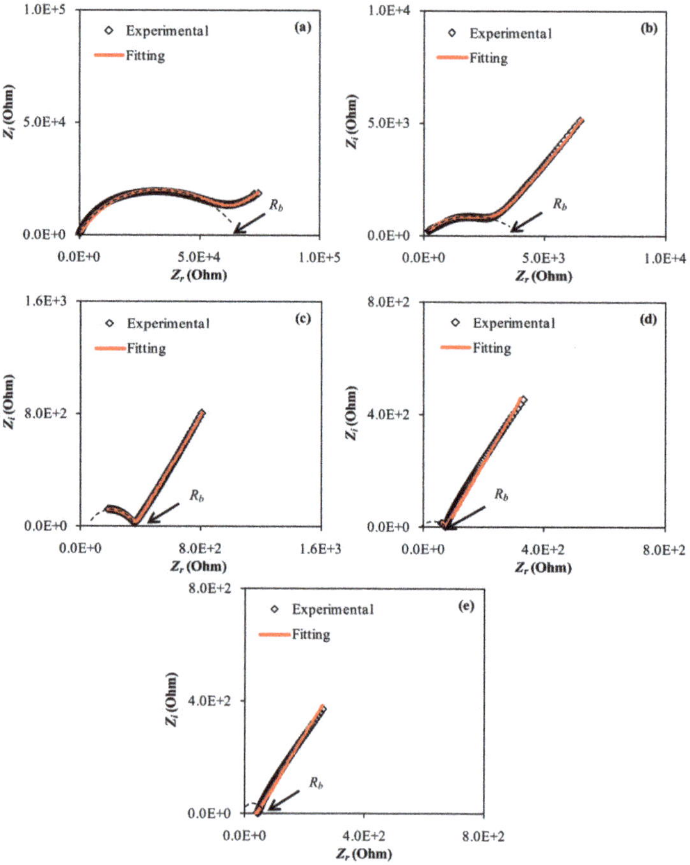

Figure 7. EIS plots for (**a**) CSPVNG11, (**b**) CSPVNG22, (**c**) CSPVNG33, (**d**) CSPVNG44 and (**e**) CSPVNG55 electrolyte films.

The facility to recognize the relaxation frequency and to separate electrode (low frequency spike) and bulk (high frequency semicircular region) affects are two of the benefits of the CIS technique. Real (Z') and imaginary (Z'') components of impedance can be obtained using the CIS approach across a large frequency range. This technique has been effectively utilized to assess DC ionic conductivity and ionic conductor activation energy in recent years [44,53]. An electrochemical sample holder must be exposed to an alternating voltage across a wide frequency range for CIS measurements [54]. The direct correlation between the system's response and the proposed equivalent circuit, which represents the system's electrical behavior, is a key element of CIS [55,56]. In physics, a dielectric material's dissipative response is characterized by a resistance (R), whereas its storing response is characterized by a capacitance (C) [56,57]. One of the outputs of electrical impedance spectroscopy is a graph of the imaginary component of the impedance vs. the real component, CIS. This graph can be used to derive information about the expected equivalent circuit. The impedance charts should have a straight line parallel to the imaginary axis at low frequencies; instead, the blocking electrode polarization effect (double layer capacitance) produces the curvature [58–60].

It will be simple to specify the mechanism of the system after fitting experimental spectra and analyzing using the electrical equivalent circuit (EEC) technique [59]. Figures 7 and 8 show the experimental impedance spectra, as well as the related EEC models for all of the electrolyte samples. In the Nyquist plots displayed in the insets of the bulk resistance (R_b) for charge carriers in electrolyte systems, as well as two constant phase elements (CPEs); CPE1 and CPE2, can be seen Figure 8. The EEC is made up of a succession of R_b and CPE1 in parallel with a second CPE2 (from the tilted spike). The impedance's ZCPE component is written as follows [55,61,62]:

$$Z_{CPE} = \frac{1}{C\omega^{P}}\left[\cos\left(\frac{\pi p}{2}\right) - i\ \sin\left(\frac{\pi p}{2}\right)\right] \tag{1}$$

where CPE capacitance and the angular frequency are symbolized as C and ω, and represents the departure of the plot from the vertical axis in complex impedance spectra. It is important to point out that CPE and capacitor are commonly used interchangeably in the context of an EEC modeling.

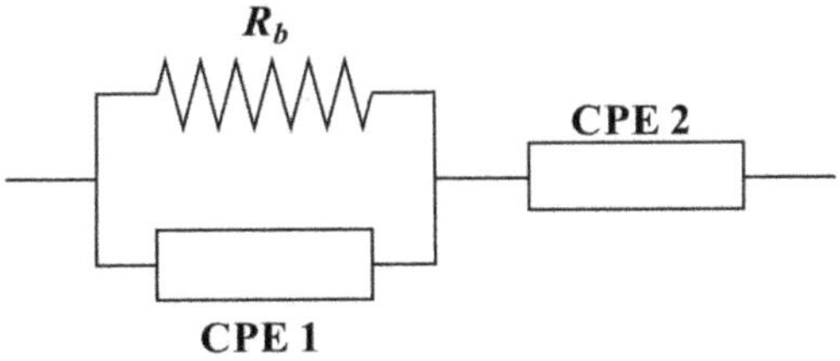

Figure 8. The equivalent electrical circuit (EEC).

The overall mathematical expression of the equivalent circuit, and the real (Z_r) and imaginary (Z_i) complex impedance (Z^*) values may be written as follows [55,61,62]:

$$Z_r = \frac{N + R_b}{M + T} + \frac{\cos\left(\frac{\pi p_2}{2}\right)}{C_2\omega^{p2}} \tag{2}$$

$$Z_i = \frac{L}{M + T} + \frac{\sin\left(\frac{\pi p_2}{2}\right)}{C_2\omega^{p2}} \tag{3}$$

where $N = R_b^2 C_1 \omega^{p1} \cos\left(\frac{\pi p_1}{2}\right)$, $M = 2R_b C_1 \omega^{p} \cos\left(\frac{\pi p}{2}\right)$, $L = R_b^2 C_1 \omega^{p1} \sin\left(\frac{\pi p_1}{2}\right)$ and $T = R_b^2 C_1^2 \omega^{2p} + 1$.

The real and imaginary sections of semicircles are attributed to the first half of Equations (8) and (9), while spike lines are attributed to the second part. Where C_1 indicates CPE1's bulk capacitance and C_2 denotes CPE2's capacitance at the electrode-electrolyte interface. In Table 3, the EEC fitting parameters are listed. In the high frequency region, the semicircle size drops significantly as the amount of GL increases, as seen in the impedance spectra. Figure 7a,b show how to model an incomplete semicircle in which the value of R_b is in series with one CPE element and parallel with another CPE element (at low frequency tail). Figure 7c–e illustrate the complete elimination of the incomplete semicircle. Given the R_b values and the film thickness, calculation of the DC conductivity can be performed using Equation (4) [63].

$$\sigma_{dc} = \left(\frac{1}{R_b}\right) \times \left(\frac{t}{A}\right) \tag{4}$$

Table 3. Circuit elements of the plasticized PBE systems.

Sample	p_1 (rad)	p_2 (rad)	CPE1 (F)	CPE2 (F)
CSPVNG11	0.71	0.52	6.67×10^{-9}	1.54×10^{-6}
CSPVNG22	0.60	0.60	1.43×10^{-7}	3.33×10^{-6}
CSPVNG33	0.73	0.68	2.22×10^{-8}	1.35×10^{-5}
CSPVNG44	0.71	0.68	5.00×10^{-8}	2.38×10^{-5}
CSPVNG55	0.64	0.67	1.00×10^{-7}	2.94×10^{-5}

Table 4 shows the computed DC conductivity and bulk resistance (R_b) values, as well as all the transport parameters of the plasticized PBE systems, including diffusion (D), mobility (μ) and charge carrier density (n). The quantity and sort of plasticizer have a big impact on the PEs' characteristics. GL is the mainly often used and least expensive plasticizer [64]. After water, it is one of the most well-known polar compounds. It is completely water soluble, has a low toxicity compared to other organic solvents, and is completely biodegradable [27]. Furthermore, renewable resources are abundant on the earth and have a low market price. The relatively high conductivity can be explained by the massive ion movement, as demonstrated by the semicircle disappearing in the high frequency area of the spectra. As a result, GL is an effective plasticizer for reducing crystallinity and increasing ion mobility, which increases overall DC conductivity. According to Kaori Kobayashi et al., the addition of GL plasticizer reduced the glass transition temperature and increased the dc conductivity (σ_{dc}) of poly (ethylene carbonate) (PEC): LiPF6Telectrolyte [65].

Table 4. Transport parameters of the plasticized PBE systems.

Sample	Rb (Ω)	σ (S cm^{-1})	D (cm^{2}s^{-1})	μ (cm^{2}V^{-1}s)	n (cm^{-3})
CSPVNG11	62,600	2.46×10^{-7}	3.48×10^{-5}	1.35×10^{-3}	1.14×10^{15}
CSPVNG22	2815	5.48×10^{-6}	1.52×10^{-5}	5.90×10^{-4}	5.79×10^{16}
CSPVNG33	369	4.18×10^{-5}	8.60×10^{-7}	3.35×10^{-5}	7.79×10^{18}
CSPVNG44	73	2.11×10^{-4}	4.06×10^{-7}	1.58×10^{-5}	8.34×10^{19}
CSPVNG55	40	3.86×10^{-4}	1.46×10^{-7}	5.67×10^{-6}	4.24×10^{20}

The heart of electrochemical devices is PEs. The majority of solid-state electrochemistry research is focused on the development of high ion-conducting materials for energy conversion and storage [66]. Massive research efforts over the past two decades have resulted in systems with enhanced conductivity and transport characteristics, and PEs fall into this category [66–68]. Transport parameters are crucial properties that should be considered in device applications. PEs with performance transport properties are the focus of many research groups.

Ion number density (n), diffusion coefficient (D), and mobility (μ) may be calculated from the impedance data of all systems by utilizing the following relations [69–71]:

$$D = \left\{ \frac{(K_2 \varepsilon_0 \varepsilon_r A)^2}{\tau_2} \right\} \quad (5)$$

where ε_r and ε_0 signify the dielectric constant and the permittivity of the space, respectively. The reciprocal of ω is represented by τ_2, which is matching to the smallest value in Z_i. From Equation (5), it is noticeable that three parameters influence the value of D, which are K_2, ε_r and τ_2. Thus, it is difficult to obtain D and μ values follow the trend of n value.

$$\mu = \left\{ \frac{eD}{K_b T} \right\} \quad (6)$$

where k_b and T refer to the Boltzmann constant and absolute temperature, respectively. DC conductivity (σ_{dc}) and Ion number density (n) are presented by the following equations, respectively:

$$\sigma_{dc} = ne\mu \quad (7)$$

$$n = \left\{ \frac{\sigma_{dc} K_b T \varepsilon \tau_2}{(eK_2 \varepsilon_0 \varepsilon_r A)^2} \right\} \quad (8)$$

From Table 4, is clear that the number density improved from 1.14×10^{15} cm^{-3} to 4.24×10^{20} cm^{-3}. A large network of ion conduction may be formed by the incorporation of plasticizer molecules into the host polymer, as shown in Figure 6a [72]. According to recent research, the conductivity of thin films may be improved by increasing the amount of existing hydroxyl groups [73]. To lower intermolecular tensions, increase the mobility of its polymeric chains and enhance mechanical properties such as the extensibility of the resultant film, GL was essential [28,74].

2.4. Dielectric Properties

For determining the electrical characteristics of many materials, including glasses, semiconductors, polymers, and transition metal oxides, impedance spectroscopy may be a valuable instrument [9,40]. CIS relies on a number of other measured or calculated parameters. The measurement, analysis, and charting of some or all of the four impedance-related functions (Z*, Y*, M*, and ε*) in the complex plane is referred to as impedance spectroscopy [75]. Dielectric measurements (ε*), such as dielectric constant (ε′) and dielectric loss (ε″), disclose a lot about polymer chemical and structural behavior. Using Equations (9) and (10), the real and imaginary parts of dielectric function were evaluated [76]. Because of the electrode polarization (EP) effect [77], both of them (ε′ and ε″) are quite high at low frequencies, as seen in Figures 9 and 10. The inclusion of another polymer or a dopant in the polymer has a significant impact on these properties (ε′ and ε″) [56]. In polymer electrolyte systems the polarization can be resulted from both permanent dipole of host matrix functional groups and space charge polarization at the electrode/electrolyte interface. As can be seen in Figure 10, the dielectric loss has recorded high value at low frequency region due to the plenty of time that is provided by the applied field for re-orientation. In addition, the increasing plasticizer has improved the rotation freedom of the site groups to response quickly to the applied field. These two factors have caused an enhancement in dielectric loss value. However, at high frequency regions such a high value is not observed due to the fast switching of the applied field.

$$\varepsilon' = \frac{Z_i}{\omega C_0\left(Z_r^2 + Z_i^2\right)} \tag{9}$$

$$\varepsilon'' = \frac{Z_r}{\omega C_0\left(Z_r^2 + Z_i^2\right)} \tag{10}$$

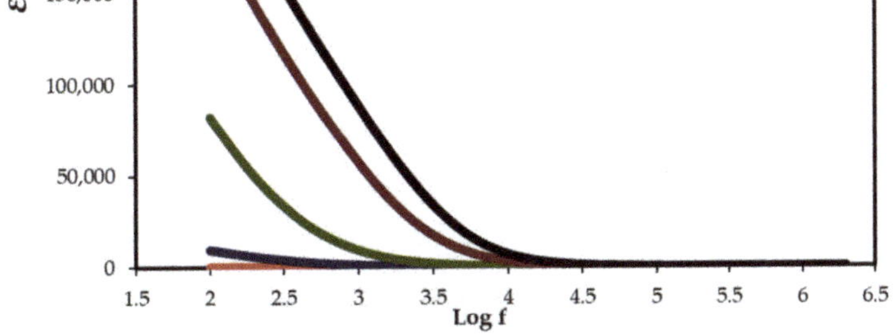

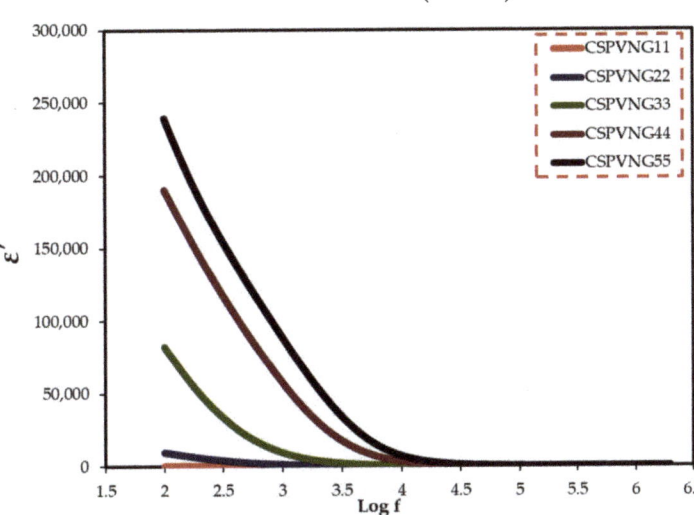

Figure 9. Dielectric constant versus log(f) for the electrolyte samples.

The vacuum capacitance, C_o, is provided by the formula $\varepsilon_o A/t$, where ε_o is the free space permittivity, which is to 8.85×10^{-12} F/m. If f is the applied field frequency, the angular frequency (ω) is equal to $2\pi f$.

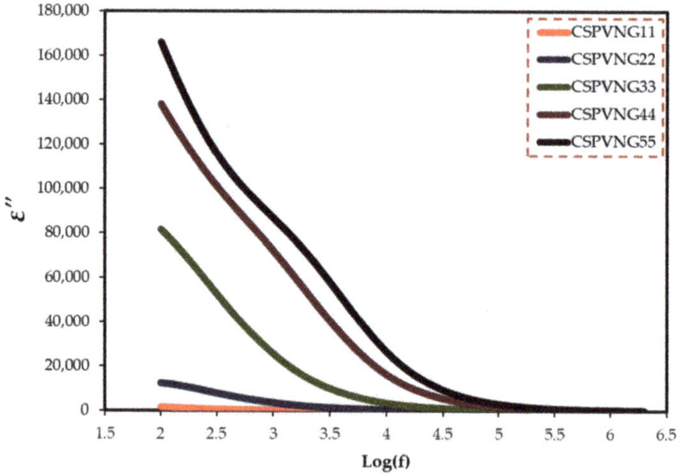

Figure 10. Dielectric loss versus log(f) for the electrolyte samples.

At low frequencies, ion migration and polarization have a significant impact on dielectric properties. For polymer/filler composites, this phenomenon is not possible. Due to the delayed dielectric relaxation of various polarizations in polymer composites, the permittivity falls gradually with increasing frequency. Permittivity has a low frequency dependency, which is advantageous across a large frequency range [78].

The migration of ions from one site to another in solid PEs will cause the electric potential of the environment to be perturbed. The perturb potential will have an impact on the movement of the other ions in this location. Non-exponential decay, or conduction processes with a dispersion of relaxation time, will occur as a result of such cooperative ion motion [79]. The behavior of bulk dielectric constant and bulk DC conductivity is essentially identical with GL content, as can be seen. This finding demonstrates that DC conductivity is greatly influenced by the dielectric constant. It is important to note that as the plasticizer concentration rises, the dielectric dispersion switches to the high frequency region. Both DC conductivity and ε' are clearly lower for Gl content of 10 to 20 Gl. When crystallites cover a significant percentage of the polymer, the conductivity and dielectric constant are dramatically reduced, which is an undesirable circumstance [80]. Most of the dielectric constant is reduced by structural densification (i.e., the rise in crystallinity between the provided lamellae). Charge carrier mobility is hindered during the remainder of the amorphous phase as a consequence. The conduction channel may become unstable even if the amorphous phase composition changes slightly in such a compact arrangement. There is also the possibility that electrode contact loss and sample stiffness are to blame for such a large conductivity decline or resistivity rise [81]. Usually, the dominance of the EP effect can be attributed to significant charge carrier buildup at high GL concentration or at high temperature at the electrode/electrolyte interface because at these situations the semicircle disappears in CIS plots [82].

When an electric field is given to a solid substance, it causes ion conduction between favorable binding sites. The tail at low frequencies may be owing to the electrode/electrolyte interfaces blocking mobile ions and/or equipment constraints in viewing the low-frequency electrode polarization region [83]. The magnitude of ionic conductivity is described in theory as

$$\sigma = \sum q_i n_i \mu_i \tag{11}$$

where n_i denotes the number of carriers, and q_i and μ_i denote the carriers' charge and mobility, respectively. To achieve good ionic conductivity, you will need a lot of ion carriers with a large mobility. As a result, it is critical to discover materials that match these requirements [84]. The concentration of charge carriers' growths when the dielectric constant rises $\left(n_i = n_o \exp\left(-\frac{U}{\varepsilon' K_B T}\right)\right)$, where U denotes the dissociation energy, resulting in anincrease in DC conductivity $(\sigma = \sum q_i n_i \mu_i)$.

2.5. Tanδ Study

Broadband dielectric spectroscopy has been revealed to be a very efficient method for studying polymeric system relaxation processes. The complex dielectric utility, ε*, which consists of the dielectric constant and loss, is a material parameter that is affected by frequency, temperature, and structure, and so the tan, which is the ratio of ε'' to ε', is likewise frequency dependent. The plot of tanδ dielectric relaxation peaks as a function of frequency at room temperature for all samples is explored in this section. The aim of this study is to learn more about the relaxation processes and structure of polymeric-based materials. The study of dielectric relaxation can be a useful technique for understanding dipole relaxation in PEs. The dielectric relaxation processes are usually linked to one or more of the material's polarization processes. The dipolar polarization and polarization owing to migrating charges are the two major components of polymer dielectric response. In frequencies less than 10^9 Hz, dipolar and migrating charge polarizations can be identified [85]. To our understanding, there is no literature that discusses the use of loss tangent (Tanδ) relaxation peaks in determining polymeric structural attributes. The mechanism of ion transport is presently under debate among scientists. As a result, the study's primary contribution is to provide the first experimental data and insights into the tan relaxation peaks and structure identification. The dielectric loss peak has been recognized as useful in investigating relaxations such as, α, β and γ, which are connected to dipole rotation in the crystalline phase, dipole orientation in amorphous areas, and side group or end-group movement in the amorphous phase, respectively [86]. α relaxation occurs at low frequencies most of the time, but β relaxation can be seen at intermediate to higher frequencies, and it transitions to higher frequencies as the amorphous phases increases [55]. The height of the peak associated with the amorphous phase at high frequency is clearly increased as the GL content increases. The results of the morphological and impedance analyses are exactly in line with these conclusions. As a result, the loss tangent peak can be used to determine structural features of materials in a sensitive manner. PEs are known to be heterogeneous materials, as both amorphous and crystalline phases exist. At high GL concentrations, the single peak disappears, implying system homogeneity [87]. The findings in this work suggest that tanδ relaxation peaks can be used to distinguish between crystalline and amorphous polymer phases. Furthermore, one peak formed at 10 wt.% of GL, while two loss peaks appeared at 20 wt.% of GL, as shown in Figure 11a. In both crystalline and amorphous regions, the peaks are connected to dipole orientation. The α -relaxation loss peak was discovered to be narrower and asymmetrical than the β -relaxation peak. The α -relaxation is partially filtered at low frequencies due to ionic conductivity [55,88]. While Figure 11b shows that α -relaxation loss peak has been disappeared, and β -relaxation peak has dominated due to the transfer of these systems from crystalline phase to fully amorphous phase, which is supported by XRD results (Figure 4).

2.6. Electric Modulus and Relaxation Study

The material is exposed to an alternating electric field, which is generated by applying a sinusoidal voltage, in dielectric measurements; this process causes dipoles in the material to align, resulting in polarization. Dipoles on the side chain of the polymer backbone can cause dipolar polarization in polymeric materials, as well as the presence of ion translational diffusion [89]. Because of the huge conductivity effects, dielectric parameters should be expressed in terms of the complex electric modulus ($M^* = 1/\varepsilon^*$) [90]. It can be deduced from the electric modulus study that ion transport happens either by polymer segmental relaxation or conductivity relaxation [91]. Electrical relaxation phenomena in PE are known to be caused by phase transitions, interfacial effects, and polarization or conductivity mechanisms [92]. Macedo et al. devised the electric modulus formalism to limit the effect of electrode polarization. Equations (12) and (13) were used to compute the real and imaginary components of electric modulus [93]. The peak maximum in the imaginary part of the electric modulus indicates that the sample is an ion conductor. Figure 12 depicts the frequency dependence of M′ for various GL plasticizer concentrations. Figure 13 shows that a distinct relaxation peak can be detected in the imaginary part of modulus (M″) spectra, which is associated with conductivity processes, however no peak can be found in the dielectric loss spectra. This indicates that ionic and polymer segmental movements are highly connected, as evidenced by a single peak in the M″ spectra and no equivalent characteristic in dielectric loss spectra (Figure 13) [94]. As a result, charge migration of ions between coordinated sites of the polymer and segmental relaxation of the polymer are both involved in conduction in PEs.

$$M' = \omega C_0 Z_i \quad (12)$$

$$M'' = \omega C_0 Z_r \quad (13)$$

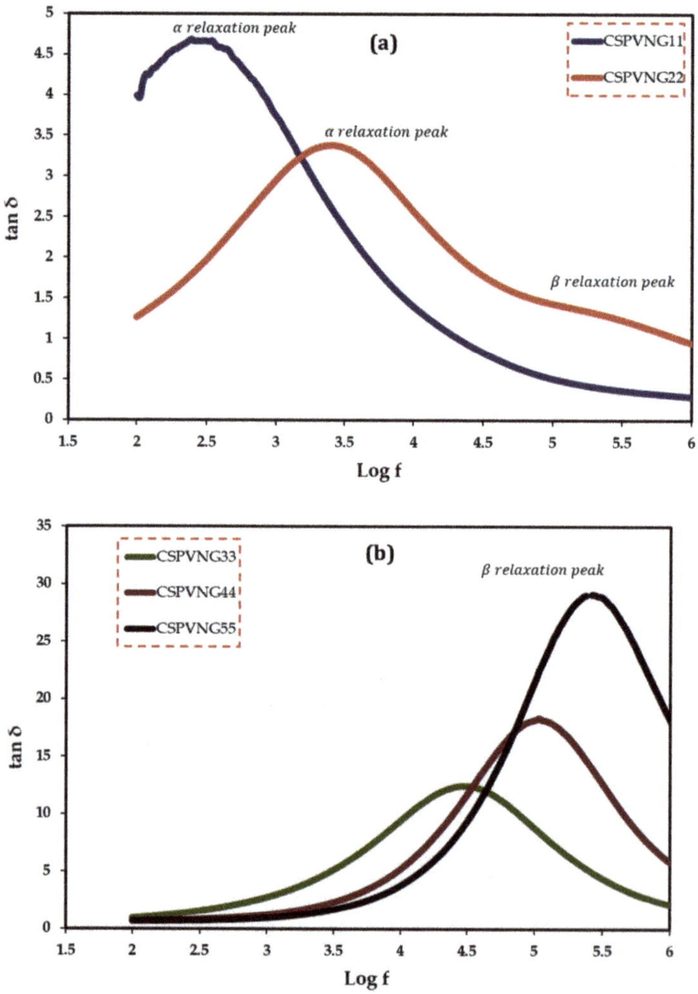

Figure 11. Loss tangent versus log (f) for (**a**) CSPVNG11 and CSPVNG12, and (**b**) CSPVNG33, CSPVNG44 and CSPVNG55 the electrolyte samples.

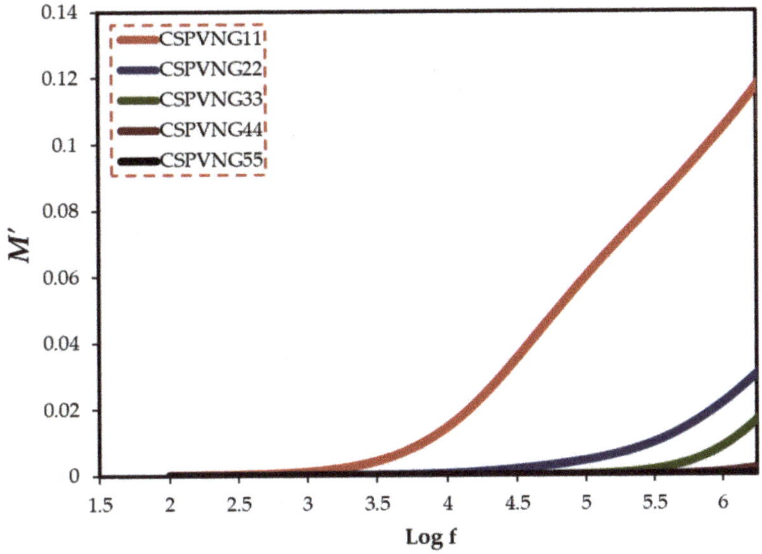

Figure 12. The real part of the electric modulus versus log (f) for the electrolyte samples.

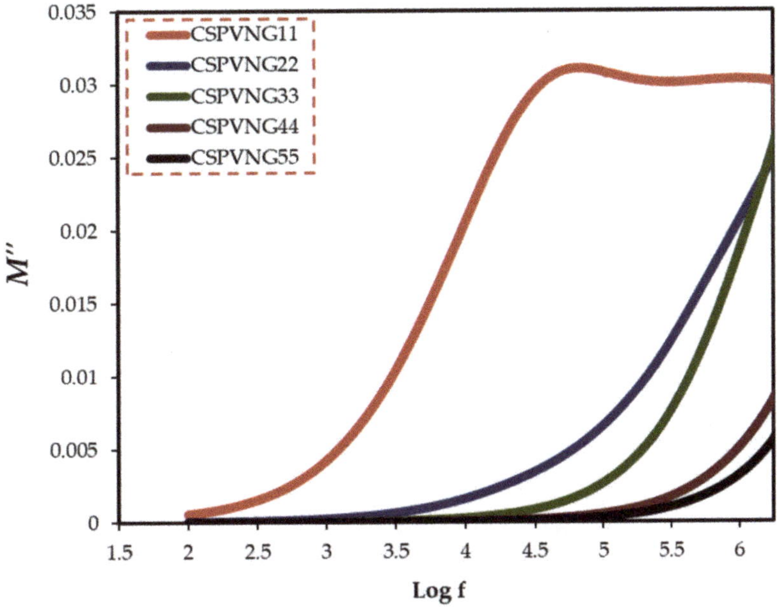

Figure 13. The imaginary part of the electric modulus versus log (f) for all electrolyte samples.

The use of Argand (M″ vs. M′) plots to decide whether a relaxation dynamic fits in to viscoelastic or conductivity relaxation is a well-known approach. The type of relaxation processes in the current PEs can be revealed. The Argand curves for all of the samples are shown in Figure 14 at room temperature. According to [95], if the Argand plot between the imaginary component (M″) and the real part (M′) of the electric modulus displays semicircular behavior, the relaxation is due to conductivity relaxation; otherwise, viscoelastic relaxation (or polymer molecular relaxation) is responsible [96]. The circle diameters do not line up with the real axis. This means that ion transport happens in all samples via viscoelastic relaxation. The conductivity relaxation differs significantly from the viscoelastic relaxations found in polymers. The conductivity relaxation corresponds to a single-relaxation-time Debye model process, whereas viscoelastic relaxations are known to have a dispersion of relaxation times [97].

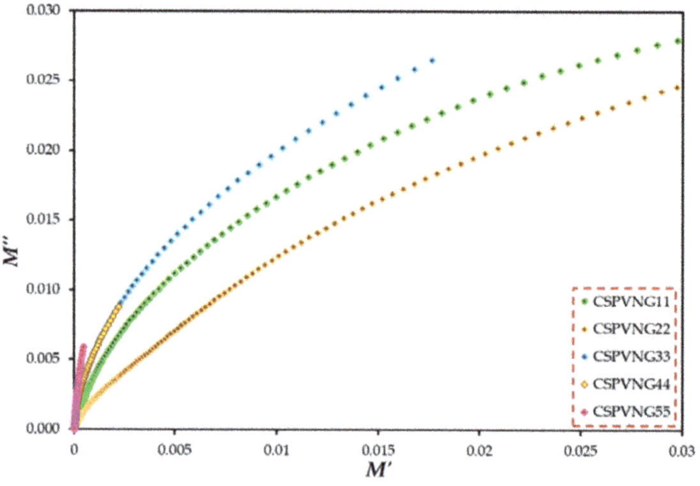

Figure 14. M″-M′ plot for all plasticized SPE films.

2.7. AC Conductivity Analysis (σ_{ac})

Even though experiments have been carried out on a broad range of semiconducting and insulating materials, the literature on AC conductivity in disorder solids and its characteristics with composition is valuable. In general, frequency-dependent conductivity follows a power-law behavior ($\sigma \sim \omega^s$) [98]. The properties of crystalline semiconductors and insulators have been determined by several solid state theories, but when these theories are applied to data on polymers, the predictions they produce are not sufficiently unique, making it surprisingly difficult to determine all the important characteristics of electrical conduction in polymers even today [99]. The σ_{ac} were evaluated from the real (Z_r) and imaginary (Z_i) parts of complex impedance (Z^*) using Equation (14).

$$\sigma'_{ac} = \left[\frac{Z_r}{Z_r^2 + Z_i^2}\right] \times \left(\frac{t}{A}\right) \tag{14}$$

Figure 15 shows the AC conductivity spectra of plasticized PVA : CS : NaBr sheets at ambient temperatures. At high frequencies, ac conductivity in disordered solids is highly dispersed, making it one of the most distinctive properties of electrical conduction in these materials. AC methods are widely used to study electrical properties of ionically conducting materials, which would need the construction of non-blocking electrodes for DC research [100]. Frequency dependent data may be used to highlight the contributions of bulk materials (high frequency semicircle area), grain boundaries, and electrode-electrolyte effects (low frequency spike region) [101]. In the GL range of 11 to 33 wt.%, the ac conductivity spectra (see Figure 15) may be separated into three discrete regions. The electrode-electrolyte interfacial phenomenon known as electrode polarization (EP) effect caused the low frequency area (region I) to appear as a spike [102]. It is worth noting that a linear relation between the spike region and GL content can be observed. A plateau of ac conductivity is identified in the intermediate frequency band (region II), which correlates to DC conductivity. As a result of increased electrode polarization (EP) and a shift to the higher frequency side, this region (plateau region) grows dramatically with increasing GL content. At high GL concentrations, the disappearance of the high frequency semicircle in the impedance plot (Z_i-Z_r) is closely related to the lowering of dispersion region (Figure 7a–e). These dispersion areas are critical for determining the ion conduction type in solid PEs.

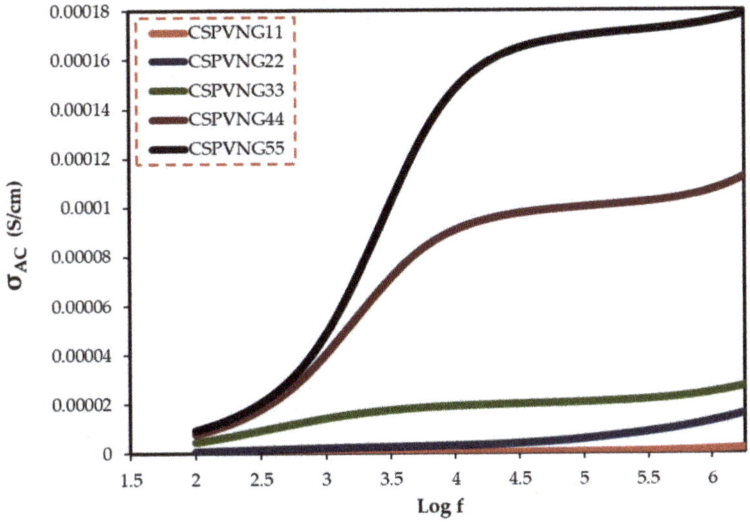

Figure 15. AC conductivity spectra.

The Jonscher's relation is a good way to describe the link between ac conductivity and charge carrier motion [103]:

$$\sigma_{dc}(\omega) = \sigma_{dc} + A\omega^{s}\ (0 < s < 1) \quad (15)$$

where σ_{dc} denotes the frequency-independent dc conductivity, A denotes a temperature-dependent constant, and s denotes the charge carrier interactions during hopping processes [104]. The second component is caused by accumulated interfacial charges and permanent/induced polarization (restricted mobility) of the dipoles. As frequency rises, the total conductivity of the second component due to polarization increases, according to Equation (15) [105]. $\sigma'_{ac}(\omega)$ represents the charge transport mechanism and many-body interactions among charge carriers [104,106].

3. Conclusions

Finally, the casting approach was employed to make plasticized PBEs based on polyvinyl alcohol (PVA): chitosan (CS) polymers. The samples structural and electrical properties were investigated using a variety of techniques including XRD, FTIR and EIS. There is interaction at the molecular level, as shown by shifting peaks and changes in the intensity of FTIR bands. PVA and CS polymers behave semicrystallinly, as shown by the XRD data. It was determined that the samples inserted with high content of GL were amorphous due to the XRD pattern exhibiting a wide hum. The GL plasticizer fastened the dissociation of NaBr salt and reduced the crystallinity of polymers. For a sample containing 55 wt% GL, the CIS approach revealed a maximum ionic conductivity of 3.8×10^{-4} S/cm. Each electrolyte's CIS data were fitted with EECs to have a comprehensive understanding of the ion-conducting systems' full electrical properties. The DC conductivity and carrier density increased with increasing GL ratio. With the help of EEC modeling, the transport parameters associated with dissociated ions are calculated. The conductivity measurement was connected to the dielectric characteristics. High dielectric constant and dielectric loss were recorded at low frequencies. The relaxation behaviors of the samples are examined using loss tangent and electric modulus graphs. The peak detected in the spectra of tanδ and M″ plots and the distribution of data points are asymmetric beside the peaks. This confirms that the movement of ions and their contribution to conductivity belongs to the viscoelastic relaxation type. The high ionic conductivity and improved transport properties (n, μ and D) revealed the suitableness of the films for energy storage device applications.

4. Materials and Methods

4.1. Material Components

PVA (MW 89,000–98,000 g/mol, 99+% hydrolyzed) and chitosan (CS) (MW 310,000–375,000 g/mol, Sigma-Aldrich, St. Louis, MO, USA) are the polymeric materials. Other raw materials employed included sodium bromide (NaBr) (MW 102.89 g/mol, Sigma-Aldrich) salt as a dopant, acetic acid (CH3COOH) solution and distill water as a solvent, and GL (C3H8O3) (MW 92.09 g/mol, Sigma-Aldrich) as a plasticizer. No additional purification was necessary.

4.2. Electrolyte Preparation Using Polymer Blends

Acetic acid solvents were prepared by adding 1 mL of glacial acetic acid into 99 mL of distilled water. 0.5 g of CS was dissolved in 70 mL of the prepared (1%) acetic acid at ambient temperature and 0.5 g of PVA was dissolved in 30 mL distilled water at 90 °C under stirring. After cooling to room temperature, the PVA solution was mixed with dissolved CS. Using a magnetic stirrer, the solution was continuously stirred for 24 h to obtain an identical PVA:CS solution. To make ion-conducting PEs, sodium salt was used. For this purpose, 35 wt.% NaBr salt was added to the PVA:CS solutions under stirring and then addition of varying concentrations (11, 22, 33, 44, and 55 wt.%) of GL as plasticizers into the PVA:CS:NaBr solution were carried out. In order to obtain a dry film free of solvent at room temperature, the solutions were deposited in numerous plastic Petri dishes (8 cm in diameter). This procedure is seen in the Figure 16. The 0.023 mm-thick flexible films were then created.

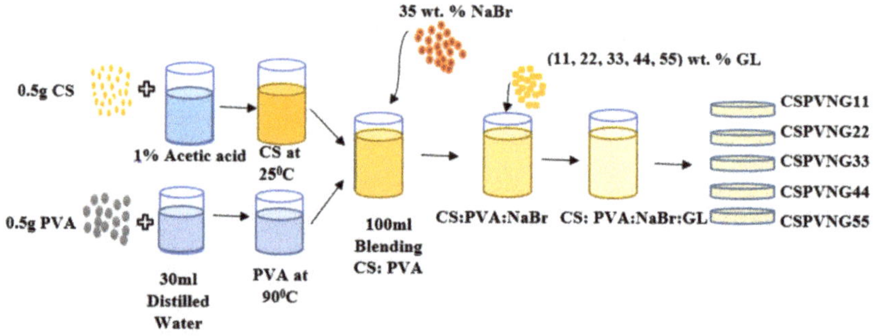

Figure 16. Schematic processes for the fabrication of the plasticized PE samples.

It was determined that CSPVNG0, CSPVNG11, CSPVNG22, CSPVNG33, CSPVNG44, and CSPVNG55 were the appropriate coding for the PVA:CS:NaBr mix systems doped with varied concentrations of GL. Table 5 depicts the film's chemical composition.

Table 5. Solid PE based on PVA:CS:NaBr:GL.

Designation	CS : PVA : NaBr (g)	Glycerol Wt.%	Glycerol (g)
CSPVNG0	(0.5 : 0.5 : 0.538)	0	0
CSPVNG11	(0.5 : 0.5 : 0.538)	11	0.190
CSPVNG22	(0.5 : 0.5 : 0.538)	11	0.4339
CSPVNG33	(0.5 : 0.5 : 0.538)	22	0.758
CSPVNG44	(0.5 : 0.5 : 0.538)	33	1.209
CSPVNG55	(0.5 : 0.5 : 0.538)	44	1.880

4.3. XRD and FTIR Spectra Analysis

Bruker D8 ADVANCE X-ray powder diffractometer (Bruker, Berlin, Germany) with CuKα radiation sources (k = 1.54 A°) is used to record XRD patterns of pure and composition polymer films in the region of 10–80° in order to provide information about their crystal structure. The interaction between the composite components was studied using a Perkin Elmer FTIR spectrometer with a range of 500 to 4000 cm^{-1} and a resolution of 2 cm^{-1}.

4.4. Electrical Impedance Spectroscopy (EIS)

Complex impedance spectroscopy may be used to determine the electrical properties of materials and their interaction with electrically conducting electrodes. SPE films were cut into 2 cm diameter compact discs and placed between two stainless steel electrodes under spring pressure as shown in Figure 17.

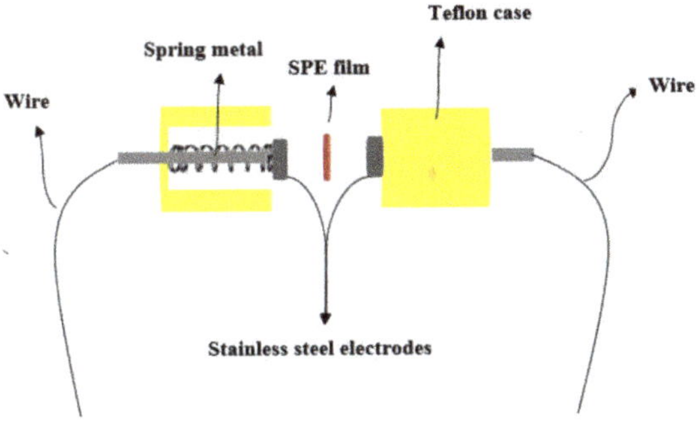

Figure 17. Schematic the stainless-steel electrodes.

HIOKI 3531 Z Hi-tester, linked to a computer, was used to measure the impedance of the films throughout a frequency range of 100 Hz to 2 MHz. Z_r and Z_i are calculated by the software and controlled by it throughout the measuring process. An intersection of the plot and real impedance axis yielded bulk resistance (R_b), and the Z_r and Z_i values were presented as a Nyquist plot. The equation below may be used to determine conductivity [107,108].

In Equation (1), there are three parameters to consider: film thickness (t), bulk resistance (R_b), and active area (A). It is also possible to determine the real and imaginary components of dielectric and electric modulus using Z_r and Z_i by the Equations (9) and (10), (12) and (13), which are based on complex impedance (Z^*) [107,108].

Author Contributions: Conceptualization, S.B.A. and M.F.Z.K.; formal analysis, N.M.S.; investigation, N.M.S.; Methodology, N.M.S.; project administration, S.B.A. and M.F.Z.K.; supervision, S.B.A. and M.F.Z.K.; validation, N.M.S., S.B.A. and M.F.Z.K.; writing—original draft, N.M.S.; writing—review & editing, S.B.A. and M.F.Z.K. All authors have read and agreed to the published version of the manuscript.

Funding: This research received no external funding.

Institutional Review Board Statement: Not applicable.

Informed Consent Statement: Not applicable.

Acknowledgments: The authors gratefully acknowledge the financial support for this study from the Ministry of Higher Education and Scientific Research-Kurdish National Research Council (KNRC), Kurdistan Regional Government/Iraq. The financial support from the University of Sulaimani, and Komar University of Science and Technology are greatly appreciated.

Conflicts of Interest: The authors declare that they have no known competing financial interests or personal relationships that could have appeared to influence the work reported in this paper.

References

1. Zhou, Y.; Wang, P.; Ruan, G.; Xu, P.; Ding, Y. Synergistic Effect of P[MPEGMA-IL] Modified Graphene on Morphology and Dielectric Properties of PLA/PCL Blends. *ES Mater. Manuf.* **2020**, *11*, 20–29. [CrossRef]
2. Wang, Z.; He, S.; Nguyen, V.; Riley, K.E. Ionic liquids as "Green solvent and/or electrolyte" for energy interface. *Eng. Sci.* **2020**, *11*, 3–18. [CrossRef]
3. Wang, Y.P.; Gao, X.H.; Li, H.K.; Li, H.J.; Liu, H.G.; Guo, H.X. Effect of active filler addition on the ionic conductivity of PVDF-PEG polymer electrolyte. *J. Macromol. Sci. Part A Pure Appl. Chem.* **2009**, *46*, 461–467. [CrossRef]
4. Angell, C.A. Polymer electrolytes—Some principles, cautions, and new practices. *Electrochim. Acta* **2017**, *250*, 368–375. [CrossRef]
5. Elayappan, V.; Murugadoss, V.; Fei, Z.; Dyson, P.J.; Angaiah, S. Influence of polypyrrole incorporated electrospun poly (vinylidene fluoride-cO–Hexafluoropropylene) nanofibrous composite membrane electrolyte on the photovoltaic performance of dye sensitized solar cell. *Eng. Sci.* **2020**, *10*, 78–84. [CrossRef]
6. Srivastava, M.; Surana, K.; Singh, P.K.; Ch, R. Nickel Oxide embedded with Polymer Electrolyte as Efficient Hole Transport Material for Perovskite Solar Cell. *Eng. Sci.* **2021**, *17*, 216–223. [CrossRef]
7. Wu, X.; Zhong, J.; Zhang, H.; Liu, H.; Mai, J.; Shi, S.; Deng, Q.; Wang, N. Garnet Li7La3Zr2O12 Solid-State Electrolyte: Environmental Corrosion. *Countermeas. Appl. ES Energy Environ.* **2021**, *14*, 22–33. [CrossRef]
8. Maitra, A.; Heuer, A. Understanding correlation effects for ion conduction in polymer electrolytes. *J. Phys. Chem. B* **2008**, *112*, 9641–9651. [CrossRef]
9. Aziz, S.B. Li^+ ion conduction mechanism in poly (ε-caprolactone)-based polymer electrolyte. *Iran. Polym. J.* **2013**, *22*, 877–883. [CrossRef]
10. Hassan, M.; Rafiuddin, R. Ionic Conductivity and Phase Transition Behaviour in 4AgI-(1-)-2CuI System. *Res. Lett. Phys.* **2008**, *2008*, 249402. [CrossRef]
11. Wiśniewski, Z.; Górski, L.; Zasada, D. Investigation of structure and conductivity of superionic conducting materials obtained on the basis of silver iodide. *Acta Phys. Pol. A* **2008**, *113*, 1231–1236. [CrossRef]
12. Mourya, V.K.; Inamdar, N.N. Chitosan-modifications and applications: Opportunities galore. *React. Funct. Polym.* **2008**, *68*, 1013–1051. [CrossRef]
13. Aziz, S.B.; Rasheed, M.A.; Abidin, Z.H.Z. Optical and Electrical Characteristics of Silver Ion Conducting Nanocomposite Solid Polymer Electrolytes Based on Chitosan. *J. Electron. Mater.* **2017**, *46*, 6119–6130. [CrossRef]
14. Aguirre-chagala, Y.E.; Pavón-pérez, L.B.; Altuzar, V.; Domínguez-chávez, J.G.; Muñoz-aguirre, S. Comparative Study of One-Step Cross-Linked Electrospun Chitosan-Based Membranes. *J. Nanomater.* **2017**, *2017*, 1980714. [CrossRef]
15. Fernandes, D.M.; Andrade, J.L.; Lima, M.K.; Silva, M.F.; Andrade, L.H.C.; Lima, S.M.; Hechenleitner, A.A.W.; Pineda, E.A.G. Thermal and photochemical effects on the structure, morphology, thermal and optical properties of PVA/Ni0.04Zn0.96O and PVA/Fe 0.03Zn0.97O nanocomposite films. *Polym. Degrad. Stab.* **2013**, *98*, 1862–1868. [CrossRef]
16. Li, X.; Chen, K.; Ji, X.; Yuan, X.; Lei, Z.; Ullah, M.W.; Xiao, J.; Yang, G. Microencapsulation of poorly water-soluble finasteride in polyvinyl alcohol/chitosan microspheres as a long-term sustained release system for potential embolization applications. *Eng. Sci.* **2021**, *13*, 105–120. [CrossRef]
17. Rao, C.V.S.; Ravi, M.; Raja, V.; Bhargav, P.B.; Sharma, A.K.; Rao, V.V.R.N. Preparation and characterization of PVP-based polymer electrolytes for solid-state battery applications. *Iran. Polym. J.* **2012**, *21*, 531–536. [CrossRef]
18. Nofal, M.M.; Aziz, S.B.; Hadi, J.M.; Karim, W.O.; Dannoun, E.M.A.; Hussein, A.M.; Hussen, S.A. Polymer composites with 0.98 transparencies and small optical energy band gap using a promising green methodology: Structural and optical properties. *Polymers* **2021**, *13*, 1648. [CrossRef] [PubMed]
19. Borgohain, M.M.; Joykumar, T.; Bhat, S.V. Studies on a nanocomposite solid polymer electrolyte with hydrotalcite as a filler. *Solid State Ion.* **2010**, *181*, 964–970. [CrossRef]
20. Kumar, J.; Rodrigues, S.J.; Kumar, B. Interface-mediated electrochemical effects in lithium/polymer-ceramic cells. *J. Power Sources* **2010**, *195*, 327–334. [CrossRef]
21. Aziz, S.B.; Abidin, Z.H.Z. Electrical Conduction Mechanism in Solid Polymer Electrolytes: New Concepts to Arrhenius Equation. *J. Soft Matter* **2013**, *2013*, 323868. [CrossRef]
22. Yusof, Y.M.; Illias, H.A.; Kadir, M.F.Z. Incorporation of NH_4Br in PVA-chitosan blend-based polymer electrolyte and its effect on the conductivity and other electrical properties. *Ionics* **2014**, *20*, 1235–1245. [CrossRef]
23. Buraidah, M.H.; Arof, A.K. Characterization of chitosan/PVA blended electrolyte doped with NH 4I. *J. Non. Cryst. Solids* **2011**, *357*, 3261–3266. [CrossRef]
24. Kadir, M.F.Z.; Majid, S.R.; Arof, A.K. Plasticized chitosan-PVA blend polymer electrolyte based proton battery. *Electrochim. Acta* **2010**, *55*, 1475–1482. [CrossRef]
25. Kadir, M.F.Z.; Arof, A.K. Application of PVA-chitosan blend polymer electrolyte membrane in electrical double layer capacitor. *Mater. Res. Innov.* **2011**, *15* (Suppl. 2), s217–s220. [CrossRef]
26. Hadi, J.M.; Aziz, S.B.; Nofal, M.M.; Hussen, S.A.; Hamsan, M.H.; Brza, M.A.; Abdulwahid, R.T.; Kadir, M.F.Z.; Woo, H.J. Electrical, dielectric property and electrochemical performances of plasticized silver ion-conducting chitosan-based polymer nanocomposites. *Membranes* **2020**, *10*, 151. [CrossRef]
27. Jung, H.-E.; Shin, M. Surface-Roughness-Limited Mean Free Path in Si Nanowire FETs. 2013. Available online: http://arxiv.org/abs/1304.5597 (accessed on 11 May 2015).

28. Ayala, G.; Agudelo, A.; Vargas, R. Effect of glycerol on the electrical properties and phase behavior of cassava starch biopolymers. *Dyna* **2012**, *79*, 138–147.
29. Seoane, N.; Martinez, A.; Brown, A.R.; Asenov, A. Study of surface roughness in extremely small Si nanowire MOSFETs using fully-3D NEGFs. In Proceedings of the 2009 Spanish Conference Electron Devices, Santiago de Compostela, Spain, 11–13 February 2009; pp. 180–183. [CrossRef]
30. Bhargav, P.B.; Sarada, B.A.; Sharma, A.K.; Rao, V.V.R.N. Electrical conduction and dielectric relaxation phenomena of PVA based polymer electrolyte films. *J. Macromol. Sci. Part A Pure Appl. Chem.* **2010**, *47*, 131–137. [CrossRef]
31. Mohan, V.M.; Bhargav, P.B.; Raja, V.; Sharma, A.K.; Rao, V.V.R.N. Optical and electrical properties of pure and doped PEO polymer electrolyte films. *Soft Mater.* **2007**, *5*, 33–46. [CrossRef]
32. Olewnik-Kruszkowska, E.; Gierszewska, M.; Jakubowska, E.; Tarach, I.; Sedlarik, V.; Pummerova, M. Antibacterial films based on PVA and PVA-chitosan modified with poly(hexamethylene guanidine). *Polymers* **2019**, *11*, 2093. [CrossRef] [PubMed]
33. Sharma, P.; Mathur, G.; Goswami, N.; Sharma, S.K.; Dhakate, S.R.; Chand, S.; Mathur, A. Evaluating the potential of chitosan/poly(vinyl alcohol) membranes as alternative carrier material for proliferation of Vero cells. *E-Polymers* **2015**, *15*, 237–243. [CrossRef]
34. Tretinnikov, O.N.; Zagorskaya, S.A. Determination of the degree of crystallinity of poly(Vinyl alcohol) by ftir spectroscopy. *J. Appl. Spectrosc.* **2012**, *79*, 521–526. [CrossRef]
35. Aziz, S.B.; Nofal, M.M.; Ghareeb, H.O.; Dannoun, E.M.A.; Hussen, S.A.; Hadi, J.M.; Ahmed, K.K.; Hussein, A.M. Characteristics of poly(Vinyl alcohol) (PVA) based composites integrated with green synthesized Al^{3+}-metal complex: Structural, optical, and localized density of state analysis. *Polymers* **2021**, *13*, 1316. [CrossRef] [PubMed]
36. Kadir, M.F.Z.; Aspanut, Z.; Majid, S.R.; Arof, A.K. FTIR studies of plasticized poly(vinyl alcohol)-chitosan blend doped with NH4NO3 polymer electrolyte membrane. *Spectrochim. Acta—Part A Mol. Biomol. Spectrosc.* **2011**, *78*, 1068–1074. [CrossRef]
37. Wei, D.; Sun, W.; Qian, W.; Ye, Y.; Ma, X. The synthesis of chitosan-based silver nanoparticles and their antibacterial activity. *Carbohydr. Res.* **2009**, *344*, 2375–2382. [CrossRef] [PubMed]
38. Abdullah, O.G.; Aziz, S.B.; Omer, K.M.; Salih, Y.M. Reducing the optical band gap of polyvinyl alcohol (PVA) based nanocomposite. *J. Mater. Sci. Mater. Electron.* **2015**, *26*, 5303–5309. [CrossRef]
39. Krithiga, N.; Rajalakshmi, A.; Jayachitra, A. Green Synthesis of Silver Nanoparticles Using Leaf Extracts of Clitoria ternatea and Solanum nigrum and Study of Its Antibacterial Effect against Common Nosocomial Pathogens. *J. Nanosci.* **2015**, *2015*, 928204. [CrossRef]
40. Salleh, N.S.; Aziz, S.B.; Aspanut, Z.; Kadir, M.F.Z. Electrical impedance and conduction mechanism analysis of biopolymer electrolytes based on methyl cellulose doped with ammonium iodide. *Ionics* **2016**, *22*, 2157–2167. [CrossRef]
41. Amran, N.N.A.; Manan, N.S.A.; Kadir, M.F.Z. The effect of $LiCF_3SO_3$ on the complexation with potato starch-chitosan blend polymer electrolytes. *Ionics* **2016**, *22*, 1647–1658. [CrossRef]
42. Aziz, S.B. Modifying Poly(Vinyl Alcohol) (PVA) from Insulator to Small-Bandgap Polymer: A Novel Approach for Organic Solar Cells and Optoelectronic Devices. *J. Electron. Mater.* **2016**, *45*, 736–745. [CrossRef]
43. Ricciardi, R.; Auriemma, F.; De Rosa, C.; Lauprêtre, F. X-ray Diffraction Analysis of Poly(vinyl alcohol) Hydrogels, Obtained by Freezing and Thawing Techniques. *Macromolecules* **2004**, *37*, 1921–1927. [CrossRef]
44. Aziz, S.B.; Abidin, Z.H.Z.; Arof, A.K. Effect of silver nanoparticles on the DC conductivity in chitosansilver triflate polymer electrolyte. *Phys. B Condens. Matter* **2010**, *405*, 4429–4433. [CrossRef]
45. Yusuf, S.N.F.; Azzahari, A.D.; Yahya, R.; Majid, S.R.; Careem, M.A.; Arof, A.K. From crab shell to solar cell: A gel polymer electrolyte based on N-phthaloylchitosan and its application in dye-sensitized solar cells. *RSC Adv.* **2016**, *6*, 27714–27724. [CrossRef]
46. Malathi, J.; Kumaravadivel, M.; Brahmanandhan, G.M.; Hema, M.; Baskaran, R.; Selvasekarapandian, S. Structural, thermal and electrical properties of PVA-LiCF3SO3 polymer electrolyte. *J. Non. Cryst. Solids* **2010**, *356*, 2277–2281. [CrossRef]
47. Gasperini, A.; Wang, G.J.N.; Molina-Lopez, F.; Wu, H.C.; Lopez, J.; Xu, J.; Luo, S.; Zhou, D.; Xue, G.; Tok, J.B.H.; et al. Characterization of Hydrogen Bonding Formation and Breaking in Semiconducting Polymers under Mechanical Strain. *Macromolecules* **2019**, *52*, 2476–2486. [CrossRef]
48. Stephan, A.M.; Kumar, T.P.; Kulandainathan, M.A.; Lakshmi, N.A. Chitin-incorporated poly(ethylene oxide)-based nanocomposite electrolytes for lithium batteries. *J. Phys. Chem. B* **2009**, *113*, 1963–1971. [CrossRef]
49. Honary, S.; Orafai, H. The effect of different plasticizer molecular weights and concentrations on mechanical and thermomechanical properties of free films. *Drug Dev. Ind. Pharm.* **2002**, *28*, 711–715. [CrossRef] [PubMed]
50. Honary, S.; Golkar, M. Effect of Polymer Grade and Plasticizer Molecular Weights on Viscoelastic Behavior of Coating Solutions. *Iran. J. Pharm. Res.* **2003**, *2*, 125–127. [CrossRef]
51. Asami, K. Characterization of heterogeneous systems by dielectric spectroscopy. *Prog. Polym. Sci.* **2002**, *27*, 1617–1659. [CrossRef]
52. Polu, A.R.; Kumar, R. Impedance spectroscopy and FTIR studies of PEG—Based polymer electrolytes. *E-J. Chem.* **2011**, *8*, 347–353. [CrossRef]
53. Mahato, D.K.; Dutta, A.; Sinha, T.P. Impedance spectroscopy analysis of double perovskite Ho_2NiTiO_6. *J. Mater. Sci.* **2010**, *45*, 6757–6762. [CrossRef]
54. Fortunato, R.; Branco, L.C.; Afonso, C.A.M.; Benavente, J.; Crespo, J.G. Electrical impedance spectroscopy characterisation of supported ionic liquid membranes. *J. Memb. Sci.* **2006**, *270*, 42–49. [CrossRef]

55. Aziz, S.B.; Abdullah, R.M. Crystalline and amorphous phase identification from the tanδ relaxation peaks and impedance plots in polymer blend electrolytes based on [CS:AgNt]x:PEO(x-1) ($10 \leq x \leq 50$). *Electrochim. Acta* **2018**, *285*, 30–46. [CrossRef]
56. Benavente, J.; García, J.M.; Riley, R.; Lozano, A.E.; De Abajo, J. Sulfonated poly(ether ether sulfones): Characterization and study of dielectrical properties by impedance spectroscopy. *J. Memb. Sci.* **2000**, *175*, 43–52. [CrossRef]
57. Benavente, J.; Zhang, X.; Valls, R.G. Modification of polysulfone membranes with polyethylene glycol and lignosulfate: Electrical characterization by impedance spectroscopy measurements. *J. Colloid Interface Sci.* **2005**, *285*, 273–280. [CrossRef] [PubMed]
58. Aziz, S.B.; Abdullah, R.M.; Rasheed, M.A.; Ahmed, H.M. Role of Ion Dissociation on DC Conductivity and Silver Nanoparticle Formation in PVA:AgNt Based Polymer Electrolytes: Deep Insights to Ion Transport Mechanism. *Polymers* **2017**, *9*, 338. [CrossRef] [PubMed]
59. Jacob, M.M.E.; Prabaharan, S.R.S.; Radhakrishna, S. Effect of PEO addition on the electrolytic and thermal properties of PVDF-LiClO4 polymer electrolytes. *Solid State Ion.* **1997**, *4*, 267–276. [CrossRef]
60. Qian, H.; Wang, Y.; Fang, Y.; Gu, L.; Lu, R.; Sha, J. High-performance ZnO nanowire field-effect transistor with forming gas treated SiO2 gate dielectrics. *J. Appl. Phys.* **2015**, *117*, 164308. [CrossRef]
61. Shukur, M.F.; Ithnin, R.; Kadir, M.F.Z. Electrical characterization of corn starch-LiOAc electrolytes and application in electrochemical double layer capacitor. *Electrochim. Acta* **2014**, *136*, 204–216. [CrossRef]
62. Teo, L.P.; Buraidah, M.H.; Nor, A.F.M.; Majid, S.R. Conductivity and dielectric studies of Li_2SnO_3. *Ionics* **2012**, *18*, 655–665. [CrossRef]
63. Asnawi, A.S.F.M.; Aziz, S.B.; Brevik, I.; Brza, M.A.; Yusof, Y.M.; Alshehri, S.M.; Ahamad, T.; Kadir, M.F.Z. The Study of Plasticized Sodium Ion Conducting Polymer Blend Electrolyte Membranes Based on Chitosan/Dextran Biopolymers: Ion Transport, Structural, Morphological and Potential Stability. *Polymers* **2021**, *13*, 383. [CrossRef]
64. Paluch, M.; Ostrowska, J.; Tyński, P.; Sadurski, W.; Konkol, M. Structural and Thermal Properties of Starch Plasticized with Glycerol/Urea Mixture. *J. Polym. Environ.* **2021**, *30*, 728–740. [CrossRef]
65. Kobayashi, K.; Pagot, G.; Vezzù, K.; Bertasi, F.; di Noto, V.; Tominaga, Y. Effect of plasticizer on the ion-conductive and dielectric behavior of poly(ethylene carbonate)-based Li electrolytes. *Polym. J.* **2021**, *53*, 149–155. [CrossRef]
66. Dinoto, V.; Negro, E.; Lavina, S.; Vittadello, M. Hybrid inorganiC–Organic polymer electrolytes. In *Polymer Electrolytes*; Woodhead Publishing: Sawston, UK, 2010; pp. 219–277. [CrossRef]
67. Muldoon, J.; Bucur, C.B.; Boaretto, N.; Gregory, T.; di Noto, V. Polymers: Opening doors to future batteries. *Polym. Rev.* **2015**, *55*, 208–246. [CrossRef]
68. di Noto, V.; Lavina, S.; Giffin, G.A.; Negro, E.; Scrosati, B. Polymer electrolytes: Present, past and future. *Electrochim. Acta* **2011**, *57*, 4–13. [CrossRef]
69. Arof, A.K.; Amirudin, S.; Yusof, S.Z.; Noor, I.M. A method based on impedance spectroscopy to determine transport properties of polymer electrolytes. *Phys. Chem. Chem. Phys.* **2014**, *16*, 1856–1867. [CrossRef] [PubMed]
70. Shukur, M.F.; Ithnin, R.; Kadir, M.F.Z. Electrical properties of proton conducting solid biopolymer electrolytes based on starch-chitosan blend. *Ionics* **2014**, *20*, 977–999. [CrossRef]
71. Yusof, Y.M.; Shukur, M.F.; Hamsan, M.H.; Jumbri, K.; Kadir, M.F.Z. Plasticized solid polymer electrolyte based on natural polymer blend incorporated with lithium perchlorate for electrical double-layer capacitor fabrication. *Ionics* **2019**, *25*, 5473–5484. [CrossRef]
72. Asnawi, A.S.F.M.; Aziz, B.S.; Nofal, M.M.; Hamsan, M.H.; Brza, M.A.; Yusof, Y.M.; Abdilwahid, R.T.; Muzakir, S.K.; Kadir, M.F.Z. Glycerolized Li^+ Ion Conducting Chitosan-Based Polymer Electrolyte for Energy Storage EDLC Device Applications with Relatively High Energy Density. *Polymers* **2020**, *12*, 1433. [CrossRef] [PubMed]
73. Lu, J.; Deng, Y.; Song, J.; Hu, Y.; Deng, Z.; Cui, Q.; Lou, Z. Role of Hydroxyl on Conductivity Switching of Poly(ethylene oxide)/TiO_2 Electrical Bistable Devices. *Phys. Status Solidi Appl. Mater. Sci* **2019**, *216*, 1900443. [CrossRef]
74. Müller, C.M.O.; Yamashita, F.; Laurindo, J.B. Evaluation of the effects of glycerol and sorbitol concentration and water activity on the water barrier properties of cassava starch films through a solubility approach. *Carbohydr. Polym.* **2008**, *72*, 82–87. [CrossRef]
75. Impedance Spectroscopy. In *Gas Adsorption Equilibria*; Springer: Boston, MA, USA, 2005; pp. 287–357. [CrossRef]
76. Aziz, S.B. Role of dielectric constant on ion transport: Reformulated Arrhenius equation. *Adv. Mater. Sci. Eng.* **2016**, *2016*, 2527013. [CrossRef]
77. Okutan, M.; Şentürk, E. β Dielectric relaxation mode in side-chain liquid crystalline polymer film. *J. Non. Cryst. Solids* **2008**, *354*, 1526–1530. [CrossRef]
78. Sun, L.; Liang, L.; Shi, Z.; Wang, H.; Xie, P.; Dastan, D.; Sun, K.; Fan, R. Optimizing strategy for the dielectric performance of topological-structured polymer nanocomposites by rationally tailoring the spatial distribution of nanofillers. *Eng. Sci.* **2020**, *12*, 95–105. [CrossRef]
79. Pradhan, D.K.; Choudhary, R.N.P.; Samantaray, B.K. Studies of dielectric relaxation and AC conductivity behavior of plasticized polymer nanocomposite electrolytes. *Int. J. Electrochem. Sci.* **2008**, *3*, 597–608.
80. Marzantowicz, M.; Dygas, J.R.; Krok, F.; Łasińska, A.; Florjańczyk, Z.; Zygadło-Monikowska, E. In situ microscope and impedance study of polymer electrolytes. *Electrochim. Acta* **2006**, *51*, 1713–1727. [CrossRef]
81. Marzantowicz, M.; Dygas, J.R.; Krok, F.; Łasińska, A.; Florjańczyk, Z.; Zygadło-Monikowska, E.; Affek, A. Crystallization and melting of PEO:LiTFSI polymer electrolytes investigated simultaneously by impedance spectroscopy and polarizing microscopy. *Electrochim. Acta* **2005**, *50*, 3969–3977. [CrossRef]

82. Aziz, S.B.; Abidin, Z.H.Z. Electrical and morphological analysis of chitosan:AgTf solid electrolyte. *Mater. Chem. Phys.* **2014**, *144*, 280–286. [CrossRef]
83. Hurd, J.A.; Vaidhyanathan, R.; Thangadurai, V.; Ratcliffe, C.I.; Moudrakovski, I.L.; Shimizu, G.K.H. Anhydrous proton conduction at 150 °C in a crystalline metal-organic framework. *Nat. Chem.* **2009**, *1*, 705–710. [CrossRef]
84. Bureekaew, S.; Horike, S.; Higuchi, M.; Mizuno, M.; Kawamura, T.; Tanaka, D.; Yanai, N.; Kitagawa, S. One-dimensional imidazole aggregate in aluminium porous coordination polymers with high proton conductivity. *Nat. Mater.* **2009**, *8*, 831–836. [CrossRef]
85. Nicolau, A.; Nucci, A.M.; Martini, E.M.A.; Samios, D. Electrical impedance spectroscopy of epoxy systems II: Molar fraction variation, resistivity, capacitance and relaxation processes of 1,4-butanediol diglycidyl ether/succinic anhydride and triethylamine as initiator. *Eur. Polym. J.* **2007**, *43*, 2708–2717. [CrossRef]
86. Wang, W.; Alexandridis, P. Composite Polymer Electrolytes: Nanoparticles Affect Structure and Properties. *Polymers* **2016**, *8*, 387. [CrossRef] [PubMed]
87. Aziz, S.B.; Abdullah, O.G.; Rasheed, M.A. Structural and electrical characteristics of PVA:NaTf based solid polymer electrolytes: Role of lattice energy of salts on electrical DC conductivity. *J. Mater. Sci. Mater. Electron.* **2017**, *28*, 12873–12884. [CrossRef]
88. Marzantowicz, M.; Dygas, J.R.; Krok, F.; Florjaczyk, Z.; Zygado-Monikowska, E. Conductivity and dielectric properties of polymer electrolytes PEO:LiN(CF3SO2)2 near glass transition. *J. Non. Cryst. Solids* **2007**, *353*, 4467–4473. [CrossRef]
89. Mohomed, K.; Moussy, F.; Harmon, J.P. Dielectric analyses of a series of poly(2-hydroxyethyl methacrylate-co-2,3-dihydroxypropyl methacylate) copolymers. *Polymer* **2006**, *47*, 3856–3865. [CrossRef]
90. Kamal, A.; Rafiq, M.A.; Rafiq, M.N.; Usman, M.; Waqar, M.; Anwar, M.S. Structural and impedance spectroscopic studies of CuO-doped (K0.5Na0.5Nb0.995Mn0.005O3) lead-free piezoelectric ceramics. *Appl. Phys. A Mater. Sci. Process.* **2016**, *122*, 1037. [CrossRef]
91. Rayssi, C.; El Kossi, S.; Dhahri, J.; Khirouni, K. Frequency and temperature-dependence of dielectric permittivity and electric modulus studies of the solid solution Ca0.85Er0.1Ti1-: XCo4 x /3O3 ($0 \leq x \leq 0.1$). *RSC Adv.* **2018**, *8*, 17139–17150. [CrossRef]
92. Smaoui, H.; Arous, M.; Guermazi, H.; Agnel, S.; Toureille, A. Study of relaxations in epoxy polymer by thermally stimulated depolarization current (TSDC) and dielectric relaxation spectroscopy (DRS). *J. Alloys Compd.* **2010**, *489*, 429–436. [CrossRef]
93. Agrawal, S.L.; Singh, M.; Asthana, N.; Dwivedi, M.M.; Pandey, K. Dielectric and ion transport studies in [PVA: $LiC_2H_3O_2$]: $Li_2Fe_5O_8$ polymer nanocomposite electrolyte system. *Int. J. Polym. Mater. Polym. Biomater.* **2011**, *60*, 276–289. [CrossRef]
94. Ayesh, A.S. Dielectric relaxation and thermal stability of polycarbonate doped with $MnCl_2$ salt. *J. Thermoplast. Compos. Mater.* **2008**, *21*, 309–322. [CrossRef]
95. Ayesh, A.S. Electrical and optical characterization of PMMA doped with Y 0.0025Si0.025Ba0.9725 (Ti(0.9)Sn0.1)O3 ceramic. *Chin. J. Polym. Sci.* **2010**, *28*, 537–546. [CrossRef]
96. Aziz, S.B.; Karim, W.O.; Brza, M.A.; Abdulwahid, R.T.; Saeed, S.R.; Al-Zangana, S.; Kadir, M.F.Z. Ion Transport Study in CS: POZ Based Polymer Membrane Electrolytes Using Trukhan Model. *Int. J. Mol. Sci.* **2019**, *20*, 5265. [CrossRef] [PubMed]
97. Mohomed, K.; Gerasimov, T.G.; Moussy, F.; Harmon, J.P. A broad spectrum analysis of the dielectric properties of poly(2-hydroxyethyl methacrylate). *Polymer* **2005**, *46*, 3847–3855. [CrossRef]
98. Gommans, H.H.P. Charge Transport and Interface Phenomena in Semiconducting Polymers. Ph.D.Thesis, Technische Universiteit Eindhoven, Eindhoven, The Netherlands, September 2005.
99. Barker, R.E. Mobility and Conduction of Ions in and into Polymeric Solids. *Pure & Appi. Chem.* **1976**, *46*, 157–170. [CrossRef]
100. Rozanskia, S.A.; Kremer, F.; Kijberle, P. Relaxation and charge transport in mixtures of zwitterionic polymers and inorganic salts. *Macromol. Chem. Phys.* **1995**, *890*, 877–890. [CrossRef]
101. Bassiouni, M.E.; Al-shamy, F.; Madi, N.K.; Kassem, M.E. Temperature and electric field effects on the dielectric dispersion of modified polyvinyl chloride. *Mater. Lett.* **2003**, *57*, 1595–1603. [CrossRef]
102. Hema, M.; Selvasekerapandian, S.; Sakunthala, A.; Arunkumar, D.; Nithya, H. Structural, vibrational and electrical characterization of PVA-NH4Br polymer electrolyte system. *Phys. B Condens. Matter* **2008**, *403*, 2740–2747. [CrossRef]
103. Yang, J.; Meng, X.J.; Shen, M.R.; Fang, L.; Wang, J.L.; Lin, T.; Sun, J.L.; Chu, J.H. Hopping conduction and low-frequency dielectric relaxation in 5 mol % Mn doped (Pb, Sr) TiO_3 films. *J. Appl. Phys.* **2008**, *104*, 113. [CrossRef]
104. Migahed, M.D.; Ishra, M.; Fahmy, T.; Barakat, A. Electric modulus and AC conductivity studies in conducting PPy composite films at low temperature. *J. Phys. Chem. Solids* **2004**, *65*, 1121–1125. [CrossRef]
105. Ravi, M.; Pavani, Y.; Kumar, K.K.; Bhavani, S.; Sharma, A.K.; Rao, V.V.R.N. Studies on electrical and dielectric properties of PVP: $KBrO_4$ complexed polymer electrolyte films. *Mater. Chem. Phys.* **2011**, *130*, 442–448. [CrossRef]
106. Bhadra, S.; Singha, N.K.; Khastgir, D. Dielectric properties and EMI shielding efficiency of polyaniline and ethylene 1-octene based semi-conducting composites. *Curr. Appl. Phys.* **2009**, *9*, 396–403. [CrossRef]
107. Daniel, R.; Janet Priscilla, S.; Karthikeyan, S.; Selvasekarapandian, S.; Madeswaran, S.; Sivaji, K. Structural and Electrical Studies on PVP—Pan Blend Polymer System for Energy Storage Devices. *Int. J. Curr. Res. Rev.* **2018**, *10*, 19–24. [CrossRef]
108. Ibrahim, S.; Yasin, S.M.M.; Nee, N.M.; Ahmad, R.; Johan, M.R. Conductivity and dielectric behaviour of PEO-based solid nanocomposite polymer electrolytes. *Solid State Commun.* **2012**, *152*, 426–434. [CrossRef]

gels

Article

Properties of the PVA-VAVTD KOH Blend as a Gel Polymer Electrolyte for Zinc Batteries

Alisson A. Iles Velez [1], Edwin Reyes [1], Antonio Diaz-Barrios [1], Florencio Santos [2], Antonio J. Fernández Romero [2,*] and Juan P. Tafur [1,2,*]

1 School of Chemical Science and Engineering, Yachay Tech University, Yachay City of Knowledge, Urcuqui 100650, Ecuador; alisson.iles@yachaytech.edu.ec (A.A.I.V.); edwin.reyes@yachaytech.edu.ec (E.R.); adiaz@yachaytech.edu.ec (A.D.-B.)

2 Grupo de Materiales Avanzados para la Producción y Almacenamiento de Energía, Universidad Politécnica de Cartagena, Aulario II, Campus de Alfonso XIII, 30203 Cartagena, Spain; florencio.santos@upct.es

* Correspondence: antonioj.fernandez@upct.es (A.J.F.R.); jtafur@yachaytech.edu.ec (J.P.T.)

Abstract: Rechargeable zinc-air batteries are promising for energy storage and portable electronic applications because of their good safety, high energy density, material abundance, low cost, and environmental friendliness. A series of alkaline gel polymer electrolytes formed from polyvinyl alcohol (PVA) and different amounts of terpolymer composed of butyl acrylate, vinyl acetate, and vinyl neodecanoate (VAVTD) was synthesized applying a solution casting technique. The thin films were doped with KOH 12 M, providing a higher amount of water and free ions inside the electrolyte matrix. The inclusion of VAVTD together with the PVA polymer improved several of the electrical properties of the PVA-based gel polymer electrolytes (GPEs). X-ray diffraction (XRD), thermogravimetric analysis (TGA), and attenuated total reflectance- Fourier-transform infrared spectroscopy (ATR-FTIR) tests, confirming that PVA chains rearrange depending on the VAVTD content and improving the amorphous region. The most conducting electrolyte film was the test specimen 1:4 (PVA-VAVTD) soaked in KOH solution, reaching a conductivity of 0.019 S/cm at room temperature. The temperature dependence of the conductivity agrees with the Arrhenius equation and activation energy of ~0.077 eV resulted, depending on the electrolyte composition. In addition, the cyclic voltammetry study showed a current intensity increase at higher VAVTD content, reaching values of 310 mA. Finally, these gel polymer electrolytes were tested in Zn–air batteries, obtaining capacities of 165 mAh and 195 mAh for PVA-T4 and PVA-T5 sunk in KOH, respectively, at a discharge current of −5 mA.

Keywords: gel polymer electrolyte; PVA blend; potassium hydroxide; XRD and ATR-FTIR; TGA-DTG; ionic conductivity; cyclic voltammetry study; Zn–air battery

Citation: Velez, A.A.I.; Reyes, E.; Diaz-Barrios, A.; Santos, F.; Fernández Romero, A.J.; Tafur, J.P. Properties of the PVA-VAVTD KOH Blend as a Gel Polymer Electrolyte for Zinc Batteries. *Gels* **2021**, *7*, 256. https://doi.org/10.3390/gels7040256

Academic Editor: Hiroyuki Takeno

Received: 17 November 2021
Accepted: 6 December 2021
Published: 10 December 2021

1. Introduction

Polymer electrolytes have been considered as a possible ionically active material since Fenton and coworkers discovered the poly(ethylene oxide) (PEO) complexes with alkali metal ions [1] and their potential application in batteries was discovered by Armand [2]. Polymer electrolytes (PEs) have been drawing attention because they are a safer choice than liquid electrolytes [3]. In recent decades, numerous electrolytes containing different cations such as Zn(II) [4,5], Cd(II) [6], Cu(II) [7], or Co(II) [8] have been investigated as samples of thin-film polymeric electrolytes. Nevertheless, the main drawback of PE materials is their relatively low ionic conductivity at room temperature [9,10]. Ionic conductivity increases in inverse proportion to the degree of crystallinity and the viscosity of the polymeric matrix [11]. Therefore, many efforts have been implemented to improve the ionic motion of PEs.

Thus, gel polymer electrolytes (GPEs) were proposed to improve the limited ionic conductivity of common solid PEs. GPEs are also known as plasticized PEs, and they are neither liquid nor solid. Therefore, they have properties of both, conserving the cohesive characteristics of solids together with the ion diffusive properties of liquids [12]. Usually, a polymer containing heteroatoms such as oxygen, nitrogen, or sulfur [2,12,13] acts as a hosting matrix of inorganic salts of Li^+ or Zn^{2+} to obtain the gel polymer electrolyte.

On the other hand, the GPEs applied in energy storage devices need to have high ionic conductivities, good mechanical properties, and excellent electrochemical stability at room temperature. Metal-air batteries consist of a metal as a negative electrode coupled to an air-breathing positive one, and are considered as alternatives to the conventional Li-ion batteries used nowadays [14,15] because metal-air batteries provide higher energy density thanks to the oxygen involved in the reaction being directly drawn from the surrounding air—it is not stored in the battery in advance [15,16]. Notably, alkaline zinc-air batteries, among metal-air batteries, hold enough merit to be highlighted. These devices use relatively inexpensive and environmentally friendly raw materials and provide high specific energy values (1084 Wh/kg) [17]. In addition, their low-cost, high natural abundance, low toxicity, nonflammability, and large stable potential window make them very attractive as energy storage applications [17–19].

According to the literature, polyvinyl alcohol (PVA) has been widely used due to its good film-forming ability, good mechanical strength, optical properties, biocompatibility, biodegradability, and non-toxicity [20,21]. PVA is a semicrystalline, hydrophilic, synthetic polymer material at room temperature, which has been extensively used as a host material in various PEs systems due to its polar nature, easy processability, non-corrosive nature, and low production cost [22]. Some researchers have highlighted the electrochemical and mechanical behavior of PVA upon the addition of differently concentrated salt solutions and blending with other materials to prepare GPEs. Saeed et al. [23] reported the effect of a high ammonium salt concentration on electrolytes-based PVA, and they reached an ionic conductivity of 5.17×10^{-5} S/cm at room temperature. Tiong et al. [24] reported that the highest ionic conductivity reached by the PEO/PVA blend-based gel polymer electrolytes was 5.5×10^{-3} S/cm. In addition, Merle et al. [25] improved the ionic conductivity values until 2.2×10^{-1} S/cm by crosslinking the PVA-KOH polymer with poly(ethylene glycol) diglycidyl ether.

Moreover, vinyl acrylic emulsions are important classes of polymer dispersions usually applied in architectural coating [26]. One of the most important industrial latexes is the emulsion copolymerization of vinyl acetate (VA)/butyl acrylate (BA). According to their properties, VA and BA present glass transition temperatures at $T_{g(VA)}$ = 32 °C and $T_{g(BA)}$ = −54 °C, which give a stable thermal range to provide a suitable temperature for electrochemical applications [24,25].

In addition, the KOH incorporation inside GPE membranes provides free ions, which interact with the polar groups of PVA, modifying their structural arrangement and improving the motions of the ions inside the electrolyte matrix. Lewandowski et al. [27] and Santos et al. [4] reported ionic conductivity in the order of 10^{-3}–10^{-1} S/cm for PVA-KOH-H_2O alkaline polymer electrolytes, respectively. In this work, different GPEs are prepared from PVA blended with VAVTD as the host matrix and doped with a concentrated KOH solution used as a source of ions to improve the system's ionic conductivity. The structural, electrical, and transport properties of the synthesized membranes have been analyzed as a function of polymer blend proportions and the results point to a good material to be applied as GPE in Zn batteries.

2. Results and Discussion

2.1. Structural Characterization

2.1.1. Swelling Ratio

PVA-VAVTD membranes synthesized in this article were soaked in a 12 M KOH solution to provoke the entrance of KOH and water inside the polymeric matrix and thus improve their electrical properties. The swelling ratio (SR) was determined for all GPEs following Equation (2). It was observed that the progressive increase in VAVTD in the PVA matrix improved the swelling ratio of the resulting matrix as shown in Figure 1.

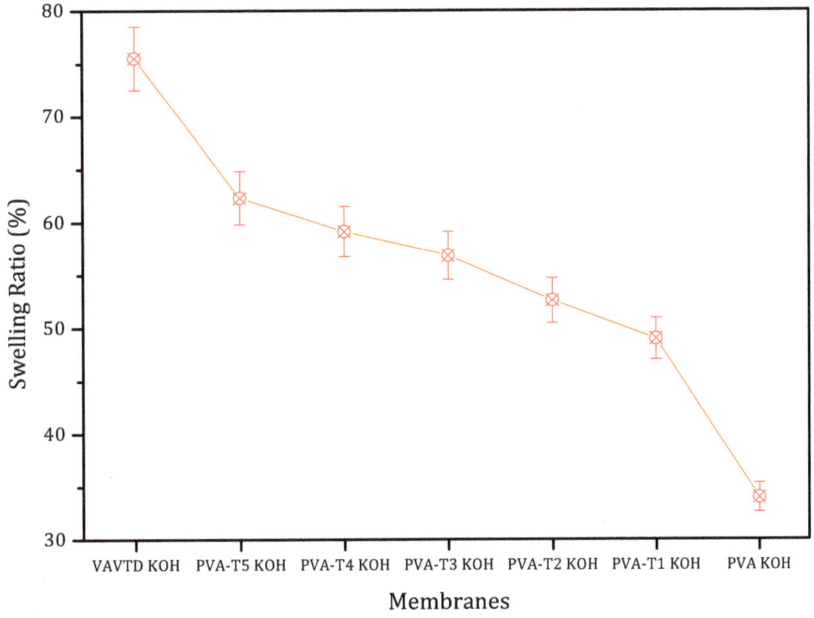

Figure 1. Swelling ratio of the PVA-VAVTD KOH matrix as a function of the VAVTD increase in membrane.

Compared with the KOH-doped PVA matrix, it was possible to observe that a minimum swelling ratio was established for the PVA KOH system. In addition, pure VAVTD reached the highest SR value—75%—which means that it retained three times its electrolyte weight. The progressive incorporation of VAVTD elevated the overall swelling ratio of the matrix due to the growth of internal sites. Although VAVTD absorbed a large quantity of KOH, its mechanical resistance was significantly affected and the resulting membrane did not keep good properties for being used as a GPE. It was thus necessary to add another polymer, namely PVA, to improve the mechanical resistance. However, VAVTD was used to increase the KOH absorption resulting in an excellent blend.

2.1.2. XRD Analysis

The XRD spectra were analyzed in order to determine the VAVTD influence on the crystalline structure of the PVA matrix. The XRD profile of pure PVA, VAVTD, PVA-VAVTD, and PVA-VAVTD KOH samples are depicted in Figure 2, where VAVTD was introduced to decrease the PVA crystallinity due to the disruption of internal hydrogen bonds generated by $(OH)^-$ arrangement. The XRD pattern of the pure PVA membranes revealed a crystalline peak at $2\theta = 19.8°$ and a shoulder at 23.01°, representing reflections from (101) and (200) from a monoclinic unit cell [28]. Similar studies have considered polymer blending as an effective technique to reduce PVA crystallinity [27–29].

On the other hand, the XRD profile of VAVTD was presented with the deconvolution method in Figure 3a. This showed the peak at $2\theta = 20°$, but with a high level of amorphousness compared with pure PVA. This peak was related to the presence of VA segments [30] long enough to form nanocrystals due to the highest participation of VA in the VAVTD synthesis.

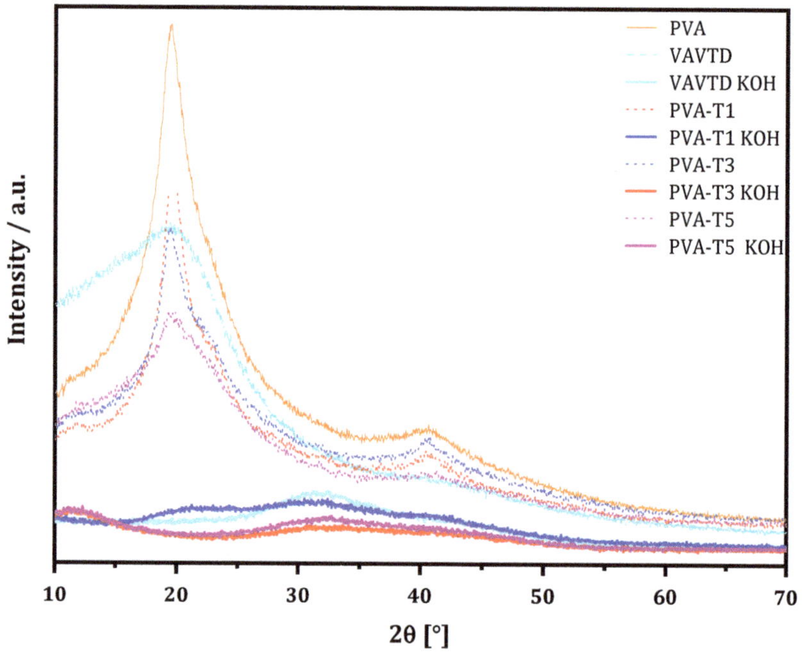

Figure 2. XRD patterns of PVA-VAVTD and PVA-VAVTD KOH systems at different VAVTD contents.

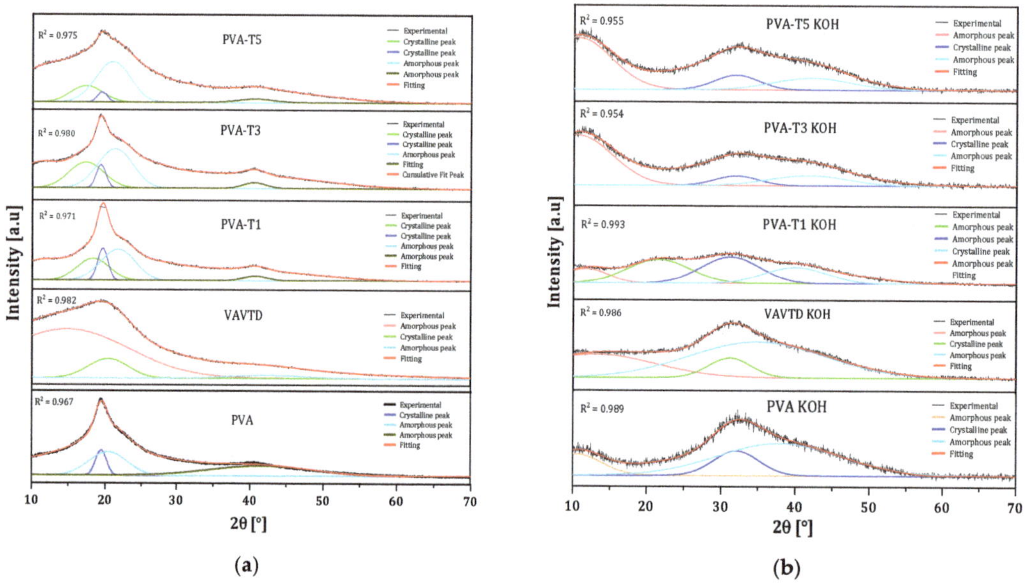

Figure 3. Comparison of the deconvoluted XRD patterns of (**a**) PVA-VAVTD and (**b**) PVA-VAVTD KOH systems at different VAVTD contents.

Moreover, the deconvoluted XRD spectra of PVA KOH, VAVTD KOH, and PVA-blend KOH systems are shown in Figure 3b. It is notable the modification of the PVA peaks, such as the reduction of the principal crystalline peak at 2θ = 20° and the wideness of peak at

$2\theta = 40°$ [31], when the VAVTD content increases. This crystalline reduction was caused by the breaking of hydrogen bonds formed between –OH groups, provoking the intensity reduction seen in the comparison spectra of Figure 2. Thus, these wider and lower intensity peaks confirm the amorphousness of the PVA blends' structure [32].

As can be seen, pure PVA deconvolution presents a crystalline peak at 20°, which is consistent with previously reported results [31,32]. This peak was observed again when VAVTD was included in the membrane, but in this case, another crystalline peak was observed due to the presence of VAVTD (Figure 3a). Similar behavior has previously been observed for PVA-based composites [31,32]. The inclusion of KOH and water inside the polymeric matrix provokes important structural changes providing very different XRD patterns and, thus, new deconvolution peaks (Figure 3b). However, the intensity of these peaks is very low, as can be seen in Figure 2.

From the deconvolution method, the Xc of each PVA blend system was obtained using Equation (3), and all data are summarized in Table 1, where it can be seen that Xc decreases with the amount of VAVTD and the inclusion of KOH and water. Thus, the electrolyte with the most amorphous region corresponds to PVA-T5 when KOH is inserted, with $Xc = 5.06\%$. These results confirm the favored amorphous nature of the polymer electrolytes. In addition, it is notable that, in PVA-VAVTD GPEs, the area of the crystalline peak (1 0 1) decreases as the VAVTD content increases, and this peak entirely disappears when the membranes are immersed in KOH (Figure 3b). Moreover, the degree of crystallinity of PVA and PVA KOH matches with the values reported in the literature [31,33].

Table 1. The Xc for each electrolyte calculated by the deconvoluted method.

Electrolyte	***Xc* (%)**	
	Dried	**Soaked in KOH**
PVA	41.97	17.81
VAVTD	9.773	1.612
PVA-T1	25.48	11.90
PVA-T3	19.01	7.16
PVA-T5	15.95	5.06

2.1.3. ATR-FTIR Analysis

The ATR-FTIR spectra (Figure 4A) show the characteristic bands of PVA and VAVTD polymers, as well as of PVA-VAVTD blends. Regarding the PVA structure, bands associated with the C–O, C–O–C and C=O vibrational motions of acetate groups were identified at 945–1086, 1241 and 1722 cm^{-1}, respectively. The VAVTD spectrum shows the same characteristic bands, but they are shifted to 1020, 1225 and 1734 cm^{-1}, respectively. In addition, there is no C=C band at 1600–1680 cm^{-1}, which corroborates both polymers' complete polymerization [29]. All of the spectra show an intense band between 3100 and 3500 cm^{-1}, which corresponds to the $(OH)^-$ stretching vibration. The bands at 2925 and 2884 cm^{-1} arise from the stretching of CH_3–and–CH_2–groups [34].

With regard to PVA-VAVTD systems (PVA-T), the spectra display similar vibrational frequencies for all of them, although differences in their intensities were found. Furthermore, the combination of the PVA and VAVTD terpolymer can be appreciated because the characteristic absorption bands of both polymers are preserved. In addition, the interaction between the two polymeric chains is confirmed by the shift in the main bands. PVA-T1 presents a similar ATR-FTIR spectrum to that of PVA, but the increase in the VAVTD amount inside the polymer provides spectra close to that of VAVTD terpolymer.

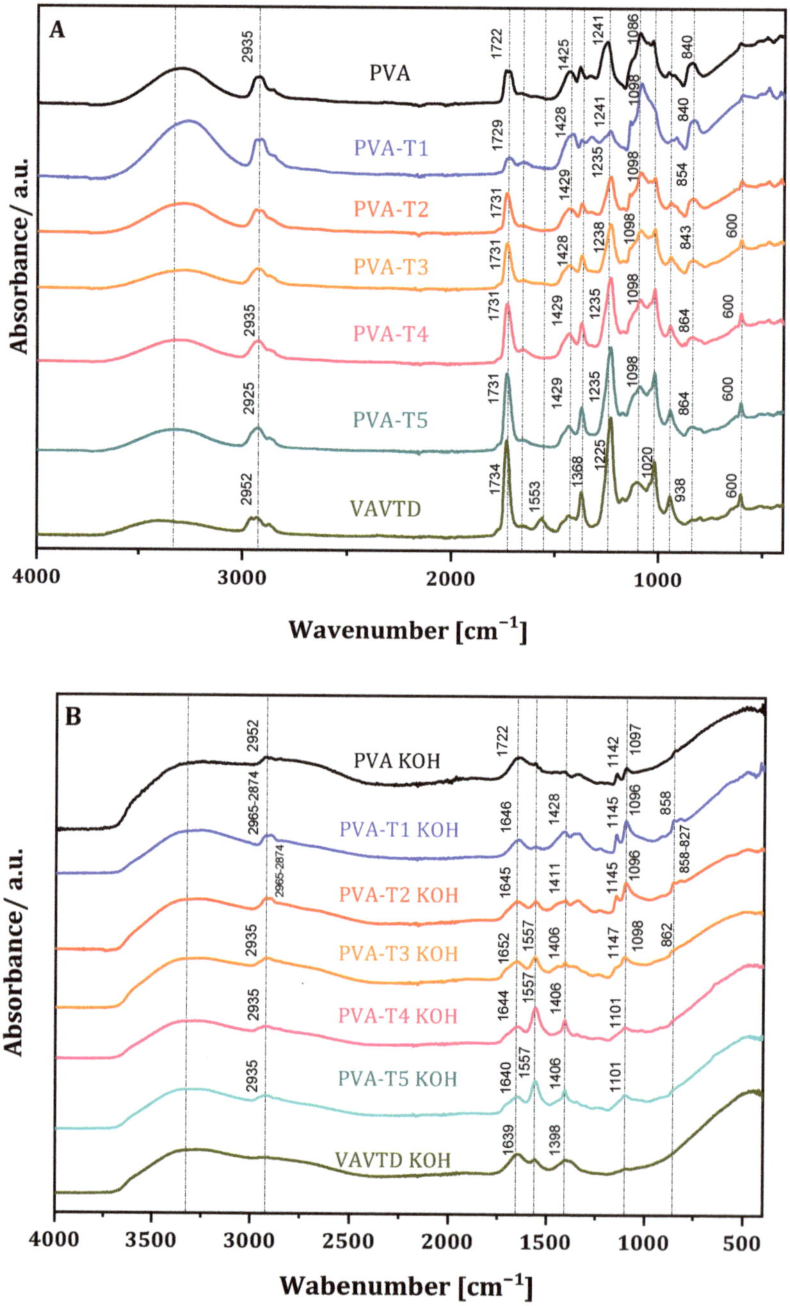

Figure 4. Comparison of the ATR-FTIR spectra of (**A**) PVA, VAVTD, PVA-VAVTD; and (**B**) PVA-VAVTD KOH membranes at different VAVTD proportions.

On the other hand, the incorporation of KOH and H_2O molecules inside the PVA-VAVTD polymeric matrix broke its semicrystalline structure and induced several changes in the ATR-FTIR spectra (Figure 4B). The band at υ 1645 cm^{-1} confirms the bending mode frequency of water [35]. In addition, υ 1733, 945 and 1241–1225 cm^{-1} related to carboxylate, C–O and C–O–C vibrational motions, disappeared at all spectra, which may confirm the hydrolysis of acetyl groups by $(OH)^-$ groups of the KOH [4]. On the other hand, a new band is observed at 1569 cm^{-1}, which may be assigned to the asymmetrical stretching vibration absorption of –C=O–(–O–K), as it has already been reported [4,31]. In fact, the oxygen atoms of –OH and the carbonyl groups have lone pairs available to coordinate with K^+ ions and form C=O–K^+/C–O–K^+ complexes [4].

Furthermore, the intensity of this band increased with the amount of VAVTD in the blend due to the higher number of carboxylate groups included in the polymeric matrix. Finally, the peak at 1143 cm^{-1} associated with the C–O stretching mode is mostly attributed to the remaining crystallinity of the PVA because it is able to form some domains [32]. This band depends on the new intermolecular hydrogen bondings that the samples can build, which were lost for PVA-T3 KOH, PVA-T4 KOH and PVA-T5 KOH membranes showing the dominance of the amorphous region as the VAVTD content increases.

The fundamental –OH absorption region presents a broader peak between 3100 and 3500 cm^{-1} associated with hydrogen interactions. Additionally, the PVA-VAVTD KOH blend has wider and more intense bands than PVA-VAVTD, indicating a higher amount and stronger hydrogen bonds inside the polymeric matrix. Furthermore, the introduction of KOH and water molecules produces changes in the chain network, breaking some H-bonds and forming new H-interactions between the hydroxyl anions, water, and acetate groups of the polymer chains.

In this section, it is worth noting that the synthesis of the polymer blend, casting, and its immersion in KOH solution was performed under ambient conditions without any atmosphere control. Thus, CO_2 from the air could always be in contact with the blend. However, it is known that the KOH solution is used to capture CO_2 from the air, generating K_2CO_3 [36]. Furthermore, PVA-KOH GEPS previously synthesized by our group had to be prepared in a controlled atmosphere to prevent carbon dioxide forming during the casting and when the membrane was immersed in KOH 12 M, preventing the formation of carbonates inside the polymer [4]. The presence of carbonate in the electrolyte is a concern for Zn/air batteries because it can form K_2CO_3, which may be deposited on the air electrode, blocking the oxygen transfer and resulting in the earlier performance decline of the Zn/air battery [37]. Additionally, the Zn electrode may also be passivated by precipitating insoluble $ZnCO_3$ on the electrode surface [38].

Figure 5 shows the ATR-FTIR spectra of PVA-KOH and PVA-KOH immersed in 12 M KOH solution, synthesized in the presence and in the absence of carbon dioxide. As can be seen, when CO_2 is present, an intense peak assigned to K_2CO_3 was observed at υ 1370 cm^{-1} [39]. This band appeared in the spectra of both PVA-KOH and that soaked in KOH 12 M membranes. However, this band disappeared for membranes synthesized when the atmosphere was controlled, avoiding the absorption of CO_2.

Contrarily, PVA-VAVTD and PVA-VAVTD soaked in KOH 12 M solution were synthesized in the presence of atmospheric CO_2, but the ATR-FTIR spectra do not show the 1370 cm^{-1} peak. Furthermore, this is a clear advantage of the PVA-VAVTD-KOH blend with respect to the PVA-KOH ones.

2.1.4. Thermal Analysis

Figure 6 compares the TGA analysis obtained from pure VAVTD, PVA-VAVTD, and PVA-VAVTD soaked in the KOH solution [4,40,41]. The VAVTD terpolymer (black line) presents three degradation steps. The first one occurs in the water region due to the slow loss of internal water, which suppose a 10% of weight loss. The second degradation step starts at ~300 °C until ~380 °C, where 60% of the mass is lost. The last degradation step occurs at ~420 °C and is associated with the breaking backbone. Finally, at ~480 °C, a stable behavior is observed, with a remaining residue of ~5% wt. Similar behavior has been reported for PVA [4,36] and PVA with BA/VAc [42] or PMMA blends [43].

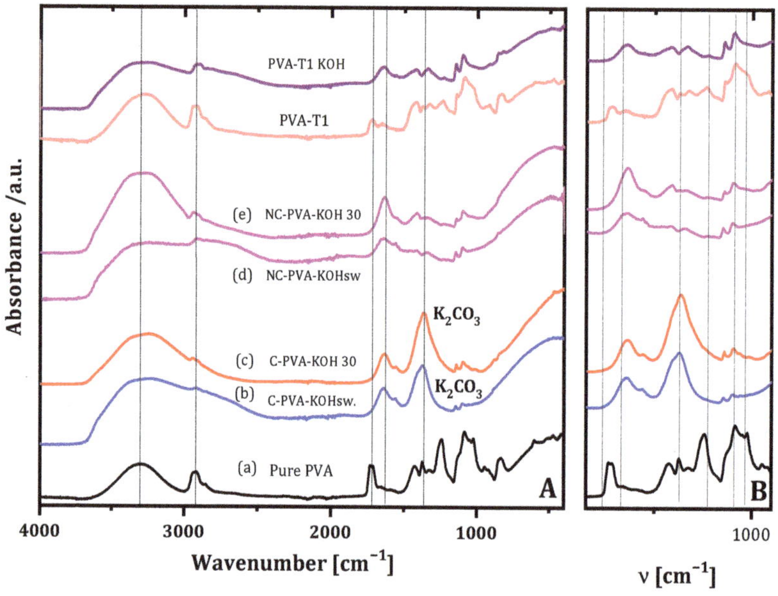

Figure 5. (**A**) ATR-FTIR spectra of pure PVA (**a**) PVA-KOH (**b**) and PVA-KOH immersed in 12 M KOH solution (**c**) synthesized in the presence of carbon dioxide; PVA-KOH (**d**) and PVA-KOH immersed in 12 M KOH solution (**e**) synthesized in the absence of carbon dioxide; and PVA-T1 and PVA-T1 KOH membranes synthesized in the presence of carbon dioxide; (**B**) depicts the K_2CO_3 vibration region. Sw indicates that PVA-KOH was swelled in KOH 12 M solution.

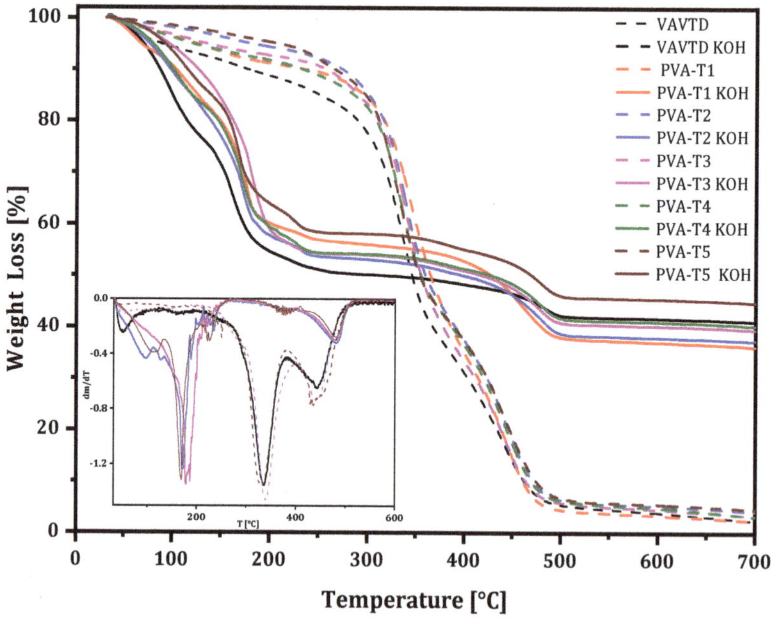

Figure 6. TG curves of PVA-VAVTD and PVA-VAVTD KOH membranes. Ins: DTG curves of electrolytes.

PVA-VAVTD blends show TGA curves very close to those of VAVTD and PVA polymers, and they do not show changes with the VAVTD amount included in the polymeric matrix. However, PVA-VAVTD KOH displays very different TGA curves indicating strong structural changes. This behavior is very similar to the results obtained for PVA-KOH soaked in 12 M KOH solution [4]. The first mass loss occurs between 30 °C and 150 °C, which is associated with the elimination of water. Then, it reaches a loss of 20 wt%, 10% higher than those of VAVTD and VAVTD-KOH films, indicating the presence of more water inside the polymer due to the swelling of the membrane which occurred during its immersion in KOH solution.

A second weight loss occurs with an onset at 150 °C. It is worth noting that this temperature onset is much lower than that observed for VAVTD-PVA membranes, at ~300 °C, confirming the lower thermal stability of the membranes due to KOH solution uptake. This fact is clearly shown in the dm/dT vs. T graph included in the inset of Figure 6, where the peaks corresponding to the second step are found at 342 °C for PVA-VAVTD and at 170 °C for PVA-VAVTD KOH membranes. It should be noted that the TGA curves of PVA-VAVTD KOH are independent of the VAVTD quantity included in the GPEs. This fact points to the fact that the shift to lower temperatures may be associated with the high entrance of KOH and water [44], which interacts with $(OH)^-$ and C=O groups which are mainly present in PVA chains. Furthermore, these interactions can explain the reduction in thermal stability observed for membranes soaked in KOH solution, agreeing with the higher amorphous behavior observed in XRD.

The entrance of water molecules during the swelling process was confirmed by the increasing weight loss at a temperature lower than 150 °C. Additionally, the entrance of KOH inside the polymer was demonstrated with the residue obtained at 700 °C. As can be seen, the GPEs preserve 40% of their original mass, which can be associated with the formation of K_2O by the decomposition of KOH [45] according to Equation (1):

$$2KOH_{(s)} \rightarrow K_2O_{(s)} + H_2O_{(g)} \quad (1)$$

Once again, it has to be noted that the percentage of weight remaining at 700 °C is the same for all PVA-VAVTD KOH GPEs, independently of the amount of VAVTD inside the polymer. Thus, KOH should mainly be included in the PVA portion of the polymer.

As a result, the TGA analysis confirms how incorporating KOH and H_2O molecules inside the GPEs contributes to the weakening of the thermal resistance due to their interactions with the pendant groups of the membranes. This result supports the ATR-FTIR and XRD analyses. When the KOH-H_2O penetrates the hydrophilic GPEs matrix due to PVA segments, it increases the amorphous behavior and the distance between the polymer chains, providing more free volume for molecular movement [44], decreasing its melting temperature, and allowing a more accessible ionic motion and transport.

2.2. *Electrochemical Characterization*

2.2.1. The Influence of VAVTD Content on Ionic Conductivity

Ionic conductivity is an important factor in determining the applicability of a membrane as a polymer electrolyte in batteries. The gel polymer electrolytes used in this study are composed of a host polymer, VAVTD or PVA-VAVTD, soaked in 12 KOH solution to incorporate $(OH)^-$, K^+, and water molecules inside the polymeric matrix. Furthermore, the amount of KOH and water inside the polymer host will determine the number of ionic carriers, and thus, the ionic conductivity values.

Figure 7 shows the dependence of ionic conductivity with the temperature for GPEs based on PVA at different concentrations of VAVTD and soaked in KOH solution. The ionic conductivity values were obtained from Equation (4). The plots follow a linear fitting, confirming the Arrhenius behavior, as can be deduced from Equation (5), proving that the conductivity is thermally assisted [46].

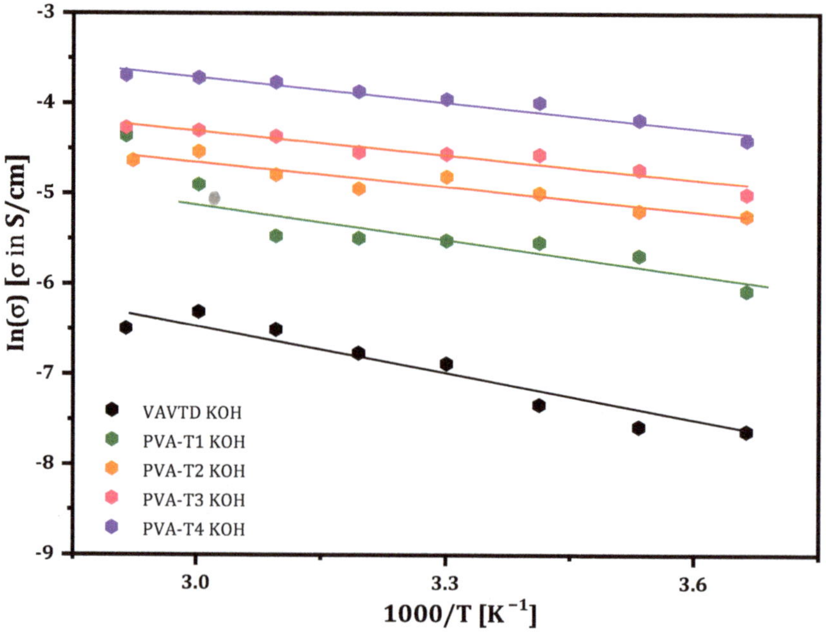

Figure 7. Ionic conductivity of PVA-based electrolytes at different temperatures.

From the slope of the linear fitting, the Activation Energy (Ea) values for the GPEs studied can be calculated, which are shown in the inset of Figure 8. As can be seen, Ea values are higher for GPEs with a low amount of VAVTD, but this reaches a constant value of ~0.033 eV for the membrane with a higher VAVTD quantity.

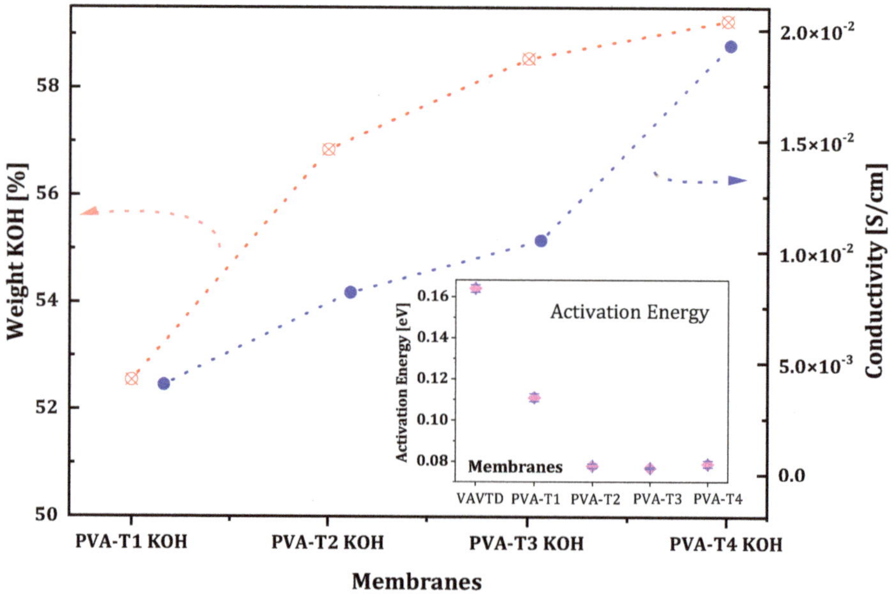

Figure 8. Comparison between the ionic conductive values and KOH absorption percentage (SR) of PVA-VAVTD-based electrolytes.

Table 2 and Figure 1 demonstrate that PVA-VAVTD polymers have a higher swelling ratio, i.e., they absorb a larger amount of KOH and water, with the increase in VAVTD quantity added to the PVA homopolymer. In addition, conductivity values follow the same tendency of increasing their value with the amount of VAVTD. This fact indicates that more KOH and water molecules are stabilized inside the polymer matrix due to the interaction with a higher number of acetate groups of the VAVTD, as was deduced from ATR-FTIR results. Thus, the ionic conductivity values increase as a consequence of the higher ionic concentration inside the polymer as well as the greater amorphousness and the higher chains separation, as previously deduced. Figure 8 shows the increase in the swelling ratio and conductivity values with the amount of VAVTD inside the blend.

Table 2. Swelling ratio (SR), activation energy (Ea), and the conductivity (σ) values of PVA-VAVTD KOH electrolytes. σ values were obtained at T = 20 °C.

Electrolyte	SR (%)	Ea (eV)	σ (S/cm)
VAVTD KOH	75.2	0.164	0.001
PVA-T1 KOH	52.5	0.111	0.004
PVA-T2 KOH	56.8	0.078	0.008
PVA-T3 KOH	58.6	0.077	0.011
PVA-T4 KOH	59.3	0.079	0.019
PVA-T5 KOH	62.7	-	0.019

As confirmed by XRD analysis, the crystallization of the PVA-VAVTD GPE membrane was broken by doping with the KOH solution. In addition, this effect improves when the temperature increases, generating a more structural relaxation of the polymer chains (amorphous phase) and expanding the free volume. As a result, this decrease the energy barrier to ionic transport and promotes the fast ion migration [47]. The maximum ionic conductivity, 1.93×10^{-2} S/cm, was found when the VAVTD content was 80 wt%, and it reduced to 4.0×10^{-3} S/cm when the VAVTD content was 50 wt%. Table 2 summarizes the ionic conductivity and Ea values for VAVTD and the PVA-VAVTD KOH analyzed.

The conductivity of PVA-VAVTD systems could be explained by the mechanism of ions transferring through polymer molecular chains, implying an association–dissociation process between the ions and polar groups of the PVA [47,48] and acrylic groups of the VAVTD matrix, presumably following a Grotthuss mechanism, as has been described before for PVA-KOH GPEs [4]. This phenomenon can be attributed to the increase in the swelling ratio when the VAVTD content increases, as shown in Table 2. This means that a higher amount of KOH and H_2O molecules penetrates the polymer, favoring the ionic conductivity. In addition, the increase in ion mobility inside the electrolyte is related to the segmental motion of the chains that creates a large free volume and improves the pathway of ionic species. In addition, all blends of PVA-VAVTD KOH have lower Ea values than the VAVTD KOH membrane, which decrease with the amount of VAVTD included in the GPE, until reach a minimum value of ~0.077 eV. This fact indicates that membranes with a lower amount of VAVTD restrict ion mobility, whereas in GPEs with higher VAVTD contents, the energy necessary for providing the ionic movement decreases. The spatial arrangement changes of the electrolytes were demonstrated by the XRD and ATR-FTIR techniques, removing the crystalline domains and favoring the amorphous region.

It was also observed that the membranes of VAVTD pure have lower ionic conductivity and the highest activation energy. However, they present a higher KOH absorption than the other membranes shown in Table 2. This apparent contradictory result is explained by the poor mechanical resistance of pure VAVTD film. This result proves that the VAVTD terpolymer is a poor electrolyte, and it cannot work alone, making necessary the inclusion of PVA to form a sufficiently mechanically resistant blend.

2.2.2. Cyclic Voltammetry

Cyclic voltammetry analysis was performed using a Zn/GPE/Zn cell to confirm the ionic transport in the GPEs. Voltammograms of PVA-VAVTD (inset) and PVA-VAVTD-KOH

GPEs with different contents of VAVTD are shown in Figure 9. The cyclic voltammograms of PVA-VAVTD membranes without immersion in KOH salt show capacitive curves which are due to the absence of mobile ions inside the membrane, hindering the ionic transport to the electrode to balance the charge changes during the redox reactions. Thus, these films behave as an inadequate electrolyte.

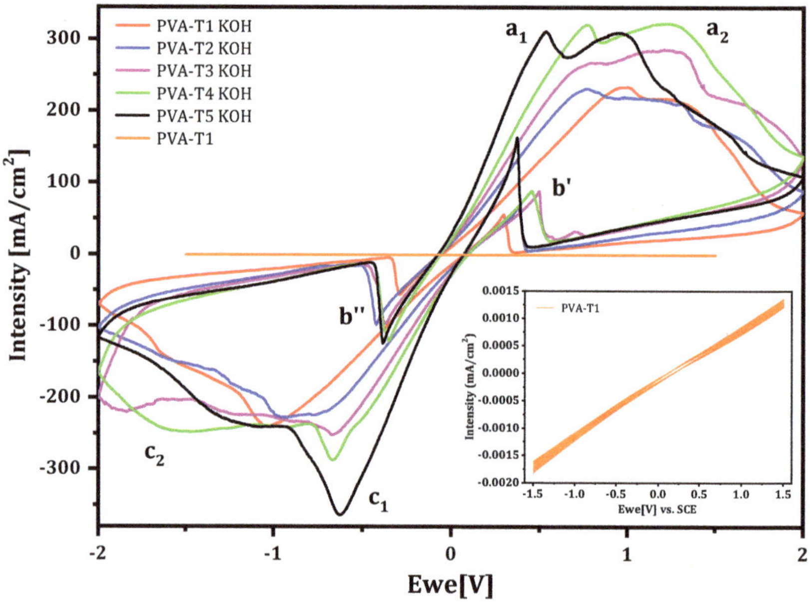

Figure 9. Comparison of the cyclic voltammetry of PVA-VAVTD and GPEs immersed in KOH 12 M at different VAVTD contents. Inset: cyclic voltammetry of PVA-T1 membrane. Peak labels are explained in the text.

Contrarily, when PVA-VAVTD KOH GPEs (Figure 9) were used in the cell, a quasi-reversible behavior of the Zn^{2+}/Zn oxidation/reduction processes was obtained. As can be seen in Figures 9 and 10, two peaks (a_1 and a_2) are observed at the anodic branch. Similar peaks have been previously reported by Cai et al. [49] in an electrochemical study of the Zn-electrode in alkaline solution. They assigned the a_1 peak to the oxidation of Zn to $Zn(OH)_4{}^{2-}$, whereas the peak a_2 was associated with the oxidation of Zn to $Zn(OH)_3{}^{-}$, which is due to the depletion of $(OH)^-$ anions in the proximity of the electrode surface, forming a prepassive layer at a more positive potential than that of peak a_1. These authors only observed the double peak in the anodic branch because they used a usual three-electrode cell. However, we found the peak split in both anodic (a_1 and a_2) and cathodic (c_1 and c_2) branches because we used an Zn/PVA-VAVTD KOH/Zn cell.

Moreover, Figure 9 shows the voltammograms obtained using PVA-VAVTD KOH GPEs synthesized with different amounts of VAVTD, where the peak charges and intensities increase with the concentration of VAVTD inside the gel. This behavior agrees with increase in the swelling of the KOH solution with the amount of VAVTD, which reduces the crystallinity of the GPEs and facilitates the ion transport, as was mentioned previously. In addition, the inset in Figure 10 presents 30 consecutive cycles using a PVA-T4 KOH film, which demonstrates the GPE stability.

On the other hand, the inverse peak b′ appearing in the cathodic branch is due to the oxidation of Zn after the dissolution the passive film deposited on the Zn-electrode surface, which come off during the cathodic scan. The same happens for the peak b″ in the anodic scan [4,49].

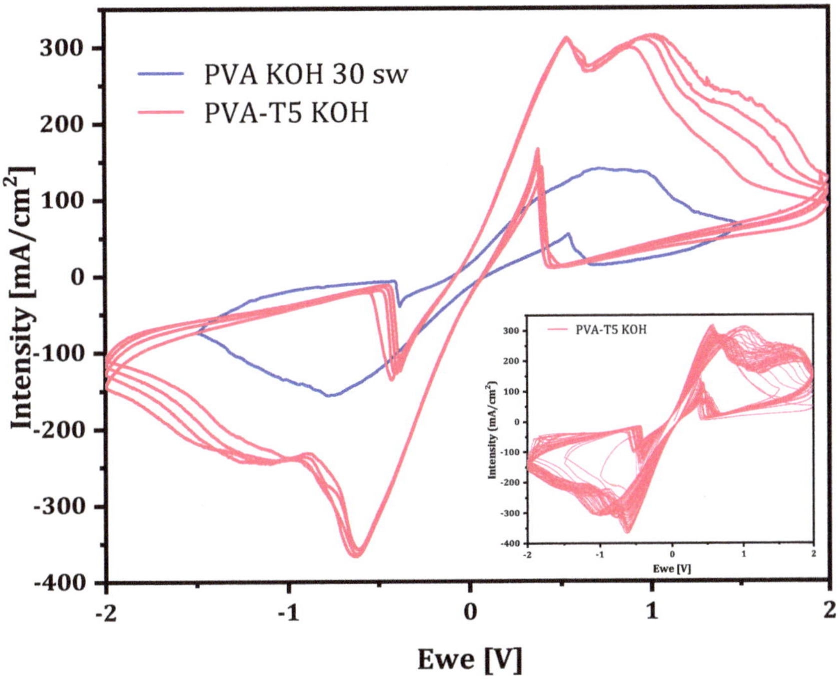

Figure 10. Cyclic voltammograms of the PVA-T4 KOH and PVA-KOH 30 swollen membranes [4]. Inset: 30 consecutive cycles of the PVA-T4 KOH film.

A comparison between PVA-KOH soaked in KOH solution and PVA-VAVTD KOH GPEs is shown in Figure 10. The current intensity of the PVA-KOH system presents a maximum close to 150 mA/cm^2 when immersed in KOH 12 M [4]. However, this value corresponds to half of that obtained when VAVTD was added, 321 mA/cm^2. As the intensity values result depends on the number of electrons transferred between the redox species and the electrode, which depends on the ions' movement, the current increase may be related to the improvement of the fast-ionic motion across the electrolyte matrix. Furthermore, PVA-VAVTD KOH electrolytes are presented as an alternative to be applied in energy storage devices.

2.3. Zn/PVA-VAVTD KOH/Air Battery

The electrochemical performances of PVA-T4 KOH and PVA-T5 KOH GPEs were examined as electrolytes in Zn–air batteries. Zn powder and a commercial Air E4B electrode were used as negative and positive electrodes. PVA-VAVTD-based membranes were used as electrolyte, placing them between Zn and air electrodes. The discharge current density was -5 mA/cm^2 and the cut-off voltage was 0 V (Figure 11).

Zn–air batteries using PVA-VAVTD GPEs present a capacity of 135 mAh/g and 150 mAh/g for PVA-T4 KOH and PVA-T5 KOH, respectively, with a cut-off voltage of 0.9 V. However, when the discharge is carried out until a cut off of 0 V, the maximum capacity reached was 165 mAh/g and 195 mAh/g for PVA-T4 KOH and PVA-T5 KOH, respectively During the discharge process, a stable potential between 1.1 and 1.2 V until reaching 95 mAh/g was maintained when using PVA-T4 KOH GPE, whereas with PVA-T5, KOH maintains a potential of 1.2–1.3 V until reaching 150 mAh/g, confirming the positive effect of VAVTD on the discharge performance, in agreement with the cyclic voltammetry results. It is normally accepted that Zn oxidation needs enough $(OH)^-$ anions to form soluble species such as $Zn(OH)_4{}^{2-}$ or $Zn(OH)_3{}^-$, previously to be deposited as a passive film of ZnO. Furthermore, the increase in KOH and water swelling during the immersion of the membranes in 12 M KOH solution is essential to enable the battery to function properly. Thus, these results agree with the values obtained in the CV and conductivity measurements, where high carrier charges are transported through PVA-VAVTD KOH membranes.

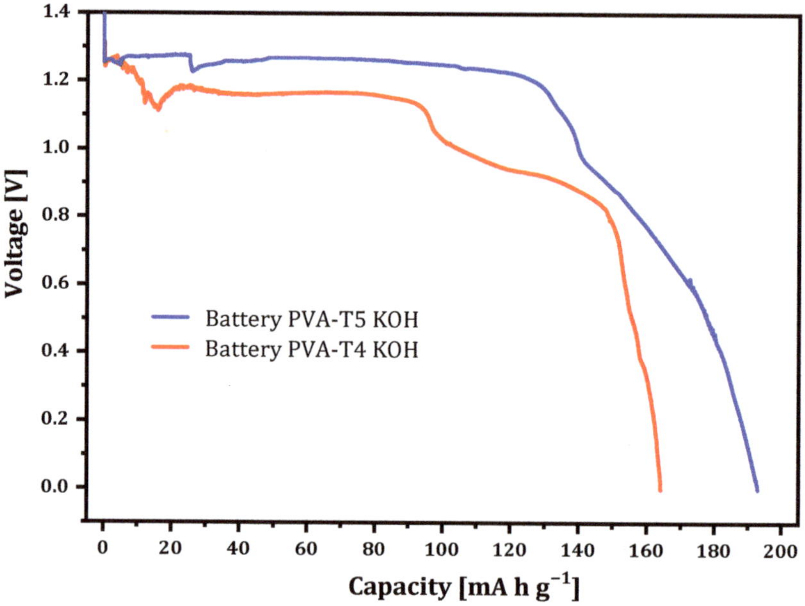

Figure 11. Discharge curves of Zn/PVA-T4 KOH/Air, Zn/PVA-T5 KOH/Air.

In addition, the best and more stable battery performance was found when the PVA-T5 KOH GPE was used. This result is in accordance with the structural characterization results, which confirmed a larger amorphous phase and higher amount of KOH solution uptake in the GPE, with the amount of VAVTD incorporated to the blend. Thus, once again, the battery results agree with the CV and conductivity measurements.

3. Conclusions

In this article, we proved the good electrical properties of the PVA-VAVTD blends as a consequence of the synergetic interactions of PVA and VAVTD polymers, separately improving the individual properties of each polymer. The VAVTD terpolymer has a high swelling ratio when it is soaked in 12 M KOH solution. However, their poor mechanical properties provoke its failure as a gel polymer electrolyte. This fact makes it necessary that it is blended with another polymer with better mechanical properties. For this aim, PVA was chosen due to the well-known high performance of this polymer when used as host of a GPE.

All PVA-VAVTD blends prepared present higher swelling ratios, as values of ~50–60% are obtained, than PVA-KOH polymers, which arise with maximum values of 36%, when they are soaked in KOH solution. This behavior provides a higher quantity of KOH and H_2O molecules inside the polymeric matrix of PVA-VAVTD blends. In addition, the swelling ratio increased with the amount of VAVTD included in the membrane.

On the other hand, the structural characterization carried out by TGA, ATR-FTIR, and XRD techniques revealed the decrease in the crystallinity inside the polymer blend when VAVTD was mixed with PVA. In addition, the amorphousness of the blend raised when it was soaked in KOH solution.

Regarding the electrochemical characterization, both the conductivity and voltametric intensity values increased with the amount of VAVTD incorporated into the GPE. Ionic conductivity values demonstrate an Arrhenius behavior with the temperature and a maximum value of 0.019 S/cm at 20 °C was found, which is somewhat less than the value obtained for PVA-KOH. However, it must be noted that the Activation Energy values, ~0.077 eV, are one magnitude order lower than those measured for PVA-KOH GEPs. In addition, a quasi-reversible voltametric behavior was observed for all PVA-VAVTD blends, whose intensity peaks increase with the amount of VAVTD into the polymer and reaching intensity values approximately three times higher than those found for PVA-KOH gels, thus confirming the high ionic transfer through the gel polymer to favor the Zn/Zn^{2+} redox processes. Finally, PVA-VAVTD KOH GPEs were tested in Zn/PVA-VAVTD-KOH/air batteries, showing remarkable capacity values.

As concluding remarks, the blend of PVA and VAVTD polymers and its immersion in KOH solution provide high-performance gel polymer electrolytes, with high mechanical and electrical properties, as a consequence of the extended amorphous regions and the free volume existent inside the polymeric matrix, caused by the interaction between the VAVTD and PVA chain, as well as the high KOH and water molecules uptake. Furthermore, this GPE is a good candidate to be used in all-solid energy storage devices.

4. Materials and Methods

Poly (vinyl alcohol) (PVA), MOWIOL 18-88 (MW 130,000), and potassium hydroxide (KOH), Mw = 56.12 g/mol, 85% of purity were purchased from Sigma-Aldrich. Distilled water was used as a solvent in the polymers blending.

For the electrochemical characterization, zinc (Zn) powder, 98.7% purity, and Pt foils, 99.95%, were provided by Goodfellow (Hamburg, Germany), air E4B by Electric Fuel Ltd. (Beit Shemesh, Israel), and Al_2O_3 Polishing Suspension, (1 and 0.05) micron, Buehler (Lake Bluff, IL, USA).

4.1. *VAVTD Synthesis*

Poly (vinyl acetate)-co-poly (butyl acrylate)-co-poly (vinyl tert-decanoato) (VAVTD), of a particle size in the range of 150–300 nm [50], was synthetized with the same chemicals and procedure previously reported [29]. Vinyl acetate (Tg = 35 °C and water solubility of 2.5 g/100g at 20 °C), butyl acrylate (Tg = −53 °C and water solubility of 0.16 g/100g at 20 °C), and vinyl tert-decanoate (Tg = −3 °C and water solubility of 0.001 g/100g at 20 °C) with an initial concentration of 70/15/15 w/w, respectively, were the starting monomers of the emulsion polymerization. Ammonium persulfate and tert-butyl hydroperoxide were the radical initiators. The reaction was carried out at 353 K for 5 h.

4.2. *Preparation of PVA-VAVTD GPEs*

The polymer electrolyte was prepared by a solution casting method. One gram of PVA was dissolved in 15 mL of deionized water at 90 °C under continuous stirring for 2 h until PVA was completely dissolved. When the solution reached room temperature, different amounts of VAVTD (1,2,3,4,5) g were added under continuous stirring. Table 3 list the sample code used in the article. The mix was placed in a Petri dish and allowed to cast at ambient room temperature for 1 week. When the water was completely removed, 0.15 ± 0.05 cm-thick membranes were obtained. These films were immersed in a 12 M KOH solution for 1 day. To name the last blends, 'KOH' was added to the sample codes of Table 3.

Table 3. Sample code of polymer electrolytes used in this article.

Electrolyte	Sample Code
Poly (vinyl alcohol)	PVA
PVA + VAVTD (1:1)	PVA-T1
PVA + VAVTD (1:2)	PVA-T2
PVA + VAVTD (1:3)	PVA-T3
PVA + VAVTD (1:4)	PVA-T4
PVA + VAVTD (1:5)	PVA-T5

4.3. Swelling Behavior of GPEs

For swelling ratio calculations, the samples were weighed before and after 24 h of being immersed in KOH. After that, the KOH absorbed was determined using Equation (2):

$$SR = \frac{W_t - W_0}{W_0} \tag{2}$$

where W_t and W_0 are the swollen gel weights and the initial sample, respectively.

4.4. Structural, Thermal, and Electrochemical Characterization Techniques

4.4.1. ATR-FTIR and XRD Methods

Attenuated total reflectance Fourier transformed infrared spectroscopy (ATR-FTIR) measurements were obtained using a Thermo Nicolet 5700 Infrared Spectrometer (Waltham, MA, USA) in the wavenumber range of 4000–400 cm^{-1}, with a resolution < 0.5 cm^{-1}. X-ray diffraction patterns were collected using a Bruker D8 Advance laboratory diffractometer (Billerica, MA, USA), operated in the reflection Bragg–Brentano geometry and configured in the θ/θ mode to always maintain a horizontal sample position. The data were collected at room temperature, using Cu-Kα (λ = 1.5418 Å). The degree of crystallinity was calculated by applying Equation (3):

$$X_c = \frac{A_c}{A_T} \times 100\% \tag{3}$$

where A_c is the total crystallinity area and A_T represents the total area of the XRD deconvoluted plot (the sum of the amorphous and crystalline area) using OriginPro software. The Gaussian function mode was employed to fit the XRD spectra.

4.4.2. Thermal Analysis

Thermogravimetric analysis was performed on samples of 5–10 mg using a Mettler-Toledo TGA/DSC 1HT (Columbus, OH, USA), from room temperature up to 700 °C at a heating rate of 10 °C/min and under N_2 atmosphere.

4.4.3. Conductivity and Voltammetry Studies

Symmetric Zn/GPE/Zn cells were used to carried out the cyclic voltammetry measurements by a Biologic VSP Modular 5 channels potentiostat/galvanostat (Seyssinet-Pariset, France). The zinc electrodes' area was 1 cm^2. The scan rate was 50 mV/s, in a potential window between −2 V and +2 V. Ionic conductivity was determined by AC impedance technique using the same potentiostat/galvanostat in the frequency range from 1 kHz to 10 mHz. The sample thickness was measured with a micrometer. The temperature was set by a Julabo (Seelbach, Germany) F25-D cryothermostat in the range from 5 °C to 70 °C with a variation of ±1 °C. Two blocking platinum electrodes of 1 cm^2 area were used in a Pt/GPE/Pt cell. Ionic conductivity, σ, was calculated using Equation (4):

$$\sigma = \frac{l}{A * R_b} \tag{4}$$

where l, A, and R_b represent the film thickness, the Pt electrodes area, and the bulk resistance, respectively, obtained from the intersections of impedance curves with the X axis, as shown in Figure 12. Three impedance measurements were carried out for each membrane.

To determine the activation energy (E_a) of each electrolyte, the Arrhenius Equation (5), was used with a linear fitting by plotting a logarithmic relationship between log(σT) and 1000/T:

$$\sigma = \sigma_0 \exp\left(-\frac{\mathrm{Ea}}{\mathrm{K_b(T)}}\right) \tag{5}$$

where K_b is the Boltzmann's constant, T is the absolute temperature, and σ_0 is a pre-exponential factor [51].

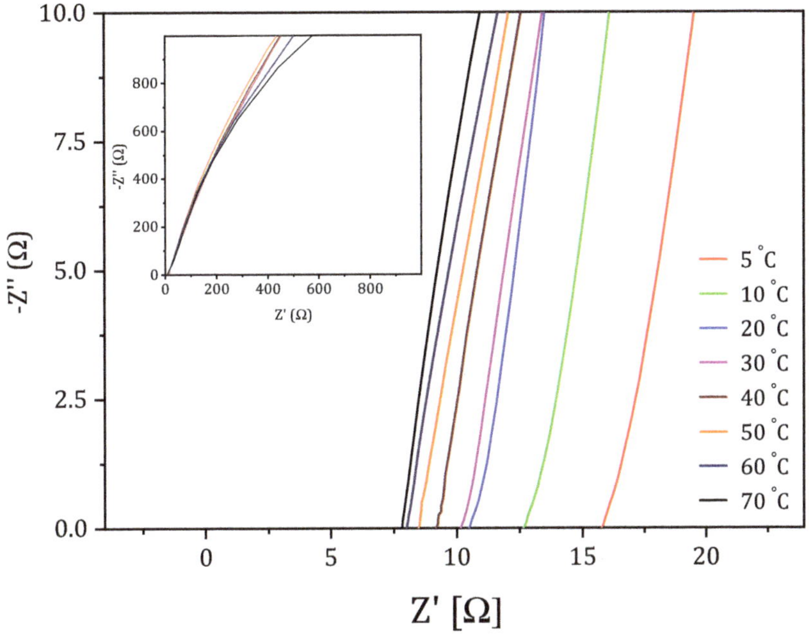

Figure 12. Nyquist plots obtained to calculate the bulk resistance values of the PVA-T4 KOH membrane at different temperatures.

4.5. Electrode Preparation and Battery Setup

4.5.1. Zn Electrode Preparation

The electrodes were prepared with 1 g of zinc powder, which were pressed at a pressure of 10-ton cm^{-2}. After that, 1 μm and 0.05 μm polishing suspensions were used to eliminate irregularities on the electrode surface. Finally, the electrodes were washed with distilled water to remove any polish trace.

4.5.2. Zn/PVA-VAVTD/Air Battery Fabrication

To prepare the Zn–air batteries, the GPE was placed between 1 g of Zn powder placed in a steel capsule and the air E4B cathode. Nickel meshes were utilized as electrode collectors. Finally, the battery was placed on Teflon support to maintain constant contact during the test. Two screws located at each end of the bracket provided stability.

Author Contributions: Conceptualization, methodology, formal analysis, data curation, investigation, software, A.A.I.V. and E.R.; writing—original draft preparation, A.A.I.V.; writing—review and editing, A.J.F.R., F.S., A.D.-B. and J.P.T.; conceptualization, methodology, supervision, A.J.F.R., F.S., A.D.-B. and J.P.T. project administration, funding acquisition, A.J.F.R. and F.S. All authors have read and agreed to the published version of the manuscript.

Funding: This research was funded by FUNDACIÓN SÉNECA (Región de Murcia, Spain), grant number: 20985/PI/18 and SPANISH AGENCIA ESTATAL DE INVESTIGACIÓN, grant number PID2019-104272RB-C55/AEI/10.13039/501100011033. The APC was funded by FUNDACIÓN SÉNECA (Región de Murcia, Spain).

Data Availability Statement: No new data were created or analyzed in this study. Data sharing is not applicable to this article.

Acknowledgments: We would like to acknowledge all support for this work provided by Fundación Séneca (Región de Murcia, Spain; Ref: 20985/PI/18) and the Spanish Agencia Estatal de Investigación

(PID2019-104272RB-C55/AEI/10.13039/501100011033. In addition, we would like to acknowledge Yachay Tech University and the Technical University of Cartagena.

Conflicts of Interest: The authors declare no conflict of interest.

References

1. Fenton, D.E.; Parker, J.M.; Wright, P.V. Complexes of alkali metal ions with poly(ethylene oxide). *Polymer* **1973**, *14*, 589. [CrossRef]
2. Michel, A. Polymers with Ionic Conductivity. *Adv. Mater.* **1990**, *2*, 278–286.
3. Aziz, S.B.; Abidin, Z.H.Z. Ion-transport study in nanocomposite solid polymer electrolytes based on chitosan: Electrical and dielectric analysis. *J. Appl. Polym. Sci.* **2015**, *132*, 1–10. [CrossRef]
4. Santos, F.; Tafur, J.P.; Abad, J.; Fernández Romero, A.J. Structural modifications and ionic transport of PVA-KOH hydrogels applied in Zn/Air batteries. *J. Electroanal. Chem.* **2019**, *850*, 113380. [CrossRef]
5. Guisao, J.P.T.; Romero, A.J.F. Interaction between Zn^{2+} cations and n-methyl-2-pyrrolidone in ionic liquid-based Gel Polymer Electrolytes for Zn batteries. *Electrochim. Acta* **2015**, *176*, 1447–1453. [CrossRef]
6. Prakash, R.; Maiti, P. Functionalized thermoplastic polyurethane gel electrolytes for cosensitized TiO_2/CdS/CdSe photoanode solar cells with high efficiency. *Energy Fuels* **2020**, *34*, 16847–16857. [CrossRef]
7. Maria, M.B.; Sivasubramanian, G.; Kulasekaran, P.; Deivanayagam, P. Sulfonated polystyrene-block-poly(ethylene-ran-butylene)-block-polystyrene based membranes containing CuO@g-C_3N_4 embedded with 2,4,6-triphenylpyrylium tetrafluoroborate for fuel cell applications. *Soft Matter.* **2021**, *17*, 8387–8393. [CrossRef]
8. Ni, J.; Zhou, J.; Bing, J.; Guan, X. Recycling the cathode materials of spent Li-ion batteries in a H-Shaped neutral water electrolysis cell. *Sep. Purif. Technol.* **2021**, *278*, 119485. [CrossRef]
9. Asnawi, A.S.F.M.; Aziz, S.B.; Nofal, M.M.; Yusof, Y.M.; Brevik, I.; Hamsan, M.H.; Brza, M.A.; Abdulwahid, R.T.; Kadir, M.F.Z. Metal complex as a novel approach to enhance the amorphous phase and improve the EDLC performance of plasticized proton conducting chitosan-based polymer electrolyte. *Membranes* **2020**, *10*, 132. [CrossRef]
10. Li, Q.; Itoh, T.; Imanishi, N.; Hirano, A.; Takeda, Y.; Yamamoto, O. All solid lithium polymer batteries with a novel composite polymer electrolyte. *Solid State Ionics* **2003**, *159*, 97–109. [CrossRef]
11. Ngai, K.S.; Ramesh, S.; Ramesh, K.; Juan, J.C. A review of polymer electrolytes: Fundamental, approaches and applications. *Ionics* **2016**, *22*, 1259–1279. [CrossRef]
12. Zhou, D.; Shanmukaraj, D.; Tkacheva, A.; Armand, M.; Wang, G. Polymer Electrolytes for Lithium-Based Batteries: Advances and Prospects. *Chem* **2019**, *5*, 2326–2352. [CrossRef]
13. Aziz, S.B.; Woo, T.J.; Kadir, M.F.Z.; Ahmed, H.M. A conceptual review on polymer electrolytes and ion transport models. *J. Sci. Adv. Mater. Devices.* **2018**, *3*, 1–17. [CrossRef]
14. Hamsan, M.H.; Nofal, M.M.; Aziz, S.B.; Brza, M.A.; Dannoun, E.M.A.; Murad, A.R.; Kadir, M.F.Z.; Muzakir, S.K. Plasticized Polymer Blend Electrolyte Based on Chitosan for Energy Storage Application: Structural, Circuit Modeling, Morphological and Electrochemical Properties. *Polymers* **2021**, *13*, 8. [CrossRef]
15. Hosseini, S.; Lao-atiman, W.; Han, S.J.; Arpornwichanop, A.; Yonezawa, T.; Kheawhom, S. Discharge Performance of Zinc-Air Flow Batteries Under the Effects of Sodium Dodecyl Sulfate and Pluronic F-127. *Sci. Rep.* **2018**, *8*, 14909. [CrossRef]
16. Santos, F.; Urbina, A.; Abad, J.; López, R.; Toledo, C.; Fernández Romero, A.J. Environmental and economical assessment for a sustainable Zn/air battery. *Chemosphere* **2020**, *250*, 126273. [CrossRef]
17. Suren, S.; Kheawhom, S. Development of a High Energy Density Flexible Zinc-Air Battery. *J. Electrochem. Soc.* **2016**, *163*, A846–A850. [CrossRef]
18. Han, X.; Li, X.; White, J.; Zhong, C.; Deng, Y.; Hu, W.; Ma, T. Metal–Air Batteries: From Static to Flow System. *Adv. Energy Mater.* **2018**, *8*, 1801396. [CrossRef]
19. Santos, F.; Fernández Romero, A.J. Hydration as a solution to zinc batteries. *Nat. Sustain.* **2021**, *2021*, 1–2. [CrossRef]
20. Fu, J.; Cano, Z.P.; Park, M.G.; Yu, A.; Fowler, M.; Chen, Z. Electrically Rechargeable Zinc–Air Batteries: Progress, Challenges, and Perspectives. *Adv. Mater.* **2017**, *297*, 1604685. [CrossRef]
21. Amunátegui, B.; Ibáñez, A.; Sierra, M.; Pérez, M. Electrochemical energy storage for renewable energy integration: Zinc-air flow batteries. *J. Appl. Electrochem.* **2018**, *48*, 627–637. [CrossRef]
22. Abdullah, O.G.; Aziz, S.B.; Rasheed, M.A. Incorporation of NH_4NO_3 into MC-PVA blend-based polymer to prepare proton-conducting polymer electrolyte films. *Ionics* **2018**, *24*, 777–785. [CrossRef]
23. Saeed, M.A.M.; Abdullah, O.G. Effect of high ammonium salt concentration and temperature on the structure, morphology, and ionic conductivity of proton-conductor solid polymer electrolytes based pva. *Membranes* **2020**, *10*, 262. [CrossRef] [PubMed]
24. Tiong, T.S.; Buraidah, M.H.; Teo, L.P.; Arof, A.K. Conductivity studies of poly(ethylene oxide)(PEO)/poly(vinyl alcohol) (PVA) blend gel polymer electrolytes for dye-sensitized solar cells. *Ionics* **2016**, *22*, 2133–2142. [CrossRef]
25. Merle, G.; Wessling, M.; Nijmeijer, K. Anion exchange membranes for alkaline fuel cells: A review. *J. Memb. Sci.* **2011**, *377*, 1–35. [CrossRef]
26. Asua, J.M. *Polymer Reaction Engineering*; Blackwell Publishing Ltd.: Oxford, UK, 2008; ISBN 1405144424.
27. Lewandowski, A.; Skorupska, K.; Malinska, J. Novel poly(vinyl alcohol)-KOH-H O alkaline polymer electrolyte. *Solid State Ion.* **2000**, *133*, 265–271. [CrossRef]

28. Gupta, S.; Pramanik, A.K.; Kailath, A.; Mishra, T.; Guha, A.; Nayar, S.; Sinha, A. Composition dependent structural modulations in transparent poly(vinyl alcohol) hydrogels. *Colloids Surf. B Biointerfaces.* **2009**, *74*, 186–190. [CrossRef] [PubMed]
29. Díaz-Barrios, A.; González, G.; Reinoso, C.; Santiana, J.; Quiroz, F.; Chango, J.I.; Vera, C.C.; Caniglia, L.; Salazar, V.; Fernández-Delgado, M. In situ synthesis and long-term stabilization of nanosilver in poly(vinyl acetate-co-butyl acrylate-co-neodecanoate) matrix for antibacterial applications. *Mater. Chem. Phys.* **2020**, *255*, 123476. [CrossRef]
30. Ismail, A.S.; Darwish, M.S.A.; Ismail, E.A. Synthesis and characterization of hydrophilic chitosan-polyvinyl acetate blends and their sorption performance in binary methanol–water mixture. *Egypt. J. Pet.* **2017**, *26*, 17–22. [CrossRef]
31. Aziz, S.B.; Marf, A.S.; Dannoun, E.M.A.; Brza, M.A.; Abdullah, R.M. The study of the degree of crystallinity, electrical equivalent circuit, and dielectric properties of polyvinyl alcohol (PVA)-based biopolymer electrolytes. *Polymers* **2020**, *12*, 2184. [CrossRef] [PubMed]
32. Brza, M.A.; Aziz, S.B.; Anuar, H.; Dannoun, E.M.A.; Ali, F.; Abdulwahid, R.T.; Al-Zangana, S.; Kadir, M.F.Z. The study of EDLC device with high electrochemical performance fabricated from proton ion conducting PVA-based polymer composite electrolytes plasticized with glycerol. *Polymers* **2020**, *12*, 1896. [CrossRef]
33. Tretinnikov, O.N.; Zagorskaya, S.A. Determination of the degree of crystallinity of poly(Vinyl alcohol) by ftir spectroscopy. *J. Appl. Spectrosc.* **2012**, *79*, 521–526. [CrossRef]
34. Qiao, J.; Fu, J.; Lin, R.; Ma, J.; Liu, J. Alkaline solid polymer electrolyte membranes based on structurally modified PVA/PVP with improved alkali stability. *Polymer* **2010**, *51*, 4850–4859. [CrossRef]
35. Seki, T.; Chiang, K.Y.; Yu, C.C.; Yu, X.; Okuno, M.; Hunger, J.; Nagata, Y.; Bonn, M. The bending mode of water: A powerful probe for hydrogen bond structure of aqueous systems. *J. Phys. Chem. Lett.* **2020**, *11*, 8459–8469. [CrossRef] [PubMed]
36. Pak, J.; Han, S.J.; Wee, J.H. Precipitation of potassium-based carbonates for carbon dioxide fixation via the carbonation and re-carbonation of KOH dissolved aqueous ethanol solutions. *Chem. Eng. J.* **2022**, *427*, 131669. [CrossRef]
37. Chen, P.; Zhang, K.; Tang, D.; Liu, W.; Meng, F.; Huang, Q.; Liu, J. Recent Progress in Electrolytes for Zn–Air Batteries. *Front. Chem.* **2020**, *8*, 1–7. [CrossRef] [PubMed]
38. Zhao, Z.; Fan, X.; Ding, J.; Hu, W.; Zhong, C.; Lu, J. Challenges in Zinc Electrodes for Alkaline Zinc-Air Batteries: Obstacles to Commercialization. *ACS Energy Lett.* **2019**, *4*, 2259–2270. [CrossRef]
39. Jackson, P.; Robinson, K.; Puxty, G.; Attalla, M. In situ Fourier Transform-Infrared (FT-IR) analysis of carbon dioxide absorption and desorption in amine solutions. *Energy Procedia* **2009**, *1*, 985–994. [CrossRef]
40. Gilman, J.W.; VanderHart, D.L.; Kashiwagi, T. *Thermal Decomposition Chemistry of Poly(Vinyl Alcohol)*; Gordon, L.N., Ed.; Florida Institute of Technology: Melbourne, VIC, Australia, 1995; Volume 599, pp. 161–185.
41. Restrepo, I.; Medina, C.; Meruane, V.; Akbari-Fakhrabadi, A.; Flores, P.; Rodríguez-Llamazares, S. The effect of molecular weight and hydrolysis degree of poly(vinyl alcohol)(PVA) on the thermal and mechanical properties of poly(lactic acid)/PVA blends. *Polímeros* **2018**, *28*, 169–177. [CrossRef]
42. Duquesne, S.; Lefebvre, J.; Delobel, R.; Camino, G.; LeBras, M.; Seeley, G. Vinyl acetate/butyl acrylate copolymers—Part 1: Mechanism of degradation. *Polym. Degrad. Stab.* **2004**, *83*, 19–28. [CrossRef]
43. Ludwig, B.; Harald, C.; Manfred. F.; Willi, K.; Arnold, S. *Comprehensive Polymer Science: The Synthesis, Characteri-Zation, Reactions and Applications of Polymers*, 7th ed.; Pergamon Press: Oxford, UK, 1989; ISBN 0-08-03251 6-5.
44. Zelkó, R.; Szakonyi, G. The effect of water on the solid state characteristics of pharmaceutical excipients: Molecular mechanisms, measurement techniques, and quality aspects of final dosage form. *Int. J. Pharm. Investig.* **2012**, *2*, 18. [CrossRef]
45. Strydom, C.A.; Collins, A.C.; Bunt, J.R. The influence of various potassium compound additions on the plasticity of a high-swelling South African coal under pyrolyzing conditions. *J. Anal. Appl. Pyrolysis* **2015**, *112*, 221–229. [CrossRef]
46. Hatta, F.F.; Yahya, M.Z.A.; Ali, A.M.M.; Subban, R.H.Y.; Harun, M.K.; Mohamad, A.A. Electrical Conductivity Studies on PVA/PVP-KOH Alkaline Solid Polymer Blend Electrolyte. *Ionics* **2005**, *11*, 418–422. [CrossRef]
47. Fan, L.; Wang, M.; Zhang, Z.; Qin, G.; Hu, X.; Chen, Q. Preparation and characterization of PVA alkaline solid polymer electrolyte with addition of bamboo charcoal. *Materials* **2018**, *11*, 679. [CrossRef] [PubMed]
48. Zhang, J.; Han, H.; Wu, S.; Xu, S.; Yang, Y.; Zhou, C.; Zhao, X. Conductive carbon nanoparticles hybrid PEO/P(VDF-HFP)/SiO_2 nanocomposite polymer electrolyte type dye sensitized solar cells. *Solid State Ionics* **2007**, *178*, 1595–1601. [CrossRef]
49. Cai, M.; Park, S. Spectroelectrochemical Studies on Dissolution and Passivation of Zinc Electrodes in Alkaline Solutions. *J. Electrochem. Soc.* **1996**, *143*, 2125–2131. [CrossRef]
50. Santiana, A.; Jessica, M. Desarrollo de una Emulsión Polimérica Vinil-Acrílica Con Características Hidrofóbicas. Bachelor's Thesis, Escuela Politecnica Nacional, Quito, Ecuador, 2017. Available online: http://bibdigital.epn.edu.ec/handle/15000/17319 (accessed on 17 May 2017).
51. Ramesh, S.; Upender, G.; Raju, K.C.J.; Padmaja, G.; Reddy, S.M.; Reddy, C.V. Effect of Ca on the Properties of Gd-Doped Ceria for IT-SOFC. *J. Mod. Phys.* **2013**, *04*, 859–863. [CrossRef]

gels

Review

Research Progress in the Multilayer Hydrogels

Lu Jin, Jia Xu, Youcai Xue, Xinjiang Zhang, Mengna Feng, Chengshuang Wang, Wei Yao, Jinshan Wang and Meng He *

School of Materials Science and Engineering, Yancheng Institute of Technology, Yancheng 224051, China; jinlu525@126.com (L.J.); xujia202109@163.com (J.X.); xyc1246@163.com (Y.X.); zhangxinjiang1983@163.com (X.Z.); fengmengnafwch@126.com (M.F.); wangcs@ycit.cn (C.W.); xiaoniu1981@126.com (W.Y.); wangjinshan@ycit.cn (J.W.)

* Correspondence: hemeng315@163.com

Abstract: Hydrogels have been widely used in many fields including biomedicine and water treatment. Significant achievements have been made in these fields due to the extraordinary properties of hydrogels, such as facile processability and tissue similarity. However, based on the in-depth study of the microstructures of hydrogels, as a result of the enhancement of biomedical requirements in drug delivery, cell encapsulation, cartilage regeneration, and other aspects, it is challenge for conventional homogeneous hydrogels to simultaneously meet different needs. Fortunately, heterogeneous multilayer hydrogels have emerged and become an important branch of hydrogels research. In this review, their main preparation processes and mechanisms as well as their composites from different resources and methods, are introduced. Moreover, the more recent achievements and potential applications are also highlighted, and their future development prospects are clarified and briefly discussed.

Keywords: multilayer hydrogels; composites; fabrication process; mechanisms; application

Citation: Jin, L.; Xu, J.; Xue, Y.; Zhang, X.; Feng, M.; Wang, C.; Yao, W.; Wang, J.; He, M. Research Progress in the Multilayer Hydrogels. *Gels* **2021**, *7*, 172. https://doi.org/10.3390/gels7040172

Academic Editor: Hiroyuki Takeno

Received: 27 September 2021
Accepted: 13 October 2021
Published: 17 October 2021

1. Introduction

Hydrogel is a kind of hydrophilic material with a three-dimensional crosslinked network structure, which is infiltrated with water [1]. It can absorb water quickly and retain water for a certain period without dissolving in water. The properties of hydrogels are similar to those of biological tissues and their excellent biocompatibility makes them extremely suitable for biomedical research [2–4]. According to different formation mechanisms and molecular structures, hydrogels can be generally divided into chemically crosslinked hydrogels and physically crosslinked hydrogels [5]. Chemically crosslinked hydrogels are formed through crosslinking with chemical bonds, which is an irreversible permanent crosslinking, and show extraordinary chemical stability with good properties such as good solvent resistance. Physically crosslinked hydrogels are noncovalently crosslinked and usually exhibit excellent properties such as reversibility, repairability, and high responsiveness to external stimuli, which can be modified further to endow them with other properties as "intelligent" materials [6,7]. It has been noted that hydrogels are widely used in different fields of drug delivery, tissue engineering, medical implants, wound dressings, and various mechanical and electronic devices due to their extreme mechanical properties, such as high degrees of toughness, robustness, elasticity, stickiness, and fatigue resistance [8–14].

The research and design of polymer-based hydrogels are mainly based on the overall consideration of their properties, and the resultant hydrogels are usually homogeneous materials [15]. However, it is challenging for these homogeneous hydrogels to simultaneously meet the needs of further microstructure control and different applications, including precise drug delivery and release of different drugs, bone repair and regeneration, and carriers of different cells. As a kind of heterogeneous hydrogel, multilayer hydrogels have emerged and become an important and novel branch of hydrogels. Multilayer hydrogels exhibit many excellent properties such as high ductility, unique complex internal structure, and excellent response to stimuli [16–18]. Moreover, multilayer hydrogels exhibit a variety of different shapes, such as spherical, cylindrical, spindle-like, and multilayer tubular, to satisfy different applications [19–22]. The internal structure of multilayer hydrogels is complex, which could exhibit cavities between each layer (namely inter-layer space) for the storage of drugs, microorganisms, and cells. Different layers can be prepared from different substances or methods to independently exhibit different physical and chemical properties [23].

In this review, we summarize the recent progress of the multilayer hydrogels and their composites. Particularly, we review the preparation processes and mechanisms of multilayer hydrogels and their innovative applications in different fields, especially in the biomedical field, including drug delivery, cell carrier or encapsulation, wound dressings (coatings), and bone repair. Finally, we provide a brief perspective on the future development of multilayer hydrogels, hoping to provide some theoretical guidance on broadening hydrogels' application.

2. Preparation Methods and Mechanisms of the Multilayer Hydrogels

Many different preparation methods for multilayer hydrogels have been developed in recent years according to different raw materials, crosslinking structures (chemical or physical), and potential applications. These methods can be generally divided into two pathways according to the formation direction of hydrogel layers: from the inside to the outside and from the outside to the inside.

More recently, the preparation of multilayer hydrogels from inside to outside has been widely studied, and its main preparation technology is as follows: First, suitable materials are selected to prepare inner gel cores (generally agarose) with special shapes according to application requirements. Then, a first gel layer is formed on the gel core by different crosslinking or coagulation methods through pretreating the gel core with subsequent soaking in certain solutions. Finally, the fabrication process is repeated several times to generate multilayer hydrogels with the desired number of layers. The volume of multilayer hydrogels continues to grow until the preparation process of the multilayer structure is completed by gradually wrapping or covering the gel cores with gel layers.

Compared with the pathway from inside to outside, the preparation of multilayer hydrogels from outside to inside has been studied much less, although this form appeared earlier. The main preparation process is as follows: First, suitable raw materials are selected to prepare the hydrogel shells, which can wrap the solutions corresponding to the following gel layers. Second, a first gel layer is formed on the inner wall of the shell by different coagulation or crosslinking methods. Finally, the above fabrication process is repeated several times to generate multilayer hydrogels with the desired number of layers. The total volume of multilayer hydrogels prepared from outside to inside cannot increase after the formation of the shells. All the gelation processes occur inside the shells and move from the shell to the center of the gel with a gradual layer-by-layer (LBL) formation.

Detailed preparation methods are commonly used for the above two pathways, which can be roughly divided into the lLBL method and non-LBL method, as follows.

2.1. LBL Assembling Methods for Multilayer Hydrogels

At present, LBL assembly is the most widely used method for the preparation of multilayer hydrogels from different polymers such as polysaccharides. LBL assembly can be used to prepare multilayer structures by different driving forces or crosslinking types. Different LBL assembly methods are suitable for the preparation of multilayer hydrogels with different performances and application requirements. In the following discussion, several commonly used LBL methods are analyzed from the aspects of preparation principle, advantages, and application prospects.

2.1.1. LBL by Chemical Crosslinking

Chemical crosslinking is a common method of preparing homogenous hydrogels via the use of crosslinking agents. Interestingly, multilayer hydrogels with arbitrary shapes, including onion-like, tubular, and star-like, can be readily prepared using the chemical crosslinking method (Figure 1a–d) through the LBL process. As an example, the preparation process of a chitosan multilayer hydrogel from Xiong's group is roughly shown in Figure 1e [24]. First, suitable raw materials (e.g., agarose) were selected to prepare a gel core, which was immersed in the crosslinker solutions (such as glutaraldehyde, terephthalaldehyde, and epoxy chloropropane) for an appropriate time to load the crosslinker. Second, the agarose gel core loaded with the crosslinker was soaked in the chitosan solution for a required time to crosslink the neighboring chitosan chains for the formation of the first chitosan gel layer. Third, the obtained gel core with the first chitosan gel layer was immersed in the crosslinker solution again for the desired time to ripen the chitosan gel layer and load the crosslinker for the formation of the second chitosan gel layer, followed by soaking in the chitosan solution. Onion-like CS multilayer (multi-membrane) hydrogels with the desired layers can be prepared by repeating the above-mentioned process. The chitosan layers were covalently crosslinked with chemical crosslinkers to endow them with good solvent resistance and pH sensitivity.

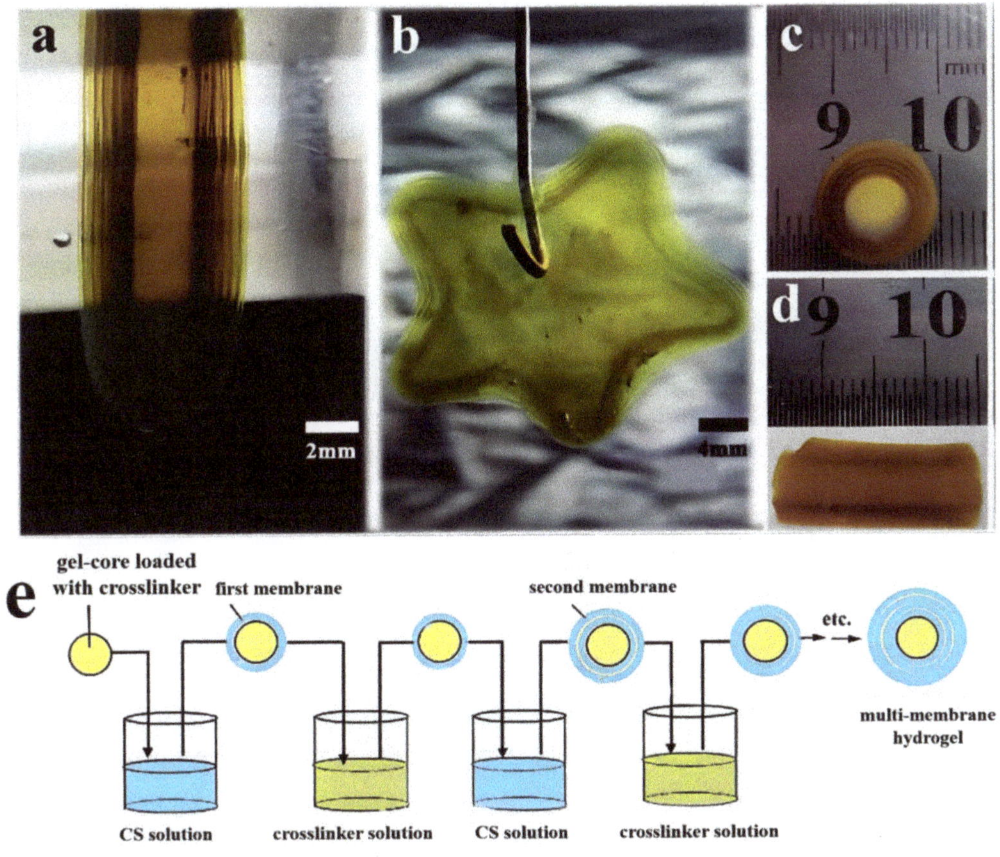

Figure 1. Chitosan multilayer hydrogels with various shapes: column (**a**), star (**b**), and tubular (**c**,**d**), and scheme of the preparation process of chitosan multilayer (multi-membrane) hydrogels by the LBL assembly method (**e**). (Reproduced with permission from [24]. Royal Society of Chemistry, 2013).

For chemically crosslinked multilayer hydrogels by the LBL process, reasonable and rapid crosslinking is essential for successful preparation. The formation and growth of each gel layer are related to the diffusion of the crosslinker. The inter-layer space canbe adjusted by changing the crosslinking degree of gel layers. Moreover, the chemically crosslinked chitosan multilayer hydrogels have a unique sub-layer structure [24]. The chitosan multilayer hydrogels have pH sensitivity and can disintegrate layer by layer, thus showing promise for applications in different fields, including drug delivery and tissue engineering, due to their unique structure.

2.1.2. LBL by Ion Crosslinking

In addition to common crosslinking agents, such as organic dibasic acids and polyols, metal ions can also be used as special crosslinking agents to promote the gelation of polymer solutions. Moreover, the introduction of different metal ions has different effects on the hydrogel structure [25–27]. For example, the preparation process and mechanism of an alginate multilayer hydrogel from Xu's group [25] is roughly shown in Figure 2. First, an egg-box structure gel core of sodium alginate crosslinked by Ca^{2+} was prepared by diffusing Ca^{2+} into the sodium alginate solution to crosslink alginate molecule chains. Second, the above gel core was immersed in the sodium alginate solution for a given time to prepare a hydrogel layer, which was further cured in the Ca^{2+} solution. Finally, alginate-based multilayer hydrogels with the desired layers could be prepared by repeating the above-mentioned process. In addition, carboxymethyl cellulose and other polyanions can be crosslinked by Ca^{2+}, Al^{3+}, and other polyvalent inorganic cations to form hydrogels [27]. This LBL assembly method can also be used to prepare carboxymethyl cellulose multilayer hydrogels using $AlCl_3$ aqueous solution as a crosslinking agent, showing good versatility [24].

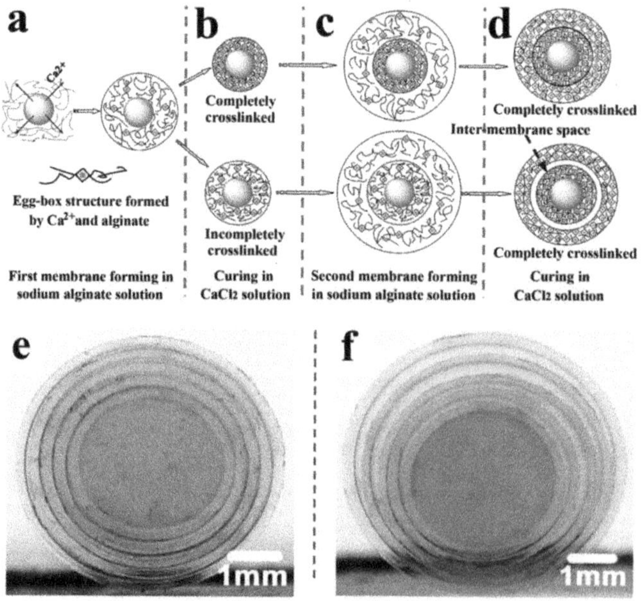

Figure 2. Schematic illustration of the preparation of alginate multilayer hydrogels. The partly crosslinked layer (membrane) formed with the egg-box structure in the sodium alginate solution (**a**). The completely or incompletely crosslinked hydrogels (**b**). The second layer formed at the periphery of either completely or incompletely crosslinked hydrogel (**c**). The finally obtained double-membrane hydrogel with or without inter-membrane space (**d**). Photographs of the multi-membrane alginate hydrogels without (**e**) and with (**f**) inter-layer space. (Reproduced with permission from [24]. Royal Society of Chemistry, 2009).

For multilayer hydrogels from the ion crosslinking method, complete or incomplete crosslinking is essential for the successful preparation of inter-layer spaces. Complete or incomplete membrane crosslinking can be readily controlled by adjusting the crosslinking time (Figure 2). A hydrogel with an inter-membrane space can be obtained by fully crosslinking incompletely crosslinked alginate hydrogel layers in $CaCl_2$ solution (Figure 2d). Every ion-crosslinked layer is independent of other layers in the hydrogels. Therefore, these multilayer hydrogels produced using the ion crosslinking method are expected to be used in investigating the co-culture of multiple cells, drug delivery, and tissue engineering due to their unique structure.

2.1.3. LBL by Electrostatic Interaction

In addition to the introduction of crosslinking agents to prepare multilayer hydrogels, polyelectrolyte-based multilayer hydrogels can also be facilely prepared through simple electrostatic interactions [28,29]. Figure 3 shows the scheme for the formation of a multilayer hydrogel by electrostatic interaction. Positively and negatively charged polyelectrolytes can form hydrogel layers alternately on the substrate through electrostatic interaction. The key to electrostatic interaction is the mutual adsorption and surface charge reversal of positive and negative polyelectrolyte-based hydrogels. The concentration, pH, and temperature of polyelectrolytes are the most important factors affecting the formation and stability of multilayer hydrogels [28]. As the most commonly used LBL deposition strategy, electrostatic interaction has been widely studied and applied. Because electrostatic LBL assembly can be carried out in aqueous solutions, it is convenient to prepare LBL multilayers automatically using a LBL deposition machine [29]. These multilayer hydrogels produced using the electrostatic interaction method are widely used in different fields, including surface modification.

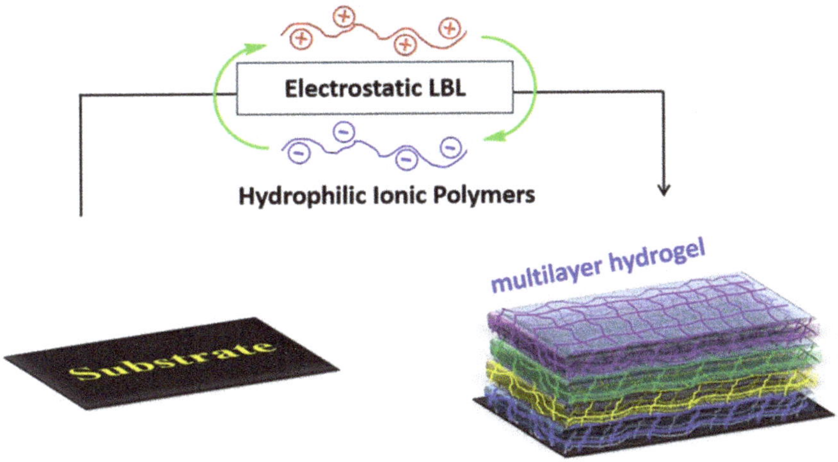

Figure 3. The scheme for formation of multilayer hydrogel from electrostatic LBL. (Reproduced with permission from [28]. John Wiley & Sons, 2020).

2.1.4. LBL through Acid-Base Neutralization

Neutralization is an effective pathway to fabricate hydrogels, especially for acid-dissolved chitosan, by converting NH_3^+ in low pH solution to NH_2 with the addition of a base solution. This can weaken the ionic repulsions between chitosan chains, resulting in physical cross-links through hydrogen bonding, hydrophobic interactions, and crystallite formation [30]. As early as in 2008, Alain Domard's group reported a chitosan multilayer hydrogel using the interrupted neutralization process (Figure 4a). A chitosan physical alcohol gel was prepared by adding 1,2-propanediol aqueous solution to chitosan/HCl solution and the subsequent evaporation process. Then, NaOH aqueous solution was used to neutralize acid in the chitosan alcohol gel and form the first chitosan layer and inter-layer space for a given time. Finally, onion-like chitosan multilayer hydrogels with the desired layers (Figure 4b) were prepared by repeating the above-mentioned process (namely LBL).

In contrast to the above chitosan multilayer hydrogel, Shi's group fabricated alginate/chitosan composite multilayer hydrogels via the interrupted neutralization of the as-prepared fluid-filled capsules with a polyelectrolyte shell layer [31]. First, a single drop of a chitosan solution was added to an alginate solution, which was further incubated to form a fluid-filled capsule with a chitosan/alginate layer through electrostatic attraction effect. Then, the capsule was neutralized with alkaline solution for some time to form a chitosan layer through the gelation of chitosan solution in the capsule. The corresponding multilayer hydrogels could be fabricated layer by layer through the repeating of the intermittent neutralization.

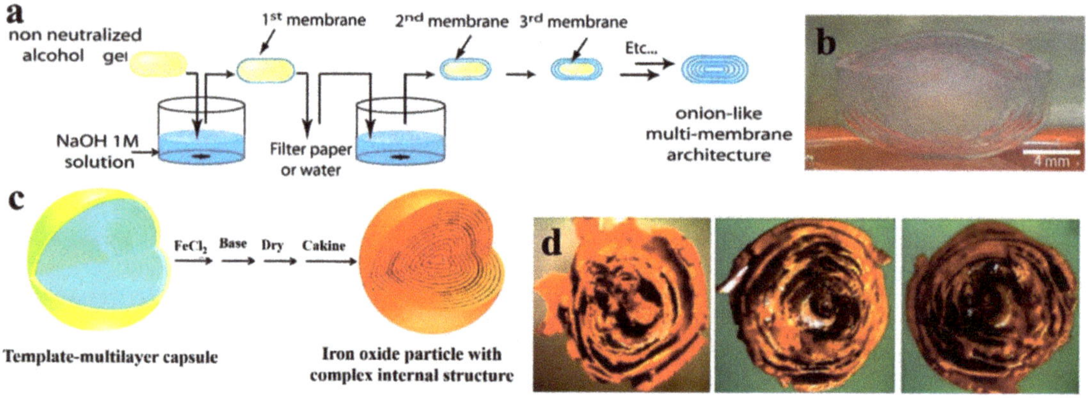

Figure 4. Process diagram of the preparation of chitosan multilayer hydrogel by LBL process through neutralization (**a**) and the photograph of the corresponding chitosan multilayer hydrogel (**b**). Schematic illustrating the iron oxide templating procedure (**c**) and photographs revealing the internal structure of iron oxide particles generated from the multilayer template) (**d**). (Reproduced with permission from [30,31]. Nature publishing group, 2008 and American Chemical Society, 2014).

From the microscopic perspective, the semi-permeable polyelectrolyte complex shell layer from chitosan and alginate can retain chitosan solution inside the capsule, and the chitosan hydrogel layers cannot block the movement of OH^- in the alkaline aqueous solution to the inner chitosan solution. The OH^- group can convert NH_3^+ in the chitosan solution into NH_2, resulting in the further formation of the chitosan hydrogel layer [32]. The quantity, thickness, and microstructure of multilayer hydrogels can be controlled by the concentration of alkali and contact time [30–32]. Interestingly, the prepared multilayer hydrogels can be used as templates to create hard particles with a complex internal structure, such as iron oxide particles (Figure 4c,d), and composite multilayer hydrogels consisting of organic-inorganic substances can be formed accordingly. Moreover, multilayer hydrogels prepared using the neutralization method have the advantages of uniform drug bearing, controllable inter-layer space, and good biocompatibility, showing potential application in the fields of cell culture and drug delivery.

As shown above, the general method to manufacture multilayer hydrogels from outside to inside was to utilize an interrupted chain condensation and contraction of an as-prepared hydrogel to form gel layers (namely LBL). However, this method is usually time-consuming and cannot readily load drugs [33]. To solve the above problems, the acid-base neutralization interaction can also be generalized to form hydrogel layers from the inside to the outside using gel-cores and other raw materials. As an example, the preparation process of a cellulose multilayer hydrogel from our previous work [34] is roughly shown in Figure 5. First, the as-prepared agarose gel rod was immersed into an acetic acid solution to load acetic acid as a coagulant. Subsequently, the gel rod loaded with acetic acid was immersed in a NaOH/urea dissolved cellulose solution to prepare the first cellulose layer. The gel rod with the first layer was immersed again in the acetic acid solution, so the new gel rod loaded with acetic acid again could be used to form the second cellulose layer. Finally, the cellulose multilayer hydrogels were fabricated by repeating this process (LBL).

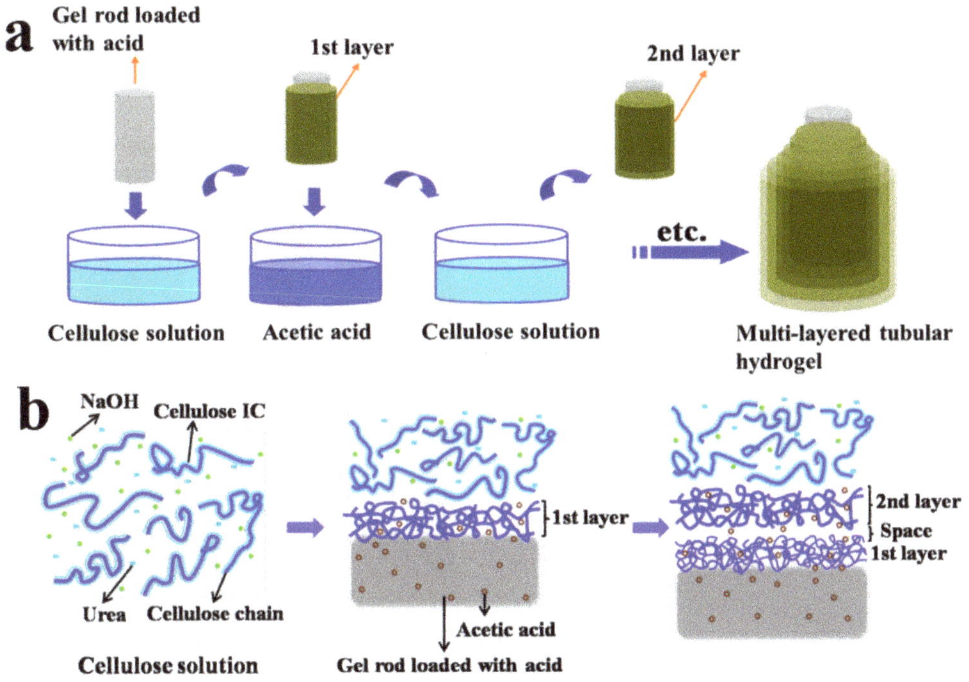

Figure 5. Preparation process of a cellulose multilayer hydrogel by a multi-step interrupted gelation (**a**) and the corresponding schematic model to describe the formation process (**b**). (Reproduced with permission from [34]. American Chemical Society, 2014).

Interestingly, a water-soluble inclusion complex (IC) associated with cellulose, NaOH, urea, and water occurs in the NaOH/urea solvent system at low temperature, which leads to cellulose dissolution [35]. When the cellulose IC was destroyed with acetic acid through the contact of acetic acid in the gel core and the cellulose solution, the strong inter-chain interactions of the exposed cellulose chains led to the rapid self-aggregation of cellulose and the formation of the first cellulose layer along the gel core (namely gelation or coagulation). Subsequently, the first cellulose layer was cured by re-immersion in acetic acid solution to load acetic acid, which could be used for the regeneration of the next cellulose hydrogel layer. Moreover, the inter-layer space was formed with the progress of the curing process. The fabrication process is facile and rapid, and the thickness and inter-layer spacing of the hydrogel can be controlled by adjusting the cellulose concentration, the diameter of the gel core, and the contact time. Multilayer cellulose hydrogels showed high compressive strength due to the dense packing of cellulose chains. The multilayer hydrogels prepared by the LBL process through acid-base neutralization have the advantages of stable gel structure, controllable shape, size, and thickness, and good biocompatibility, which are expected to be applied in cell culture and tissue engineering scaffolds [34].

Moreover, the electrochemical method can also be used to fabricate multilayer hydrogel through broadly defined acid-base neutralization interaction. Briefly, electrochemical synthesis is based on the use of electrochemical workstations to generate multilayer hydrogels through programming input electrical signals. The chitosan-based multilayer hydrogel from Shi's group is used as an example [36] and its preparation process is shown in Figure 6a. First, a chitosan solution was prepared with the pH of 5 [37]. Secondly, a stainless-steel wire was adopted as the working electrode to immerse in the above chitosan solution, and a platinum wire was adopted as the counter electrode to connect the electrochemical workstation. Finally, chitosan multilayer hydrogels with different layers and thicknesses could be fabricated layer by layer using a pulsed electrical signal under the On–Off model.

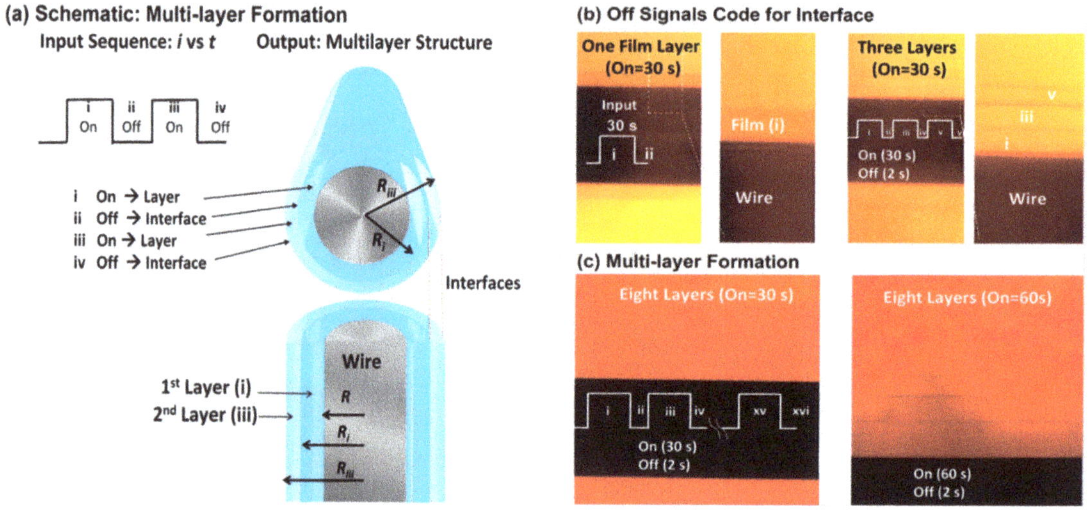

Figure 6. Chitosan multilayers generated by input sequences of "on-steps" (0.5 mA) and "off-steps" (0 mA). Schematic illustrating how electrical input controls the output multilayer structure (**a**). Off-steps (interruptions) code for interfaces (**b**). The left images of b show one layer and its enlarged photo on the wire by biasing a 30 s electrical input; the right images of b show a three-layered gel on a wire by three successive on–off sequences. The duration of on-steps controls the layer thickness (**c**). Images show eight layers with different thicknesses controlled by 30 s and 60 s on-steps, respectively. (Reproduced with permission from [36]. Royal Society of Chemistry, 2013).

This method enabled the assembly of the chitosan hydrogel in the cathode by a neutralization mechanism through the input of electrical signals [38,39]. Electrolysis can control the local pH [40], and the generation of OH^- at the cathode is believed to neutralize acidic chitosan solution and induce its localized sol–gel transition [41]. Physical crosslinks of the deposited chitosan hydrogels occurred in the crystalline regions [42]. Obviously, each interruption (off-step) generates an interface and a multilayer structure can be generated by an input sequence with multiple interrupts (Figure 6b). The duration of the on-step can control the thickness of the individual layers (Figure 6c). The above images demonstrate that the controllable multilayer hydrogels can be created using electronic input signals. Moreover, the simple inputs of the electrical signal do not change the solution compositions, making it more convenient to control and adjust [43]. Thus, this work provides an initial proof of principle that electronic codes and can be used to guide the assembly and control the hydrogel structure. Multilayer hydrogels prepared by this method provided new possibilities for tissue regeneration, multifunctional coating, and controlled drug delivery.

2.1.5. LBL by Compound Methods

It is well known that the blood vessel is a tri-layered substance with different components for each layer [44]. Layers from different raw materials are expected to have different properties for varied requirements; thus, mimicking blood-vessel like multilayer hydrogels is important and attractive. However, the fabrication of multilayer hydrogels with different layer components is difficult using a single method, as mentioned above, and the combination of several methods can be used during the LBL process to solve this problem. As an example, in one of our previous studies, alternate layered chitosan/alginate composite hydrogels (CACH) were fabricated successfully using the LBL process with the combination of acid-base neutralization for the formation of the chitosan layer and ion crosslinking for the alginate layer (Figure 7a) [45]. The CACH was constructed by repeating the alternate formation of chitosan and alginate gel layers. All the tubular CACH exhibited good appearance and controllable layers (Figure 7b–d). The layer thickness increased with the increase in chitosan or alginate concentrations and soaking time. Moreover, the CACH exhibited good architectural stability and biocompatibility towards endothelial cells, thus showing significant potential as a cell culture carrier and a matrix for the controlled release of molecules.

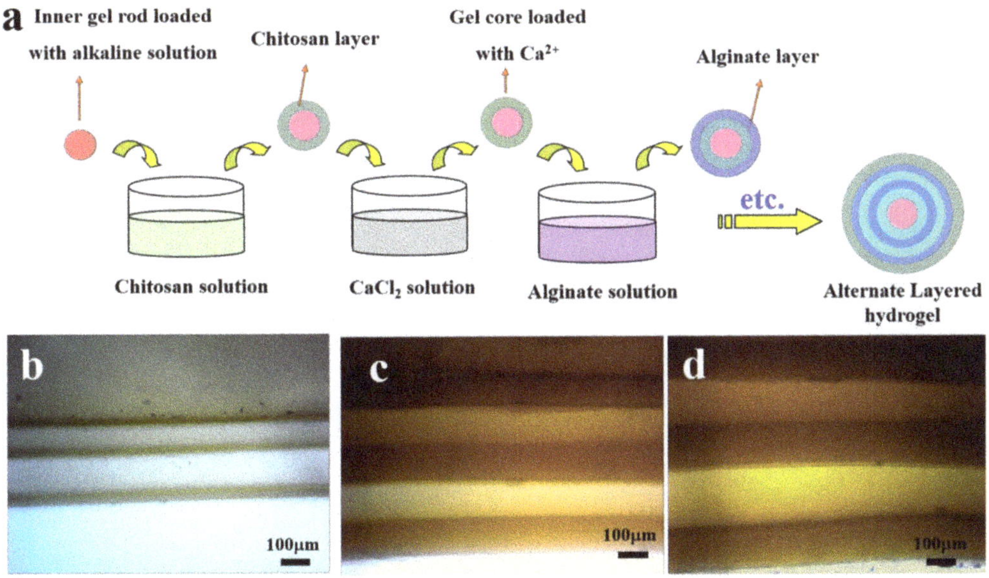

Figure 7. The preparation process of spherical alternate layered chitosan/alginate composite hydrogels (CACH) through acid-base neutralization and ion crosslinking (**a**), and photographs of the tubular CACH fabricated from different chitosan/alginate concentrations and soaking time (**b**–**d**). (Reproduced with permission from [45]. Elsevier, 2017).

2.2. Non-LBL Methods for Multilayer Hydrogels

As mentioned above, LBL has been predominantly used to prepare multilayer hydrogels in recent years. Non-LBL methods (traditionally) have also been developed to prepare multilayer hydrogels. Two non-LBL assembling methods are described in the following.

2.2.1. New Concept Welding

New concept welding is a method used to prepare anisotropic multilayer hydrogels by an ion-induced interfacial reconfiguration. Taking anisotropic cellulose multilayer hydrogels by Jeon's group as an example [46], the design principle is shown in Figure 8a. First, cellulose was dissolved in a lithium chloride/N,N'-dimethylacetamide (LiCl/DMAc) mixed solution to obtain a cellulose solution, which was then cast on a mold to form an organogel layer through intermolecular hydrogen bonds (H-bonds). The resulting cellulose organogel was transformed into an isotropic cellulose hydrogel by immersion in water. The anisotropic cellulose hydrogel film was formed using axial force on the above isotropic hydrogel sheet, where the highly aligned polymer chains were fixed by H-bonds. Ion-induced welding by adding a LiCl/DMAc mixture was then used between the adjacent hydrogel layers through the realization of the intermolecular H-bond exchange at the interface. Finally, anisotropic cellulose multilayer hydrogels with different morphologies were prepared accordingly. Four different forms of anisotropic multilayer hydrogels (parallel laminated (PL), orthogonally laminated (OL), axially rolled (AR), and concentrically rolled (CR) multilayered hydrogels) were prepared through the hierarchical programming of cellulose chain orientation in hydrogels (Figure 8b,c), indicating the versatility of new concept welding.

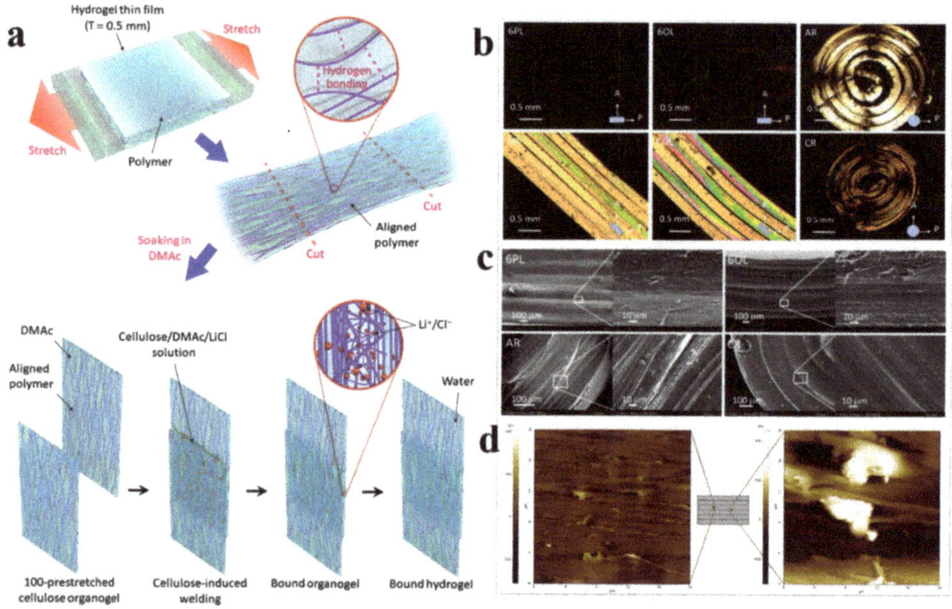

Figure 8. Design of weldable anisotropic cellulose multilayer hydrogels (**a**), POM images (taken in cross-polarized mode; A: analyzer; P: polarizer) (**b**) and SEM images (**c**) of the cross-section of 6PL, 6OL, AR, and CR multilayer hydrogels, and AFM images of the cross-section of 6PL hydrogel (**d**). (Reproduced with permission from [46]. Royal Society of Chemistry, 2019).

In this method, a thin layer of cellulose/LiCl/DMAc solution was trapped at the interface of the two thin layers, and the interfacial LiCl gradually diffused into the hydrogel layer over time due to the concentration gradient, and reassembled the cellulose interface through the exchange of H-bonds. Highly aligned microfibers appeared in the bulk of a layer from 6PL gel, and the fibers in the interfacial regions of two layers were randomly distributed (Figure 8d). The hydrogel layers were completely integrated by an isotropic interfacial region with the possible occurrence of a full reconfiguration of the polymer chains. This facile method achieved the interface reconfiguration of hydrogels through ion-induced welding without an adverse effect on the highly aligned polymer orientation and the common adoption of covalent crosslinking, shedding light on the design of novel hydrogels used in the engineering and biomedical fields.

2.2.2. Metal Ions Modulation

Many polymers such as chitosan (CS) have functional groups including -NH_2, which can coordinate with numerous metal ions through the chelation effect. Metal ions modulation can be used to prepare multilayer hydrogels through strong chelation interaction. A copper-chitosan composite multilayer hydrogel using Cu^{2+} modulation was fabricated by Wang's group (Figure 9a) [47]. In this work, a Cu^{2+}-CS solution was prepared first by adding $CuCl_2$ powder to pure CS solution. Then, the above solution was filled in a single opening mold and immersed in an alkaline coagulation bath to complete the gelation process (Figure 9a). Finally, the resultant copper-CS multilayer hydrogel was formed and unloaded from the mold, which was repeatedly washed with deionized water to be neutral.

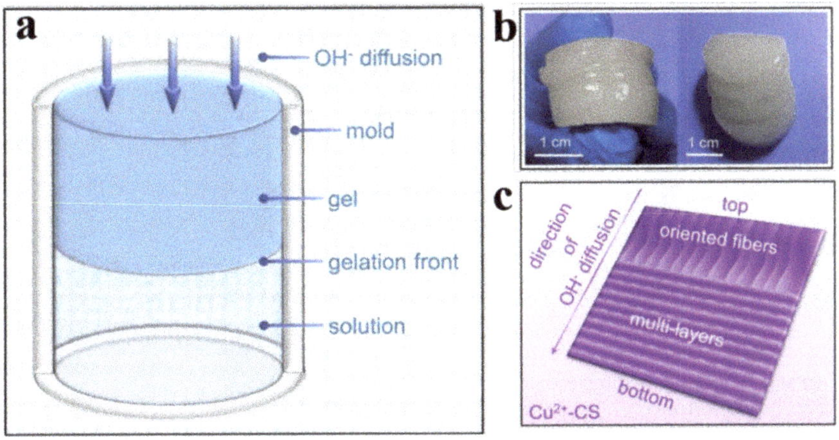

Figure 9. Schematic illustration of the formation of CS hydrogel (**a**), photograph of a CS multilayer hydrogel (**b**), and schematic illustration of the typical morphology of the copper-CS multilayer hydrogel (**c**). (Reproduced with permission from [47]. Nature publishing group, 2016).

Pure CS could form a multilayer hydrogel through the addition of OH^- to the CS solution (Figure 9a,b) as mentioned above [48]. By comparison, the mechanism for copper-CS multilayer hydrogel formation with the structural transition differs and is summarized in the following. CS chain entanglement existed on the gel-sol interface (Figure 9c). $Cu^{2+}$and CS can form a strong complex due to their strong affinity, resulting in the increased tendency of the volume of polymer zones to shrink. The gelation rate depends on the proximity of the gelation front to the system–coagulation interface. The introduction of Cu^{2+} ions increases the volume shrinkage of the CS bands, which causes a contraction at the gel–sol interface and enhances the disentanglement of macromolecules, resulting in the formation of a "clear space". Two layers can be created with the gelation process by further diffusion of OH^- (Figure 9c). Thus, the copper-CS multilayer hydrogel can be fabricated accordingly, which had potential value in applications including copper-based fungicides, redox catalysts, and urea uptake.

3. Achievements and Practical Applications of Multilayer Hydrogels

Multilayer hydrogels with internal cavities and a complex internal structure have been widely applied in biomedical fields, including drug delivery, cell carrier or encapsulation, bacteria delivery, wound dressings (coatings), and bone repair. The achievements and practical applications of multilayer hydrogels have mainly focused on the biomedical field. Moreover, multilayer hydrogels have also been used in other fields, such as tuning ice nucleation and propagation, dye adsorptions, and forward-osmosis (FO) desalination, due to their unique structure, modifiable properties, and higher surface area.

3.1. Drug and Bioactive Substances Delivery

Natural polymer-based multilayer hydrogels usually have excellent biocompatibility, and multilayer hydrogels in the forms of microspheres or capsules can effectively encapsulate and release drugs and bioactive substances. Inspired by biologic lipid bilayers [49], the development of multilayer hydrogels further deepened the study of drug delivery through the homogenization of the drug release in different layers, and by restricting the migration and diffusion of different drugs. The outer layer without drugs can effectively isolate the external environment and prevent the drug precipitating from the hydrogel surface, or the burst release of the drugs, which can prolong the drug release [50–52]. After arriving at the targeted sites, multilayer hydrogels can be induced to degrade by different stimuli-response pathways to achieve the precise drug release. Table 1 shows several kinds of multilayer hydrogels from different resources for the precise delivery and release of drugs by different pathways.

Table 1. Precise delivery and release of drugs by different multilayer hydrogels.

Drug Carriers	Drug Species	Release Pathways	Advantages	Reference
Poly(methacrylic acid)/poly(N-vinylpyrrolidone) multilayer hydrogel capsules	Nucleic acids	Ultrasound-triggered release	Higher effective loading capacity and controlled delivery of sensitive biomolecules	[53]
Poly(N-vinylcaprolactam) multilayer hydrogels	Sodium diclofenac	Increase in temperature	Multiple drug delivery	[54]
Chitosan multilayer hydrogel capsules	Doxorubicin	Adjustment of pH	Significant inhibition of the burst release and good biocompatibility,	[55]
Polycarboxymethyl-β-cyclodextrin (polyCM-β-CD)/polyetherimide (PEI) multilayer	Ofloxacin	Adjustment of pH	Controllable release in different media	[56]
Poly(methacrylic acid) (PMAA) multilayer hydrogel cubes	7-(benzylamino)-3,4-dihydro-pyrrolo[4,3,2-de]quinolin-8(1H)-one	Redox-triggered release	Long-term storage, combination of well-regulated drug release and shape-modulated drug delivery	[57]

Wound healing is a dynamic and complex process that comprises several sequential phases, for which a number of drugs are effective. However, most of the current drug delivery systems were designed to treat only one phase of wound repair, ignoring the fact that every stage plays a critical role in the wound healing process. In an inspiring study, Maet al. reported that an injectable sodium alginate/bioglass (SA/BG) composite hydrogel can be used to carry SA microparticles containing a conditioned medium (CM) of cells (SA_{CM}) [58]. Inside the SA_{CM} microparticles, poly(lactic-co-glycolic acid) (PLGA) microspheres containing pirfenidone (PFD) were encapsulated (PLGAPFD). This multilayer injectable hydrogel system (SA/BG-SACM-$PLGA_{PFD}$) was designed to sequentially deliver bioactive molecules for satisfying the bioactivity requirement and timeline of each wound-healing stage (Figure 10).

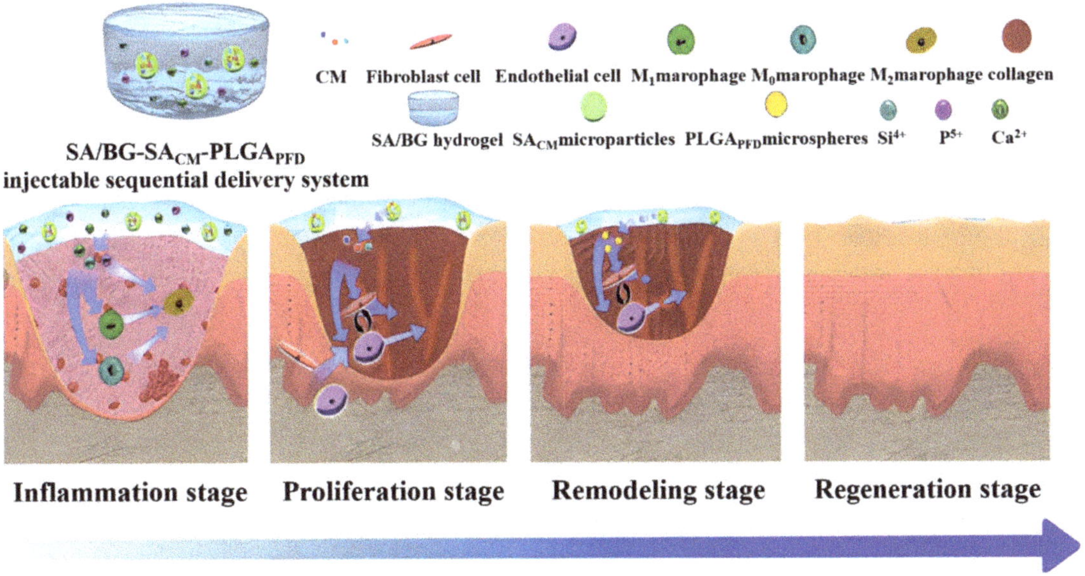

Figure 10. Scheme of a multilayer injectable hydrogel system that sequentially delivers bioactive substances for each wound-healing stage. (Reproduced with permission from [58]. American Chemical Society, 2020).

3.2. Cell Encapsulation (Carrier or Bioreactors) and Bacteria Delivery

The formation of multilayer hydrogels can sequentially and heterogeneously control the organization of cells, and the cavities between the layers can serve as good cell carriers or bioreactors [59]. The fast diffusion-induced gelation method was used by Sun's group to fabricate multilayer hydrogels with controllable layer thickness for the encapsulation of viable cells [60]. Five layers of cells marked with alternate green/red fluorescence were assembled in a LBL fashion into a tubular structure by immersing the core gel into alternating solutions of each labeled cell (Figure 11A). The cells in each layer were separated by distinct boundaries, indicating limited mixing of the gel components at each step. Heterogeneous cell-laden multilayer hydrogel tubes were fabricated with HUVECs, SMCs, and fibroblasts, which were distributed from the inside to the outside of tubes to mimic native blood vessels (Figure 11B). Moreover, all the layers from the multilayer hydrogels exhibited high cell viability (>90%) according to the live–dead staining result (Figure 11C).

It is well known that the strong acid environment of the stomach is harmful to probiotics, and oral delivery of probiotics is a significant challenge. To address this challenge, Chen's group prepared multilayer alginate hydrogel beads (MAHBs) by an emulsion method via ionic crosslinking between calcium ions and the carboxylic group of alginates, which can be used as an encapsulating material for oral delivery of a model probiotic bacterium *B. breve* [61]. MAHBs can be widely used as a carrier for probiotics oral delivery because they can significantly promote the viability of a variety of bacteria (including *B. breve*, *S. aureus*, and *E. coli*) at a low pH environment similar to that stomach, thus retaining the activity of the probiotics in the stomach. MAHBs can be utilized in the fermentation process, which is needed to release metabolite continuously and to avoid the burst release, and have been shown to be an excellent encapsulating material for oral administration.

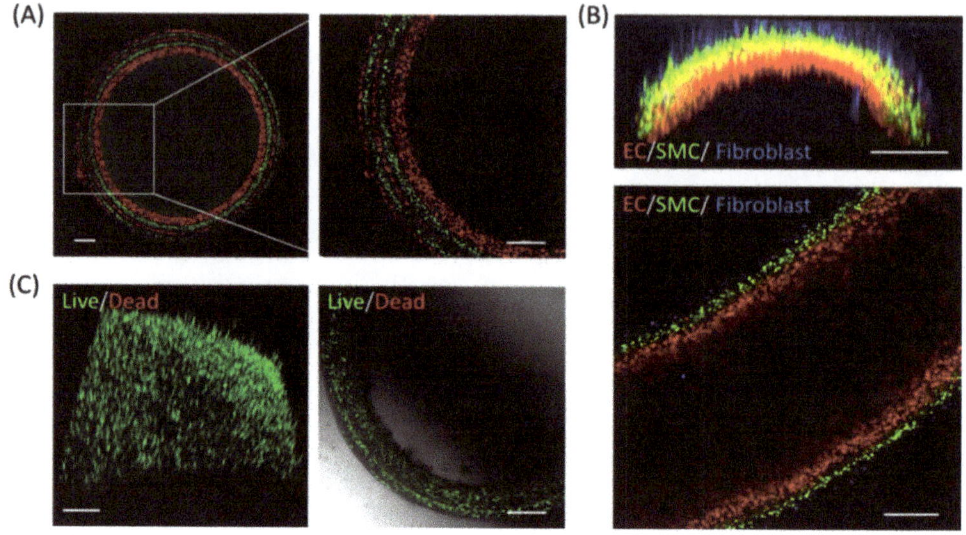

Figure 11. Cell encapsulation with multilayer hydrogels. (**A**) Images of a five-layer multilayer hydrogel tube embedded with fluorescence-tracked C2C12 cells of alternating color. (**B**) Cross-sectional (**top**) and longitudinal section (**bottom**) images of a three-layer tube embedded with HUVECs (red), SMCs (green), and fibroblasts (blue) in different layers. (**C**) 3D reconstruction (**left**) and cross-sectional (**right**) images of C2C12 cell-laden multilayer hydrogel walls stained for live (green) and dead (red) cells. Scale bar: (**A–C**) 500 µm. (Reproduced with permission from [60]. American Chemical Society, 2018).

3.3. *Cartilage Repair*

The need for bone repair materials has increased due to the complications associated with population aging. Hydrogels are often used as temporary fracture internal fixation materials due to their good mechanical properties, biocompatibility, and biodegradability [62]. Moreover, hydrogels are expected to treat cartilage diseases by mimicking the structural and functional characteristics of the natural extracellular matrix (ECM). It has been noted that multilayer hydrogels exhibit more advantages than ordinary single-layer hydrogels due to their unique structure, and layers with different and modifiable properties. Multilayer hydrogels can not only simulate the overall structure of cartilage, but also allow chondrocytes to migrate in the best form of tissue [63,64]. The internal structure of multilayer hydrogels can provide an appropriate microenvironment for the proliferation and differentiation of cells and microorganisms [65].

Recently, Nasr-Esfahani's group successfully constructed a chitosan/polycaprolactone multilayer hydrogel as a sustained Kartogenin (KGN) delivery system for cartilage engineering [66]. KGN was released from the hydrogels by three different mechanisms consisting of diffusion, swelling and erosion, or degradation (Figure 12a). KGN-conjugated multilayer materials (MLS + K) showed lower swelling ability and higher compressive modulus with gradual release of KGN in a longer retention time, which not only facilitated the effective treatment, but also provided a suitable mechanical structure for cartilage engineering and osteoarthritis treatment. Multilayer systems capable of simultaneous dual tissue formation were crucial for the regeneration of the osteochondral (OC) unit. Pereira et al. developed bi-layered hydrogel composites (BHCs) via the combination of two structurally stratified layers from nature-derived gellan-gum (GG) and hydroxyapatite (HAp) [67]. Either low acyl GG (LAGG) alone or in combination with high acyl GG (HAGG) were used for the fabrication of cartilage-like layers, and LAGG incorporating different ratios of HAp were used to prepare bone-like layers. GG-based layers and HAp reinforcement created a resilient bilayered structure with an interfacial region, which was not only able to integrate dissimilar zones, but also provided good stability during the degradation process. The BHC had good integration with surrounding tissues, and provided support for cartilage and bone-like tissue formation (Figure 12b), showing its feasibility as a osteochondral substitute with unique features for osteochondral regeneration.

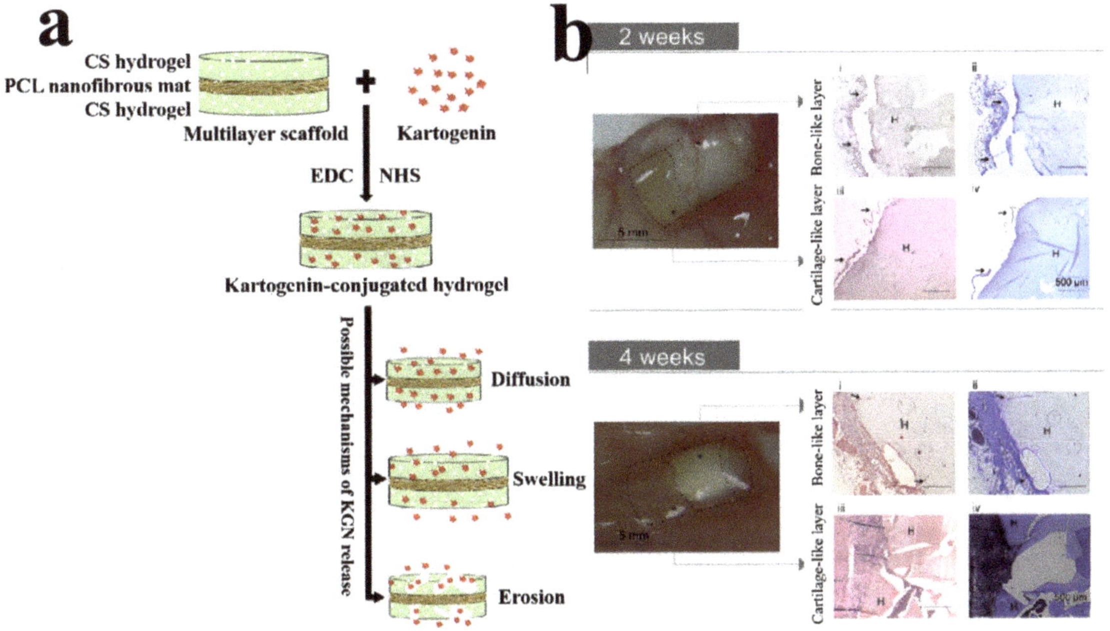

Figure 12. Schematic illustration of conjugation and mechanisms of KGN release from a MLS + K sample, showing three possible mechanisms (diffusion, swelling, and erosion or degradation) that are responsible for KGN release (**a**). Macroscopic images of the explants after implantation of (LAGG/LAGG-HAp 20%) hydrogels (H) in the dorsum of the mice for 2 weeks and 4 weeks (**b**). (Reproduced with permission from [66,67]. Elsevier, 2021, 2018).

3.4. Medical Dressings or Coatings

As mentioned above, wound healing remains a challenge in the biomedical field, which can be primarily be addressed by adopting appropriate wound care management. Wound dressings cannot only protect the wound from external damage, but also provide a suitable microenvironment for tissue regeneration. Hydrogels are suitable for the fabrication of medical dressing due to their excellent physical and chemical properties. Compared with monolayer or homogenous hydrogel wound dressings, multilayer hydrogel wound dressings can better promote wound healing because different layers can exhibit varied properties, which can satisfy different requirements of the top layer (barrier for bacterial transition and control of moist environment), middle layer (supply of controlled drug release for a long time and mass transfer limitations for drug molecules), and lower layer (absorption of the excess exudate, adhesion onto the wound surface, and support for new tissue formation) [68–71].

Recently, Tamahkar and others fabricated a new type of four-layered hydrogel (ML) antibacterial wound dressing using carboxylated polyvinyl alcohol (PVA-C), gelatin (G), hyaluronic acid (HA), and G (Figure 13a,b) [72]. The PVA-C and G upper layers provided the most control and a physical barrier for microorganisms. The HA-based middle layer served as an antibiotic-loaded layer. The G lower layer was able to be used to release antibiotics and provide the removal of excess exudate from the wound site. ML hydrogels showed unique antibacterial performance against *S. aureus* and *E. coli* (Figure 13c,d).

Moreover, the ML hydrogels showed antibacterial activity against oxacillin sensitivity, indicating that the novel wound dressings were an effective option for selective treatment of bacterial infections. Shokrollahi et al. prepared biocompatible electrospun multilayer nanofibrous dressings using PCL nanofibers as the first layer, hybrid nanofibers of chamomile/CECS/PVA and PCL as the second layer, and chamomile-loaded CECS/PVA as the third layer [73]. This multilayer dressing exhibited sufficient mechanical and swelling properties, and had excellent antibacterial efficiency due to the loading of chamomile, and could potentially be used for wound healing.

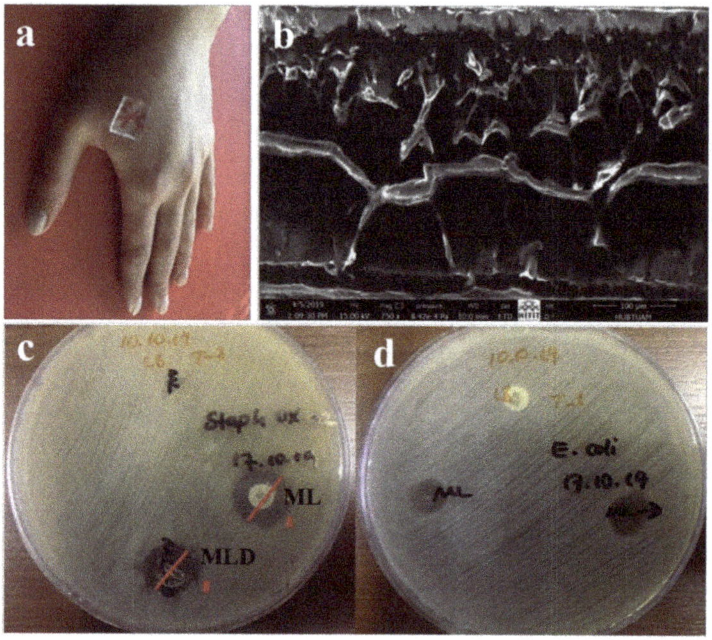

Figure 13. Photograph of a multilayer hydrogel as a wound healing dressing (**a**); the corresponding cross-sectional morphology (**b**); antimicrobial activities of ampicillin disc, ampicillin-loaded multilayer hydrogels (ML-D), and ML against oxacillin sensitive *S. aureus* (**c**); and antimicrobial activities of ampicillin disc, ML-D, and ML against *E. coli* (**d**). (Reproduced with permission from [72]. Elsevier, 2020).

Türkŏglu et al. fabricated a wheat germ oil (WGO)-loaded multilayer hydrogel dressing by cross-linking sodium alginate (SA) with poly(ethylene glycol) diglycidyl ether (PEGDGE) on textile nonwovens [74]. This multilayer hydrogel showed rapid and positive swelling properties with an interconnected network of pores, and the resultant product was able to support the treatment of burns and wounds with medium to high exudate, and thus may be a promising alternative to conventional products in the wound healing field.

In addition to dressings, coatings from organic and inorganic materials can also be considered an excellent strategy to prevent bacterial adhesion, bacterial infection, and subsequent biofilm formation [75]. Multilayer hydrogels are able to extend the application of multifunctional biomedical coatings with a long use time due to their unique structure. Zhao et al. prepared an antibacterial and biocompatible multilayer biomedical coating by alternate deposition of chitosan (CS) and sodium carboxymethyl cellulose (CMC), which can be used to heal damage [76]. This multilayer coating exhibited high antibacterial properties by adsorbing the negative charge on the surface of bacteria, and fast and efficient self-healing properties through H-bonds and electrostatic attraction under specific stimuli. These features enabled the CS/CMC multilayer polyelectrolyte coating to have an extended lifespan, showing potential as a novel functional biomedical material.

Shi's group fabricated a chitosan/silver nanoparticle (AgNP) multilayer hydrogel coating via the combination of in situ synthesis of AgNPs on a pre-deposited chitosan multilayer hydrogel [77]. The coating conferred antibacterial properties by embedding AgNPs into the chitosan hydrogel network, using the ability of chitosan to adsorb and stabilize metal salts and sterilization by silver ion diffusion. The nanocomposite multilayer hydrogel coating exhibited a staged release behavior of AgNPs based on acidic triggered dissolution of chitosan hydrogel layer by layer due to its unique layered structure (Figure 14a). The obtained AgNPs with a narrow size of ~15 nm were evenly distributed throughout the hydrogel matrix to confer the multilayer hydrogel with excellent antibacterial properties (Figure 14b). This antibacterial multilayer hydrogel showed significant potential either to be used as a new coating material for the interfacial improvement of implants or as a wound dressing.

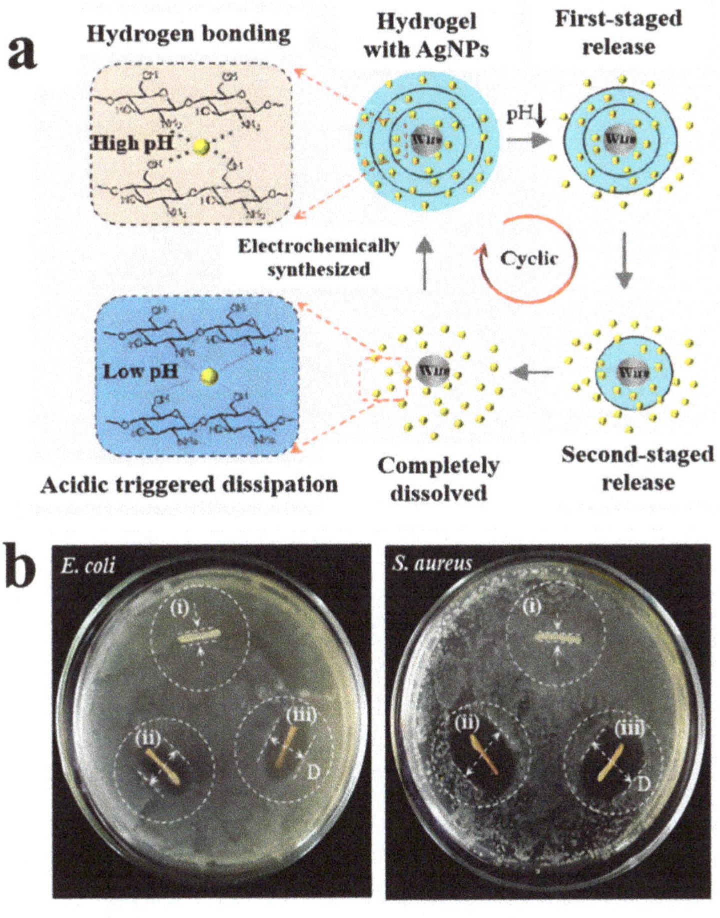

Figure 14. Schematic illustration of the staged release processes of AgNPs from a chitosan multilayer hydrogel based on acidic triggered dissolution of the hybrid coating layer by layer (**a**). Antibacterial activities of chitosan hydrogels with distinct compositions (**b**). i, bare 1-layer chitosan; ii, 1-layer chitosan loaded with AgNPs; iii, 3-layer chitosan loaded with AgNPs. (Reproduced with permission from [77]. Elsevier, 2021).

3.5. Other Fields

In addition to the large number of achievements and applications in the biomedical field, multilayer hydrogels can also be used in other fields. Three selected different application fields are introduced in the following.

The tuning of both ice nucleation and ice propagation via a simple anti-icing coating method is an important research topic, and was first investigated by Guo et al. using multilayer hydrogels [78]. Figure 15a shows the fabrication of poly(methacrylic acid) (PMAA)n multilayer hydrogels (n is the bilayer number). They first prepared a hydrogen-bonded multilayer of PMAA/poly(N-vinylpyrrolidone) (PVPON) at pH = 2.5 based on a LBL deposition approach. The neighboring PMAA layers were crosslinked with ethylenediamine (EDA), followed by the removal of the sacrificial template layers of PVPON at pH = 8.0. The ice nucleation and subsequent ice propagation on PMAA hydrogels with different counterions were investigated accordingly. The removal of dyes from effluents is also an important and urgent area of research, and hydrogels are important adsorption materials due to their advantages of low cost, high efficiency, and easy handling. Multilayer structures can increase the adsorption area for dyes such as methylene blue. In previous work, Chen et al. fabricated a novel multilayer composite hydrogel bead using alginate, acrylamide, and attapulgite for dye adsorption (Figure 15b) [79]. The multilayer hydrogels effectively adsorbed methylene blue and the maximum adsorption capacity reached 155.7 mg/g. (Figure 15c). These hydrogels are a promising adsorption material for dye-contaminated water treatment. Moreover, multilayer hydrogel capsules were also reported to load and release solutes including dyes via controlling temperature [80].

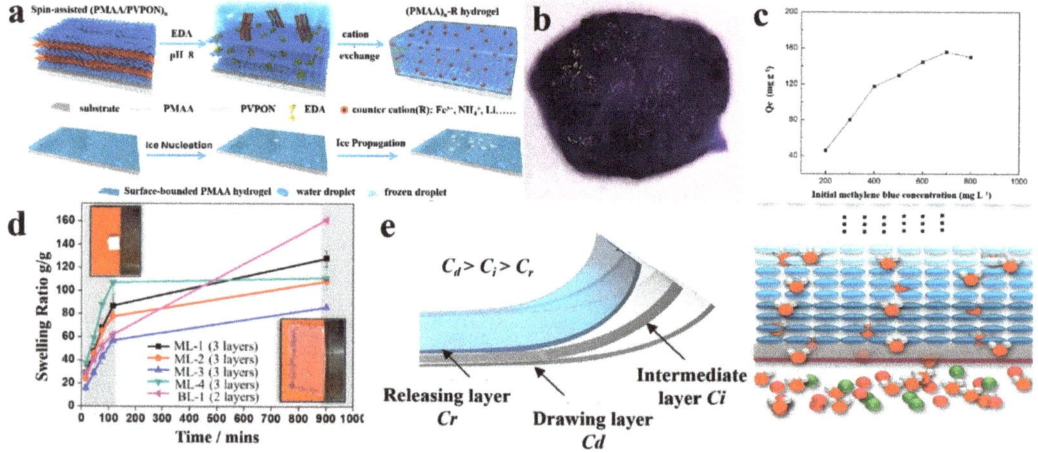

Figure 15. Schematic illustration of the fabrication of (PMAA)n-R multilayer hydrogels with different counterions ("R" denotes the type of counterion) and illustration of ice nucleation and ice propagation on (PMAA)n hydrogel surfaces (**a**); a photograph of the cross-section for the SAA2 multilayer hydrogel bead in methylene blue solution after adsorption of 72 h (**b**); and the corresponding effect of initial concentration of methylene blue on its adsorption capacity (**c**). Multilayer hydrogels with different configurations of layers along the direction of water transport; the inserted figures represent the ML-1 sample at t = 0 (top left) and t = 900 min (bottom right), respectively (**d**), and multilayer design with gradual reduction of SA concentration along the water transport pathway (**e**). (Reproduced with permission from [78,79,81]. American Chemical Society 2018, Tech Science Press 2019, Elsevier 2019).

To solve the bottleneck of the lack of suitable draw agents in the development of FO desalination, Zeng et al. developed a multilayer temperature-responsive hydrogel on the basis of poly(N-isopropylacrylamide-co-sodium acrylate) (P(NIPAAm-co-SA)) [81]. The multilayer hydrogel was completely dry and white before the test (t = 0), which then swelled and became transparent (Figure 15d, inserts). The corresponding swelling curves were delineated into the initial fast swelling stage and the subsequent steady stages after 150 min (Figure 15d), indicating that the multilayer configuration did not affect the intrinsic swelling property of the P(NIPAAm-co-SA). The multilayer hydrogels showed a favorable performance for water storage with a reasonable mass transfer rate. The multilayer hydrogels consisted of a drawing layer with a high SA concentration for high osmotic pressure in the FO process, a releasing P-NIPAAm layer for fast water release, and intermediate layers for the reduction of the mass transfer resistance (Figure 15e). After dewatering and then cooling below the LCST, the P-NIPAAm releasing layer was expected to draw the water molecules from the intermediate layer more easily when compared with the bi-layer hydrogels. The multilayer hydrogel yielded a high capacity of water absorption and high permeable flux, which was very important for the development of hydrogel-based energy-efficient FO desalination.

4. Conclusions and Prospects

As a result of the continuous improvement in people's living standards and the deepening of research in various fields, hydrogels have been widely developed and studied. Although many homogenous hydrogels with different excellent properties have been developed and improved continuously in recent decades, significant room remains for further development of hydrogels, because it is a challenge for these homogeneous hydrogels to simultaneously meet different needs due to the restrictions of their structure. Fortunately, novel multilayer hydrogels from different resources have emerged as required and become a new branch of hydrogels. Multilayer hydrogels have attracted significant attention and been studied and utilized in various fields due to their unique structure and excellent properties, as outlined above. Most preparation techniques (roughly classified as LBL and non-LBL) and related mechanisms of different multilayer hydrogels were cited and systematically discussed in this review. These impressive works will not only have a significant impact on the construction of multilayer hydrogels in the future, but also shed light on industrial processing for exploiting multilayer hydrogels in daily life.

However, in general, recent research and applications of multilayer hydrogels remain insufficient and lack maturity. Moreover, commercial products have not yet emerged because the existing methods for the production of multilayer hydrogels are relatively complex and difficult to apply in industrial settings. In addition, it is a challenge to implement the main biomedical applications, particularly in the clinical phase, because caution is needed in the evaluation of biomedical materials. Thus, further effort is required to develop novel methods and accelerate the evaluation of the long-term biocompatibility of multilayer hydrogels as implants. As a result of the continuous development of science and technology, prolific creativity, and effective cooperation, we believe that the preparation process of multilayer hydrogels will be optimized, and their properties will be continuously improved. Therefore, the issues relating to industrialization and biomedical applications of multilayer hydrogels may be solved based on the in-depth study of abundant resources, modifiable layers, and advanced technologies, and multilayer hydrogels can be eventually applied to every aspect of our lives.

Author Contributions: Conceptualization, M.H., J.W., W.Y. and C.W.; writing—original draft preparation, L.J., J.X., and M.H.; writing—review and editing, M.H., Y.X. and X.Z.; supervision, M.H., M.F. and C.W.; funding acquisition, M.H., M.F. and C.W. All authors have read and agreed to the published version of the manuscript.

Funding: This research was funded by National Natural Science Foundation of China under the grant numbers of 51903127 and 51503177, the Natural Science Foundation of the Jiangsu Higher Education Institutions of China under the grant number of No. 20KJB430032.

Conflicts of Interest: The authors declare no conflict of interest.

References

1. Wichterle, O.; Lim, D. Hydrophilic gels for biological use. *Nature* **1960**, *185*, 117–118. [CrossRef]
2. Xue, K.; Wang, X.; Yong, P.W.; Young, D.J.; Wu, Y.L.; Li, Z.; Loh, X.J. Hydrogels as emerging materials for translational biomedicine. *Adv. Ther.* **2018**, *2*, 1800088. [CrossRef]
3. Correa, S.; Grosskopf, A.K.; Lopez Hernandez, H.; Chan, D.; Yu, A.C.; Stapleton, L.M.; Appel, E.A. Translational applications of hydrogels. *Chem. Rev.* **2021**, *121*, 11385–11457. [CrossRef] [PubMed]
4. Zhang, Y.S.; Khademhosseini, A. Advances in engineering hydrogels. *Science* **2017**, *356*, eaaf3627. [CrossRef] [PubMed]
5. Maitra, J.; Shukla, V.K. Cross-linking in hydrogels—A review. *Am. J. Polym. Sci.* **2014**, *4*, 25–31.
6. Webber, M.J.; Appel, E.A.; Meijer, E.W.; Langer, R. Supramolecular biomaterials. *Nat. Mater.* **2016**, *15*, 13–26. [CrossRef]
7. Lim, J.Y.C.; Lin, Q.; Xue, K.; Loh, X.J. Recent advances in supramolecular hydrogels for biomedical applications. *Mater. Today Adv.* **2019**, *3*, 100021. [CrossRef]
8. Peppas, N.A.; Bures, P.; Leobandung, W.S.; Ichikawa, H. Hydrogels in pharmaceutical formulations. *Eur. J. Pharm. Biopharm.* **2000**, *50*, 27–46. [CrossRef]
9. Qiu, Y.; Park, K. Environment-sensitive hydrogels for drug delivery. *Adv. Drug Deliv. Rev.* **2001**, *53*, 321–339. [CrossRef]
10. Peppas, N.A.; Hilt, J.Z.; Khademhosseini, A.; Langer, R. Hydrogels in biology and medicine: From molecular principles to bionanotechnology. *Adv. Mater.* **2006**, *18*, 1345–1360. [CrossRef]
11. Li, J.; Mooney, D.J. Designing hydrogels for controlled drug delivery. *Nat. Rev. Mater.* **2016**, *1*, 16071. [CrossRef] [PubMed]
12. Liu, Y.; Liu, J.; Chen, S.; Lei, T.; Kim, Y.; Niu, S.; Wang, H.; Wang, X.; Foudeh, A.M.; Tok, J.B.H.; et al. Soft and elastic hydrogel-based microelectronics for localized low-voltage neuromodulation. *Nat. Biomed. Eng.* **2019**, *3*, 58–68. [CrossRef] [PubMed]
13. Lu, B.; Yuk, H.; Lin, S.; Jian, N.; Qu, K.; Xu, J.; Zhao, X. Pure pedot: Pss hydrogels. *Nat. Commun.* **2019**, *10*, 1043. [CrossRef]
14. Gao, Y.; Song, J.; Li, S.; Elowsky, C.; Zhou, Y.; Ducharme, S.; Chen, Y.M.; Zhou, Q.; Tan, L. Hydrogel microphones for stealthy underwater listening. *Nat. Commun.* **2016**, *7*, 12316. [CrossRef]
15. Elisseeff, J. Structure starts to gel. *Nat. Mater.* **2008**, *7*, 271–273. [CrossRef]
16. Wu, J.; Ren, Y.; Sun, J.; Feng, L. Carbon nanotube-coated macroporous poly (N-isopropylacrylamide) hydrogel and its electrosensitivity. *ACS Appl. Mater. Interfaces* **2013**, *5*, 3519–3523. [CrossRef]
17. Johnson, L.M.; DeForest, C.A.; Pendurti, A.; Anseth, K.S.; Bowman, C.N. Formation of three-dimensional hydrogel multilayers using enzyme-mediated redox chain initiation. *ACS Appl. Mater. Interfaces* **2010**, *2*, 1963–1972. [CrossRef]
18. Zarzar, L.D.; Kim, P.; Aizenberg, J. Bio-inspired design of submerged hydrogel-actuated polymer microstructures operating in response to pH. *Adv. Mater.* **2011**, *23*, 1442–1446. [CrossRef]
19. Kozlovskaya, V.; Chen, J.; Zavgorodnya, O.; Hasan, M.B.; Kharlampieva, E. Multilayer hydrogel capsules of interpenetrated network for encapsulation of small molecules. *Langmuir* **2018**, *34*, 11832–11842. [CrossRef] [PubMed]
20. Nie, J.; Lu, W.; Ma, J.; Yang, L.; Wang, Z.; Qin, A.; Hu, Q. Orientation in multi-layer chitosan hydrogel: Morphology, mechanism and design principle. *Sci. Rep.* **2015**, *5*, 7635. [CrossRef] [PubMed]
21. Sun, L.; Wu, W.; Yang, S.; Zhou, J.; Hong, M.; Xiao, X.; Ren, F.; Jiang, C. Template and silica interlayer tailorable synthesis of spindle-like multilayer α-Fe_2O_3/Ag/SnO_2 ternary hybrid architectures and their enhanced photocatalytic activity. *ACS Appl. Mater. Interfaces* **2014**, *6*, 1113–1124. [CrossRef] [PubMed]
22. Chiriac, A.P.; Nistor, M.T.; Nita, L.E. An investigation on multi-layered hydrogels based on poly (N, N-Dimethylacrylamide-co-3, 9-Divinyl-2, 4, 8, 10-tetraoxaspiro (5.5) undecane). *Rev. Roum. Chim.* **2014**, *59*, 1059–1068.
23. Dhanasingh, A.; Groll, J. Polysaccharide based covalently linked multi-membrane hydrogels. *Soft Matter* **2012**, *8*, 1643–1647. [CrossRef]
24. Duan, J.; Hou, R.; Xiong, X.; Wang, Y.; Wang, Y.; Fu, J.; Yu, Z. Versatile fabrication of arbitrarily shaped multi-membrane hydrogels suitable for biomedical applications. *J. Mater. Chem. B* **2013**, *1*, 485–492. [CrossRef]
25. Dai, H.J.; Li, X.F.; Long, Y.H.; Wu, J.J.; Liang, S.M.; Zhang, X.L.; Zhao, N.; Xu, J. Multi-membrane hydrogel fabricated by facile dynamic self-assembly. *Soft Matter* **2009**, *5*, 1987–1989. [CrossRef]
26. Liu, L.; Wu, F.; Ju, X.J.; Xie, R.; Wang, W.; Niu, C.H.; Chu, L.Y. Preparation of monodisperse calcium alginate microcapsules via internal gelation in microfluidic-generated double emulsions. *J. Colloid Interface Sci.* **2013**, *404*, 85–90. [CrossRef]
27. Mohammadi, G.; Barzegar-Jalali, M.; Shadbad, M.S.; Azarmi, S.; Barzegar-Jalali, A.; Rasekhian, M.; Adibkia, K.; Danesh-Bahreini, M.; Kiafar, F.; Zare, M.; et al. The effect of inorganic cations Ca^{2+} and Al^{3+} on the release rate of propranolol hydrochloride from sodium carboxymethylcellulose matrices. *DARU* **2009**, *17*, 131–138.
28. Seidi, F.; Zhao, W.; Xiao, H.; Jin, Y.; Zhao, C. Layer-by-layer assembly for surface tethering of thin-hydrogel films: Design strategies and applications. *Chem. Rec.* **2020**, *20*, 857–881. [CrossRef]
29. Zhang, X.; Chen, H.; Zhang, H. Layer-by-layer assembly: From conventional to unconventional methods. *Chem. Commun.* **2007**, *14*, 1395–1405. [CrossRef] [PubMed]
30. Ladet, S.; David, L.; Domard, A. Multi-membrane hydrogels. *Nature* **2008**, *452*, 76–79. [CrossRef] [PubMed]
31. Xiong, Y.; Yan, K.; Bentley, W.E.; Deng, H.; Du, Y.; Payne, G.F.; Shi, X.W. Compartmentalized multilayer hydrogel formation using a stimulus-responsive self-assembling polysaccharide. *ACS Appl. Mater. Interfaces* **2014**, *6*, 2948–2957. [CrossRef] [PubMed]
32. Li, B.; Gao, Y.; Feng, Y.; Ma, B.; Zhu, R.; Zhou, Y. Formation of concentric multi-layer chitosan hydrogel loaded with isoniazid. *J. Control. Release* **2011**, *152*, e45–e47. [CrossRef] [PubMed]

33. Li, B.; Gao, Y.; Feng, Y.; Ma, B.; Zhu, R.; Zhou, Y. Formation of concentric multilayers in a chitosan hydrogel inspired by Liesegang ring phenomena. *J. Biomater. Sci. Polym. Ed.* **2011**, *22*, 2295–2304. [CrossRef]
34. He, M.; Zhao, Y.; Duan, J.; Wang, Z.; Chen, Y.; Zhang, L. Fast contact of solid–liquid interface created high strength multi-layered cellulose hydrogels with controllable size. *ACS Appl. Mater. Interfaces* **2014**, *6*, 1872–1878. [CrossRef]
35. Cai, J.; Zhang, L.; Liu, S.; Liu, Y.; Xu, X.; Chen, X.; Chu, B.; Guo, X.; Xu, J.; Cheng, H. Dynamic self-assembly induced rapid dissolution of cellulose at low temperatures. *Macromolecules* **2008**, *41*, 9345–9351. [CrossRef]
36. Yan, K.; Ding, F.; Bentley, W.E.; Deng, H.; Du, Y.; Payne, G.F.; Shi, X.W. Coding for hydrogel organization through signal guided self-assembly. *Soft Matter* **2013**, *10*, 465–469. [CrossRef]
37. Wei, X.Q.; Payne, G.F.; Shi, X.W.; Du, Y. Electrodeposition of a biopolymeric hydrogel in track-etched micropores. *Soft Matter* **2013**, *9*, 2131–2135. [CrossRef]
38. Wu, L.Q.; Gadre, A.P.; Yi, H.; Kastantin, M.J.; Rubloff, G.W.; Bentley, W.E.; Payne, G.F.; Ghodssi, R. Voltage-dependent assembly of the polysaccharide chitosan onto an electrode surface. *Langmuir* **2002**, *18*, 8620–8625. [CrossRef]
39. Pang, X.; Zhitomirsky, I. Electrodeposition of composite hydroxyapatite–chitosan films. *Mater. Chem. Phys.* **2005**, *94*, 245–251. [CrossRef]
40. Maerten, C.; Jierry, L.; Schaaf, P.; Boulmedais, F. Review of electrochemically triggered macromolecular film buildup processes and their biomedical applications. *ACS Appl. Mater. Interfaces* **2017**, *9*, 28117–28138. [CrossRef]
41. Cheng, Y.; Luo, X.; Betz, J.; Buckhout-White, S.; Bekdash, O.; Payne, G.F.; Bentley, W.E.; Rubloff, G.W. In situ quantitative visualization and characterization of chitosan electrodeposition with paired sidewall electrodes. *Soft Matter* **2010**, *6*, 3177–3183. [CrossRef]
42. Cheng, Y.; Gray, K.M.; David, L.; Royaud, I.; Payne, G.F.; Rubloff, G.W. Characterization of the cathodic electrodeposition of semicrystalline chitosan hydrogel. *Mater. Lett.* **2012**, *87*, 97–100. [CrossRef]
43. Jin, Z.; Harvey, A.M.; Mailloux, S.; Halámek, J.; Bocharova, V.; Twiss, M.R.; Katz, E. Electrochemically stimulated release of lysozyme from an alginate matrix cross-linked with iron cations. *J. Mater. Chem.* **2012**, *22*, 19523–19528. [CrossRef]
44. Javanmard, S.H.; Anari, J.; Kharazi, A.Z.; Vatankhah, E. Invitro hemocompatibility and cytocompatibility of a three-layered vascular scaffold fabricated by sequential electrospinning of PCL, collagen, and PLLA nanofibers. *J. Biomater. Appl.* **2016**, *31*, 438–449. [CrossRef]
45. He, M.; Zhang, X.; Yao, W.; Wang, C.; Shi, L.; Zhou, P. Construction of alternate layered chitosan/alginate composite hydrogels and their properties. *Mater. Lett.* **2017**, *200*, 43–46. [CrossRef]
46. Mredha, M.T.I.; Le, H.H.; Trtik, P.; Cui, J.; Jeon, I. Anisotropic tough multilayer hydrogels with programmable orientation. *Mater. Horiz.* **2019**, *6*, 1504–1511. [CrossRef]
47. Nie, J.; Wang, Z.; Hu, Q. Chitosan hydrogel structure modulated by metal ions. *Sci. Rep.* **2016**, *6*, 36005. [CrossRef] [PubMed]
48. Droz, M. Recent theoretical developments on the formation of Liesegang patterns. *J. Stat. Phys.* **2000**, *101*, 509–519. [CrossRef]
49. Mito, K.; Haque, M.A.; Nakajima, T.; Uchiumi, M.; Kurokawa, T.; Nonoyama, T.; Gong, J.P. Supramolecular hydrogels with multi-cylindrical lamellar bilayers: Swelling-induced contraction and anisotropic molecular diffusion. *Polymer* **2017**, *128*, 373–378. [CrossRef]
50. Zhao, Q.; Han, B.; Wang, Z.; Gao, C.; Peng, C.; Shen, J. Hollow chitosan-alginate multilayer microcapsules as drug delivery vehicle: Doxorubicin loading and in vitro and in vivo studies. *Nanomed. Nanotechnol.* **2007**, *3*, 63–74. [CrossRef]
51. Nita, L.E.; Chiriac, A.P.; Nistor, M.T.; Tartau, L. Upon some multi-membrane hydrogels based on poly (N, N-dimethyl-acrylamide-co-3, 9-divinyl-2, 4, 8, 10-tetraoxaspiro (5.5) Undecane): Preparation, characterization and in vivo tests. *J. Mater. Sci. Mater. Med.* **2014**, *25*, 1757–1768. [CrossRef]
52. Veronovski, A.; Knez, Ž.; Novak, Z. Preparation of multi-membrane alginate aerogels used for drug delivery. *J. Supercrit. Fluids* **2013**, *79*, 209–215. [CrossRef]
53. Alford, A.; Tucker, B.; Kozlovskaya, V.; Chen, J.; Gupta, N.; Caviedes, R.; Gearhart, J.; Graves, D.; Kharlampieva, E. Encapsulation and ultrasound-triggered release of G-quadruplex DNA in multilayer hydrogel microcapsules. *Polymers* **2018**, *10*, 1342. [CrossRef]
54. Zavgorodnya, O.; Carmona-Moran, C.A.; Kozlovskaya, V.; Liu, F.; Wick, T.M.; Kharlampieva, E. Temperature-responsive nanogel multilayers of poly (N-vinylcaprolactam) for topical drug delivery. *J. Colloid Interface Sci.* **2017**, *506*, 589–602. [CrossRef]
55. Zhang, W.; Jin, X.; Li, H.; Wei, C.X.; Wu, C.W. Onion-structure bionic hydrogel capsules based on chitosan for regulating doxorubicin release. *Carbohyd. Polym.* **2019**, *209*, 152–160. [CrossRef]
56. Chen, P.; Wang, X.; Pang, J.; Dong, Y.; Ma, X.; Hu, X. Drug release behaviour from drug loaded LBL multilayer on hydrogel. *Mater. Technol.* **2015**, *30*, 159–161. [CrossRef]
57. Xue, B.; Wang, W.; Qin, J.J.; Nijampatnam, B.; Murugesan, S.; Kozlovskaya, V.; Zhang, R.; Velu, S.E.; Kharlampieva, E. Highly efficient delivery of potent anticancer iminoquinone derivative by multilayer hydrogel cubes. *Acta Biomater.* **2017**, *58*, 386–398. [CrossRef]
58. Ma, Z.; Song, W.; He, Y.; Li, H. Multilayer injectable hydrogel system sequentially delivers bioactive substances for each wound healing stage. *ACS Appl. Mater. Interfaces* **2020**, *12*, 29787–29806. [CrossRef] [PubMed]
59. Ladet, S.G.; Tahiri, K.; Montembault, A.S.; Domard, A.J.; Corvol, M.T. Multi-membrane chitosan hydrogels as chondrocytic cell bioreactors. *Biomaterials* **2011**, *32*, 5354–5364. [CrossRef] [PubMed]
60. Ouyang, L.; Burdick, J.A.; Sun, W. Facile biofabrication of heterogeneous multilayer tubular hydrogels by fast diffusion-induced gelation. *ACS Appl. Mater. Interfaces* **2018**, *10*, 12424–12430. [CrossRef]

61. Li, Y.; Feng, C.; Li, J.; Mu, Y.; Liu, Y.; Kong, M.; Cheng, X.; Chen, X. Construction of multilayer alginate hydrogel beads for oral delivery of probiotics cells. *Int. J. Biol. Macromol.* **2017**, *105*, 924–930. [CrossRef]
62. Hu, Q.; Li, B.; Wang, M.; Shen, J. Preparation and characterization of biodegradable chitosan/hydroxyapatite nanocomposite rods via in situ hybridization: A potential material as internal fixation of bone fracture. *Biomaterials* **2004**, *25*, 779–785. [CrossRef]
63. Ohashi, K.; Yokoyama, T.; Yamato, M.; Kuge, H.; Kanehiro, H.; Tsutsumi, M.; Amanuma, T.; Iwata, H.; Yang, J.; Okano, T.; et al. Engineering functional two-and three-dimensional liver systems in vivo using hepatic tissue sheets. *Nat. Med.* **2007**, *13*, 880–885. [CrossRef] [PubMed]
64. Bae, H.; Puranik, A.S.; Gauvin, R.; Edalat, F.; Carrillo-Conde, B.; Peppas, N.A.; Khademhosseini, A. Building vascular networks. *Sci. Transl. Med.* **2012**, *4*, 30–34. [CrossRef] [PubMed]
65. Sill, T.J.; Von Recum, H.A. Electrospinning: Applications in drug delivery and tissue engineering. *Biomaterials* **2008**, *29*, 1989–2006. [CrossRef] [PubMed]
66. Houreh, A.B.; Masaeli, E.; Nasr-Esfahani, M.H. Chitosan/polycaprolactone multilayer hydrogel: A sustained Kartogenin delivery model for cartilage regeneration. *Int. J. Biol. Macromol.* **2021**, *177*, 589–600. [CrossRef]
67. Pereira, D.R.; Canadas, R.F.; Silva-Correia, J.; da Silva Morais, A.; Oliveira, M.B.; Dias, I.R.; Mano, J.F.; Marques, A.P.; Reis, R.L.; Oliveira, J.M. Injectable gellan-gum/hydroxyapatite-based bilayered hydrogel composites for osteochondral tissue regeneration. *Appl. Mater. Today* **2018**, *12*, 309–321. [CrossRef]
68. Ding, L.; Shan, X.; Zhao, X.; Zha, H.; Chen, X.; Wang, J.; Cai, C.; Wang, X.; Li, G.; Hao, J.; et al. Spongy bilayer dressing composed of chitosan–Ag nanoparticles and chitosan–Bletilla striata polysaccharide for wound healing applications. *Carbohydr. Polym.* **2017**, *157*, 1538–1547. [CrossRef]
69. Shemesh, M.; Zilberman, M. Structure–property effects of novel bioresorbable hybrid structures with controlled release of analgesic drugs for wound healing applications. *Acta Biomater.* **2014**, *10*, 1380–1391. [CrossRef]
70. Priya, S.G.; Gupta, A.; Jain, E.; Sarkar, J.; Damania, A.; Jagdale, P.R.; Chaudhari, B.P.; Gupta, K.C.; Kumar, A. Bilayer cryogel wound dressing and skin regeneration grafts for the treatment of acute skin wounds. *ACS Appl. Mater. Interfaces* **2016**, *8*, 15145–15159. [CrossRef]
71. Guo, Y.; Pan, S.; Jiang, F.; Wang, E.; Miinea, L.; Marchant, N.; Cakmak, M. Anisotropic swelling wound dressings with vertically aligned water absorptive particles. *RSC Adv.* **2018**, *8*, 8173–8180. [CrossRef]
72. Tamahkar, E.; Özkahraman, B.; Süloğlu, A.K.; İdil, N.; Perçin, I. A novel multilayer hydrogel wound dressing for antibiotic release. *J. Drug Deliv. Sci. Technol.* **2020**, *58*, 101536. [CrossRef]
73. Shokrollahi, M.; Bahrami, S.H.; Nazarpak, M.H.; Solouk, A. Multilayer nanofibrous patch comprising chamomile loaded carboxyethyl chitosan/poly (vinyl alcohol) and polycaprolactone as a potential wound dressing. *Int. J. Biol. Macromol.* **2020**, *147*, 547–559. [CrossRef]
74. Türkoğlu, G.C.; Sariişik, M.; Karavana, S.Y.; Köse, F.A. Production of wheat germ oil containing multilayer hydrogel dressing. *Carbohydr. Polym.* **2021**, *269*, 118287. [CrossRef]
75. Syafiuddin, A.; Salim, M.R.; Beng Hong Kueh, A.; Hadibarata, T.; Nur, H. A review of silver nanoparticles: Research trends, global consumption, synthesis, properties, and future challenges. *J. Chin. Chem. Soc.* **2017**, *64*, 732–756. [CrossRef]
76. Zhao, Y.; Liang, Y.; Zou, Q.; Ma, L.; Wang, Y.; Zhu, Y. An antibacterial and biocompatible multilayer biomedical coating capable of healing damages. *RSC Adv.* **2020**, *10*, 32011–32015. [CrossRef]
77. Yan, K.; Xu, F.; Wei, W.; Yang, C.; Wang, D.; Shi, X. Electrochemical synthesis of chitosan/silver nanoparticles multilayer hydrogel coating with pH-dependent controlled release capability and antibacterial property. *Colloids Surf. B* **2021**, *202*, 111711. [CrossRef] [PubMed]
78. Guo, Q.; He, Z.; Jin, Y.; Zhang, S.; Wu, S.; Bai, G.; Xue, H.; Liu, Z.; Jin, S.; Zhao, L.; et al. Tuning ice nucleation and propagation with counterions on multilayer hydrogels. *Langmuir* **2018**, *34*, 11986–11991. [CrossRef] [PubMed]
79. Chen, X.; Zhu, J. Alginate composite hydrogel bead with multilayer flake structure for dye adsorptions. *J. Renew. Mater.* **2019**, *7*, 983–996. [CrossRef]
80. Zarket, B.C.; Raghavan, S.R. Onion-like multilayered polymer capsules synthesized by a bioinspired inside-out technique. *Nat. Commun.* **2017**, *8*, 193. [CrossRef] [PubMed]
81. Zeng, J.; Cui, S.; Wang, Q.; Chen, R. Multi-layer temperature-responsive hydrogel for forward-osmosis desalination with high permeable flux and fast water release. *Desalination* **2019**, *459*, 105–113. [CrossRef]

MDPI
St. Alban-Anlage 66
4052 Basel
Switzerland
Tel. +41 61 683 77 34
Fax +41 61 302 89 18
www.mdpi.com

Gels Editorial Office
E-mail: gels@mdpi.com
www.mdpi.com/journal/gels

www.ingramcontent.com/pod-product-compliance
Lightning Source LLC
LaVergne TN
LVHW070240100526
838202LV00015B/2157
* 9 7 8 3 0 3 6 5 6 4 2 7 2 *